ADVANCED BIOMETRIC TECHNOLOGIES

Edited by **Girija Chetty** and **Jucheng Yang**

INTECHOPEN.COM

Advanced Biometric Technologies
Edited by Girija Chetty and Jucheng Yang

Published by InTech
Janeza Trdine 9, 51000 Rijeka, Croatia

Edition 2014

Publishing Process Manager Mirna Cvijic

Technical Editor Teodora Smiljanic

Cover Designer InTech Design Team

Image Copyright pio3, 2010. Used under license from Shutterstock.com

Additional hard copies can be obtained from orders@intechopen.com

Advanced Biometric Technologies, Edited by Girija Chetty and Jucheng Yang
p. cm.
ISBN 978-953-307-487-0

INTECH

open science | open minds

Contents

Preface

The methods for human identity authentication based on biometrics – the physiological and behavioural characteristics of a person have been evolving continuously and seen significant improvement in performance and robustness over the last few years. However, most of the systems reported perform well in controlled operating scenarios, and their performance deteriorates significantly under real world operating conditions, and far from satisfactory in terms of robustness and accuracy, vulnerability to fraud and forgery, and use of acceptable and appropriate authentication protocols. To address these challenges, and the requirements of new and emerging applications, and for seamless diffusion of biometrics in society, there is a need for development of novel paradigms and protocols, and improved algorithms and authentication techniques.

This book volume on *"Advanced Biometric Technologies"* is dedicated to the work being pursued by researchers around the world in this area, and includes some of the recent findings and their applications to address the challenges and emerging requirements for biometric based identity authentication systems. The book consists of 18 Chapters and is divided into four sections namely novel approaches, .advanced algorithms, emerging applications and multimodal fusion.

Chapter 1 to 4 group some novel biometric traits and computational approaches for identity recognition task. In Chapter 1, authors examine the effect of ambient temperatures on infra-red face recognition performance, and propose novel normalization techniques to alleviate the effect of ambient temperature variations for thermal images. In Chapter 2, the authors show that it quite possible to use spectral biometrics as a complementary information to prevent spoofing of existing biometric technology. They propose an aliveness detection method based on spectral biometrics that ensures the identity decision obtained through primary biometrics comes from a living authentic person, Chapter 3 presents another novel biometric trait to recognizing identity - a low resolution contactless palm print and palm vein biometric for identity recognition. To obtain useful representation of the palm print and vein modalities, authors propose a new technique called directional coding. This method represents the biometric features in bit string format which enable speedy matching and convenient storage. In addition, authors propose a new image quality measure which can be incorporated to improve the performance of the recognition system.

Chapter 4 examines the importance of liveness detection for fingerprint recognition systems, and authors present a detailed treatment of liveness detection for fingerprints here.

Chapter 5 to 9 report some advanced computational algorithms for authenticating identity. In Chapter 5, authors propose a novel fast minutiae-based matching algorithm for fingerprint recognition. In Chapter 6, gender classification problem using facial images is considered, and authors propose several pre-processing algorithms based on PCA, JADE-ICA and an LS-SVM. In Chapter 8, the problem of security in mobile devices is considered and authors propose an interesting technique based on the use of handwritten biometric signatures adapted to mobiles. The technique is based on recognizing an identifying gesture carried out in the air. In Chapter 9, authors evaluate PCA-based face recognition algorithms in various color spaces and show how color information can be beneficial for face recognition with SV, YCbCr, and YCg'Cr' color spaces as the most appropriate spaces for authenticating identity.

Chapter 10 to 13 is a collection of works on emerging biometric applications. In Chapter 10, the authors introduced three novel ideas for perceived age estimation from face images: taking into account the human age perception for improving the prediction accuracy , clustering based active learning for reducing the sampling cost, and alleviating the influence of lighting condition change. Chapter 11 is an interesting emerging biometric application, and the authors here proposed a novel stress detection system using only two physiological signals (HR and GSR) providing a precise output indicating to what extent a user is under a stressing stimulus. Main characteristics of this system is an outstanding accuracy in detecting stress when compared to previous approaches in literature. In Chapter 12, authors proposed a direct authentication and an additive coding scheme using mathematical model for the DNA measurements. Finally, in Chapter 13, the authors develop several methods for measuring and identifying cells involved in a variety of experiments, including cell cultures. The focus was on obtaining models of the sensor system employed for data acquisition, and on using them to extract relevant information such as cell size, density, growth rate, dosimetry, etc.

The final section of the book, includes some innovative algorithms and its applications based on fusion of multiple biometric modalities and includes Chapter 14 to 18. In Chapter 14, authors propose a set of algorithms to fuse the information from multi-spectral palmprint images where fusion is performed at the matching score level to generate a unique score which is then used for recognizing a palmprint image. Authors examined several fusion rules including SUM, MIN, MAX and WHT for the fusion of the multi-spectral palmprint at the matching score level. The authors in Chapter 15 further reinforced the benefits achieved by fusion of multiple biometric modalities with a detailed treatment on fusion techniques and normalization. The authors conclude that the improvement in error rates are directly linked to the number of biometric features being combined. In Chapter 16, the authors present multimodal fusion of face and ECG biometrics. The work reported by authors in Chapter 17 is motivated by the fact that the audio-visual identity verification systems are still far from being commercially feasible for forensic and real time applications. They examine the vulnerability of audio and visual biometrics to forgery and fraudulent attacks. Chapter 18 includes some work on how multi-biometric fusion can address the requirements of next and future generation biometric systems

The book was reviewed by editors Dr. Girija Chetty and Dr. Jucheng Yang We deeply appreciate the efforts of our guest editors: Dr. Norman Poh, Dr. Loris Nanni, Dr. Jianjiang Feng, Dr. Dongsun Park and Dr. Sook Yoon, as well as a number of anonymous reviewers.

Girija Chetty, PhD
Asst. Prof. And Head of Software Engineering
Faculty of Information Sciences and Engineering
University of Canberra
Australia

Dr. Jucheng Yang
Professor
School of Information Technology
Jiangxi University of Finance and Economics
Nanchang, Jiangxi province
China

Part 1

Biometric Fusion

1

Multimodal Fusion for Robust Identity Authentication: Role of Liveness Checks

Girija Chetty and Emdad Hossain
Faculty of Information Sciences and Engineering, University of Canberra, Australia

1. Introduction

Most of the current biometric identity authentication systems currently deployed are based on modeling the identity of a person based on unimodal information, i.e. face, voice, or fingerprint features. Also, many current interactive civilian remote human computer interaction applications are based on speech based voice features, which achieve significantly lower performance for operating environments with low signal-to-noise ratios (SNR). For a long time, use of acoustic information alone has been a great success for several automatic speech processing applications such as automatic speech transcription or speaker authentication, while face identification systems based visual information alone from faces also proved to be of equally successful. However, in adverse operating environments, performance of either of these systems could be suboptimal. Use of both visual and audio information can lead to better robustness, as they can provide complementary secondary clues that can help in the analysis of the primary biometric signals (Potamianos et al (2004)). The joint analysis of acoustic and visual speech can improve the robustness of automatic speech recognition systems (Liu et al (2002), Gurbuz et al (2002).

There have been several systems proposed on use of joint face-voice information for improving the performance of current identity authentication systems. However, most of these state-of-the-art authentication approaches are based on independently processing the voice and face information and then fusing the scores - the score fusion (Chibelushi et al (2002), Pan et al (2000), Chaudari et. al.(2003)). A major weakness of these systems is that they do not take into account fraudulent replay attack scenarios into consideration, leaving them vulnerable to spoofing by recording the voice of the target in advance and replaying it in front of the microphone, or simply placing a still picture of the target's face in front of the camera. This problem can be addressed with liveness verification, which ensures that biometric cues are acquired from a live person who is actually present at the time of capture for authenticating the identity. With the diffusion of Internet based authentication systems for day-to-day civilian scenarios at a astronomical pace (Chetty and Wagner (2008)), it is high time to think about the vulnerability of traditional biometric authentication approaches and consider inclusion of liveness checks for next generation biometric systems. Though there is some work in finger print based liveness checking techniques (Goecke and Millar (2003), Molhom et al (2002)), there is hardly any work in liveness checks based on user-

friendly biometric identifiers (face and voice), which enjoy more acceptability for civilian Internet based applications requiring person identity authentication.
A significant progress however, has been made in independent processing of face only or voice only based authentication approaches (Chibelushi et al (2002), Pan et al (2000), Chaudari et. al.(2003)), in which until now, inherent coupling between jointly occurring primary biometric identifiers were not taken into consideration. Some preliminary approaches such as the ones described in (Chetty and Wagner (2008), Goecke and Millar (2003)), address liveness checking problem by using the traditional acoustic and visual speech features for testing liveness. Both these approaches, neither considered an inherent coupling between speech and orafacial articulators (lips, jaw and chin) during speech production, nor used a solid pattern recogntion based evaluation framework for the validating the performance of co-inertia features.
In this Chapter we propose a novel approach for extraction of audio-visual correlation features based on cross-modal association models, and formulate a hybrid fusion framework for modelling liveness information in the identity authentication approach. Further, we develop a sound evaluation approach based on Bayesian framework for assessing the vulnerability of system at different levels of replay attack complexity. The rest of the Chapter is organized as follows. Section 2 describes the motivation for using the proposed approach, and the details the cross-modal association models are described in Section 3. Section 4 describes the hybrid fusion approach for combining the correlation features with loosely couple and mutually independent face-speech components. The data corpora used and the experimental setup for evaluation of the proposed features is described in Section 5. The experimental results, evaluating proposed correlation features and hybrid fusion technique is discussed in Section 6. Finally, Section 7 summarises the conclusions drawn from this work and plans for further research.

2. Motivation for cross modal association models

The motivation to use cross-modal association models is based on the following two observations: The first observation is in relation to any video event, for example a speaking face video, where the content usually consists of the co-occurring audio and the visual elements. Both the elements carry their contribution to the highest level semantics, and the presence of one has usually a "priming" effect on the other: when hearing a dog barking we expect the image of a dog, seeing a talking face we expect the presence of her voice, images of a waterfall usually bring the sound of running water etc. A series of psychological experiments on the cross-modal influences (Molhom et al (2002), MacDonald and McGurk (1978)) have proved the importance of synergistic fusion of the multiple modalities in the human perception system. A typical example of this kind is the well-known McGurk effect (MacDonald and McGurk (1978)). Several independent studies by cognitive psychologists suggest that the type of multi-sensory interaction between acoustic and orafacial articulators occurring in the McGurk effect involves both the early and late stages of integration processing (MacDonald and McGurk (1978)). It is likely that a human brain uses a hybrid form of fusion that depends on the availability and quality of different sensory cues.
Yet, in audiovisual speech and speaker verification systems, the analysis is usually performed separately on different modalities, and the results are brought together using different fusion methods. However, in this process of separation of modalities, we lose

valuable cross-modal information about the whole event or the object we are trying to analyse and detect. There is an inherent association between the two modalities and the analysis should take advantage of the synchronised appearance of the relationship between the audio and the visual signal. The second observation relates to different types of fusion techniques used for joint processing of audiovisual speech signals. The late-fusion strategy, which comprises decision or the score fusion, is effective especially in case the contributing modalities are uncorrelated and thus the resulting partial decisions are statistically independent. Feature level fusion techniques, on the other hand, can be favoured (only) if a couple of modalities are highly correlated. However, jointly occurring face and voice dynamics in speaking face video sequences, is neither highly correlated (mutually dependent) nor loosely correlated nor totally independent (mutually independent). A complex and nonlinear spatiotemporal coupling consisting of highly coupled, loosely coupled and mutually independent components may exist between co-occurring acoustic and visual speech signals in speaking face video sequences (Jiang et al(2002), Yehia et al (1999)). The compelling and extensive findings by authors in Jiang et al (2002), validate such complex relationship between external face movements, tongue movements, and speech acoustics when tested for consonant vowel (CV) syllables and sentences spoken by male and female talkers with different visual intelligibility ratings. They proved that the there is a higher correlation between speech and lip motion for C/a/ syllables than for C/i/ and C/u/ syllables. Further, the degree of correlation differs across different places of articulation, where lingual places have higher correlation than bilabial and glottal places. Also, mutual coupling can vary from talker to talker; depending on the gender of the talker, vowel context, place of articulation, voicing, and manner of articulation and the size of the face. Their findings also suggest that male speakers show higher correlations than female speakers. Further, the authors in Yehia et al (1999), also validate the complex, spatiotemporal and non-linear nature of the coupling between the vocal-tract and the facial articulators during speech production, governed by human physiology and language-specific phonetics. They also state that most likely connection between the tongue and the face is indirectly by way of the jaw. Other than the biomechanical coupling, another source of coupling is the control strategy between the tongue and cheeks. For example, when the vocal tract is shortened the tongue does not get retracted.

Due to such a complex nonlinear spatiotemporal coupling between speech and lip motion, this could be an ideal candidate for detecting and verifying liveness, and modelling the speaking faces by capturing this information can make the biometric authentication systems less vulnerable to spoof and fraudulent replay attacks, as it would be almost impossible to spoof a system which can accurately distinguish the artificially manufactured or synthesized speaking face video sequences from the live video sequences. Next section briefly describes the proposed cross modal association models based on cross-modal association models.

3. Cross-modal association models

In this section we describe the details of extracting audio-visual features based on cross-modal association models, which capture the nonlinear correlation components between the audio and lip modalities during speech production. This section is organised as follows: The details of proposed audio-visual correlation features based on different cross modal association techniques: Latent Semantic Analysis (LSA) technique, Cross-modal Factor Analysis (CFA) and Canonical Correlation Analysis (CCA) technique is described next.

3.1 Latent Semantic Analysis (LSA)

Latent semantic analysis (LSA) is used as a powerful tool in text information retrieval to discover underlying semantic relationships between different textual units e.g. keywords and paragraphs (Li et al(2003), Li et al(2001)). It is possible to detect the semantic correlation between visual faces and their associated speech based on the LSA technique. The method consists of three major steps: the construction of a joint multimodal feature space, the normalization, the singular value decomposition (SVD), and the semantic association measurement.

Given n visual features and m audio features at each of the t video frames, the joint feature space can be expressed as:

$$X = [V_1,\ldots,V_i,\ldots,V_n,A_1,\ldots,A_i,\ldots A_m] \tag{1}$$

where

$$V_i = (v_i(1), v_i(2),\ldots,v_i(t))^T \tag{2}$$

and

$$A_i = (a_i(1), a_i(2),\ldots,a_i(t))^T \tag{3}$$

Various visual and audio features can have quite different variations. Normalization of each feature in the joint space according to its maximum elements (or certain other statistical measurements) is thus needed and can be expressed as:

$$\hat{X}_{ij} = \frac{X_{kl}}{\max\limits_{k,l}(abs(X_{kl})} \qquad \forall k,l \tag{4}$$

After normalisation, all elements in the normalised matrix $\hat{X}$ have values between –1 and 1. SVD can then be performed as follows:

$$\hat{X} = S \cdot V \cdot D^T \tag{5}$$

where S and D are matrices composed of left and right singular vectors and V is the diagonal matrix of singular values in descending order.

Keeping only the first k singular vectors in S and D, we can derive an optimal approximation of with reduced feature dimensions, where the semantic correlation information between visual and audio features is mostly preserved. Traditional Pearson correlation or mutual information calculation (Li et al (2003), Hershey and Movellan (1999), Fisher et al(2000)) can then be used to effectively identify and measure semantic associations between different modalities. Experiments in Li et al(2003), have shown the effectiveness of LSA and its advantages over the direct use of traditional correlation calculation.

The above optimization of $\hat{X}$ in the least square sense can be expressed as:

$$\hat{X} \cong \tilde{X} = \tilde{S}.\tilde{V}.\tilde{D} \tag{6}$$

Where $\tilde{S}, \tilde{V}, and\, \tilde{D}$ consist of the first k vectors in S, V, and D, respectively.

The selection of an appropriate value for *k* is still an open issue in the literature. In general, *k* has to be large enough to keep most of the semantic structures. Eqn. 6 is not applicable for applications using off-line training since the optimization has to be performed on the fly directly based on the input data. However, due to the orthogonal property of singular vectors, we can rewrite Eqn. 6 in a new form as follows:

$$\hat{X} \cong \tilde{X} = \tilde{S} \cdot V \cdot \tilde{D}^T \tag{7}$$

Now we only need the $\tilde{D}$ matrix in the calculation, which can be trained in advance using ground truth data. This derived new form is important for those applications that need off-line trained SVD results.

3.2 Cross Modal Factor Analysis (CMA)

LSA does not distinguish features from different modalities in the joint space. The optimal solution based on the overall distribution, which LSA models, may not best represent the semantic relationships between the features of different modalities, since distribution patterns among features from the same modality will also greatly impact the results of the LSA.

A solution to the above problem is to treat the features from different modalities as two separate subsets and focus only on the semantic patterns between these two subsets. Under the linear correlation model, the problem now is to find the optimal transformations that can best represent or identify the coupled patterns between the features of the two different subsets. We adopt the following optimization criterion to obtain the optimal transformations:

Given two mean-centred matrices X and Y, which consist of row-by-row coupled samples from two subsets of features, we want orthogonal transformation matrices A and B that can minimise the expression:

$$\|XA - YB\|_F^2 \tag{8}$$

where

$A^T A = I$ and $B^T B = I$.

$\|M\|_F$ denotes the Frobenius norm of the matrix M and can be expressed as:

$$\|M\|_F = \left(\sum_i \sum_j |m_{ij}|^2 \right)^{1/2} \tag{9}$$

In other words, *A* and *B* define two orthogonal transformation spaces where coupled data in *X* and *Y* can be projected as close to each other as possible.

Since we have:

$$\begin{aligned} \|XA - YB\|_F^2 &= trace\left((XA - XB).(YA - YB)^T\right) \\ &= trace\left(XAA^T X^T + YBB^T Y^T - XAB^T Y^T - YBA^T X^T\right) \\ &= trace\left((XX^T) + trace(YY^T) - 2 \cdot trace(XAB^T Y^T)\right) \end{aligned} \tag{10}$$

where the trace of a matrix is defined to be the sum of the diagonal elements. We can easily see from above that matrices A and B which maximise *trace* (XAB^TY^T) will minimise (10). It can be shown (Li et al(2003)), that such matrices are given by:

$$\begin{cases} A = S_{xy} \\ B = D_{xy} \end{cases} \quad \text{where } X^TY = S_{xy} \cdot V_{xy} \cdot D_{xy} \tag{11}$$

With the optimal transformation matrices A and B, we can calculate the transformed version of X and Y as follows:

$$\begin{cases} \tilde{X} = X \cdot A \\ \tilde{Y} = Y \cdot B \end{cases} \tag{12}$$

Corresponding vectors in $\tilde{X}$ and $\tilde{Y}$ are thus optimised to represent the coupled relationships between the two feature subsets without being affected by distribution patterns within each subset. Traditional Pearson correlation or mutual information calculation (Li et al (2003), Hershey and Movellan(1999), Fisher et al(2000)) can then be performed on the first and most important k corresponding vectors in $\tilde{X}$ and $\tilde{Y}$, which similar to those in LSA preserve the principal coupled patterns in much lower dimensions.
In addition to feature dimension reduction, feature selection capability is another advantage of CFA. The weights in A and B automatically reflect the significance of individual features, clearly demonstrating the great feature selection capability of CFA, which makes it a promising tool for different multimedia applications including audiovisual speaker identity verification.

3.3 Canonical Correlation Analysis (CCA)

Following the development of the previous section, we can adopt a different optimization criterion: Instead of minimizing the projected distance, we attempt to find transformation matrices A and B that maximise the correlation between X_A and Y_B. This can be described more specifically using the following mathematical formulations:
Given two mean centered matrices X and Y as defined in the previous section, we seek matrices A and B such that

$$correlation(XA, XB) = correlation(\tilde{X}, \tilde{Y}) = diag(\lambda_1, \cdots \lambda_i, \cdots, \lambda_l) \tag{13}$$

where $\tilde{X} = Y \cdot B$, and $1 \geq \lambda_1 \geq, \cdots, \lambda_i, \cdots, \geq \lambda_l \geq 0$. λ_i represents the largest possible correlation between the ith translated features in $\tilde{X}$ and $\tilde{Y}$. A statistical method called canonical correlation analysis (Lai and Fyfe (1998), Tabanick and Fidell (1996)] can solve the above problem with additional norm and orthogonal constraints on translated features:

$$E\{\tilde{X}^T \cdot \tilde{X}\} = I \text{ and } E\{\tilde{Y} \cdot \tilde{Y}\} = I \tag{14}$$

The CCA is described in further details in Hotelling (1936) and Hardoon et al(2004). The optimization criteria used for all three cross modal associations CFA, CCA and LSA exhibit a high degree of noise tolerance. Hence the correlation features extracted perform better as compared to normal correlation analysis against noisy environmental conditions.

4. Hybrid audiovisual fusion

In this Section, we describe the fusion approach used for combing the extracted audio-lip correlated components with mutually independent audio and visual speech features.

4.1 Feature fusion of correlated components

The algorithm for fusion of audiovisual feature extracted using the cross modal association (CMA) models (a common term being used here to represent LSA, CFA or CCA analysis methods) can be described as follows:

Let f_A and f_L represent the audio MFCC and lip-region eigenlip features respectively. A and B represent the CMA transformation matrices (LSA, CFA or CMA matrices). One can apply CMA to find two new feature sets $f_A^{'} = A^T f_A$ and $f_L^{'} = B^T f_L$ such that the between-class cross-modal association coefficient matrix of $f_A^{'}$ and $f_L^{'}$ is diagonal with maximised diagonal terms. However, maximised diagonal terms do not necessarily mean that all the diagonal terms exhibit strong cross modal association. Hence, one can pick the maximally correlated components that are above a certain correlation threshold θ_k. Let us denote the projection vector that corresponds to the diagonal terms larger than the threshold θ_k by $\widetilde{w}_A$. Then the corresponding projections of f_A and f_L are given as:

$$\tilde{f}_A = \tilde{w}_A^T . f_A \text{ and } \tilde{f}_A = \tilde{w}_L^T . f_A \tag{15}$$

Here $\tilde{f}_A$ and $\tilde{f}_L$ are the correlated components that are embedded in f_A and f_L. By performing feature fusion of correlated audio and lip components, we obtained the CMA optimised feature fused audio-lip feature vector:

$$\tilde{f}_{AL}^{LSA} = \left[\tilde{f}_A^{LSA} \quad \tilde{f}_L^{LSA} \right] \tag{16}$$

$$\tilde{f}_{AL}^{CFA} = \left[\tilde{f}_A^{CFA} \quad \tilde{f}_L^{CFA} \right] \tag{17}$$

$$\tilde{f}_{AL}^{CCA} = \left[\tilde{f}_A^{CCA} \quad \tilde{f}_L^{CCA} \right] \tag{18}$$

4.2 Late fusion of mutually independent components

In the Bayesian framework, late fusion can be performed using the product rule assuming statistically independent modalities, and various methods have been proposed in the literature as alternatives to the product rule such as max rule, min rule and the reliability-based weighted summation rule (Nefian et al(2002), Movellan and Mineiro(1997)). In fact, the most generic way of computing the joint scores can be expressed as a weighted summation

$$\rho(\lambda_r) = \sum_{n=1}^{N} w_n \log P(f_n \mid \lambda_r) \quad for\ r = 1,2,\ldots,R \tag{19}$$

where $\rho_n(\lambda_r)$ is the logarithm of the class-conditional probability, $P(f_n|\lambda_r)$, for the nth modality f_n given class λ_r, and w_n denotes the weighting coefficient for modality n, such that $\Sigma_n w_n = 1$. Then the fusion problem reduces to a problem of finding the optimal weight coefficients. Note that when $w_n = \frac{1}{N}$ $\forall n$, Eqn. 14 is equivalent to the product rule. Since the w_n values can be regarded as the reliability values of the classifiers, this combination method is also referred to as RWS (Reliability Weighted Summation) rule (Jain et al(2005), Nefian et al(2002)).The statistical and the numerical range of these likelihood scores vary from one classifier to another. Using sigmoid and variance normalization as described in (Jain et al(2005)), the likelihood scores can be normalised to be within the (0, 1) interval before the fusion process.

The hybrid audiovisual fusion vector in this Chapter was obtained by late fusion of feature fused correlated components ($\tilde{f}_{AL}^{LSA}, \tilde{f}_{AL}^{CFA}, \tilde{f}_{AL}^{CCA}$) with uncorrelated and mutually independent implicit lip texture features, and audio features with weights selected using the an automatic weight adaptation rule and is described in the next Section.

4.3 Automatic weight adaptation

For the RWS rule, the fusion weights are chosen empirically, whereas for the automatic weight adaptation, a mapping needs to be developed between modality reliability estimate and the modality weightings. The late fusion scores can be fused via sum rule or product rule. Both methods were evaluated for empirically chosen weights, and it was found that the results achieved for both were similar. However, sum rule for fusion has been shown to be more robust to classifier errors in literature (Jain et al (2005), Sanderson (2008)), and should perform better when the fusion weights are automatically, rather than empirically determined. Hence the results for additive fusion only, are presented here. Prior to late fusion, all scores were normalised to fall into the range of [0,1], using min-max normalisation.

$$\begin{aligned} P(S_i|x_A, x_V) &= \alpha P(S_i|x_A) + \beta P(S_i|x_v) \\ P(S_i|x_A, x_V) &= P(S_i|x_A)^{\alpha} \times P(S_i|x_v)^{\beta} \end{aligned} \tag{20}$$

where

$$\alpha = \begin{Bmatrix} 0 & c \le 1 \\ 1+c & -1 < c < 0 \\ 1 & c \ge 0 \end{Bmatrix} \quad and \quad \beta = \begin{Bmatrix} 1 & c \le 0 \\ 1-c & 0 < c < 1 \\ 0 & c \ge 1 \end{Bmatrix} \tag{21}$$

where, x_A and x_V refer to the audio test utterance and visual test sequence/image respectively.

To carry out automatic fusion, that adapts to varying acoustic SNR conditions, a single parameter c, the *fusion parameter*, was used to define the weightings; the audio weight α and the visual weight β, i.e., both α and β dependent on c. Higher values of c (>0) place more emphasis on the audio module whereas lower values (<0) place more emphasis on the visual module. For $c \geq 1$, $\alpha = 1$ and $\beta = 0$, hence the audiovisual fused decision is based entirely on the audio likelihood score, whereas, for $c \leq -1$, $\alpha = 0$ and $\beta = 1$, the decision is based entirely on the visual score. So in order to account for varying acoustic conditions, only c has to be adapted.

The reliability measure was the audio log-likelihood score $\rho_n(\lambda_r)$. As the audio SNR decreases, the absolute value of this reliability measure decreases, and becomes closer to the threshold for client likelihoods. Under clean test conditions, this reliability measure increases in absolute value because the client model yields a more distinct score. So, a mapping between ρ and c can automatically vary α and β and hence place more or less emphasis on the audio scores. To determine the mapping function $c(\rho)$, the values of c which provided for optimum fusion, c_{opt}, were found by exhaustive search for the N tests at each SNR levels. This was done by varying c from -1 to $+1$, in steps of 0.01, in order to find out which c value yielded the best performance. The corresponding average reliability measures were calculated, ρ_{mean}, across the N test utterances at each SNR level.

$$c(\rho) = c_{os} + \frac{h}{\exp\left[d.(\rho + \rho_{os})\right]} \tag{22}$$

A sigmoid function was employed to provide a mapping between the c_{opt} and the ρ_{mean} values, where c_{os} and ρ_{os} represent the offsets of the fusion parameter and reliability estimate respectively; h captures the range of the fusion parameter; and d determines the steepness of the sigmoid curve. The sigmoidal parameters were determined empirically to give the best performance. Once the parameters have been determined, automatic fusion can be carried out. For each set of N test scores, the ρ value was calculated and mapped to c, using $c = c(\rho)$, and hence, α and β can be determined. This fusion approach is similar to that used in (Sanderson(2008)) to perform speech recognition. The method can also be considered to be a secondary classifier, where the measured ρ value arising from the primary audio classifier is *classified* to a suitable c value; also, the secondary classifier is *trained* by determining the parameters of the sigmoid mapping.

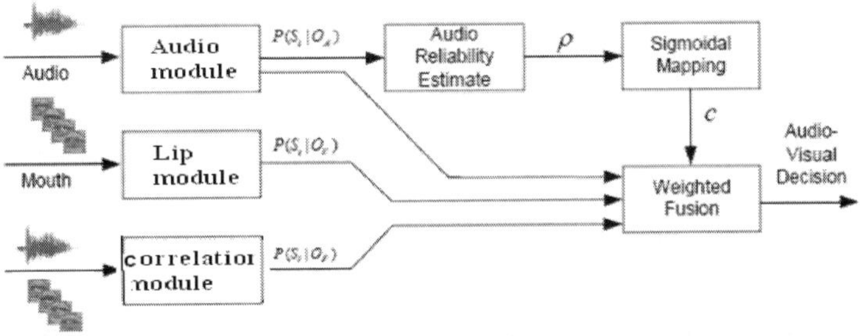

Fig. 1. System Overview of Hybrid Fusion Method

The described method can be employed to combine any two modules. It can also be adapted to include a third module. We assume here that only the audio signal is degraded when testing, and that the video signal is of fixed quality. The third module we use here is an audio-lip correlation module, which involves a cross modal transformation of feature fused audio-lip features based on CCA, CFA or LSA cross modal analysis as described in Section 3.
An overview of the fusion method described is given in Figure 1. It can be seen that the reliability measure, ρ, depends only on the audio module scores. Following the sigmoidal mapping of ρ, the fusion parameter c is passed into the fusion module along with the three scores arising from the three modules; fusion takes place to give the audiovisual decision.

5. Data corpora and experimental setup

A experimental evaluation of proposed correlation features based on cross-modal association models and their subsequent hybrid usion was carried out with two different audio-visual speaking face video corpora VidTIMIT (Sanderson(2008)) and (DaFEx (Battocchi et al (2004), Mana et al (2006)). Figure 2 show some images from the two corpora. The details of the two corpora are given in VidTIMIT (Sanderson(2008), DaFEx (Battocchi et al (2004), Mana et al (2006)).
The pattern recognition experiments with the data from the two corpora and the correlation features extracted from the data involved two phases, the training phase and the testing phase. In the training phase a 10-mixture Gaussian mixture model λ of a client's audiovisual feature vectors was built, reflecting the probability densities for the combined phonemes and visemes (lip shapes) in the audiovisual feature space. In the testing phase, the clients' live test recordings were first evaluated against the client's model λ by determining the log likelihoods $log\ p(X\,|\,\lambda)$ of the time sequences X of audiovisual feature vectors under the usual assumption of statistical independence of successive feature vectors.
For testing replay attacks, we used a two level testing, a different approach from traditional impostor attacks testing used in identity verification experiments. Here the impostor attack is a surreptitious replay of previously recorded data and such an attack can be simulated by synthetic data. Two different types of replay attacks with increasing level of sophistication and complexity were simulated: the *"static"* replay attacks and the *"dynamic"* replay attacks.

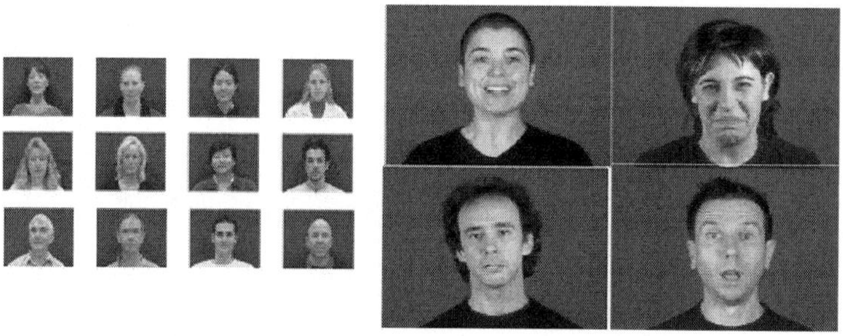

(a) VidTIMIT corpus (b) DaFeX corpus

Fig. 2. Sample Images from VidTIMIT and DaFeX corpus

For testing *"static"* replay attacks, a number of "fake" or synthetic recordings were constructed by combining the sequence of audio feature vectors from each test utterance with ONE visual feature vector chosen from the sequence of visual feature vectors and keeping that visual feature vector constant throughout the utterance. Such a synthetic sequence represents an attack on the authentication system, carried out by replaying an audio recording of a client's utterance while presenting a still photograph to the camera. Four such fake audiovisual sequences were constructed from different still frames of each client test recording. Log-likelihoods $log\ p(X'|\lambda)$ were computed for the fake sequences X' of audiovisual feature vectors against the client model λ. In order to obtain suitable thresholds to distinguish live recordings from fake recordings, detection error trade-off (DET) curves and equal error rates (EER) were determined.

For testing *"dynamic"* replay attacks, an efficient photo-realistic audio-driven facial animation technique with near-perfect lip-synching of the audio and several image key-frames of the speaking face video sequence was done to create a artificial speaking character for each person (Chetty and Wagner(2008), Sanderson(2008).

In Bayesian framework, the liveness verification task can be essentially considered as a two class decision task, distinguishing the test data as a genuine client or an impostor. The impostor here is a fraudulent replay of client specific biometric data. For such a two-class decision task, the system can make two types of errors. The first type of error is a False Acceptance Error (FA), where an impostor (fraudulent replay attacker) is accepted. The second error is a False Rejection (FR), where a true claimant (genuine client) is rejected. Thus, the performance can be measured in terms of False Acceptance Rate (FAR) and False Reject Rate (FRR), as defined as (Eqn. 23):

$$FAR\,\% = \frac{I_A}{I_T} \times 100\,\% \qquad\qquad FRR\,\% = \frac{C_R}{C_T} \times 100\,\% \tag{23}$$

where I_A is the number of impostors classified as true claimants, I_T is the total number of impostor classification tests, C_R is the number of true claimants classified as impostors, and C_T is the total number of true claimant classification tests. The implications of this is minimizing the FAR increases the FRR and vice versa, since the errors are related. The trade-off between FAR and FRR is adjusted using the threshold θ, an experimentally determined speaker-independent global threshold from the training/enrolment data. The trade-off between FAR and FRR can be graphically represented by a Receiver Operating Characteristics (ROC) plot or a Detection Error Trade-off (DET) plot. The ROC plot is on a linear scale, while the DET plot is on a normal-deviate logarithmic scale. For DET plot, the FRR is plotted as a function of FAR. To quantify the performance into a single number, the Equal Error Rate (EER) is often used. Here the system is configured with a threshold, set to an operating point when FAR % = FRR %.

It must be noted that the threshold θ can also be adjusted to obtain a desired performance on test data (data unseen by the system up to this point). Such a threshold is known as the aposteriori threshold. However, if the threshold is fixed before finding the performance, the threshold is known as the apriori threshold. The apriori threshold can be found via experimental means using training/enrolment or evaluation data, data which has also been unseen by the system up to this point, but is separate from test data.

Practically, the a priori threshold is more realistic. However, it is often difficult to find a reliable apriori threshold. The test section of a database is often divided into two sets: evaluation data and test data. If the evaluation data is not representative of the test data, then the apriori threshold will achieve significantly different results on evaluation and test data. Moreover, such a database division reduces the number of verification tests, thus decreasing the statistical significance of the results. For these reasons, many researchers prefer to use the aposteriori and interpret the performance obtained as the expected performance.

Different subsets of data from the VidTIMIT and DaFeX were used. The gender-specific universal background models (UBMs) were developed using the training data from two sessions, Session 1 and Session 2, of the VidTIMIT corpus, and for testing Session 3 was used. Due to the type of data available (test session sentences differ from training session sentences), only text-independent speaker verification experiments could be performed with VidTIMIT. This gave 1536 (2*8*24*4) seconds of training data for the male UBM and 576 (2*8*19*4) seconds of training data for the female UBM. The GMM topology with 10 Gaussian mixtures was used for all the experiments. The number of Gaussian mixtures was determined empirically to give the best performance. For the DaFeX database, similar gender-specific universal background models (UBMs) were obtained using training data from the text-dependent subsets corresponding to neutral expression. Ten sessions of the male and female speaking face data from these subsets were used for training and 5 sessions for testing.

For all the experiments, the global threshold was set using test data. For the male only subset of the VidTIMIT database, there were 48 client trials (24 male speakers x 2 test utterances in Session 3) and 1104 impostor trials (24 male speakers x 2 test utterances in Session 3 x 23 impostors/client), and for the female VidTIMIT subset, there were 38 client trials (19 male speakers x 2 test utterances in Session 3) and 684 impostor trials (19 male speakers x 2 test utterances in Session 3 x 18 impostors/client). For the male only subset for DaFeX database, there were 25 client trials (5 male speakers x 5 test utterances in each subset) and 100 impostor trials (5 male speakers x 5 test utterances x 4 impostors/client), and for the female DaFeX subset, there were similar numbers of the client and impostor trials as in the male subset as we used 5 male and 5 female speakers from different subsets.

Different sets of experiments were conducted to evaluate the performance of the proposed correlation features based on cross modal association models (LSA, CCA and CMA), and their subsequent fusion in terms of DET curves and equal error rates (EER). Next Section discusses the results from different experiments.

6. Experimental results

Figure 3 plots the maximised diagonal terms of the between class correlation coefficient matrix after the LSA, CCA and CFA analysis of audio MFCC and lip-texture (f_{eigLip}^{N}) features. Results for the CFA analysis technique for the VidTIMIT male subset are only discussed here. As can be observed from Figure 3, the maximum correlation coefficient is around 0.7 and 15 correlation coefficients out of 40 are higher than 0.1.

Table 1 presents the EER performance of the feature fusion of correlated audio-lip fusion features (cross modal features) for varying correlation coefficient threshold θ. Note that, when all the 40 transformed coefficients are used, the EER performance is 6.8%. The EER performance is observed to have a minimum around 4.7% for threshold values from 0.1 to 0.4. The optimal threshold that minimises the EER performance and the feature dimension is found to be 0.4.

	EER(%) at (θ, dim)						
Θ	0.0	0.1	0.2	0.3	0.4	0.5	0.6
Dim	40	15	12	10	8	6	4
$\tilde{f}_{AL}^{CFA}$	6.8	4.7	5.3	5.0	4.7	7.4	10.3
$\tilde{f}_{AL}^{CCA}$	7.5	5.18	5.84	5.5	5.18	8.16	11.36
$\tilde{f}_{AL}^{LSA}$	11.7	8.09	9.12	8.6	8.09	12.74	17.74

Table 1. Results for correlation features based in CMA models: EERs at varying correlation coefficient threshold values (θ) with the corresponding projection dimension (dim)

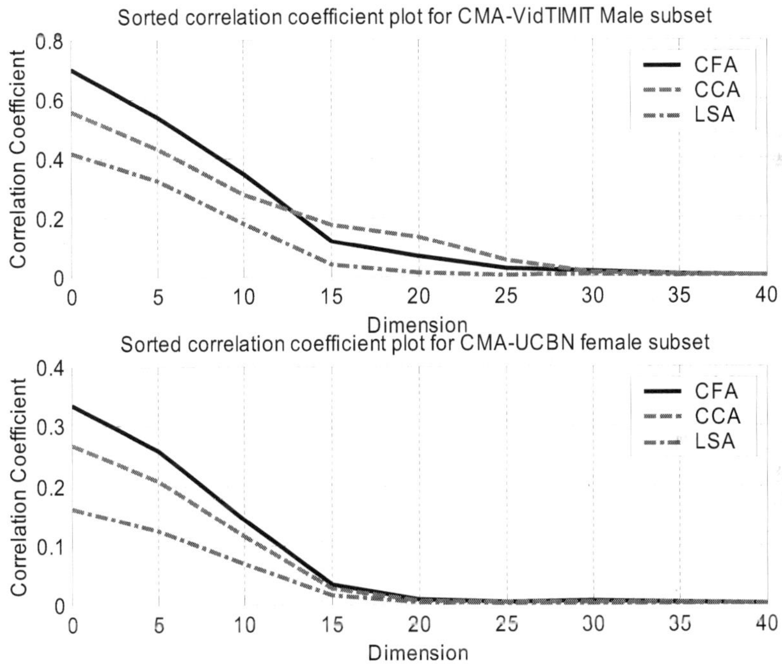

Fig. 3. Sorted correlation coefficient plot for audio and lip texture cross modal analysis

As can seen in Table 2 and Figure 4, for static replay attack scenarios (from the last four rows in Table 2), the nonlinear correlation components between acoustic and orafacial articulators during speech production is more efficiently captured by hybrid fusion scheme involving late fusion of audio f_{mfcc} features, f_{eigLip} lip features, and feature-level fusion of correlated audio-lip $\tilde{f}_{mfcc-eigLip}$ features). This could be due to modelling of identity specific mutually independent, loosely coupled and closed coupled audio-visual speech components with this approach, resulting in an enhancement in overall performance.

Modality	VidTIMIT male subset			DaFeX male subset		
	CFA EER (%)	CCA EER (%)	LSA EER (%)	CFA EER (%)	CCA EER (%)	LSA EER (%)
f_{mfcc}	4.88	4.88	4.88	5. 7	5. 7	5. 7
f_{eigLip}	6.2	6.2	6.2	7.64	7.64	7.64
$f_{mfcc-eigLip}$	7.87	7.87	7.87	9.63	9.63	9.63
$\tilde{f}_{mfcc-eigLip}$	3.78	2.3	2.76	4.15	2.89	3.14
$f_{mfcc} + f_{mfcc-eigLip}$	2.97	2.97	2.97	3.01	3.01	3.01
$f_{mfcc} + \tilde{f}_{mfcc-eigLip}$	0.56	**0.31**	0.42	0.58	0.38	0.57
$f_{mfcc} + f_{eigLip} + f_{mfcc-eigLip}$	6.68	6.68	6.68	7.75	7.75	7.75
$f_{mfcc} + f_{eigLip} + \tilde{f}_{mfcc-eigLip}$	0.92	0.72	0.81	0.85	0.78	0.83

Table 2. EER performance for static replay attack scenario with late fusion of correlated components with mutually independent components: (+) represents RWS rule for late fusion, (-) represents feature level fusion)

Though all correlation features performed well, the CCA features appear to be the best performer for static attack scenario, with an EER of 0.31%. This was the case for all the subsets of data shown in Table 2. Also, the EERs for hybrid fusion experiments with $\tilde{f}_{mfcc-eigLip}$ correlated audio lip features performed better as compared to ordinary feature fusion of $f_{mfcc-eigLip}$ features. EERs of 0.31% and 0.72% were achieved for $f_{mfcc} + \tilde{f}_{mfcc-eigLip}$ and $f_{mfcc} + f_{eigLip} + \tilde{f}_{mfcc-eigLip}$ features, the hybrid fusion types involving CMA optimised correlated features, as compared to an EER of 2.97% for $f_{mfcc} + f_{mfcc-eigLip}$ features and 6.68% for $f_{mfcc} + f_{eigLip} + f_{mfcc-eigLip}$ features, which are hybrid fusion types based on ordinary feature fusion of uncorrelated audio-lip features. This shows that correlation features based on proposed cross-modal association models can extract the intrinsic nonlinear temporal correlations between audio-lip features and could be more useful for checking liveness.
The EER table in Table 3 shows the evaluation of hybrid fusion of correlated audio-lip features based on cross modal analysis (CFA, CCA and LSA) for dynamic replay attack scenario. As can be seen, the CMA optimized correlation features perform better as compared to uncorrelated audio-lip features for complex dynamic attacks. Further, for the VidTIMIT male subset, it was possible to achieve the best EER of 10.06% for $f_{mfcc} + f_{eigLip} + \tilde{f}_{mfcc-eigLip}$ features, a hybrid fusion type involving feature fusion of correlated audio-lip features based on CCA analysis.

7. Conclusion

In this Chapter, we have proposed liveness verification for enhancing the robustness of biometric person authentication systems against impostor attacks involving fraudulent replay of client data. Several correlation features based on novel cross-modal association models have been proposed as an effective countermeasure against such attacks. These new

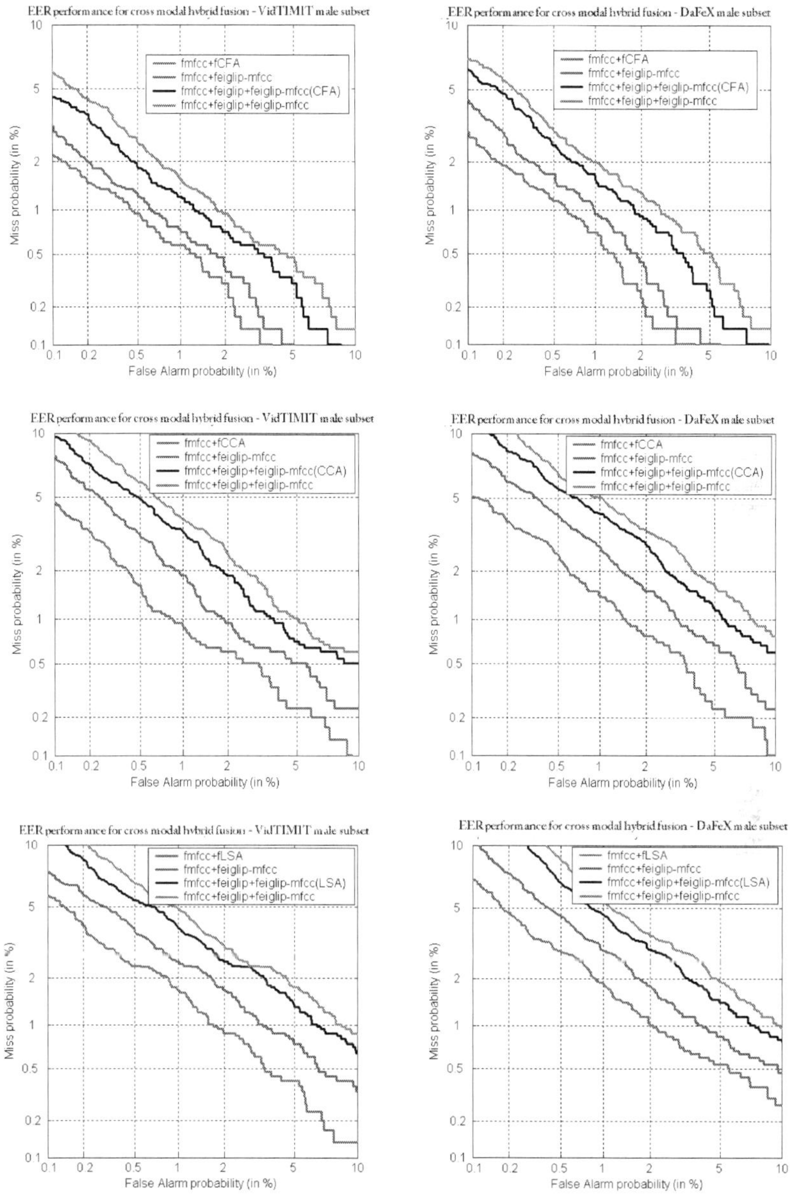

Fig. 4. DET curves for hybrid fusion of correlated audio-lip features and mutually independent audio-lip features for static replay attack scenario

correlation measures model the nonlinear acoustic-labial temporal correlations for the speaking faces during speech production, and can enhance the system robustness against replay attacks.
Further, a systematic evaluation methodology was developed, involving increasing level of difficulty in attacking the system - moderate and simple static replay attacks, and, sophisticated and complex dynamic replay attacks, allowing a better assessment of system vulnerability against attacks of increasing complexity and sophistication. For both static and dynamic replay attacks, the EER results were very promising for the proposed correlation features, and their hybrid fusion with loosely coupled (feature-fusion) and mutually independent (late fusion) components, as compared to fusion of uncorrelated features. This suggests that it is possible to perform liveness verification in authentication paradigm. and thwart replay attacks on the system. Further, this study shows that, it is difficult to beat the system, if underlying modelling approach involves efficient feature extraction and feature selection techniques, that can capture intrinsic biomechanical properties accurately.

	VidTIMIT male subset			DaFeX male subset		
Modality	CFA EER (%)	CCA EER (%)	LSA EER (%)	CFA EER (%)	CCA EER (%)	LSA EER (%)
f_{eigLip}	36.58	36. 58	36. 58	37.51	37. 51	37. 51
$f_{mfcc-eigLip}$	27.68	27.68	27.68	28.88	28.88	28.88
$\tilde{f}_{mfcc-eigLip}$	24.48	22.36	23.78	26.43	24.67	25.89
$f_{mfcc} + f_{mfcc-eigLip}$	22.45	22.45	22.45	23.67	23.67	23.67
$f_{mfcc} + \tilde{f}_{mfcc-eigLip}$	17.89	16.44	19.48	18.46	17.43	20.11
$f_{mfcc} + f_{eigLip} + f_{mfcc-eigLip}$	21.67	21.67	21.67	25.42	25.42	25.42
$f_{mfcc} + f_{eigLip} + \tilde{f}_{mfcc-eigLip}$	14.23	10.06	12.27	16.68	12.36	13.88

Table 3. EER performance for dynamic replay attack scenario with late fusion of correlated components with mutually independent components

However, though the EER performance appeared to be very promising for static replay attack scenarios (EER of 0.31 % for CCA features), the deterioration in performance for more sophisticated - dynamic replay attack scenario (EER of 10.06 % for CCA features), suggests that, there is an urgent need to investigate more robust feature extraction, feature selection, and classifier approaches, as well as sophisticated replay attack modelling techniques. Further research will focus on these two aspects.

8. References

[1] Battocchi, A, Pianesi, F(2004) DaFEx: Un Database di Espressioni Facciali Dinamiche. In Proceedings of the SLI-GSCP Workshop, Padova (Italy).
[2] Chaudhari U.V, Ramaswamy G.N, Potamianos G, and Neti C.(2003) Information Fusion and Decision Cascading for Audio-Visual Speaker Recognition Based on Time-

Varying Stream Reliability Prediction. In IEEE International Conference on Multimedia Expo., volume III, pages 9 - 12, Baltimore, USA.

[3] Chibelushi C.C, Deravi F, and Mason J(2002) A Review of Speech-Based Bimodal Recognition. IEEE Transactions on Multimedia, 4(1):23–37.

[4] Chetty G., and Wagner M(2008), Robust face-voice based speaker identity verification using multilevel fusion, Image and Vision Computing, Volume 26, Issue 9, Pages 1249-1260.

[5] Fisher III J. W, Darrell T, Freeman W. T, Viola P (2000), Learning joint statistical models for audio-visual fusion and segregation, Advances in Neural Information Processing Systems (NIPS), pp. 772-778.

[6] Gerasimos Potamianos, Chalapathy Neti, Juergen Luettin, and Iain Matthews. Audio-Visual Automatic Speech Recognition: An Overview. Issues in Visual and Audio-Visual Speech Processing, 2004.

[7] Goecke R. and Millar J.B.(2003). Statistical Analysis of the Relationship between Audio and Video Speech Parameters for Australian English. In J.L. Schwartz, F. Berthommier, M.A. Cathiard, and D. Sodoyer (eds.), Proceedings of the ISCA Tutorial and Research Workshop on Auditory-Visual Speech Processing AVSP 2003, pages 133-138, St. Jorioz, France.

[8] Gurbuz S, Tufekci Z, Patterson T, and Gowdy J.N (2002) Multi-Stream Product Modal Audio-Visual Integration Strategy for Robust Adaptive Speech Recognition. In Proc. IEEE International Conference on Acoustics, Speech and Signal Processing, Orlando.

[9] Hershey J. and Movellan J (1999) Using audio-visual synchrony to locate sounds, Proc. Advances in Neural Information Processing Systems (NIPS), pp. 813-819.

[10] Hotelling H (1936) Relations between two sets of variates Biometrika, 28:321 377.

[11] Hardoon D. R., Szedmak S. and Shawe-Taylor J (2004) Canonical Correlation Analysis: An Overview with Application to Learning Methods, in Neural Computation Volume 16, Number 12, Pages 2639–2664.

[12] Jain A, Nandakumar K, and Ross A (2005) Score Normalization in Multimodal Biometric Systems, Pattern Recognition.

[13] Jiang J, Alwan A, Keating P.A., Auer Jr. E.T, Bernstein L. E (2002) On the Relationship between Face Movements, Tongue Movements, and Speech Acoustics, EURASIP Journal on Applied Signal Processing :11, 1174–1188.

[14] Lai P. L., and Fyfe C (1998), Canonical correlation analysis using artificial neural networks, Proc. European Symposium on Artificial Neural Networks (ESANN).

[15] Li M, Li D, Dimitrova N, and Sethi I K(2003), Audio-visual talking face detection, Proc. International Conference on Multimedia and Expo (ICME), pp. 473-476, Baltimore, MD.

[16] Li D, Wei G, Sethi I. K, Dimitrova N(2001), Person Identification in TV programs, Journal on Electronic Imaging, Vol. 10, Issue. 4, pp. 930-938.

[17] Liu X, Liang L, Zhaa Y, Pi X, and Nefian A.V(2002) Audio-Visual Continuous Speech Recognition using a Coupled Hidden Markov Model. In Proc. International Conference on Spoken Language Processing.

[18] MacDonald J, & McGurk H (1978), "Visual influences on speech perception process". Perception and Psychophysics, 24, 253-257.

[19] Mana N, Cosi P, Tisato G, Cavicchio F, Magno E. and Pianesi F(2006) An Italian Database of Emotional Speech and Facial Expressions, In Proceedings of "Workshop on Emotion: Corpora for Research on Emotion and Affect", in association with "5th International Conference on Language, Resources and Evaluation (LREC2006), Genoa.

[20] Molholm S et al (2002) Multisensory Auditory-visual Interactions During Early Sensory Processing in Humans: a high-density electrical mapping study, Cognitive Brain Research, vol. 14, pp. 115-128.

[21] Movellan, J. and Mineiro, P(1997), "Bayesian robustification for audio visual fusion". In Proceedings of the Conference on Advances in Neural information Processing Systems 10 (Denver, Colorado, United States). M. I. Jordan, M. J. Kearns, and S. A. Solla, Eds. MIT Press, Cambridge, MA, 742-748.

[22] Nefian V, Liang L. H, Pi X, Liu X, and Murphy K. (2002) Dynamic Bayesian Networks for Audio-visual Speech Recognition, EURASIP Journal on Applied Signal Processing, pp. 1274-1288.

[23] Pan H, Liang Z, and Huang T(2000)A New Approach to Integrate Audio and Visual Features of Speech. In Proc. IEEE International Conference on Multimedia and Expo., pages 1093 – 1096.

[24] Potamianos G, Neti C, Luettin J, and Matthews I (2004) Audio-Visual Automatic Speech Recognition: An Overview. Issues in Visual and Audio-Visual Speech Processing.

[25] Sanderson C (2008). Biometric Person Recognition: Face, Speech and Fusion. VDM-Verlag. ISBN 978-3-639-02769-3.

[26] Tabachnick B, and Fidell L. S (1996), Using multivariate statistics, Allyn and Bacon Press.

[27] Yehia H. C, Kuratate T, and Vatikiotis-Bateson E (1999), Using speech acoustics to drive facial motion, in Proc. the 14th International Congress of Phonetic Sciences, pp. 631–634, San Francisco, Calif, USA.

2

Multimodal Biometric Person Recognition System Based on Multi-Spectral Palmprint Features Using Fusion of Wavelet Representations

Abdallah Meraoumia[1], Salim Chitroub[1] and Ahmed Bouridane[2,3]
[1]*University of Science and Technology of Houari Boumedienne (USTHB), Algiers,*
[2]*Department of Computer Science, King Saud University Riyadh*
[3]*School of Computing, Engineering and Information Sciences, Northumbria University Pandon Building Newcastle upon Tyne,*
[1]*Algeria*
[2]*Saudi Arabia*
[3]*UK*

1. Introduction

The rapid development in many applications for some different areas, such as public security, access control and surveillance, requires reliable and automatic personal recognition for effective security control. Traditionally, passwords (knowledge-based security) and ID cards (token-based) have been used. However, security can be easily breached when a password is divulged or a card is stolen; further, simple passwords are easy to guess and difficult passwords may be hard to recall (Kong & Zhang; 2002). At present, applications of biometrics are rapidly increasing due to inconveniences in using traditional identification method. Biometrics refers to the technologies that can automate the recognition of persons based on one or more intrinsic physical or behavioral traits.

Currently, a number of biometric based technologies have been developed and hand-based person identification is one of these technologies. This technology provides a reliable, low cost and user-friendly solution for a range of access control applications (Kumar & Zhang; 2002). In contrast to other modalities, like face and iris, hand based biometric recognition offers some advantages. First, data acquisition is simple using off the shelf low-resolution cameras, and its processing is also relatively simple. Second, hand based access systems are very suitable for several usages. Finally, hand features are more stable over time and are not susceptible to major changes (Sricharan & Reddy; 2006). Some features related to a human hand are relatively invariant and distinctive to an individual. Among these features, palmprint modality has been systematically used for human recognition using the palm patterns. The rich texture information of palmprint offers one of the powerful means in personal identification (Fang & Maylor; 2004).

Several studies for palmprint-based personal recognition have focused on improving the performance of palmprint images captured under visible light. However, during the past

few years, some researchers have considered multi-spectral images to improve the effect of these systems. Multi spectral imaging is a new technique in remote sensing, medical imaging and machine vision that generate several images corresponding to different wavelengths. This technique can be give different information from the same scene using an acquisition device to capture the palmprint images under visible and infrared light resulting into four color bands (*RED, BLUE, GREEN* or *Near-IR* (*NIR*) (Zhang & Guo; 2010). The idea is to employ the resulting information in these color bands to improve the performance of palmprint recognition. This paper work presents a novel technique by using information from palmprint images captured under different wavelengths, for palmprint recognition using the multivariate Gaussian Probability Density Function (*GPDF*) and Multi resolution analysis. In this method, a palmprint image (color band) is firstly decomposed into frequency sub-bands with different levels of decomposition using different techniques. We adopt as features for the recognition problem, the transform coefficients extracted from some sub-bands. Subsequently, we use the GPDF for modeling the feature vector of each color band. Finally, Log-likelihood scores are used for the matching.
In this work, a series of experiments were carried out using a multi spectral palmprint database. To evaluate the efficiency of this technique, the experiments were designed as follow: the performances under different color bands were compared to each other, in order to determine the best color band at which the palmprint recognition system performs. We also present a multi spectral palmprint recognition system using fused levels which combines several sub-bands at different decomposition levels.

2. System design

Fig. 1 illustrates the various modules of our proposed multi-spectral palmprint recognition system (single band). The proposed system consists of preprocessing, feature extraction, matching and decision stages. To enroll into the system database, the user has to provide a set of training multi-spectral palmprint images (each image is formed by: *RED, BLUE, GREEN* or *Near-IR* (*NIR*)). Typically, a feature vector is extracted from each band which describes certain characteristics of the palmprint images using multi-resolution analysis and modeling using Gaussian probability density function. Finally, the models parameters are stored as references models. For recognition (identification/verification), the same features vectors are extracted from the test palmprint images and the log-likelihood is computed using all of models references in the database. For the multi-modal system, each sub-system computes its own matching score and these individual scores are finally combined into a total score (using fusion at the matching score level), which is used by the decision module. Based on this matching score a decision about whether to accept or reject a user is made.

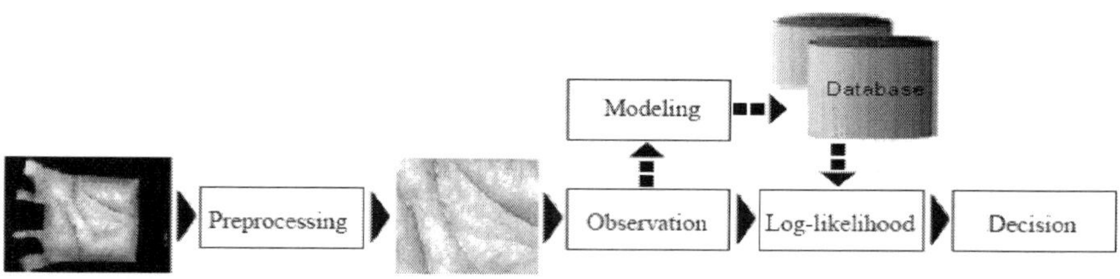

Fig. 1. Block-diagram of a multi-spectral palmprint recognition system based on the Gaussian probability density function modeling.

3. Region of interest extraction

From the whole image of the palmprint (each color band) only some characteristics are useful. Therefore, each color band images may have variable size and orientation. Moreover, the region of non useful interest may affect accurate processing and thus degrade the identification performance. Therefore, image preprocessing {Region Of Interest extraction (ROI)} is a crucial and necessary part before feature extraction. Thus, a palmprint region is extracted from each original palmprint image (each color band). In order to extract the center part of palmprint, we employ the method described in (Zhang & Kong; 2003). In this technique, the tangent of these two holes are computed and used to align the palmprint. The central part of the image, which is 128x128, is then cropped to represent the whole palmprint. The pre-processing steps are shown in Fig. 2. The basic steps to extract the ROI are summarized as follows: First, apply a low pass filter, such as Gaussian smoothing, to the original palmprint image. A threshold, T_p, is used to convert the filtered image to a binary image, then, the boundary tracking algorithm used to obtain the boundaries of the binary image. This boundary is processed to determine the points F_1 and F_2 for locating the ROI pattern and, based on the relevant points (F_1 and F_2), the ROI pattern is located on the original image. Finally, the ROI is extracted.

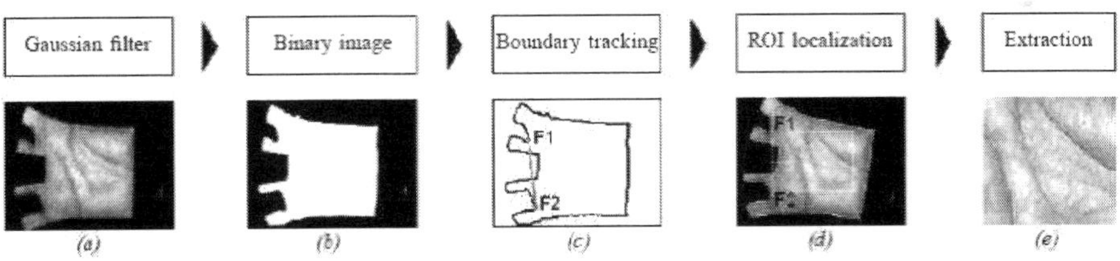

Fig. 2. Various steps in a typical region of interest extraction algorithm. (a) The filtered image, (b) The binary image, (c) The boundaries of the binary image and the points for locating the ROI pattern, (d) The central portion localization, and (e) The pre-processed result (ROI).

4. Feature extraction and modeling

The feature extraction module processes the acquired biometric data (each color band) and extracts only the salient information to form a new representation of the data. Ideally, this new representation should be unique for each person. In our method, the color band is typically analyzed using a multi-resolution analysis. After the decomposition transform of the ROI sub-image, some of the bands are selected to construct feature vectors (observation vectors). Since the Gaussian distribution of observation vectors is computed.

4.1 Feature extraction

A multi-resolution analysis of the images has better space-frequency localization. Therefore, it is well suited for analyzing images where most of the informative content is represented by components localized in space (such as edges and borders) and by information at different scales or resolutions, with large and small features. Several methods were used for obtained the multi-resolution representation such as two dimensional discrete wavelet transform (*2D-DWT*) and two dimensional block based discrete cosine transform with reordering the coefficients to come to multi-resolution representation (*2D-RBDCT*).

4.1.1 DWT decomposition

Wavelets can be used to decompose the data in the color band into components that appear at different resolutions. Wavelets have the advantage over traditional Fourier transform in that the frequency data is localized, allowing the computation of features which occur at the same position and resolution (Antonini & Barlaut; 1992). The discrete wavelet transform (*DWT*) is a multi-resolution representation. Fig. 3 shows an implementation of a one-level forward *DWT* based on a two quadrature mirror filter bank, where $h_o(n)$ and $h_i(n)$ are low-pass and high-pass analysis filters, respectively, and the block ↓2 represents the down-sampling operator by a factor 2. Thus, for *1D-DWT*, the signal is convolved with these two filters and down-sampled by a factor of two to separate it into an approximation and a representation of the details (Noore & Singh; 2007). A perfect reconstruction of the signal is possible by up-sampling (↑2) the approximation and the details and convolving with reversed filters ($g_o(n)$ and $g_i(n)$).

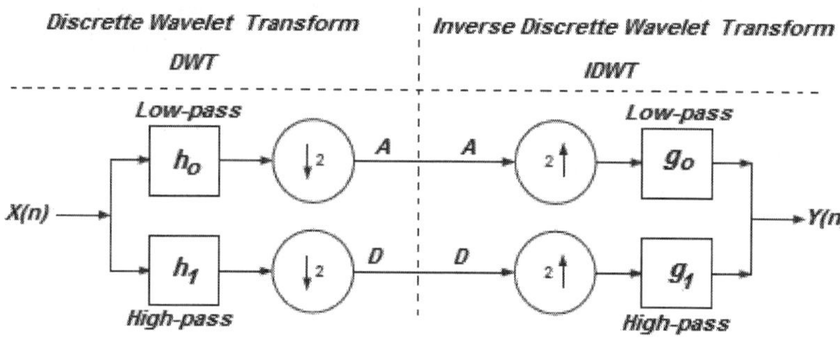

Fig. 3. Implementation of a one-level forward DWT and its inverse IDWT.

For two-dimensional signals, such as images (color band), the decomposition is applied consecutively on both dimensions, e.g. first to the rows and then to the columns. This yields four types of lower-resolution coefficient images: the approximation produced by applying two low-pass filters (*LL*), the diagonal details, computed with two high-pass filters (*HH*), and the vertical and horizontal details, output of a high-pass/low-pass combination (*LH* and *HL*). In Fig. 4 an example of two levels wavelet decomposition is reported. At the first level, the original image, A_0, is decomposed in four sub-band leading to: A_1, the scaling component containing global low-pass information, and H_1, V_1, D_1, three transform components corresponding, respectively, to the horizontal, vertical and diagonal details. In the second level, the approximation, A_1, is decomposed in four sub-bands leading to: A_2, the scaling component containing global low-pass information, and H_2, V_2, D_2, three transform components corresponding, respectively.

4.1.2 DCT decomposition

Discrete Cosine Transform (*DCT*) is a powerful transform to extract proper features for palmprint recognition. The *DCT* is the most widely used transform in image processing

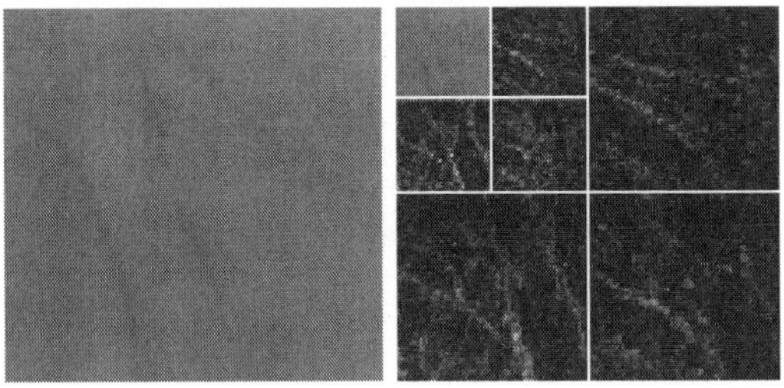

Fig. 4. Two level wavelet decomposition

algorithms, such as image/video compression and pattern recognition. Its popularity is due mainly to the fact that it achieves a good data compaction, that is, it concentrates the information content in a relatively few transform coefficients (Dabbaghchian & Ghaemmaghami; 2010). In the two dimensional block based discrete cosine transform (*2D-BDCT*) formulation, the input image is first divided into *N x N* blocks, and the *2D-DCT* of each block is determined. The *2D-DCT* can be obtained by performing a 1D-DCT on the columns and a *1D-DCT* on the rows. Given an image f, where HxW represent their size, the DCT coefficients of the spatial block are then determined by the following formula:

$$F_{ij}(u,v) = \alpha(u)\alpha(v)\sum_{m=0}^{N-1}\sum_{n=0}^{N-1} f_{ij}(n,m)\psi(n,m,u,v) \quad (1)$$

$$\psi(n,m,u,v) = \cos\left[\frac{(2n+1)u\pi}{2N}\right]\cos\left[\frac{(2m+1)v\pi}{2N}\right]$$

$u; v = 0, 1,.....,N-1, i = 0, 1,......., (H/N)-1\,,\ j = 0, 1,......., (W/N)-1$. Where $F_{ij}(u,v)$ are the *DCT* coefficients of the B_{ij} block, $f_{ij}(n;m)$ is the luminance value of the pixel (n,m) of the B_{ij} block, HxW are the dimensions of the image, and

$$C(u) = \begin{cases} \frac{1}{\sqrt{2}} & if \quad u = 0 \\ 1 & if \quad u \neq 0 \end{cases}$$

The *DCT* coefficients reflect the compact energy of different frequencies. The first coefficient $F_0 = F(0, 0)$, called DC, is the mean of visual gray scale value of pixels of a block. The AC coefficients of upper left corner of a block represent visual information of lower frequencies, whereas the higher frequency information is gathered at the right lower corner of the block (Chen & Tai; 2004).

DCT theory can provides a multi-resolution representation for interpreting the image information with the multilevel decomposition. After applying the *2D-BDCT*, the coefficients are reordered resulting in a multi-resolution representation. Therefore, if the size of the block transform, *N*, is equal to 2, each coefficient is copied into a one-band (See Fig. 5). *2D-BDCT* concentrates the information content in a relatively few transform coefficients at the top-left zone of the block. As such, the coefficients where the information is concentred, tend to be grouped together at the approximation band.

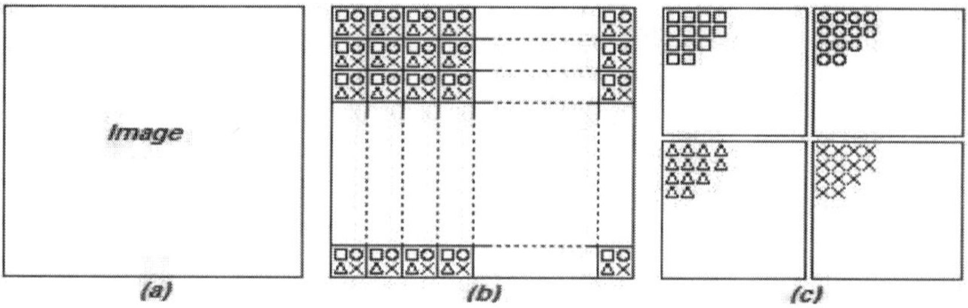

Fig. 5. Multi-resolution representation using *2D-DCT* transform with reordering these coefficients. (a) Image to be transformed, (b) *2D-BDCT* with a block size 2x2, and (c) *2D-RBDCT* decomposition.

4.2 Feature vector

To create an observation vector, the color band image is transformed into a multi-resolution form as shown in Fig. 6. Then the palmprint feature vectors are created by combining the horizontal detail (H_i), global low-pass information (Approximation: A_i) and the vertical detail (V_i) extracted using multi-resolution analysis. Three feature vectors can be extracted using three levels of decomposition for each color band (*RED, BLUE, GREEN* and *NIR*) (See Fig. 7).

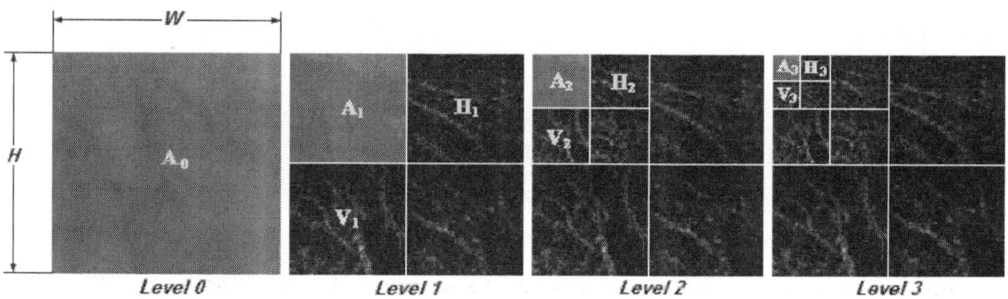

Fig. 6. Three levels decomposed into multi-resolution representation.

Let ψ_x represent a HxW palmprint ROI image (color band) and x = {R, B, G, N}, thus
$\Psi_R = RED$ $\psi_B = BLUE$ $\psi_G = GREEN$ $\psi_N = NIR$
Let *F* the applied transform method: *F: 2D-DWT* or *F: 2D-RBDCT*

- One level: $F(\psi_x) \rightarrow [A_1, H_1, V_1, D_1] / A_1, H_1, V_1, D_1$: with *H/2 x W/2* coefficients.
- Tow level: $F(A1) \rightarrow [A_2, H_2, V_2, D_2] / A_2, H_2, V_2, D_2$: with *H/4 x W/4* coefficients.
- Three level: $F(A2) \rightarrow [A_3, H_3, V_3, D_3] / A_3, H_3, V_3, D_3$: with *H/8 x W/8* coefficients.

Those three feature vectors (*observation*) are shown in Fig. 7.

$$O_1 = \begin{bmatrix} V_1 \\ A_1 \\ H_1 \end{bmatrix} \qquad O_2 = \begin{bmatrix} V_2 \\ A_2 \\ H_2 \end{bmatrix} \qquad O_3 = \begin{bmatrix} V_3 \\ A_3 \\ H_3 \end{bmatrix}.$$

Where the size of O_1 is *(3H/2)*(W/2)* coefficients, O_2 is *(3H/4)*(W/4)* coefficients and O_3 is *(3H/8)*(W/8)* coefficients, respectively. As results, the color band image as a single template (Feature vectors) as follows:

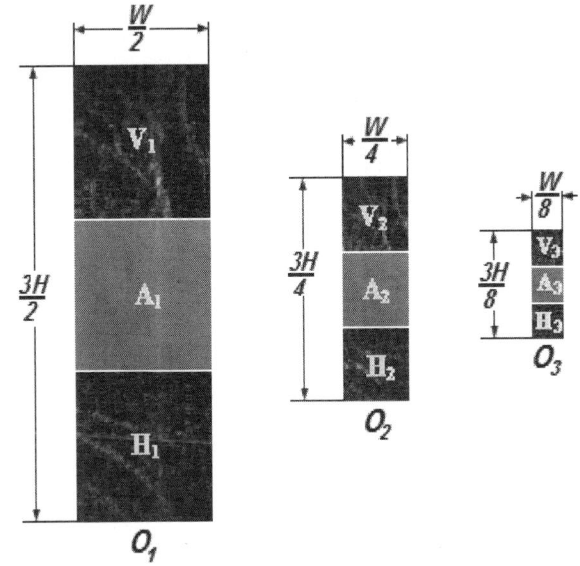

Fig. 7. The observation vector.

$$O_j = \begin{bmatrix} o_{11} & o_{12} & o_{13} & \cdots & o_{1L} \\ o_{21} & o_{22} & o_{23} & \cdots & o_{2L} \\ \vdots & \vdots & \vdots & \cdots & \vdots \\ o_{M1} & o_{M2} & o_{Mm3} & \cdots & o_{ML} \end{bmatrix}$$

Where: $j = 1 \cdots 3$ for *Level 1*, *Level* 2 and *Level* 3. $L = W/k$ and $M = 3H/k$ for $k = 2, 4, 8$. If the size of original color band is 128x128 pixels, the size of O_1 is $MxL=192x64$ coefficients, O_2 is equal to *96x32* coefficients and O_3 is *48x16* coefficients.

4.3 Modeling process: Gaussian Probability Density Function (GPDF)

In our system, the observation probabilities have been modeled as multi-variate Gaussian distributions. Arguably the single most important PDF is the Gaussian probability distribution function. It is one of the most studied and one of the most widely used distributions (Varchol & Levicky; 2007). The Gaussian has two parameters: the mean μ, and the variance σ^2. The mean specifies the centre of the distribution, and the variance tells us how "spread-out" the PDF is. For a d-dimensional vector O, the Gaussian is written

$$P(O_j / \mu, \Sigma) = \frac{1}{\sqrt{(2\pi)^d |\Sigma|}} \exp\left[-\frac{1}{2}(O_j - \mu)^T \Sigma^{-1} (O_j - \mu) \right]$$

Where μ is the mean vector, and Σ is the $d \times d$ covariance matrix. Gaussian distributions can also be written using the notation: $\eta(O;\ \mu,\ \Sigma)$ The covariance matrix Σ of a Gaussian must be symmetric and positive definite.

After the feature extraction, we now consider the problem of learning a Gaussian distribution from the vector samples O_i. Maximum likelihood learning of the parameters μ and Σ entails maximizing the likelihood (Fierrez & Ortega-Garcia; 2007). Since we assume that the data points come from a Gaussian:

$$P(O_j / \mu,\Sigma) = \prod_{i=1}^{L} P(o_i / \mu,\Sigma) = \prod_{i=1}^{L} \frac{1}{\sqrt{(2\pi)^d |\Sigma|}} \exp\left[-\frac{1}{2}(o_i - \mu)^T \Sigma^{-1} (o_i - \mu)\right]$$

It is somewhat more convenient to minimize the negative log-likelihood, LH:

$$LH(O_j,\mu,\Sigma) \equiv -\ln P(O_j / \mu,\Sigma) = -\sum_i \ln P(o_i / \mu,\Sigma)$$

$$= -\sum_i (o_i - \mu)^T \Sigma^{-1} (o_i - \mu) / 2 + \frac{L}{2}\ln|\Sigma| + \frac{Ld}{2}\ln(2\pi)$$

Solving for μ and Σ by setting $\partial LH(O_j,\mu, \Sigma)/\partial \mu = 0$ and $\partial LH(O_j, \mu, \Sigma)/\partial\Sigma = 0$ (subject to the constraint that Σ is symmetric) gives the maximum likelihood estimates (Cardinaux & Sanderson; 2004):

$$\hat{\mu} = \frac{1}{L}\sum_i o_i$$

$$\hat{\Sigma} = \frac{1}{T}\sum (o_i - \hat{\mu})(o_i - \hat{\mu})^T$$

The complete specification of the modeling process requires determining two model parameters (μ and Σ). For convenience, the compact notation $\lambda\ (\mu, \Sigma)$ is used to represent a model.

5. Feature matching

5.1 Matching process

During the identification process, the characteristics of the test color band image are analyzed by the *2D-DWT* (*2D-RBDCT*) corresponding to each person. Then the Log-likelihood score of the feature vectors given each GPDF model is computed. Therefore, the score vector is given by:

$$Lh(O_j) = \left[LH(O_j,\mu_1,\Sigma_1) \quad LH(O_j,\mu_2,\Sigma_2) \quad LH(O_j,\mu_3,\Sigma_3) \quad \cdots \quad LH(O_j,\mu_s,\Sigma_s)\right]$$

Where S represents the size of model database.

5.2 Normalization and decision process

In a verification mode, our normalization rule is formulated as $D_o = -10^{-5}\, LH(O_j, \lambda_i)$, where D_o denotes the normalized Log-likelihood scores. In an identification mode, prior to finding the decision, a *Min-Max* normalization, (Savic & Pavesic; 2002), scheme was employed to transform the Log-likelihood scores computed into similarity scores in the same range.

$$Lh_N = \frac{Lh - \min(Lh)}{\max(Lh) - \min(Lh)}$$

Where Lh_N denotes the normalized Log-likelihood scores. Therefore, these scores are compared, and the highest score is selected. Therefore, the best score is D_o and its equal to:

$$D_0 = \max(Lh_N)$$

This score, D_o, is compared with the decision threshold T_o. When $D_o \geq T_o$, the claimed identity is accepted; otherwise it is rejected.

5.3 Fusion process

The goal of the fusion process is to investigate the systems performance when the information from some color bands of a person is fused. In fact, in such a case the system works as a kind of multi-modal system with a single biometric trait but with multiple units. Therefore, the information presented by different bands (*BLUE, GREEN, RED,* and *NIR*) is fused to make the system efficient.

Fusion at the matching-score level is preferred in the field of biometric recognition because there is sufficient information content and it is easy to access and combine the matching scores (Ross & Jain; 2001). In our system we adopted the combination approach, where the individual matching scores are combined to generate a single scalar score, which is then used to make the final decision. During the system design we experiment four different fusion schemes: *Sum-score, Min-score, Max-score* and *Sum-weighting-score* (Singh & Vatsa; 2008). Suppose that the quantity D_{0i} represents the score of the ith matcher ($i = 1; 2; 3; 4$) for different palmprint color bands (*BLUE, GREEN, RED,* and *NIR*) and D_F represents the fusion score. Therefore, D_F is given by:

- *SUM* $D_F = \sum_{i=1}^{n} D_{oi}$
- *MIN* $D_F = \min\{D_{oi}\}$
- *MAX* $D_F = \max\{D_{oi}\}$
- *WHT* $D_F = \sum_{i=1}^{n} w_i D_{oi}$

$$w_i = \frac{1 / \sum_{j=1}^{n} (1 / EER_j)}{EER_i}$$

Where w_i denotes the weight associated with the matcher i, with $\Sigma w_i = 1$, and EER_i is the equal error rate of matcher i, respectively.

6. Experimental results and discussion

6.1 Experimental database

Experiments were performed using the multi-spectral palmprint database from the Hong Kong polytechnic university (PolyU) (PolyU Database; 2003). The database contains images captured with visible and infrared light. Multi-spectral palmprint images were collected from 250 volunteers, including 195 males and 55 females. The age distribution is from 20 to 60 years old. It has a total of 6000 images obtained from about 500 different palms. These samples were collected in two separate sessions. In each session, the subject was asked to provide 6 images for each palm. Therefore, 24 images of each illumination from 2 palms were collected from each subject. The average time interval between the first and the second sessions was about 9 days.

6.2 Evaluation criteria

The measure of utility of any biometric recognition system for a particular application can be explained by two values (Connie & Teoh; 2003). The value of the FAR criterion, which is the ratio of the number of instances of different feature pairs of the traits found do match to the total number of counterpart attempts, and the value of the FRR criterion, which is the ratio of the number of instances of same feature pairs of the traits found do not match to the total number of counterpart attempts. It is clear that the system can be adjusted to vary the values of these two criteria for a particular application. However, decreasing one involves increasing the other and vice versa. The system threshold value is obtained using EER criteria when FAR = FRR. This is based on the rationale that both rates must be as low as possible for the biometric system to work effectively.
Another performance measurement is obtained from FAR and FRR, which is the Genuine Acceptance Rate (GAR). It represents the identification rate of the system. In order to visually describe the performance of a biometric system, Receiver Operating Characteristics (ROC) curves are usually given. A ROC curve shows how the FAR values are changed relatively to the values of the GAR and vice-versa (Jain & Ross; 2004). Biometric recognition systems generate matching scores that represent the degree of similarity (or dissimilarity) between the input and the stored template.
Cumulative Match Curves (CMC) is another method of showing the measured accuracy performance of a biometric system operating in the closed-set identification task. Templates are compared and ranked based on their similarity. The CMC shows how often the individual's template appears in the ranks based on the match rate.

6.3 Performance of verification algorithm

In a verification mode, the log-likelihood score (LH) of the input template given each GPDF model of the claimed person is computed. To obtain the accuracy, each of the color band images was matched with all of the models in the database. In our experiment, we use randomly three color band images of each person as the training samples (enrollment) and the remainder as the test samples. The total number of matching is 562500. The number of comparisons that have corrected matching is 2250 and 560250 incorrect matching. The verification experiments were performed by using each of the Blue, Green, Red and NIR features, as well as the fusion of them at the matching score level.

6.3.1 Verification in case of uni-modal system

The goal of this experiment was to evaluate the system performance when using the information from each modality (each color band). For this, we found the performance under different modalities (Blue, Green, Red and NIR). The performances in terms of the EER of proposed method using the 2D-RBDCT technique for different threshold values T_o is shown in Table 1.

	Level 1		Level 2		Level 3	
	To	EER (%)	To	EER (%)	To	EER (%)
BLUE	0.5861	4.2586	0.1609	3.6383	0.0457	6.8941
GREEN	0.5909	6.3815	0.1638	4.7971	0.0466	9.7511
RED	0.4190	3.6407	0.1433	3.5613	0.0510	5.4266
NIR	0.3950	4.9093	0.1328	4.2071	0.0469	6.2473

Table 1. Verification test result for 2D-RBDCT (Uni-modal case)

From these results, one can clearly observe advantages of using the color band *RED* than the *BLUE, NIR* and *GREEN* in terms of EER for all levels. For example, in level 2, if the *BLUE* feature is used a GAR of 96.3617% is obtained In the case of using the *GREEN*, GAR was 95.2029 %. When *NIR* is used, the EER was 95.7929% while a *RED* feature improves the result to 96.4387 % for a database containing 250 persons. Therefore, from these test results one can conclude that an integration of color band RED with level 2 does result in an improvement of verification accuracy. Fig. 8.*(a)* compares the performance of the system for deferent levels using *RED* color band. Finally, the genuine and impostor distributions are plotted in Fig. 8.*(a)* and the ROC curves, shown in Fig. 8.*(c)*, depicts the performance of the system.

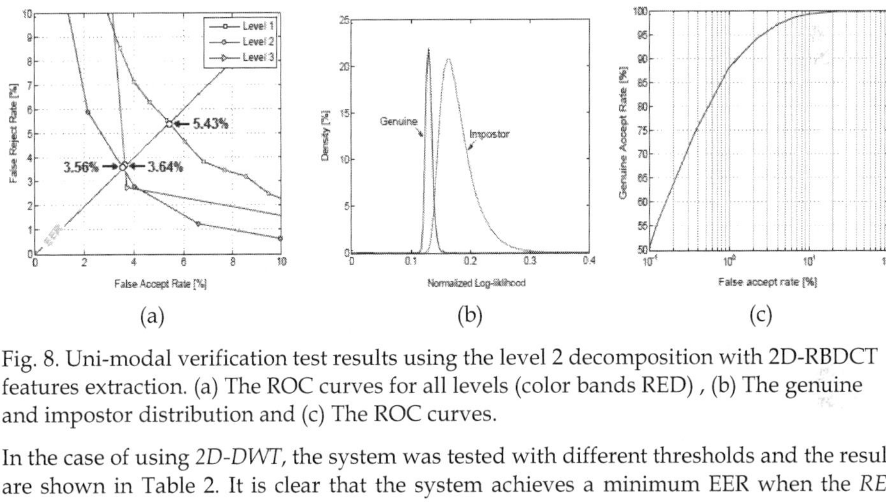

(a) (b) (c)

Fig. 8. Uni-modal verification test results using the level 2 decomposition with 2D-RBDCT features extraction. (a) The ROC curves for all levels (color bands RED) , (b) The genuine and impostor distribution and (c) The ROC curves.

In the case of using *2D-DWT*, the system was tested with different thresholds and the results are shown in Table 2. It is clear that the system achieves a minimum EER when the *RED* band with level 2 is used. In level 1, the system achieves a maximum GAR of 94.5740% if the *RED* band is used. For level 3, the system operates with a maximum GAR of 94.6539% at the *NIR* band. Fig. 9.*(a)* compares the performance of the system for deferent levels using *RED* color band. Fig. 9.*(b)* shows the two distributions for *RED* band using level 2 and the system performance at all thresholds can be depicted in the form of a receiver operating characteristic (ROC) curve, see Fig. 9.*(c)*.

6.3.2 Verification in case of multi-modal system

The objective of this section is to investigate the combination of all color bands features in order to achieve higher performances that may not be possible with uni-modal biometric

	Level 1		Level 2		Level 3	
	To	EER (%)	To	EER (%)	To	EER (%)
BLUE	0.5007	5.4339	0.1183	3.6438	0.0298	6.0094
GREEN	0.5074	6.5484	0.1212	4.8002	0.0302	6.0468
RED	0.4253	5.4260	0.1006	3.5485	0.0264	5.8528
NIR	0.3845	6.2639	0.0900	3.9351	0.0238	5.3461

Table 2. Verification test result for 2D-*DWT* (Uni-modal case)

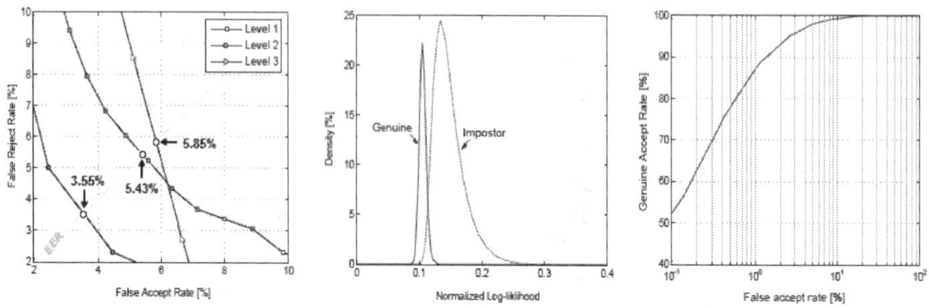

(a) (b) (c)

Fig. 9. Uni-modal verification test results using the level 2 decomposition with *2D-DWT* features extraction. (a) The ROC curves for all levels (color bands *RED*), (b) The genuine and impostor distribution and (c) The ROC curves.

alone. Thus, in order to see the performance of the system, we have evaluated different fusions of color bands and Table 3 summarizes the equal error rates for these experiments. From Table 3, we can observe the advantages of using the *RGBN* fusion modalities at level 2. For example, a fusion of *RGB* at level 1 gives a minimum EER equal to 3.0978 % at T_o = 1.6919 by using *SUM* rule fusion. This system can achieve a minimum EER of 3.8145 % for T_o = 0.4733 in the case of level 2 fusion with *SUM* rule. A fusion at level 3 results in an EER of 26.430 % at T_o = 0.1416 with *SUM* rule fusion. In the case of using the *RGBN*, an EER of 2.6848 % is achieved at a threshold T_o = 2.1067 at level 1 by *SUM* rule. In the case of level 3, EER was 16.110% at the threshold T_o = 0.1782 by using SUM rule fusion. Finally, the system can operate at a 1.8442 % EER, and the corresponding threshold is T_o = 0.6051 at level 2 by using the *WHT* rule. The experimental results show that fusion of all color bands at level 2 with WHT rule fusion is much higher than the individual color bands. The multimodal verification test results using fusion of *RGBN* bands at the level 2 decomposition with *2D-RBDCT* and all fusion schemes are shown in Fig. 10.*(a)*. The genuine and imposter distributions are shown in Fig. 10.*(b)*. Fig. 10.*(c)* depicts the ROC curve. Compared with the methods described in (Sun & Qiu; 2006), (Kumar & C.M; 2002) our system achieves better results expressed in terms of the EERs.

In the case of the *2D-DWT*, Table 4 depicts the minimum EER obtained from test database. It can be observed that the proposed scheme can recognize palmprints more accurately as a minimum EER of 2.5785 % has been obtained from *RGBN* fusion at level 2 by using *SUM* rule fusion. These results are similar as *2D-RBDCT* except that fusion takes place at level 3

EER	Fusion of level 1				Fusion of level 2				Fusion of level 3			
	SUM	MIN	MAX	WHT	SUM	MIN	MAX	WHT	SUM	MIN	MAX	WHT
RGB	3.098	4.605	5.228	3.933	3.815	16.14	9.162	9.032	26.43	42.50	48.67	46.52
RGBN	2.685	4.520	8.083	3.166	2.143	15.53	25.12	1.844	16.11	41.42	49.74	47.56

Table 3. Verification test result for *2D-RBDCT* (Multi-modal case)

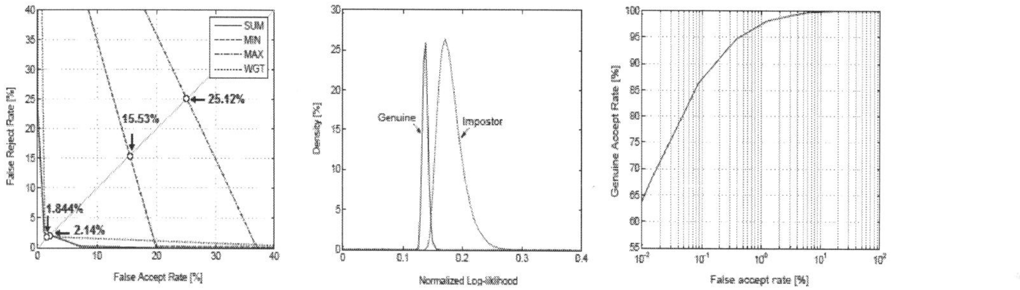

(a) (b) (c)

Fig. 10. Multimodal verification test results using fusion of RGBN bands at the level 2 decomposition with *2D-RBDCT* and all fusion schemes. (a) The ROC curves for all fusion schemes, (b) The genuine and impostor distribution and (c) The ROC curves for the level 2.

EER	Fusion of level 1				Fusion of level 2				Fusion of level 3			
	SUM	MIN	MAX	WHT	SUM	MIN	MAX	WHT	SUM	MIN	MAX	WHT
RGB	3.249	5.047	5.798	4.883	3.948	6.025	17.53	7.783	6.789	7.024	21.06	44.14
RGBN	2.777	4.944	7.526	3.835	2.579	5.981	13.54	3.902	10.56	6.994	34.52	11.11

Table 4. Verification test result for *2D-DWT* (Multi-modal case)

where the minimum EER was 6.994% with To = 0.0317 by using *MIN* rule fusion. The test results using fusion of *RGBN* bands at the decomposition level 2 with *2D-DWT* and all fusion schemes are shown in Fig. 11.*(a)*. The genuine and impostor distributions are estimated and are illustrated Fig. 11.*(b)*. Fig. 11.*(c)* presents the verification test results and show the ROC curve.

6.4 Performance of identification algorithm

In an identification mode the recognition system examines whether the user is one of enrolled candidates. Therefore, the biometric data is collected and compared to all the templates in the database. Identification is closed-set if the person is assumed to exist in the database. In open-set identification, the person is not guaranteed to exist in the database. In our work, the proposed method was tested for the two modes of identification.

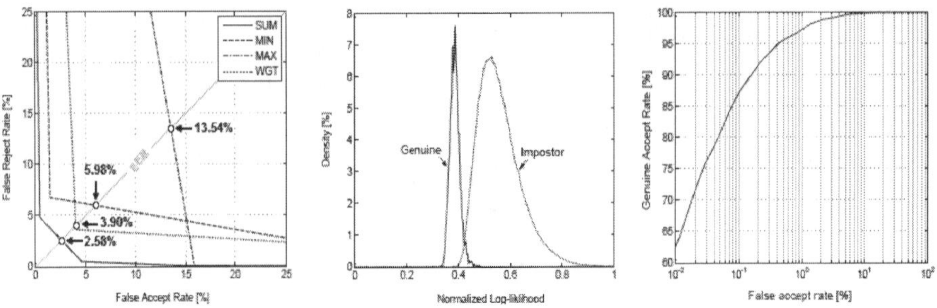

(a) (b) (c)

Fig. 11. Multimodal verification test results using fusion of *RGBN* bands at the level 2 decomposition with *2D-DWT* and all fusion schemes. (a) The ROC curves for all fusion schemes, (b) The genuine and impostor distribution and (c) The ROC curves for the level 2.

6.4.1 Identification in the case of uni-modal system

6.4.1.1 Open set identification

The open set identification result is clearly shown in Table 5 in the case of using *2D-RBDCT* features. From Table 5, one can observe a good performance and acceptability when the BLUE band is used at level 2 fusion. BLUE band open set identification system based on level 2 produces 0.0734 % accuracy, while identification system based on 'level 1' and system based on 'level 3' produce 0.2729 % and 0.1388 % accuracies, respectively. In order to show the effectiveness of the BLUE band, we have plotted ROC curves for all levels, (see Fig. 12.*(a)*). The genuine and impostor distributions are estimated and are shown in Fig. 12.*(b)*. Fig. 12.*(c)* presents the identification test results including the ROC curve. *2D-RBDCT* feature based identification method also outperforms other methods presented in the literatures, such as (Prasad & Govindan; 2009), (Varchol & Levicky; 2007), (Dai & Bi; 2004).

	Level 1		Level 2		Level 3	
	To	EER (%)	To	EER (%)	To	EER (%)
BLUE	0.9171	0.2729	0.9465	0.0734	0.9460	0.1388
GREEN	0.8960	0.5036	0.9087	0.1980	0.9268	0.2526
RED	0.9258	0.5666	0.9395	0.2340	0.9432	0.3892
NIR	0.9263	0.9851	0.9307	0.3154	0.9503	0.3437

Table 5. Open set identification test result for *2D-RBDCT* (Uni-modal case)

A performance comparison of all color bands using *2D-DWT* features are made in Table 6. The matching is employed for the proposed methods and the results shows that the system performance is found to be superior (0.0734%) with *BLUE* band when it is compared with the other three bands methods based on the level 2. It is also observed from the Table 6 that changing the *2D-RBDCT* with *2D-DWT* does not provide any improvement in the EER. The ROC curves for all levels (color bands *BLUE*), are shown in Figure 13.*(a)*. Finally, the genuine and impostor distributions are plotted in Fig. 13.*(b)* and the ROC curves, shown in Fig. 13.*(c)*, depicts the performance of the system.

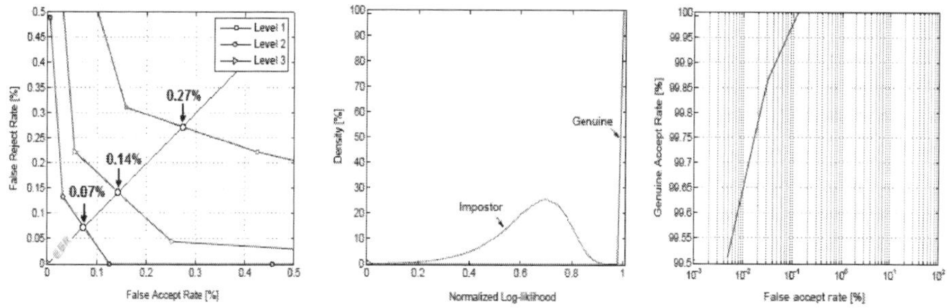

(a) (b) (c)

Fig. 12. Unimodal identification test results using the level 2 decomposition with *2D-RBDCT* features. (a) The ROC curves for all levels (color bands *BLUE*), (b) The genuine and impostor distribution and (c) The ROC curves.

	Level 1		Level 2		Level 3	
	To	EER (%)	To	EER (%)	To	EER (%)
BLUE	0.9171	0.2729	0.9465	0.0734	0.9466	0.1466
GREEN	0.8982	0.4314	0.9087	0.1980	0.9268	0.2526
RED	0.9258	0.5666	0.9395	0.2340	0.9432	0.3892
NIR	0.9261	0.9811	0.9307	0.3153	0.9503	0.3437

Table 6. Open set identification test result for 2D-DWT (Uni-modal case)

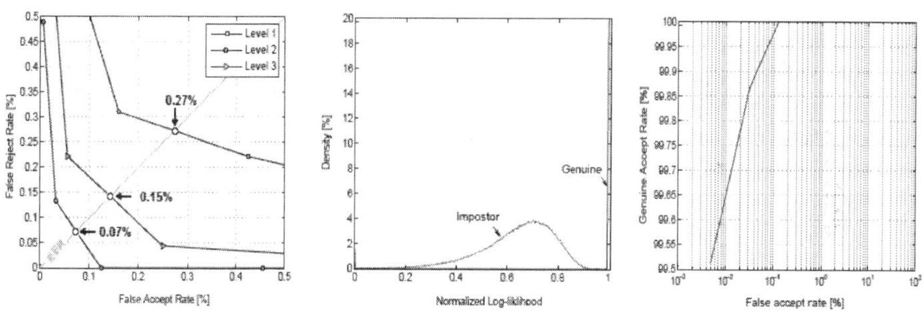

(a) (b) (c)

Fig. 13. Unimodal identification test results using the level 2 decomposition with *2D-DDWT* features. (a) The ROC curves for all levels (color bands *BLUE*), (b) The genuine and impostor distribution and (c) The ROC curves.

6.4.1.2 Closed set identification

In the case of a closed set identification, a series of experiments were carried out to select the best color band and the corresponding decomposition level (1, 2, 3). This has been done by comparing all bands using all decomposition levels (*i.e.*, 1, 2 3) and finding the color band that gives the best identification rate. Table 7 and 8 presents the experiments results obtained for all color bands for *2D-RBCD* and *2D-DWT*, respectively. From Table 7 (Table 8), the best results of rank-one identification for the *BLUE*, *GREEN*, *RED*, and *NIR* produce 98.0889 % with lowest rank (Rank of perfect rate) of 37, 99.2889% with lowest rank of 13 and 98.9333 % with lowest rank of 28, respectively. As the Table 7 (Table 8) shows, by using the *BLUE* band biometric with level 2, the performance of the system is increased. Finally, the CMC curves for *2D-RBDCT* based system and 2D-DWT based system are plotted in Fig. 14.

	Level 1		Level 2		Level 3	
	Rank-one ident [%]	lowest rank	Rank-one ident [%]	lowest rank	Rank-one identi [%]	lowest rank
BLUE	98.0889	37	99.2889	13	98.9333	28
GREEN	97.1110	119	99.2000	78	98.0000	128
RED	96.6670	160	98.8000	162	98.7556	80
NIR	94.7111	108	98.7556	69	98.4889	85

Table 7. Closed set identification test result for *2D-RBDCT* (Uni-modal case)

	Level 1		Level 2		Level 3	
	Rank-one ident [%]	lowest rank	Rank-one ident [%]	lowest rank	Rank-one identi [%]	lowest rank
BLUE	98.0889	37	99.2889	13	98.9333	28
GREEN	97.6000	121	99.2000	81	98.0000	141
RED	96.6667	161	98.8000	181	98.7552	81
NIR	94.7111	121	98.4889	101	98.7511	81

Table 8. Closed set identification test result for *2D-DWT* (Uni-modal case)

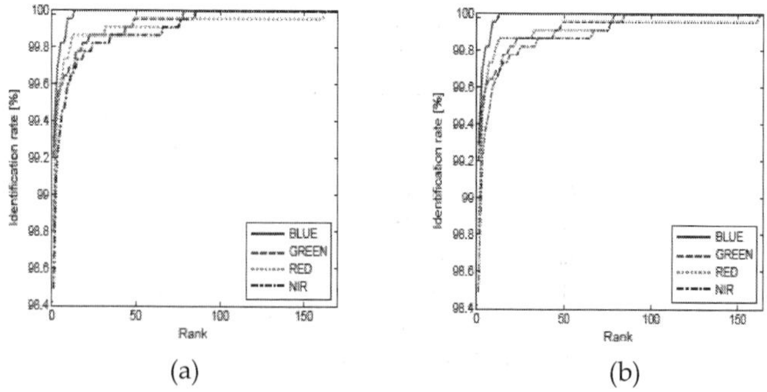

Fig. 14. Uni-modal closed-set identification test for all bands at level 2. (a) *2D-RBDCT* based system and (b) *2D-DWT* based system.

6.4.2 Identification in case of multi-modal system

6.4.2.1 Open set identification

Table 9 shows the performance of fusion using different schemes when using *2D-RBDCT* method. Compared with the performance of individual biometrics showen in Table 5, the performance of all individual biometrics in this experiment decreases to some extent. For fusion based on *RGB*, the best performance is also achieved by the *WHT* rule at a decomposition level 2 with a minimum EER of 0.0301% and a threshold T_o = 0.9235, followed by the *SUM* rule on level 3 and the *SUM* rule on level 1 at 0.0507 % and T_o = 0.9121 and the Sum rule on level 1 at 0.0889% and T_o = 0.9152. For RGBN fusion, fusion based on *SUM* rule achieves a minimum EER of 0.0732 % with T_o = 0.9321 at decomposition level of 1. Level 3 results in an EER of 0.0342 % with T_o = 0.9231 when using *SUM* rule. The multimodal system error is decreases to 0.0158 % with T_o = 0.9403 when using a decomposition level of 2 with *WHT* rule. The ROC curves for all fusion schemes (*RGBN* fusion) are shown in Fig. 15.*(a)*. The genuine and impostor distributions are plotted in Fig. 15.*(b)* and the ROC curve of GAR against FAR for various thresholds is shown in Fig. 15.*(c)*.

EER	Fusion of level 1				Fusion of level 2				Fusion of level 3			
	SUM	MIN	MAX	WHT	SUM	MIN	MAX	WHT	SUM	MIN	MAX	WHT
RGB	0.089	0.413	0.509	0.089	0.044	0.186	0.151	0.030	0.051	0.346	0.236	0.052
RGBN	0.073	0.377	0.841	0.094	0.044	0.164	0.285	0.016	0.034	0.271	0.272	0.037

Table 9. Identification test result for *2D-RBDCT* (Multi-modal case)

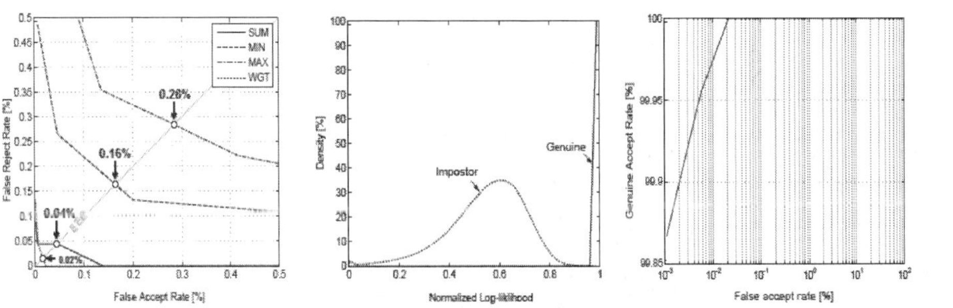

(a) (b) (c)

Fig. 15. Multimodal identification test results using fusion of *RGBN* bands at the level 2 decomposition with *2D- RBDCT*. (a) The ROC curves for all fusion schemes, (b) The genuine and impostor distribution and (c) The ROC curves for the level 2 and *WHT* rule fusion.

EER	Fusion of level 1				Fusion of level 2				Fusion of level 3			
	SUM	MIN	MAX	WHT	SUM	MIN	MAX	WHT	SUM	MIN	MAX	WHT
RGB	*0.089*	*0.387*	*0.517*	*0.107*	*0.044*	*0.186*	*0.151*	*0.021*	*0.051*	*0.346*	*0.236*	*0.054*
RGBN	*0.078*	*0.373*	*0.870*	*0.089*	*0.044*	*0.164*	*0.285*	*0.016*	*0.034*	*0.271*	*0.272*	*0.037*

Table 10. Identification test result for *2D-DWT* (Multi-modal case)

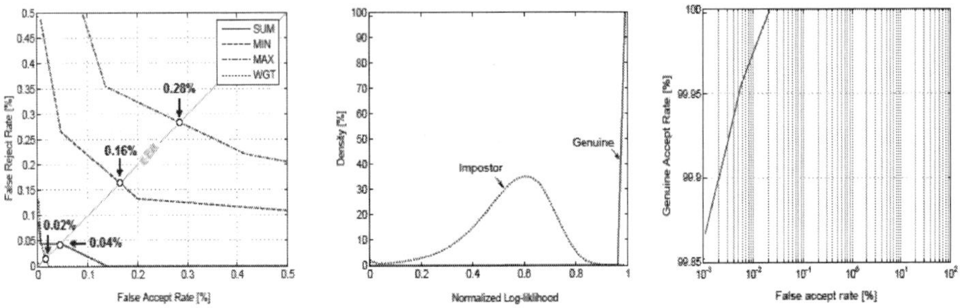

(a) (b) (c)

Fig. 16. Multimodal identification test results using fusion of RGBN bands at the level 2 decomposition with 2D-DWT. (a) The ROC curves for all fusion schemes, (b) The genuine and impostor distribution and (c) The ROC curves for the level 2 and WHT rule fusion.

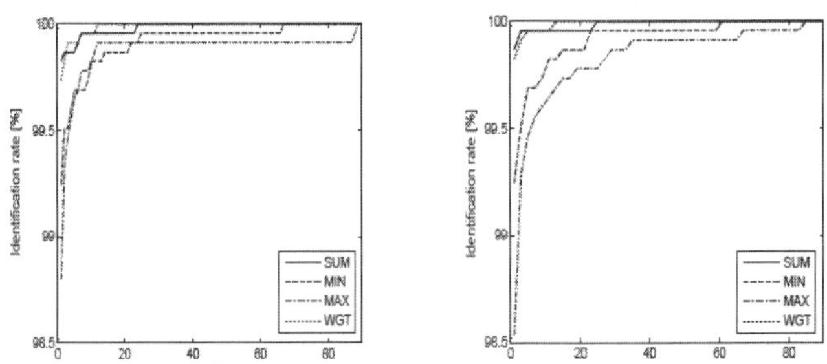

(a) (b)

Fig. 17. Uni-modal closed-set identification test for all bands at level 2. (a) 2D-RBDCT based system and (b) 2D-DWT based system.

Rank-one identification [%]

EER	Fusion of level 1				Fusion of level 2				Fusion of level 3			
	SUM	MIN	MAX	WHT	SUM	MIN	MAX	WHT	SUM	MIN	MAX	WHT
RGB	99.29	97.78	97.02	99.11	99.82	99.24	98.00	99.73	99.64	98.62	98.84	99.69
RGBN	99.33	97.82	94.93	99.02	99.87	99.24	98.53	99.73	99.85	98.89	98.89	99.64

Rank of perfect Identification rate

EER	Fusion of level 1				Fusion of level 2				Fusion of level 3			
	SUM	MIN	MAX	WHT	SUM	MIN	MAX	WHT	SUM	MIN	MAX	WHT
RGB	51	64	160	47	24	67	89	12	12	76	80	9
RGBN	7	58	108	13	24	61	85	12	5	56	70	6

Table 11. Closed set identification test result for *2D-RBDCT* (Multi-modal case)

Rank-one identification [%]

EER	Fusion of level 1				Fusion of level 2				Fusion of level 3			
	SUM	MIN	MAX	WHT	SUM	MIN	MAX	WHT	SUM	MIN	MAX	WHT
RGB	99.07	98.13	96.80	99.11	99.82	99.24	98.80	99.73	99.64	98.62	98.84	99.69
RGBN	99.07	98.18	94.89	99.06	99.87	99.24	98.53	99.82	99.85	98.89	98.89	99.83

Rank of perfect Identification rate

EER	Fusion of level 1				Fusion of level 2				Fusion of level 3			
	SUM	MIN	MAX	WHT	SUM	MIN	MAX	WHT	SUM	MIN	MAX	WHT
RGB	55	59	161	19	25	67	89	13	13	77	81	9
RGBN	45	51	109	43	25	61	85	13	5	57	71	7

Table 12. Closed set identification test result for *2D-DWT* (Multi-modal case)

In this experiment, the performance results, based on *2D-DWT* method, are shown Table 10. In this table, it can be observed that the proposed feature extraction based on *2D-DWT* has a similar performance as its *2D-RBDCT* counterpart. The ROC curves for all fusion schemes (*RGBN* fusion) are shown in Fig. 16.*(a)*. The genuine and impostor distributions are plotted in Fig. 16.*(b)* and the ROC curve of GAR against FAR for various thresholds is depicted in Fig. 16.*(c)*.

6.4.2.2 Closed set identification

To appreciate the performance differences between *2D-RBDCT* and *2D-DWT* feature extraction methods, experiments were conducted and the results are shown in Table 11 and 12. We can observe that the two feature extraction methods have marginally similar performances. Thus, the best results of rank-one identification for the *RGB*, *RGBN* are given as 99.82 % with lowest rank of 25 and 99.87 % with lowest rank of 24, respectively. As the Table 11 (Table 12) shows, by using the *RGBN* fusion with level 2, the performance of the system is increased. Finally, the CMC curves for *2D-RBDCT* based system and *2D-DWT* based system are plotted in Fig. 17.

7. Conclusion

In this chapter, we proposed algorithms to fuse the information from multi-spectral palmprint images where fusion is performed at the matching score level to generate a unique score which is used for recognizing a palmprint image. Several fusion rules including *SUM*, *MIN*, *MAX* and *WHT* are employed for the fusion of the multi-spectral palmprint at the matching score level. The features extracted from palmprint images are obtained using *2D-DCT* and *2D-DWT* methods. The algorithms are evaluated using the multi-spectral palmprint database from the Hong Kong polytechnic university (PolyU) which consists of palmprint images from *BLUE*, *GREEN*, *RED* and *NIR* color bands. Experimental results have shown that the combination of all colors bands palmprint images, *RGBN*, performs better when compared against other combinations, RGB, for both *2D-RBDCT* and *2D-DWT* extraction methods resulting in an EER of 2.1425% for verification and 0.0158% for identification. This also compares favourably against uni-modal band palmprint recognition. Experimental results also show that these proposed methods give an excellent closed set identification rate for the two extraction methods. For further improvement of the system, our future work will focus on the performance evaluation using large size database, and a combination of multi-spectral palmprint information with other biometrics such as finger-knuckle-print to obtain higher accuracy recognition performances.

8. References

Kong, W.K. & Zhang, D. (2002): *Palmprint Texture Analysis based on Low- Resolution Images for Personal Authentication*, In: 16th International Conference on Pattern Recognition, Vol. 3, pp. 807-810, August 2002.

Ajay Kumar & David Zhang. (2010). *Improving Biometric Authentication Performance from the User Quality*, In: IEEE transactions on instrumentation and measurement, vol. 59, no. 3, march 2010. pp: 730-735

K Kumar Sricharan, A Aneesh Reddy & A G Ramakrishnan. (2006). *Knuckle based Hand Correlation for User Authentication, In: Biometric Technology for Human Identification* III, Proc. of SPIE, Vol. 6202, 62020X, (2006).

Fang Li, Maylor K.H. Leung & Xaozhou Yu. (2010). *Palmprint Identi_cation Using Hausdorff Distance,* In: International Workshop on Biomedical Circuits & Systems (BioCAS'04),2004.

David Zhang, Zhenhua Guo, Guangming Lu, Lei Zhang & Wangmeng Zuo. (2010). *An Online System of Multi-spectral Palmprint Verification,* In: IEEE transactions on instrumentation and measurement, Vol. 59, No. 2, February 2010, pp 48-490.

D. Zhang, W. Kong, J. You & M. Wong. (2003). *On-line Palmprint Identification, In: IEEE transactions On pattern analysis and machine intelligence,* Vol. 25, No. 9, September 2003. pp: 1041-1050

M. Antonini, M. Barlaud, P. Mathieu & I. Daubechies. (1992). *Image coding using the wavelet transform,* In: IEEE transactions on Image Processing (2), pp 205-220.

Afzel Noore, Richa Singh & Mayank Vatsa. (2007). *Robust memory-efficient data level information fusion of multi-modal biometric images,* In: Information Fusion 8, pp. 337-346. 2007. Vol 8 N 4, pp: 337-346, October 2007

Saeed Dabbaghchian, Masoumeh P.Ghaemmaghami & Ali Aghagolzadeha. (2010). *Feature Extraction Using Discrete Cosine Transform and Discrimination Power Analysis With a Face Recognition Technology,* In: Pattern Recognition, 43, pp 1431-1440. April 2010

Yen-Yu Chen & Shen-Chuan Tai. (2004). *Embedded Medical Image Compression Using DCT Based Subband Decomposition and Modified SPIHT Data Organization,* In: Proceedings of the Fourth IEEE Symposium on Bioinformatics and Bioengineering (BIBE'04). Taichung, Taiwan, pp: 167-174, may 2004

Peter Varchol & Dusan Levicky. (2007). *Using of Hand Geometry in Biometric Security Systems, In: Radio engineering,* vol. 16, no. 4, December 2007. Julian Fierrez, Pp: 82-87

Javier Ortega-Garcia & Daniel Ramos. (2007). *HMM-Based On-Line Signature Verification: Feature Extraction and Signature Modeling,* In: Pattern recognition letters, vol. 28, no. 16, pp.2325-2334, December 2007.

Fabien Cardinaux, Conrad Sanderson & Samy Bengio. (2004). *Face Verification Using Adapted Generative Models,* In: The 6th IEEE International Conference Automatic Face and Gesture Recognition (AFGR), Seoul, 2004, pp. 825-830.

Tadej Savic & Nikola Pavesic. (2007). *Personal recognition based on an image of the palmar surface of the hand, In: Pattern Recognition,* 40, pp: 3152-3163, 2007.

A. Ross, A. Jain & J-Z. Qian. (2001). *Information Fusion in Biometrics, In: Audio and video-Based Biometric Person Authentication,* pp. 354 359, 2001.

Richa Singh, Mayank Vatsa & Afzel Noore. (2008). *Hierarchical fusion of multispectral face images for improved recognition performance,* In: Science Direct, Information Fusion 9 , pp. 200-210, 2008. Vol 9 N2, April 2008

PolyU Database. The Hong Kong Polytechnic University (PolyU) *Multispectral Palmprint Database, available* at: http://www.comp.polyu.edu.hk/ biometrics /Multispectral Palmprint/MSP.htm.

T. Connie, A. Teoh, M. Goh & D. Ngo. (2003). *Palmprint Recognition with PCA and ICA,* In: Palmerston North, 2003. Conference of image and vision computing new Zealand 2003, pp: 227-232.

Dongmei Sun, Zhengding Qiu & Qiang Li. (2006). *Palmprint Identification using Gabor Wavelet Probabilistic Neural Networks,* In: IEEE International conference on signal processing (ICSP 2006), 2006.

Ajay Kumar, David C.M. Wong, Helen C. Shen & Anil K. Jain. (2006). *Personal authentication using hand images,* In: Pattern Recognition Letters 27, pp.1478-1486, 2006.

S. M. Prasad, V. K. Govindan & P. S. Sathidevi. (2009). *Palmprint Authentication Using Fusion of Wavelet Based Representations,* In: IEEE Xplore, World Congress on Nature & Biologically Inspired Computing (NaBIC), 2009. Coimbatore, India, pp: 520-525, December 2009

A. K. Jain, A. Ross & S. Prabhakar. (2004). An Introduction to Biometric Recognition, In: IEEE Transactions on Circuits and Systems for Video Technology, Vol. 14, N. 1, pp 4-20, January 2004.

Qingyun Dai, Ning Bi, Daren Huang, DvaiJ Zhang & Feng Li. (2004). *M-Band Wavelets Application To Palmprint Recognition Based On Texture Features,* In: IEEEexplorer International Conference on Image Processing (ICIP), pp :893-896, Singapore, October 2004

3

Audio-Visual Biometrics and Forgery

Hanna Greige and Walid Karam
University of Balamand
Lebanon

1. Introduction

With the emergence of smart phones and third and fourth generation mobile and communication devices, and the appearance of a "first generation" type of mobile PC/PDA/phones with biometric identity verification, there has been recently a greater attention to secure communication and to guaranteeing the robustness of embedded multi-modal biometric systems. The robustness of such systems promises the viability of newer technologies that involve e-voice signatures, e-contracts that have legal values, and secure and trusted data transfer regardless of the underlying communication protocol. Realizing such technologies require reliable and error-free biometric identity verification (IV) systems.

Biometric IV systems are starting to appear on the market in various commercial applications. However, these systems are still operating with a certain measurable error rate that prevents them from being used in a full automatic mode and still require human intervention and further authentication. This is primarily due to the variability of the biometric traits of humans over time because of growth, aging, injury, appearance, physical state, and so forth.

Imposture can be a real challenge to biometric IV systems. It is reasonable to assume that an impostor has knowledge of the biometric authentication system techniques used on one hand, and, on the other hand, has enough information about the target client (face image, video sequence, voice recording, fingerprint pattern, etc.) A deliberate impostor attempting to be authenticated by an IV system could claim someone else's identity to gain access to privileged resources. Taking advantage of the non-zero false acceptance rate of the IV system, an impostor could use sophisticated forgery techniques to imitate, as closely as possible, the biometric features of a genuine client.

The robustness of a biometric IV system is best evaluated by monitoring its behavior under impostor attacks. This chapter studies the effects of deliberate forgery on verification systems. It focuses on the two biometric modalities people use most to recognize naturally each other: face and voice.

The chapter is arranged as follows. **Section 2** provides a motivation of the research investigated in this chapter, and justifies the need for audio-visual biometrics. **Section 3** introduces audio-visual identity verification and imposture concepts. **Section 4** then reviews automated techniques of audio-visual (A/V) IV and forgery. Typically, an A/V IV system uses audio and video signals of a client and matches the features of these signals with stored templates of features of that client.

Next, **section 5** describes imposture techniques on the visual and audio levels: face animation and voice conversion. The video sequence of the client can be altered at the audio and the visual levels. At the audio level, a voice transformation technique is employed to change the perceived speaker identity of the speech signal of the impostor to that of the target client. Techniques of voice transformation are surveyed. Such techniques were not originally developed for forgeries but have other applications. Voice conversion has been effectively used in text-to-speech systems to produce many different new voices. Other applications include dubbing movies and TV shows and the creation of virtual characters or even a virtual copy of a person's A/V identity. In this work, the main interest is to use voice conversion techniques in forgery.

At the visual level, a face transformation of the impostor to that of the client is required. This necessitates initially face detection and tracking, followed by face transformation. Face animation is described, which allows a 2-D face image of the client to be animated. This technique employs principal warps to deform defined MPEG-4 facial feature points based on determined facial animation parameters (FAP).

Evaluation and experimental results are then provided in **section 6**. Results of forgery are reported on the BANCA A/V database to test the effects of voice and face transformation on the IV system. The proposed A/V forgery is completely independent from the baseline A/V IV system, and can be used to attack any other A/V IV system. The Results drawn from the experiments show that state-of-the-art IV systems are vulnerable to forgery attacks, which indicate more impostor acceptance, and, for the same threshold, more genuine client denial. This should drive more research towards more robust IV systems, as outlined in the conclusion of **section 7**.

2. Why audio-visual biometrics?

The identification of a person is generally established by one of three schemes: Something the person owns and has (e.g. an identity card-ID, a security token, keys), something the person knows (e.g. username, password, personal identification number-PIN, or a combination of these), or something the person is or does, i.e. his/her anatomy, physiology, or behavior (e.g. fingerprint, signature, voice, gait). The latter is the more natural scheme and relies on biometric traits of a person for identification and authentication. An identity card can be lost, stolen or forged; a password or a PIN can be forgotten by its owner, guessed by an impostor, or shared by many individuals. However, biometric traits are more difficult to reproduce. Modern identity authentication systems have started to use biometric data to complement or even to replace traditional IV techniques.

The identity of a person is primarily determined visually by his or her face. A human individual can also fairly well be recognized by his or her voice. These two modalities, i.e. face and voice, are used naturally by people to recognize each other. They are also employed by many biometric identity recognition systems to automatically verify or identify humans for commercial, security and legal applications, including forensics. The combination of auditive and visual recognition is yet more efficient and improves the performance of the identity recognition systems.

Other biometric modalities have traditionally been used for identity recognition. Handwritten signatures have long been used to provide evidence of the identity of the signatory. Fingerprints have been used for over a hundred years to identify persons. They have also been successfully used in forensic science to identify suspects, criminals, and victims at a crime site. Other biometric features that help in identity recognition include the iris, the retina, hand geometry, vein pattern of hand, gait, electrocardiogram, ear form, DNA profiling, odor, keystroke dynamics, and mouse gestures. All of these biometric features, to the exception of handwritten signatures, are considered intrusive; they require the cooperation of the user, careful exposure to biometric sensors, and might necessitate the repetition of feature capture for correctness and more accuracy. Some of these features, e.g. odor, keystroke dynamics, are not stable and vary from time to time, and thus are not reliable as an IV technique. DNA, on the other hand, is infallible but cannot be used instantly as it requires laboratory analysis.

Consequently, the adoption of the two non-intrusive, psychologically neutral modalities, i.e. face and voice, for automatic identity recognition is a natural choice, and is expected to mitigate rejection problems that often restrain the social use of biometrics in various applications, and broaden the use of A/V IV technologies.

In recent years, there has been a growing interest and research in A/V IV systems. Several approaches and techniques have been developed. These techniques are surveyed below. The robustness of these schemes to various quality conditions has also been investigated in the literature. However, the effect of deliberate imposture on the performance of IV systems has not been extensively reported. The robustness of a biometric IV system is best evaluated by monitoring its behavior under impostor attacks. Such attacks may include the transformation of one, many, or all of the biometric modalities.

3. Audio-visual identity verification and imposture

Typically, an automatic A/V IV system uses audio and video signals of a client and matches the features of these signals with stored templates of features of that client. A decision, in terms of a likelihood score, is made on whether to accept the claimed identity or to reject it. Fig. 1(a) depicts the concept. To be authenticated on a biometric system, an impostor attempts to impersonate a genuine client. He uses an imposture system to convert his own audio and video signals to defeat the verification system. This process is modeled in Fig. 1(b)

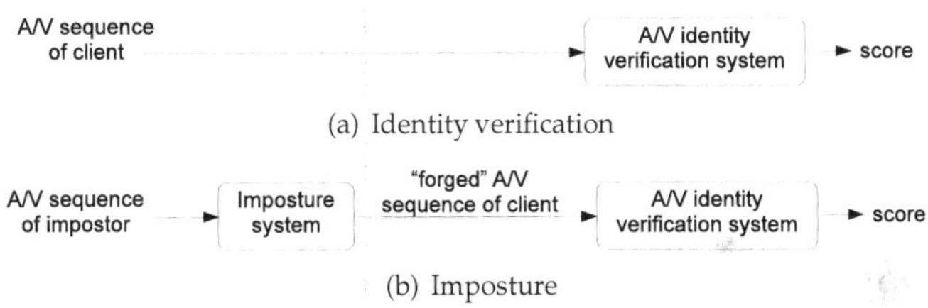

Fig. 1. Audio-visual identity verification and imposture

The imposture concept can be summarized as follows. The video sequence of the client can be altered at the audio and the visual levels. At the audio level, a voice transformation technique is employed to change the perceived speaker identity of the speech signal of the impostor to that of the target client. At the visual level, a face transformation of the impostor to that of the client is required. This necessitates initially face detection and tracking, followed by face transformation. Fig. 2 depicts the concept.

The purpose of the work described in this chapter is to build such an A/V transformation imposture system and to provide a better understanding of its effect on the performance of automatic A/V IV systems. An attempt is made to increase the acceptance rate of the impostor and to analyzing the robustness of the recognition system.

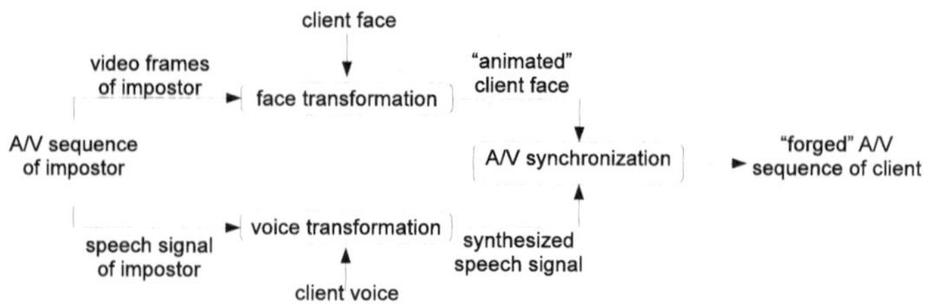

Fig. 2. The audio-visual imposture system

4. Audio-visual identity verification

An A/V IV system uses face and speech traits to verify (or to deny) a claimed identity. Face verification authenticates a person's identity by relying solely on facial information based on a set of face images (or a video sequence.) Speaker verification, on the other hand, authenticates the subject's identity based on samples of his speech. In this study, IV couples the speech and the face modalities by fusing scores of the respective verification systems.

4.1 Speaker verification

Speech carries primarily two types of information, the message conveyed by the speaker, and the identity of the speaker. In this work, analysis and synthesis of the voice of a speaker is text-independent and completely ignore the message conveyed. The focus is on the identity of the speaker.

To process a speech signal, a feature extraction module calculates relevant feature vectors from the speech waveform. On a signal window that is shifted at a regular rate a feature vector is calculated. Generally, cepstral-based feature vectors are used (section 4.1.1). A stochastic model is then applied to represent the feature vectors from a given speaker. To verify a claimed identity, new utterance feature vectors are generally matched against the claimed speaker model and against a general model of speech that may be uttered by any speaker, called the world model. The most likely model identifies if the claimed speaker has uttered the signal or not. In text independent speaker verification, the model should not reflect a specific speech structure, i.e. a specific sequence of words. State-of-the art systems use Gaussian Mixture Models (GMM) as stochastic models in text-independent mode (sections 4.1.2 and 4.1.3.)

Speaker verification encompasses typically two phases: a training phase and a test phase. During the training phase, the stochastic model of the speaker is calculated. The test phase determines if an unknown speaker is the person he claims to be. Fig. 3(a) and fig. 3(b) provide a block diagram representations of the concept.

4.1.1 Feature extraction

The first part of the speaker verification process is the speech signal analysis. Speech is inherently a non-stationary signal. Consequently, speech analysis is normally performed on short fragments of speech where the signal is presumed stationary. Typically, feature extraction is carried out on 20 to 30 ms windows with 10 to 15 ms shift between two successive windows. To compensate for the signal truncation, a weighting signal (Hamming window, Hanning windowÉ) is applied on each window of speech. The signal is also filtered with a first-order high-pass filter, called a pre-emphasis filter, to compensate for the -6dB/octave spectral slope of the speech signal. This pre-emphasis step is conventionally used before windowing.

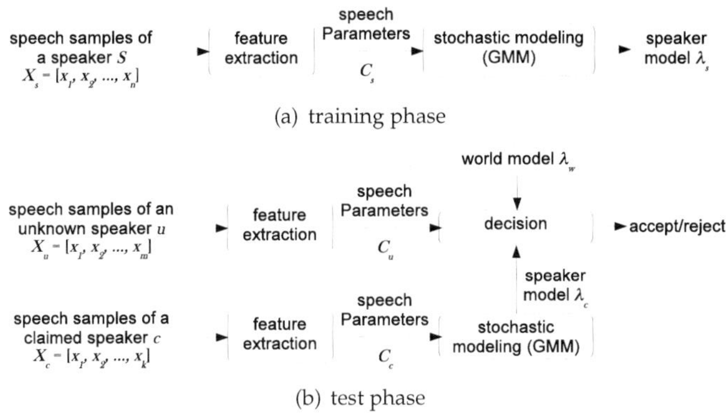

Fig. 3. The speaker verification system

Coding the truncated speech windows is achieved through variable resolution spectral analysis. Traditionally, two techniques have been employed: Filter-bank analysis, and linear-predictive analysis. Filter-bank analysis is a conventional spectral analysis technique that represents the signal spectrum with the log-energies using a filter-bank of overlapping band-pass filters. Linear predictive analysis is another accepted speech coding technique. It uses an all-pole filter whose coefficients are estimated recursively by minimizing the mean square prediction error.

The next step is cepstral analysis. The cepstrum is the inverse Fourier transform of the logarithm of the Fourier transform of the signal. A determined number of mel frequency cepstral coefficients (MFCC) are used to represent the spectral envelope of the speech signal. They are derived from either the filter bank energies or from the linear prediction coefficients. To reduce the effects of signals recorded in different conditions, Cepstral mean subtraction and feature variance normalization is used. First and second order derivatives of extracted features are appended to the feature vectors to account for the dynamic nature of speech.

4.1.2 Silence detection and removal

The silence part of the signal alters largely the performance of a speaker verification system. Actually, silence does not carry any useful information about the speaker, and its presence introduces a bias in the score calculated, which deteriorates the performance of the system. Therefore, most of the speaker recognition systems remove the silence parts from the signal before starting the recognition process. Several techniques have been used successfully for silence removal. In this work, we suppose that the energy in the signal is a random process that follows a bi-Gaussian model, a first Gaussian modeling the energy of the silence part and the other modeling the energy of the speech part. Given an utterance and more specifically the computed energy coefficients, the bi-Gaussian model parameters are estimated using the EM algorithm. Then, the signal is divided into speech parts and silence parts based on a maximum likelihood criterion. Treatment of silence detection is found in (Paoletti & Erten, 2000).

4.1.3 Speaker classification and modeling

Each speaker possesses a unique vocal signature that provides him with a distinct identity. The purpose of speaker classification is to exploit such distinctions in order to verify the

identity of a speaker. Such classification is accomplished by modeling speakers using a Gaussian Mixture Model (GMM).
Assume a given sample of speech Y, and a speaker S. Speaker verification is an attempt to determine if Y was spoken by S. The hypothesis test of equation 1 can be stated. An optimum test to decide between the null hypothesis H_0 and the alternative hypothesis H_1 is the likelihood ratio test:

$$\frac{p(Y|H_0)}{p(Y|H_1)} \begin{cases} \geq \theta & \text{accept } H_0 \\ < \theta & \text{reject } H_0 \end{cases}$$

$$\begin{cases} H_0\text{: Speech sample } Y \text{ belongs to speaker } S \\ H_1\text{: Speech sample } Y \text{ does not belong to speaker } S \end{cases} \quad (1)$$

where $p(Y \mid H_i)$, $i = 0, 1$ is the likelihood of the hypothesis H_i given the speech sample Y, and θ is the decision threshold for accepting or rejecting H_0. Figure 8 below describes speaker verification based on a likelihood ratio. A speech signal is first treated (noise reduction and linear filtering), then speaker-dependent features are extracted as described in section 4.1.1 above. The MFCC feature vectors, denoted $X = \{x_1, x_2, ..., x_n\}$, are then used to find the likelihoods of H_0 and H_1. Since speaker modeling is based on a mixture of Gaussians, as described in the next section 4.1.4, H_0 can be represented by the GMM model of the speaker λ_s, which symbolizes the mean vector and the covariance matrix parameters of the Gaussian distribution of the feature vectors of the speaker.

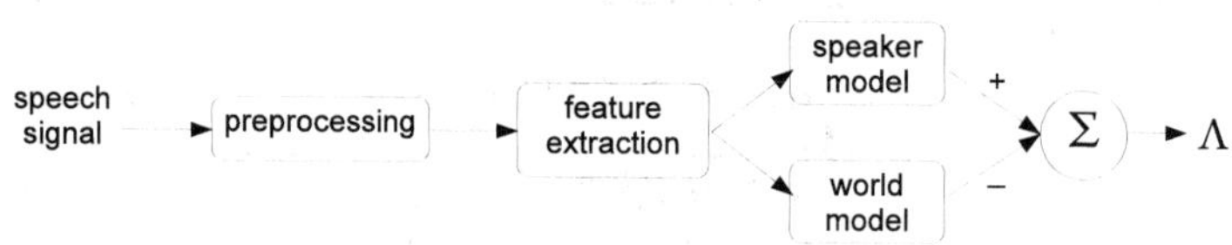

Fig. 4. Speaker verification based on a likelihood ratio

The alternative hypothesis H_1 can also be represented by a GMM model $\lambda_{\bar{s}}$, which models the entire space of alternatives to the speaker. This model is typically known as the "Universal Background Model" (UBM), or the "World Model". Alternative approaches represent the non-speaker space by a set of models representing the impostors.
The logarithm of the likelihood ratio $p(X \mid \lambda_s)/p(X \mid \lambda_{\bar{s}})$ is often computed:

$$\Lambda(X) = \log p(X \mid \lambda_s) - \log p(X \mid \lambda_{\bar{s}}) \quad (2)$$

4.1.4 Gaussian mixture models

A mixture of Gaussians is a weighted sum of M Gaussian densities $P(x \mid \lambda) = \sum_{i=1:M} \alpha_i f_i(x)$ where x is an MFCC vector, $f_i(x)$ is a Gaussian density function, and α_i the corresponding weights. Each Gaussian is characterized by its mean μ_i and a covariance matrix Σ_i. A speaker model λ is characterized by the set of parameters $(\alpha_i, \mu_i, \sum_i)_{i=1:M}$.
For each client, two GMM's are used, the first corresponds to the distribution of the training set of speech feature vectors of that client, and the second represents the distribution of the training vectors of a defined "world model". To formulate the classification concept, assume

a speaker is presented along with an identity claim C. The feature vectors $X = \{\underline{x_i}\}_{i=1}^{N}$ are extracted. The average log likelihood of the speaker having identity C is calculated as

$$\mathcal{L}(X \mid \lambda_c) = \frac{1}{N}\sum_{i=1}^{N} \log p\left(\underline{x_i} \mid \lambda_c\right) \tag{3}$$

where

$$p\left(\underline{x_i} \mid \lambda_c\right) = \sum_{j=1}^{N_G} m_j \mathcal{N}\left(\underline{x}; \underline{\mu_j}, Cov_j\right)$$

$$\lambda_c = \left\{m_j \underline{\mu_j}, Cov_j\right\}_{j=1}^{N_G} \tag{4}$$

$$\mathcal{N}\left(\underline{x}; \underline{\mu_j}, Cov_j\right) = \frac{1}{(2\pi)^{\frac{D}{2}} \left|Cov_j\right|^{\frac{1}{2}}} e^{-\frac{1}{2}(\underline{x}-\underline{\mu_j})^T Cov_j^{-1}(\underline{x}-\underline{\mu_j})}$$

is a multivariate Gaussian function with mean $\underline{\mu_j}$ and diagonal covariance matrix Cov_j, and D is the dimension of the feature space, λ_c is the parameter set for person C, N_G is the number of Gaussians, m_j = weight for Gaussian j, and $\sum_{k=1}^{N_j} m_k = 1, m_k \geqslant 0 \forall k$. With a world model of w persons, the average log likelihood of a speaker being an impostor is found as $\mathcal{L}(X \mid \lambda_w) = \frac{1}{N}\sum_{i=1}^{N_W} \log p\left(\underline{x_i} \mid \lambda_w\right)$. An opinion on the claim is then found: $\mathcal{O}(X) = \mathcal{L}(X \mid \lambda_c) - \mathcal{L}(X \mid \lambda_w)$ As a final decision to whether the face belongs to the claimed identity, and given a certain threshold t, the claim is accepted when $\mathcal{O}(X) \geqslant t$, and rejected when $\mathcal{O}(X) < t$. To estimate the GMM parameters λ of each speaker, the world model is adapted using a Maximum a Posteriori (MAP) adaptation (Gauvain & Lee, 1994). The world model parameters are estimated using the Expectation Maximization (EM) algorithm (Dempster et al., 1977).

The EM algorithm is an iterative algorithm. Each iteration is formed of two phases: the Estimation (E) phase and the Maximization (M) phase. In the E phase the likelihood function of the complete data given the previous iteration model parameters is estimated. In the M phase new values of the model parameters are determined by maximizing the estimated likelihood. The EM algorithm ensures that the likelihood on the training data does not decrease with the iterations and therefore converges towards a local optimum. This local optimum depends on the initial values given to the model parameters before training and therefore, the initialization of the model parameters is a crucial step.

GMM client training and testing is performed on the speaker verification toolkit BECARS (Blouet et al., 2004). BECARS implements GMM's with several adaptation techniques, e.g. Bayesian adaptation, MAP, maximum likelihood linear regression (MLLR), and the unified adaptation technique defined in (Mokbel, 2001).

The speaker recognition system requires two models: the model of the claimed speaker and the world model. The direct estimation of the GMM parameters using the EM algorithm requires a large amount of speech feature vectors. This can be easily satisfied for the world model where several minutes from several speakers may be collected offline. For the speaker model, this introduces a constraint, i.e. the speaker to talk for large duration. To overcome this, speaker adaptation techniques may be used (Mokbel, 2001) to refine the world model parameters λ_w into speaker specific parameters λ_s.

4.1.5 Score normalization

The last step in a speaker verification system is the decision of whether a speaker is the claimed identity (fig. 3(b)). It involves matching the claimed speaker model to speech samples by comparing a likelihood measure to a decision threshold. If the likelihood measure is smaller that the threshold, the speaker is rejected, otherwise, he is accepted. The choice of the decision threshold is not a simple task, and cannot be universally fixed due to the large score variability of various experiments. This is primarily due to the variability of conditions of speech capture, voice quality, speech sample duration, and background noise. It is also due to inter-speaker or intra-speaker differences between the enrollment speech data and the data used for testing. Score normalization is used to lessen the effect of the score variability problem. Different score normalization techniques have been proposed (Bimbot et al., 2004; Li & Porter, 1988). These include Z-norm, H-norm, T-norm, HT-norm, C-norm, D-norm, and WMAP.

Z-norm is a widely used normalization technique that has the advantage of being performed offline during the speaker training phase. Given a speech sample data X, a speaker model λ, and the corresponding score $L_\lambda(X)$. The normalized score is given by $\widetilde{L}_\lambda(X) = \frac{L_\lambda(X)-\mu_\lambda}{\sigma_\lambda}$, where μ_λ and σ_λ are the score mean and standard deviation of the speaker λ.

4.2 Face verification

The human face is a prominent characteristic that best distinguishes individuals. It forms an important location for person identification and transmits momentous information in social interaction. Psychological processes involved in face identification are known to be very complex, to be present from birth, and to involve large and widely distributed areas in the human brain. The face of an individual is entirely unique. It is determined by the size, shape, and position of the eyes, nose, eyebrows, ears, hair, forehead, mouth, lips, teeth, cheeks, chin, and skin.

Face verification is a biometric person recognition technique used to verify (confirm or deny) a claimed identity based on a face image or a set of faces (or a video sequence). Methods of face verification have been developed and surveyed in the literature (Chellappa et al., 1995; Dugelay et al., 2002; Fromherz et al., 1997; Zhang et al., 1997). (Zhao et al., 2003) classifies these methods into three categories: Holistic methods, feature-based methods, and hybrid methods. These methods are classified according to the differences in the feature extraction procedures and/or the classification techniques used.

4.2.1 Holistic methods

Holistic (global) identity recognition methods treat the image information without any localization of individual points. The face is dealt with as a whole without any explicit isolation of various parts of the face. Holistic techniques employ various statistical analysis, neural networks and transformations. They normally require a large training set but they generally perform better. However, such techniques are sensitive to variations in position, rotation, scale, and illumination and require preprocessing and normalization.

Holistic methods include Principal-component analysis (PCA) and eigenfaces (Turk & Pentland, 1991), Linear Discriminant Analysis (LDA) (Zhao et al., 1998), Support Vector Machine (SVM) (Phillips, 1998), and Independent Component Analysis (ICA) (Bartlett et al., 2002).

4.2.2 Feature-based methods

As opposed to holistic methods, feature-based techniques depend on the identification of fiducial points (reference or feature points) on the face such as the eyes, the nose, and the mouth. The relative locations of these feature points are used to find geometric association between them. Face recognition thus combines independent processing of the eyes, the nose, the mouth, and other feature points of the face. Since detection of feature points precedes the analysis, such a system is robust to position variations in the image.

Feature-based methods include graph matching (Wiskott et al., 1997), Hidden Markov Models (HMM) (Nefian & Hayes, 1998; Samaria & Young, 1994), and a Bayesian Framework (Liu & Wechsler, 1998; Moghaddam et al., 2000).

4.2.3 Hybrid and other methods

These methods either combine holistic and feature-based techniques or employ methods that do not fall in either category. These include Active Appearance Models (AAM) (Edwards et al., 1998), 3-D Morphable Models (Blanz & Vetter, 2003), 3-D Face Recognition (Bronstein et al., 2004), the trace transform (Kadyrov & Petrou, 2001; Srisuk et al., 2003), and kernel methods (Yang, 2002; Zhou et al., 2004).

The process of automatic face recognition can be thought of as being comprised of four stages: 1-Face detection, localization and segmentation, 2-Normalization, 3-Facial feature extraction, and 4-Classification (identification and/or verification). These subtasks have been independently researched and surveyed in the literature, and are briefed next.

4.2.4 Face detection and tracking in a video sequence

4.2.4.1 Face detection

Face detection is an essential part of any face recognition technique. Given an image, face detection algorithms try to answer the following questions

- Is there a face in the image?
- If there is a face in the image, where is it located?
- What are the size and the orientation of the face?

Face detection techniques are surveyed in (Hjelmas & Low, 2001). The face detection algorithm used in this work has been introduced initially by (Viola & Jones, 2001) and later developed further by (Lienhart & Maydt, 2002) at Intel Labs. It is a machine learning approach based on a boosted cascade of simple and rotated haar-like features for visual object detection.

4.2.4.2 Face tracking in a video sequence

Face tracking in a video sequence is a direct extension of still image face detection techniques. However, the coherent use of both spatial and temporal information of faces makes the detection techniques more unique. The technique used in this work employs the algorithm developed by (Lienhart & Maydt, 2002) on every frame in the video sequence. However, three types of errors are identified in a talking face video: 1-More than one face is detected, but only one actually exists in a frame, 2-A wrong object is detected as a face, and 3-No faces are detected. Fig. 5 shows an example detection from the BANCA database (Popovici et al., 2003), where two faces have been detected, one for the actual talking-face subject, and a false alarm. The correction of these errors is done through the exploitation of spatial and temporal information in the video sequence as the face detection algorithm is run on every subsequent frame. The correction algorithm is summarized as follows:

(a) More than one face area detected: The intersections of these areas with the area of the face of the previous frame are calculated. The area that corresponds to the largest calculated intersection is assigned as the face of the current frame. If the video frame in question is the first one in the video sequence, then the decision to select the proper face for that frame is delayed until a single face is detected at a later frame, and verified with a series of subsequent face detections.

(b) No faces detected: The face area of the previous frame is assigned as the face of the current frame. If the video frame in question is the first one in the video sequence, then the decision is delayed as explained in part (a) above.

(c) A wrong object detected as a face: The intersection area with the previous frame face area is calculated. If this intersection ratio to the area of the previous face is less than a certain threshold, e.g. 80%, the previous face is assigned as the face of the current frame.

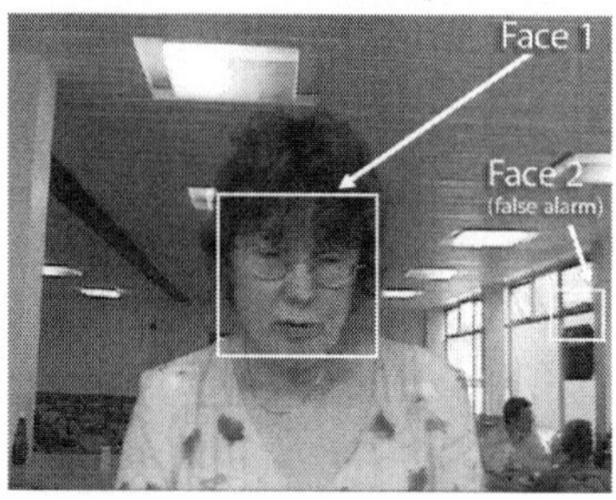

Fig. 5. Face Detection and Tracking

Fig. 6 below shows a sample log of face detection. The coordinates shown are (x, y) of the top left-hand corner of the rectangle encompassing the face area, and its width and height.

```
...
Frame 409: Found 1 face: 334 297 136 136 100.00 %
Frame 410: Found 1 face: 334 293 138 138 95.69 %
Frame 411: Found 1 face: 332 292 139 139 97.85 %
Frame 412: Found 2 faces. Previous face: 332 292 139 139
Face 0: 114 425 55 55 0.00 %
Face 1: 331 291 141 141 97.18 %
Selected face 1: 331 291 141 141

Frame 413: Found 1 face: 332 292 141 141 98.59 %
Frame 414: Found 1 face: 333 292 138 138 100.00 %
Frame 415: Found 1 face: 333 292 138 138 100.00 %
Frame 416: Found 1 face: 333 294 138 138 98.55 %
Frame 417: Found 3 faces. Previous face: 333 294 138 138
Face 0: 618 381 52 52 0.00 %
Face 1: 113 424 55 55 0.00 %
Face 2: 334 294 135 135 100.00 %
Selected face 2: 334 294 135 135
Frame 418: Found 1 face: 336 295 132 132 100.00 %
Frame 419: Found 1 face: 332 291 141 141 87.64 %
Frame 420: Found 1 face: 332 292 139 139 100.00 %
Frame 421: Found 1 face: 334 292 136 136 100.00 %
Frame 422: Found 1 face: 332 291 141 141 93.03 %
...
```

Fig. 6. Sample log of a face detection and tracking process

4.2.5 Face normalization

Normalizing face images is a required pre-processing step that aims at reducing the variability of different aspects in the face image such as contrast and illumination, scale, translation, rotation, and face masking. Many works in the literature (Belhumeur & Kriegman, 1998; Swets & Weng, 1996; Turk & Pentland, 1991) have normalized face images with respect to translation, scale, and in-plane rotation, while others have also included masking and affine warping to properly align the faces (Moghaddam & Pentland, 1997). (Craw & Cameron, 1992) have used manually annotated points around shapes to warp the images to the mean shape, leading to shape-free representation of images useful in PCA classification.

The pre-processing stage in this work includes four steps:

- Scaling the face image to a predetermined size (w_f, h_f).
- Cropping the face image to an inner-face, thus disregarding any background visual data.
- Disregarding color information by converting the face image to grayscale.
- Histogram equalization of the face image to compensate for illumination changes.

Fig. 7 below shows an example of the four steps.

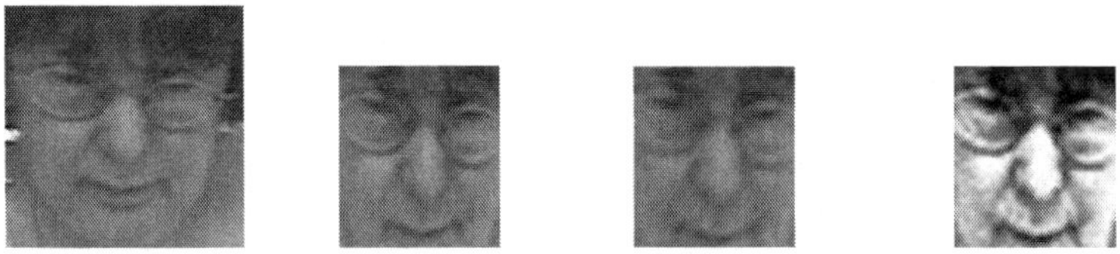

(a) detected face (b) cropped "inner" face (c) face (d) histogram-equalized face

Fig. 7. Preprocessing face images

4.2.6 Facial feature extraction

The face feature extraction technique used in this work is DCT-*mod2*, initially introduced by (Sanderson & Paliwal, 2002), who showed that their proposed face feature extraction technique outperforms PCA and 2-D Gabor wavelets in terms of computational speed and robustness to illumination changes. This feature extraction technique is briefed next. A face image is divided into overlapping $N \times N$ blocks. Each block is decomposed in terms of orthogonal 2-D DCT basis functions, and is represented by an ordered vector of DCT coefficients:

$$\left[c_0^{(b,a)} c_1^{(b,a)} \dots c_{M-1}^{(b,a)}\right]^T \tag{5}$$

where (b, a) represent the location of the block, and M is the number of the most significant retained coefficients. To minimize the effects of illumination changes, horizontal and vertical delta coefficients for blocks at (b, a) are defined as first-order orthogonal polynomial coefficients:

$$\Delta^h c_n^{(b,a)} = \frac{\sum_{k=-1}^{1} k h_k c_n^{(b,a+k)}}{\sum_{k=-1}^{1} h_k k^2} \qquad \Delta^v c_n^{(b,a)} = \frac{\sum_{k=-1}^{1} k h_k c_n^{(b+k,a)}}{\sum_{k=-1}^{1} h_k k^2} \tag{6}$$

The first three coefficients c_0, c_1, and c_2 are replaced in (5) by their corresponding deltas to form a feature vector of size M+3 for a block at (b, a):

$$\left[\Delta^h c_0 \Delta^v c_0 \Delta^h c_1 \Delta^v c_1 \Delta^h c_2 \Delta^v c_2 c_3 c_4 ... c_{M-1}\right]^T \quad (7)$$

In this study, neighboring blocks of 8 $\times$ 8 with an overlap of 50% is used. M, the number of retained coefficients is fixed at 15.

4.2.7 Classification

Face verification can be seen as a two-class classification problem. The first class is the case when a given face corresponds to the claimed identity (client), and the second is the case when a face belongs to an impostor. In a similar way to speaker verification, a GMM is used to model the distribution of face feature vectors for each person. The reader is referred back to sections 4.1.3 and 4.1.4 for a detailed treatment of identity classification and Gaussian mixture modeling.

4.3 Audio-visual data fusion

An A/V IV system uses face and voice biometric traits to verify (or to deny) a claimed identity. Face verification authenticates a person's identity by relying solely on facial information based on a set of face images (or a video sequence.) Speaker verification, on the other hand, authenticates the subject's identity based on samples of his speech. In this study, IV couples the speech and the face modalities by fusing scores of the respective verification systems.

It has been shown that biometric verification systems that combine multiple modalities outperform single biometric modality systems (Kittler, 1998). A final decision on the claimed identity of a person relies on both the speech-based and the face-based verification systems. To combine both modalities, a fusion scheme is needed. Fusion can be achieved at different levels (Ross & Jain, 2003) (Fig. 8):

- Fusion at the feature extraction level, when extracted feature vectors originating from the multiple biometric systems are combined (Fig. 8(a))
- Fusion at the matching level, when the multiple scores of each system are combined (Fig. 8(b)), and
- Fusion at the decision level, when the accept/reject decisions are consolidated (Fig. 8(c)).

Different fusion techniques have been proposed and investigated in the literature. (Ben-Yacoub et al., 1999) evaluated different binary classification approaches for data fusion, namely Support Vector Machine (SVM), minimum cost Bayesian classifier, Fisher's linear discriminant analysis, C4.5 decision classifier, and multi layer perceptron (MLP) classifier. The use of these techniques is motivated by the fact that biometric verification is merely a binary classification problem. Other fusion techniques used include the weighted sum rule and the weighted product rule. It has been shown that the sum rule and support vector machines are superior when compared to other fusion schemes (Chatzis et al., 1999; Fierrez-Aguilar et al., 2003).

In this study, fusion at the classification level is used. The weighted sum rule fusion technique is employed to fuse the scores of the face and voice classifiers. The sum rule computes the A/V score s by weight averaging: $s = w_s s_s + w_f s_f$, where w_s and w_f are speech and face score weights computed so as to optimize the equal error rate (EER) on the training set. The

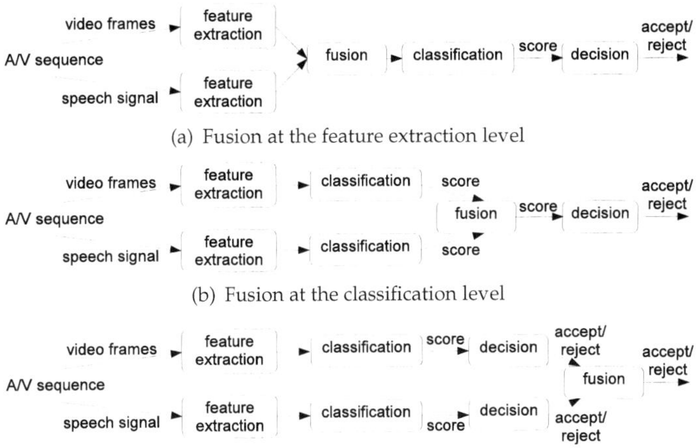

(a) Fusion at the feature extraction level

(b) Fusion at the classification level

(c) Fusion at the decision level

Fig. 8. The three levels of fusion of A/V biometric systems

speech and face scores must be in the same range (e.g. $\mu = 0, \sigma = 1$) for the fusion to be meaningful. This is achieved by normalizing the scores $s_{norm(s)} = \frac{s_s - \mu_s}{\sigma_s}$, $s_{norm(f)} = \frac{s_f - \mu_f}{\sigma_f}$.

5. Audio-visual imposture

Imposture is the act or conduct of a person (impostor) to pretend to be somebody else in an effort to gain financial or social benefits. In the context of this work, imposture is an attempt to increase the acceptance rate of one or more computer-based biometric verification systems and get authenticated to gain access to privileged resource. A/V imposture encompasses the transformation of both audio (voice) and visual (face) features, as developed below.

5.1 Voice transformation

Voice transformation, also referred to as speaker transformation, voice conversion, or speaker forgery, is the process of altering an utterance from a speaker (impostor) to make it sound as if it were articulated by a target speaker (client.) Such transformation can be effectively used by an avatar to impersonate a real human and converse with an Embodied Conversational Agent (ECA). Speaker transformation techniques might involve modifications of different aspects of the speech signal that carries the speaker's identity such as the Formant spectra i.e. the coarse spectral structure associated with the different phones in the speech signal (Kain & Macon, 1998), the Excitation function i.e. the "fine" spectral detail, the Prosodic features i.e. aspects of the speech that occur over timescales larger than individual phonemes, and the Mannerisms such as particular word choice or preferred phrases, or all kinds of other high-level behavioral characteristics. The formant structure and the vocal tract are represented by the overall spectral envelope shape of the signal, and thus are major features to be considered in voice transformation (Kain & Macon, 2001).

Several voice transformation techniques have been proposed in the literature (Abe et al. (1988); Kain & Macon (2001); Perrot et al. (2005); Stylianou & Cappe (1998); Sundermann et al. (2006); Toda (2009); Ye & Young (2004)). These techniques can be classified as text-dependent

methods and text independent methods. In text-dependent methods, training procedures are based on parallel corpora, i.e. training data have the source and the target speakers uttering the same text. Such methods include vector quantization (Abe et al. (1988); Arslan (1999)), linear transformation (Kain & Macon, 2001; Ye & Young, 2003), formant transformation (Turajlic et al., 2003), vocal tract length normalization (VTLN) (Sundermann et al., 2003), and prosodic transformation (Erro et al., 2010). In text-independent voice conversion techniques, the system trains on source and target speakers uttering different text. Techniques include text-independent VTLN (Sundermann et al., 2003), maximum likelihood adaptation and statistical techniques (Karam et al., 2009; Mouchtaris et al., 2004; Stylianou & Cappe, 1998), unit selection (Sundermann et al., 2006), and client memory indexation. (Chollet et al., 2007; Constantinescu et al., 1999; Perrot et al., 2005).
The analysis part of a voice conversion algorithm focuses on the extraction of the speaker's identity. Next, a transformation function is estimated. At last, a synthesis step is achieved to replace the source speaker characteristics by those of the target speaker.
Consider a sequence of spectral vectors uttered by a source speaker (impostor) $X_s = [x_1, x_2, ..., x_n]$, and another sequence pronounced by a target speaker $Y_t = [y_1, y_2, ..., y_n]$. Voice conversion is based on the estimation of a conversion function $\mathcal{F}$ that minimizes the mean square error $\epsilon_{mse} = E\lfloor \|y - \mathcal{F}(x)\|^2 \rfloor$, where E is the expectation. Two steps are useful to build a conversion system: training and conversion. In the training phase, speech samples from the source and the target speakers are analyzed to extract the main features. For text-dependent voice conversion, these features are time aligned and a conversion function is estimated to map the source and the target characteristics (Fig. 9(a)). In a text-independent voice conversion system, a model of the target speaker is used to estimate the mapping function as illustrated in Fig. 9(b).
The aim of the conversion is to apply the estimated transformation rule to an original speech pronounced by the source speaker. The new utterance sounds like the same speech pronounced by the target speaker, i.e. pronounced by replacing the target characteristics by those of the source voice. The last step is the re-synthesis of the signal to reconstruct the source speech voice (Fig. 10).

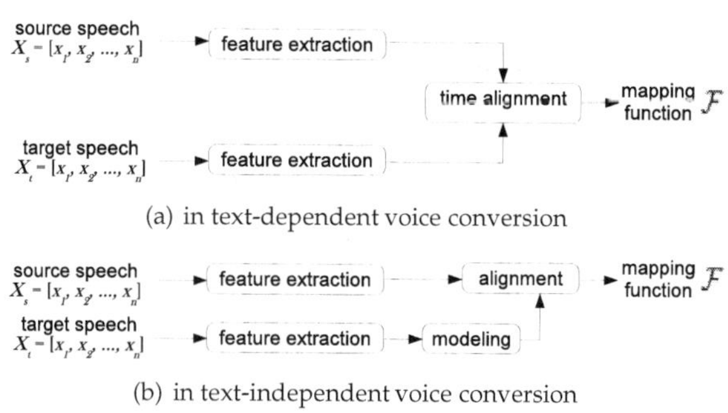

Fig. 9. Voice conversion mapping function estimation

Voice conversion can be effectively used by an avatar to impersonate a real human and hide his identity in a conversation with an ECA. This technique is complementary with face transformation in the creation of an avatar that mimics in voice and face a target real human.

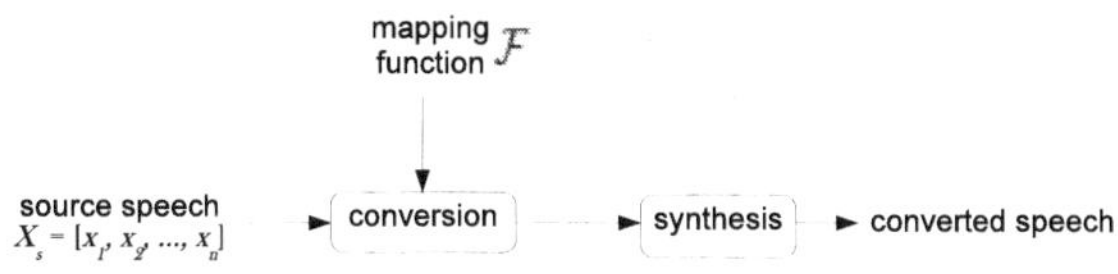

Fig. 10. Voice conversion

In this work, Mix*Trans*, initially introduced by (Karam et al., 2009), is used as a mixture-structured bias voice transformation, and it is briefed next.

5.1.1 Mix*Trans* speaker transformation

MixTrans is a text-independent speaker transformation technique that operates in the cepstral domain by defining a transformation that maps parts of the acoustical space to their corresponding time-domain signals. A speaker is stochastically represented with a GMM model. A deviation from this statistical estimate of the speaker it estimates could make the model better represent another speaker. Given a GMM model λ_t of a known target speaker t; λ_t can be obtained from a recorded speech of speaker t. The impostor s provides to the system the source speech X_s, which was never uttered by the target. Mix*Trans* makes use of X_s, its extracted features (MFCC), and the target model λ_t to compute the transformed speech X_s'. X_s' is time-aligned with, and has the same text as X_s, but appears to have been uttered by the target speaker t, as it inherits the characteristics of the source speaker s. Fig. 11 provides a block diagram of the proposed Mix*Trans* technique.

Mix*Trans* comprises several linear time-invariant filters, each of them operating in a part of the acoustical space:

$$\mathcal{T}_\gamma(\mathbf{X}) = \sum_k \mathbf{P}_k(\mathbf{X} + \mathbf{b}_k) = \sum_k \mathbf{P}_k\mathbf{X} + \sum_k \mathbf{P}_k\mathbf{b}_k = \mathbf{X} + \sum_k \mathbf{P}_k\mathbf{b}_k \tag{8}$$

where $\mathbf{b}_k$ represents the kth bias and $\mathbf{P}_k$ is the probability of being in the kth part of the acoustical space given the observation vector $\mathbf{X}$, parameter γ being the set of biases $\{\mathbf{b}_k\}$. $\mathbf{P}_k$ is calculated using a universal GMM modeling the acoustic space (Karam et al. (2009)). The parameters of the transformation are estimated such that source speech vectors are best represented by the target client GMM model λ using the maximum likelihood criterion $\hat{\gamma} = \arg\max_\gamma \mathcal{L}(\mathcal{T}_\gamma(\mathbf{X})|\lambda)$. Parameters $\{\mathbf{b}_k\}$ are calculated iteratively using the EM algorithm.

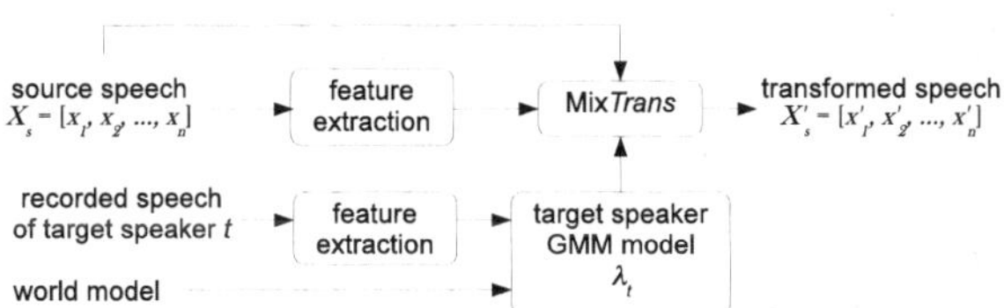

Fig. 11. Mix*Trans* block diagram

5.2 Face transformation

Face transformation is the process of converting someone's face to make it partially or completely different. Face transformation can be divided into two broad categories: Inter-person transformation, and intra-person transformation. In intra-person transformation, the face is transformed in such a way as to have the subject retain his/her identity. The application of such transformation might be

- Face Beautification, i.e. to make the face look more attractive. Such applications are used in beauty clinics to convince clients of certain treatments. Example characteristic features of a female "more attractive" face include a narrower facial shape, fuller lips, a bigger distance of eyes, darker and narrower eye brows, a narrower nose, and no eye rings.
- Age modification, i.e. to make the face look younger or older by modifying the qualities of the facial appearance. Such qualities include those of a baby face, a juvenile, a teenager, and an adult. For example, the visual facial features of a baby include a small head, a curved forehead, large round eyes, small short nose, and chubby cheeks. One application of age modification is the projection of how a person might look when he gets old.
- Expression modification, i.e. the alteration of the mental state of the subject by changing the facial expression, e.g. joy, sadness, anger, fear, disgust, surprise, or neutral.
- Personalized Avatars, i.e. a computer user's representation of himself or herself, whether in the form of a 2-D or a 3-D model that could be used in virtual reality environments and applications such as games and online virtual worlds.

In inter-person transformation, the face is transformed so as to give the impression of being somebody else. The application of such transformation might be decoy, i.e. a form of protection for political, military, and celebrity figures. This involves an impersonator who is employed (or forced) to perform during large public appearances, to mislead the public. Such act would entail real time face transformation and 3-D animation and projection on large screens as well as imitating, naturally or synthetically, the public figure's voice, and mannerism.

- *Caricatural* impression, i.e. an exaggerated imitation or representation of salient features of another person in an artistic or theatrical way. This is used by impressionists, i.e. professional comedians, to imitate the behavior and actions of a celebrity, generally for entertainment, and makes fun of their recent scandals or known behavior patterns.
- Imposture or identity theft, i.e. the act of deceiving by means of an assumed character. The face of the impostor is altered so as to resemble some other existing person whose identity and facial features are publicly known in a certain real or virtual community. This application can be used to test the robustness of A/V IV systems to fraud. This latter concept is the purpose of this work.

Face transformation involves face animation of one or several pictures of a face, and can be two or three-dimensional. Face animation is described next.

5.2.1 Face animation

Computer-based face animation entails techniques and models for generating and animating images of the human face and head. The important effects of human facial expressions on verbal and non-verbal communication have caused considerable scientific, technological, and artistic interests in the subject. Applications of computer facial animation span a wide variety of areas including entertainment (animated feature films, computer games, etc.) communication (teleconferencing), education (distance learning), scientific simulation, and agent-based systems (e.g. online customer service representative).

To complete the scenario of A/V imposture, speaker transformation is coupled with face transformation. It is meant to produce synthetically an "animated" face of a target person, given a still photo of his face and some animation parameters defined by a source video sequence.

Several commercial and experimental tools for face animation are already available for the professional and the research communities. CrazyTalk[1] is an animation studio that provides the facility to create 3-D talking characters from photos, images or illustrations, and provides automatic lip-sync animation from audio and typed text. Other computer animated talking heads and conversational agents used in the research community include Greta (Pasquariello & Pelachaud, 2001), Baldi (Massaro, 2003), MikeTalk (Ezzat & Poggio, 1998), and Video Rewrite (Bregler et al., 1997), among others.

The face animation technique used in this work is MPEG-4 compliant, which uses a very simple thin-plane spline warping function defined by a set of reference points on the target image, driven by a set of corresponding points on the source image face. This technique is described next.

5.2.2 MPEG-4 2D face animation

MPEG-4 is an object-based multimedia compression standard, which defines a standard for face animation (Tekalp & Ostermann, 2000). It specifies 84 feature points 12 that are used as references for Facial Animation Parameters (FAPs). 68 FAPs allow the representation of facial expressions and actions such as head motion and mouth and eye movements. Two FAP groups are defined, visemes (FAP group 1) and expressions (FAP group 2). Visemes (FAP1) are visually associated with phonemes of speech; expressions (FAP2) are joy, sadness, anger, fear, disgust, and surprise.

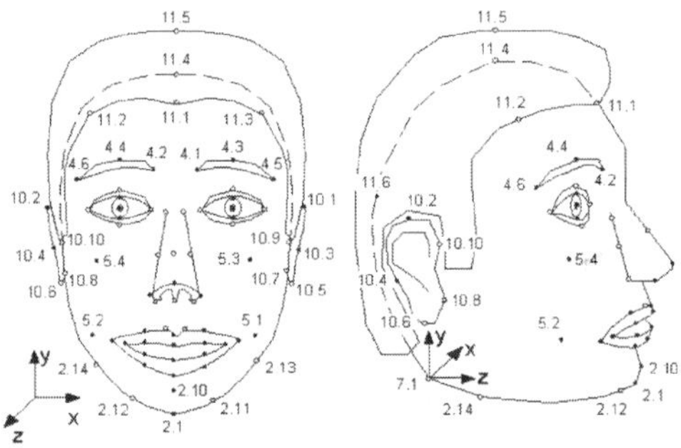

Fig. 12. MPEG-4 feature points

An MPEG-4 compliant system decodes a FAP stream and animates a face model that has all feature points properly determined. In this paper, the animation of the feature points is accomplished using a simple thin-plate spline warping technique, and is briefly described next.

5.2.3 Thin-plate spline warping

The thin-plate spline (TPS), initially introduced by Duchon (Duchon, 1976), is a geometric mathematical formulation that can be applied to the problem of 2-D coordinate

[1] http://www.reallusion.com/crazytalk/

transformation. The name thin-plate spline indicates a physical analogy to bending a thin sheet of metal in the vertical z direction, thus displacing x and y coordinates on the horizontal plane. TPS is briefed next. Given a set of data points $\{w_i, i = 1, 2, ..., K\}$ in a 2-D plane Ð for our case, MPEG-4 facial feature points Ð a radial basis function is defined as a spatial mapping that maps a location x in space to a new location $f(x) = \sum_{i=1}^{K} c_i \phi(\|x - w_i\|)$, where $\{c_i\}$ is a set of mapping coefficients and the kernel function $\phi(r) = r^2 \ln r$ is the thin-plate spline (Bookstein, 1989). The mapping function $f(x)$ is fit between corresponding sets of points $\{x_i\}$ and $\{y_i\}$ by minimizing the "bending energy" I, defined as the sum of squares of the second-order derivatives of the mapping function:

$$I[f(x,y)] = \iint_{\mathbb{R}^2} \left[\left(\frac{\partial^2 f}{\partial x^2}\right)^2 + 2\left(\frac{\partial^2 f}{\partial xy}\right)^2 + \left(\frac{\partial^2 f}{\partial y^2}\right)^2 \right] dx\, dy \tag{9}$$

Fig. 13 shows thin-plate spline warping of a 2-D mesh, where point A on the original mesh is displaced by two mesh squares upward, and point B is moved downward by 4 mesh squares. The bent mesh is shown on the right of fig. 13.

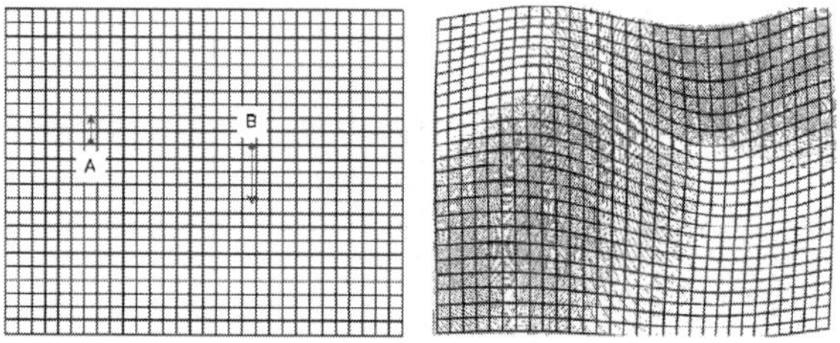

Fig. 13. TPS warping of a 2-D mesh

14 shows an example of TPS warping as it could be used to distort, and eventually animate a human face. The original face is on the left. The second and the third faces are warped by displacing the lowest feature point of the face, FDP number 2.1 as defined in the MPEG-4 standard 12. The corresponding FAP is "open_jaw". The last face on the right is a result of displacing the FDP's of the right eye so as to have the effect of a wink.

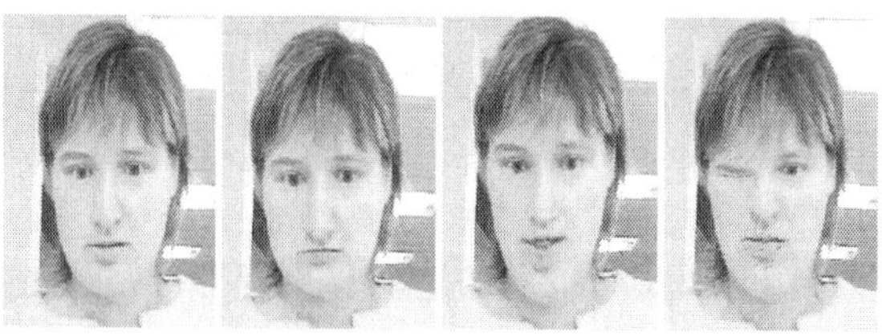

Fig. 14. Face animation using TPS warping

6. Evaluation and experimental results

To test the robustness of IV systems, a state-of-the-art baseline A/V IV system is built. This system follows the BANCA[2] "pooled test" of its evaluation protocol (Popovici et al., 2003).

[2] http://www.ee.surrey.ac.uk/CVSSP/banca/

The evaluation of a biometric system performance and its robustness to imposture is measured by the rate of errors it makes during the recognition process. Typically, a recognition system is a "comparator" that compares the biometric features of a user with a given biometric reference and gives a "score of likelihood". A decision is then taken based on that score and an adjustable defined acceptance "threshold" Θ. Two types of error rates are traditionally used: The False Acceptance Rate (FAR), i.e. the frequency that an impostor is accepted as a genuine client, and the False Rejection Rate (FRR), the frequency that a genuine client is rejected as an impostor. Results are reported in terms of an equal error rate (EER), the value where FAR=FRR. The lower the value of EER, the more performing the recognition system is. Typically, FAR and FRR can be traded off against each other by adjusting a decision threshold. The accuracy estimate of the EER is also reported by computing a confidence interval of 95% for the error rates.

6.1 Speaker verification

To process the speech signal, a feature extraction module calculates relevant feature vectors from the speech waveform. On a signal "FFT" window shifted at a regular rate, cepstral coefficients are derived from a filter bank analysis with triangular filters. A Hamming weighting window is used to compensate for the truncation of the signal. Then GMM speaker classification is performed with 128 and 256 Gaussians. The world model of BANCA is adapted using MAP adaptation, and its parameters estimated using the EM algorithm. A total of 234 true client tests and 312 "random impostor" tests per group were performed. EER's of 5.4%[$\pm$0.09] and 6.2%[$\pm$0.11] are achieved for 256 and 128 Gaussians respectively.

6.2 Face verification

Face verification is based on processing a video sequence in four stages: 1-Face detection, localization and segmentation, 2-Normalization, 3-Facial Feature extraction and tracking, and 4-Classification. The face detection algorithm used in this work is a machine learning approach based on a boosted cascade of simple and rotated haar-like features for visual object detection Lienhart & Maydt (2002). Once a face is detected, it is normalized (resized to 48x64, cropped to 36x40, gray-scaled, and histogram equalized) to reduce the variability of different aspects in the face image such as contrast and illumination, scale, translation, and rotation. The face tracking module extracts faces in all frames and retains only 5 per video for training and/or testing. The next step is face feature extraction. We use DCT-*mod*2 proposed in Sanderson & Paliwal (2002). In a similar way to speaker verification, GMM's are used to model the distribution of face feature vectors for each person. For the same BANCA "P" protocol, a total of 234 true clients and 312 "random impostor" tests are done (per group per frame, 5 frames per video.) EER's of 22.89%[$\pm$0.04] and 23.91%[$\pm$0.05] are achieved for 256 and 128 Gaussians respectively.

6.3 Speaker transformation

BANCA has total of 312 impostor attacks per group in which the speaker claims in his own words to be someone else. These attempts are replaced by the *MixTrans* transformed voices. For each attempt, MFCC analysis is performed and transformation coefficients are calculated in the cepstral domain using the EM algorithm. Then the signal transformation parameters are estimated using a gradient descent algorithm. The transformed voice signal is then reconstructed with an inverse FFT and OLA as described in Karam et al. (2009).

Verification experiments are repeated with the transformed voices. EER's of 7.88%[±0.08] and 7.96%[±0.10] are achieved for 256 and 128 Gaussians respectively.

6.4 Face transformation

Given a picture of the face of a target person, the facial feature points are first annotated as defined by MPEG-4. The facial animation parameters (FAP) used in the experiments correspond to a subset of 33 out of the 68 FAP's defines by MPEG-4. Facial actions related to head movement, tongue, nose, ears, and jaws are not used. The FAP's used correspond to mouth, eye, and eyebrow movements, e.g. horizontal displacement of right outer lip corner (stretch_r_corner_lip_o). A synthesized video sequence is generated by deforming a face from its neutral state according to determined FAP values, using the thin plate spline warping technique. Speaker verification experiments are repeated with the forged videos. EER's of 50.64%[±0.08] and 50.83%[±0.09] are achieved for 256 and 128 Gaussians respectively.

6.5 Audio-visual verification and imposture

Reporting IV results on A/V verification and A/V imposture is done simply by fusing scores of the verification of face and speaker and their transformations. In this paper, A/V scores are computed by weight averaging: $s = w_s s_s + w_f s_f$, where w_s and w_f are speech and face score weights computed so as to optimize EER on the training set, s_s and s_f being the speaker and the face scores respectively. Table 1 provides a summary of results of IV in terms of EER's, including speaker and face verification and their transformations, as well as A/V verification and transformation. The results clearly indicate an increase in EER's between the base experiments with no transformation and the experiments when either face, speaker, or both transformations are in effect. This indicates the acceptance of more impostors when any combination of voice/face transformation is employed.

		GMM size	
		256	128
speaker	no transformation	**5.40** [±0.09]	**6.22** [±0.11]
	MixTrans	7.88 [±0.08]	7.96 [±0.10]
face	no transformation	**22.89** [±0.04]	**23.91** [±0.05]
	TPS face warping	50.64 [±0.08]	50.83 [±0.09]
A/V	no transformation	**5.24** [±0.10]	**5.10** [±0.10]
	MixTrans+TPS warping	14.37 [±0.10]	15.39 [±0.10]
	MixTrans only	6.87 [±0.10]	6.60 [±0.10]
	TPS face warping only	13.38 [±0.10]	13.84 [±0.10]

Table 1. Summary of results - EER's with an accuracy estimate over a 95% interval of confidence

7. Conclusion

An important conclusion drawn from the experiments is that A/V IV systems are still far from being commercially feasible, especially for forensic or real time applications. A voice transformation technique that exploits the statistical approach of the speaker verification system can easily break that system and consent to a higher false acceptance rate. In a similar way to speaker verification systems, face verification systems that use the holistic statistical approach are vulnerable to imposture attacks that exploit the statistical approach of the verification.

After decades of research and development on face and speech recognition, such systems are yet to find their place in our lives. With error rates of 1 to 5 percent for speech recognition, and 10 to 25 for face recognition, such systems have found their way only in exhibition proof-of-concept type of demos and limited noncritical applications, such as computer games and gadgets. It has remained an open question why A/V biometrics has remained in the research laboratory and has not found its way to public use. Will the world witness a breakthrough in A/V biometrics? Will we use our face and voice instead of our passport as we walk through the security zone at the airport to authenticate and declare our passage?

8. References

Abe, M., Nakamura, S., Shikano, K. & Kuwabara, H. (1988). Voice conversion through vector quantization, *Proceedings of the IEEE International Conference on Acoustics, Speech, and Signal Processing (ICASSP'08*, pp. 655–658.

Arslan, L. M. (1999). Speaker transformation algorithm using segmental codebooks (stasc), *Speech Commun.* 28(3): 211–226.

Bartlett, M., Movellan, J. & Sejnowski, T. (2002). Face recognition by independent component analysis, *Neural Networks, IEEE Transactions on* 13(6): 1450–1464.

Belhumeur, P. N. & Kriegman, D. J. (1998). What is the set of images of an object under all possible lighting conditions, *IJCV* 28: 270–277.

Ben-Yacoub, S., Abdeljaoued, Y. & Mayoraz, E. (1999). Fusion of Face and Speech Data for Person Identity Verification, *IEEE Transactions on Neural Networks* 10(05): 1065–1074.

Bimbot, F., Bonastre, J.-F., Fredouille, C., Gravier, G., Magrin-Chagnolleau, I., Meignier, S., Merlin, T., Ortega-Garcia, J., Petrovska-Delacretaz, D. & Reynolds, D. (2004). A tutorial on text-independent speaker verification, *EURASIP J. Appl. Signal Process.* 2004(1): 430–451.

Blanz, V. & Vetter, T. (2003). Face recognition based on fitting a 3d morphable model, *IEEE Trans. Pattern Anal. Mach. Intell.* 25(9): 1063–1074.

Blouet, R., Mokbel, C., Mokbel, H., Soto, E. S., Chollet, G. & Greige, H. (2004). Becars: A free software for speaker verification, *Proc. ODYSSEY'04*, pp. 145–148.

Bookstein, F. (1989). Principal warps: Thin-plate splines and the decomposition of deformations, *IEEE Transactions on Pattern Analysis and Machine Intelligence* 11(6): 567–585.

Bregler, C., Covell, M. & Slaney, M. (1997). Video rewrite: driving visual speech with audio, *SIGGRAPH '97: Proceedings of the 24th annual conference on Computer graphics and interactive techniques*, ACM Press/Addison-Wesley Publishing Co., New York, NY, USA, pp. 353–360.

Bronstein, A., Bronstein, M., Kimmel, R. & Spira, A. (2004). 3d face recognition without facial surface reconstruction, *ECCV'04: Proceesings of the European Conference on Computer Vision*, Prague.

Chatzis, V., Bors, A. G. & Pitas, I. (1999). Multimodal decision-level fusion for person authentication, *IEEE Transactions on Systems, Man, and Cybernetics, Part A* 29(6): 674–680.

Chellappa, R., Wilson, C. & Sirohey, S. (1995). Human and machine recognition of faces: a survey, *Proceedings of the IEEE* 83(5): 705–741.

Chollet, G., Hueber, T., Bredin, H., Mokbel, C., Perrot, P. & Zouari, L. (2007). *Advances in Nonlinear Speech Processing*, Vol. 1, Springer Berlin / Heidelberg, chapter Some experiments in Audio-Visual Speech Processing, pp. 28–56.

Constantinescu, A., Deligne, S., Bimbot, F., Chollet, G. & Cernocky, J. (1999). Towards alisp: a proposal for automatic language, *Independent Speech Processing. Computational Models of Speech Pattern Processing (ed. Ponting K.), NATO ASI Series*, pp. 375–388.

Craw, I. & Cameron, P. (1992). Face recognition by computer, *Proc. British Machine Vision Conference*, Springer Verlag, pp. 498–507.

Dempster, A. P., Laird, N. M. & Rubin, D. B. (1977). Maximum likelihood from incomplete data via the em algorithm, *Journal of the Royal Statistical Society* 39(1): 1–38.

Duchon, J. (1976). Interpolation des fonctions de deux variables suivant le principe de la flexion des plaques minces, *R.A.I.R.O. Analyse numérique* 10: 5–12.

Dugelay, J.-L., Junqua, J.-C., Kotropoulos, C., Kuhn, R., Perronnin, F. & Pitas, I. (2002). Recent advances in biometric person authentication, *ICASSP 2002 , 27th IEEE International Conference on Acoustics,Speech and Signal Processing - May 13-17, 2002, Orlando, USA*.

Edwards, G. J., Cootes, T. F. & Taylor, C. J. (1998). Face recognition using active appearance models, *ECCV '98: Proceedings of the 5th European Conference on Computer Vision-Volume II*, Springer-Verlag, London, UK, pp. 581–595.

Erro, D., Navas, E., Herná andez, I. & Saratxaga, I. (2010). Emotion conversion based on prosodic unit selection, *Audio, Speech, and Language Processing, IEEE Transactions on* 18(5): 974 –983.

Ezzat, T. & Poggio, T. (1998). Miketalk: A talking facial display based on morphing visemes, *Proceedings of the Computer Animation Conference*, pp. 96–102.

Fierrez-Aguilar, J., Ortega-Garcia, J., Garcia-Romero, D. & Gonzalez-Rodriguez, J. (2003). A comparative evaluation of fusion strategies for multimodal biometric verification, *Proceedings of IAPR International Conference on Audio and Video-based Person Authentication, AVBPA*, Springer, pp. 830–837.

Fromherz, T., Stucki, P. & Bichsel, M. (1997). A survey of face recognition, *Technical report*, MML Technical Report.

Gauvain, J.-L. & Lee, C.-H. (1994). Maximum a posteriori estimation for multivariate gaussian mixture observations of markov chains, *IEEE Transactions on Speech and Audio Processing* 2: 291–298.

Hjelmas, E. & Low, B. (2001). Face detection: A survey, *Computer Vision and Image Understanding* 83: 236–274(39).

Kadyrov, A. & Petrou, M. (2001). The trace transform and its applications, *Pattern Analysis and Machine Intelligence, IEEE Transactions on* 23(8): 811–828.

Kain, A. & Macon, M. (1998). Spectral voice conversion for text-to-speech synthesis, *Acoustics, Speech and Signal Processing, 1998. Proceedings of the 1998 IEEE International Conference on* 1: 285–288 vol.1.

Kain, A. & Macon, M. (2001). Design and evaluation of a voice conversion algorithm based on spectral envelope mapping and residual prediction, *Acoustics, Speech, and Signal Processing, 2001. Proceedings. (ICASSP '01). 2001 IEEE International Conference on* 2: 813–816 vol.2.

Karam, W., Bredin, H., Greige, H., Chollet, G. & Mokbel, C. (2009). Talking-face identity verification, audiovisual forgery, and robustness issues, *EURASIP Journal on Advances in Signal Processing* 2009(746481): 18.

Kittler, J. (1998). Combining classifiers: a theoretical framework, *Pattern Analysis and Applications* 1: 18–27.

Li, K. P. & Porter, J. (1988). Normalizations and selection of speech segments for speaker recognition scoring, *In Proc. IEEE Int. Conf. Acoustics, Speech and Signal Processing*, pp. 595–598.

Lienhart, R. & Maydt, J. (2002). An extended set of haar-like features for rapid object detection, *2002 International Conference on Image Processing. 2002. Proceedings*, Vol. 1, pp. I–900–I–903.

Liu, C. & Wechsler, H. (1998). A unified bayesian framework for face recognition, *Image Processing, 1998. ICIP 98. Proceedings. 1998 International Conference on* 1: 151–155 vol.1.

Massaro, D. W. (2003). A computer-animated tutor for spoken and written language learning, *ICMI '03: Proceedings of the 5th international conference on Multimodal interfaces*, ACM, New York, NY, USA, pp. 172–175.

Moghaddam, B., Jebara, T. & Pentland, A. (2000). Bayesian face recognition, *Pattern Recognition 2000* 33(11): 1771–1782.

Moghaddam, B. & Pentland, A. (1997). Probabilistic visual learning for object representation, *IEEE Trans. Pattern Anal. Mach. Intell.* 19(7): 696–710.

Mokbel, C. (2001). Online adaptation of hmms to real-life conditions: A unified framework, *IEEE Transactions on Speech and Audio Processing* 9: 342–357.

Mouchtaris, A., der Spiegel, J. V. & Mueller, P. (2004). Non-parallel training for voice conversion by maximum likelihood constrained adaptation, *Proceedings of the IEEE International Conference on Acoustics, Speech, and Signal Processing (ICASSP'04*, Vol. 1, pp. I1–I4.

Nefian, A. & Hayes, M. I. (1998). Hidden markov models for face recognition, *Acoustics, Speech and Signal Processing, 1998. Proceedings of the 1998 IEEE International Conference on* 5: 2721–2724 vol.5.

Paoletti, D. & Erten, G. (2000). Enhanced silence detection in variable rate coding systems using voice extraction, *Circuits and Systems, 2000. Proceedings of the 43rd IEEE Midwest Symposium on* 2: 592–594 vol.2.

Pasquariello, S. & Pelachaud, C. (2001). Greta: A simple facial animation engine, *In Proc. of the 6th Online World Conference on Soft Computing in Industrial Applications*.

Perrot, P., Aversano, G., Blouet, R., Charbit, M. & Chollet, G. (2005). Voice forgery using alisp: Indexation in a client memory, *Proceedings of the IEEE International Conference on Acoustics, Speech, and Signal Processing (ICASSP'05*, Vol. 1, pp. 17–20.

Phillips, P. J. (1998). Support vector machines applied to face fecognition, *Adv. Neural Inform. Process. Syst* 1(11): 803–809.

Popovici, V., Thiran, J., Bailly-Bailliere, E., Bengio, S., Bimbot, F., Hamouz, M., Kittler, J., Mariethoz, J., Matas, J., Messer, K., Ruiz, B. & Poiree, F. (2003). The BANCA Database and Evaluation Protocol, *4th International Conference on Audio- and Video-Based Biometric Person Authentication, Guildford, UK*, Vol. 2688 of *Lecture Notes in Computer Science*, SPIE, Berlin, pp. 625–638.

Ross, A. & Jain, A. K. (2003). Information fusion in biometrics, *Pattern Recogn. Lett.* 24(13): 2115–2125.

Samaria, F. & Young, S. (1994). Hmm based architecture for face identification, *IntŠl J. Image and Vision Computing* 1(12): 537583.

Sanderson, C. & Paliwal, K. K. (2002). Fast feature extraction method for robust face verification, *IEE Electronics Letters* 38(25): 1648–1650.

Srisuk, S., Petrou, M., Kurutach, W. & Kadyrov, A. (2003). Face authentication using the trace transform, *Computer Vision and Pattern Recognition, 2003. Proceedings. 2003 IEEE Computer Society Conference on* 1: I–305–I–312 vol.1.

Stylianou, Y. & Cappe, O. (1998). A system for voice conversion based on probabilistic classification and a harmonic plus noise model, *Acoustics, Speech and Signal Processing, 1998. Proceedings of the 1998 IEEE International Conference on* 1: 281–284 vol.1.

Sundermann, D., Hoge, H., Bonafonte, A., Ney, H., Black, A. & Narayanan, S. (2006). Text-independent voice conversion based on unit selection, *Acoustics, Speech and Signal Processing, 2006. ICASSP 2006 Proceedings. 2006 IEEE International Conference on* 1: I–I.

Sundermann, D., Ney, H. & Hoge, H. (2003). Vtln-based cross-language voice conversion, *IEEE Automatic Speech Recognition and Understanding Workshop*, St. Thomas, Virgin Islands, USA, pp. 676–681.

Swets, D. L. & Weng, J. (1996). Using discriminant eigenfeatures for image retrieval, *IEEE Transactions on Pattern Analysis and Machine Intelligence* 18: 831–836.

Tekalp, A. & Ostermann, J. (2000). Face and 2-d mesh animation in mpeg-4, *Image Communication Journal* 15(4-5): 387–421.

Toda, T. (2009). Eigenvoice-based approach to voice conversion and voice quality control, *Proc. NCMMSC, International Symposium*, Lanzhou, China, pp. 492–497.

Turajlic, E., Rentzos, D., Vaseghi, S. & Ho, C. (2003). Evaluation of methods for parametric formant transformation in voice conversion, *Proceedings of the International Conference on Acoustics, Speech, and Signal Processing (ICASSP'03)*.

Turk, M. & Pentland, A. (1991). Eigenfaces for recognition, *J. Cognitive Neuroscience* 3(1): 71–86.

Viola, P. & Jones, M. (2001). Rapid object detection using a boosted cascade of simple features, *IEEE CVPR'01*.

Wiskott, L., Fellous, J.-M., Kuiger, N. & von der Malsburg, C. (1997). Face recognition by elastic bunch graph matching, *Pattern Analysis and Machine Intelligence, IEEE Transactions on* 19(7): 775–779.

Yang, M.-H. (2002). Kernel eigenfaces vs. kernel fisherfaces: Face recognition using kernel methods, *Automatic Face and Gesture Recognition, 2002. Proceedings. Fifth IEEE International Conference on* 1: 215–220.

Ye, H. & Young, S. (2003). Perceptually weighted linear transformations for voice conversion, *Proceedings of EUROSPEECH'03, Geneva*, Vol. 4, pp. 2409–2412.

Ye, H. & Young, S. (2004). Voice conversion for unknown speakers, *Proceedings of the ICSLPŠ04, Jeju Island, South Korea*, pp. 1161–1164.

Zhang, J., Yan, Y. & Lades, M. (1997). Face recognition: eigenface, elastic matching, and neural nets, *Proceedings of the IEEE*, pp. 1423–1435.

Zhao, W., Chellappa, R. & Krishnaswamy, A. (1998). Discriminant analysis of principal components for face recognition, *FG '98: Proceedings of the 3rd. International Conference on Face & Gesture Recognition*, IEEE Computer Society, Washington, DC, USA, p. 336.

Zhao, W., Chellappa, R., Phillips, P. J. & Rosenfeld, A. (2003). Face recognition: A literature survey, *ACM Comput. Surv.* 35(4): 399–458.

Zhou, S., Moghaddam, B. & Zhou, S. K. (2004). Intra-personal kernel space for face recognition, *In Proceedings of the IEEE International Automatic Face and Gesture Recognition*, pp. 235–240.

4

Face and ECG Based Multi-Modal Biometric Authentication

Ognian Boumbarov[1], Yuliyan Velchev[1], Krasimir Tonchev[1] and Igor Paliy[2]

[1]*Technical University of Sofia*
[2]*Ternopil National Economic University*
[1]*Bulgaria*
[2]*Ukraine*

1. Introduction

A biometric system is essentially a pattern recognition system. This system measures and analyses human body physiological characteristics, such as face and facial features, fingerprints, eye, retinas, irises, voice patterns or behavioral characteristic for enrollment, verification or identification (Bolle & Pankanti, 1998). Uni-modal biometric systems have poor performance and accuracy, and over last few decades the multi-modal biometric systems have become very popular. The main objective of multi biometrics is to reduce one or more false accept rate, false reject rate and failure to enroll rate. Face Recognition (FR) is still considered as one of the most challenging problems in pattern recognition. The FR systems try to recognize the human face in video sequences as 3D object (Chang et al., 2003; 2005), in unconstrained conditions, in comparison to the early attempts of 2D frontal faces in controlled conditions. Despite the effort spent on research today there is not a single, clearly defined, solution to the problem of Face Recognition, leaving it an open question. One of the key aspects of FR is its application, which also acts as the major driving force for research in that area. The applications range from law enforcement to human-computer interactions (HCI). The systems used in these applications fall into two major categories: *systems for identification* and *systems for verification* (Abate et al., 2007). The first group attempts to identify the person in a database of faces, and extract personal information. These systems are widely used, for instance, in police departments for identifying people in criminal records. The second group finds its main application in security, for example to gain access to a building, where face is used as more convenient biometric. The more general HCI systems include not only identification or verification, but also tracking of a human in a complex environment, interpretation of human behavior and understanding of human emotions.
Another biometric modality that we use in our approach is the electrocardiogram (ECG). The modern concept for ECG personal identification is to extract the signal features using transform methods, rather than parameters in time domain (amplitudes, slopes, time intervals). The proper recognition of the extracted features and the problem of combining different biometric modalities in intelligent video surveillance systems are the novel steps that we introduce in this work.

2. Recognition of facial images

2.1 Framework for face recognition

In real case scenario, human faces often appear in scenes with complex background, rather than as a single object. In addition, they have varying appearance due to different lightning conditions, changes in pose, human expressions etc. Thus, a reliable system for FR must be robust to noise, variations and be able to work in real time. To meet these requirements we are proposing a framework for recognition of facial images depicted on Fig. 1. This framework consists of three stages, namely Face Detection (FD), Subspace Projection (SP) and Classification.

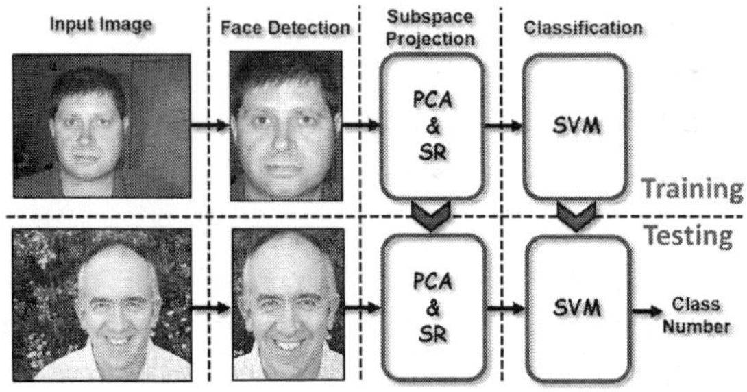

Fig. 1. Pictorial depiction of facial recognition framework

The purpose of the FD is to locate a human face in a scene and extract it as a single image. In this work, we propose a combination of two classifiers for rapid and accurate FD. The first one is faster but less precise, while the second, compensates for the imprecision of the first classifier. The second stage of the proposed framework, namely SP, is used for dimensionality reduction of the detected facial images, when represented as vectors in high-dimensional Euclidean space. Thus, it is necessary to transform them from the original high-dimensional space to a low dimensional one for alleviating the *curse of dimensionality*. The SP is based on Principal Component Analysis (PCA) and Spectral Regression (SR) algorithms. The PCA discovers the subspace which contains all vectors of facial images and we use it mainly to remove noises. PCA also preserves Euclidean distances between vector pairs in the subspace. Based on this, further dimensionality reduction is done by using the SR algorithm. This algorithm is robust with respect to the variation of lightning conditions and human expressions. Finally we perform classification using Support Vector Machines classifier in the subspace. In the following, the three stages of the proposed FR framework we will be discussed in details.

2.2 Face detection

A combination of rapid cascaded classifier and accurate monolithic one is used as a two level Face Detection algorithm. The first level is represented by the Haar-like features' cascade of weak classifiers, which is responsible for fast detection of face-like objects. The second level is a Convolutional Neural Network (CNN) used for filtration of falsely detected faces. The Haar-like features' cascade of weak classifiers allows detecting face candidates very quickly. It consists of a cascade of one or more weak classifiers. The weak classifier's input is represented by Haar-like feature with a value (Viola & Jones, 2004):

$$Feat(x) = s_w \times SUM_w + s_b \times SUM_b \tag{1}$$

where x is input image's sub-window, s_w and s_b - whole rectangle's and its black part's weights accordingly, SUM_w and SUM_b - whole rectangle's and its black part's sums of pixels. A weak classifier's output value is:

$$h(x) = \begin{cases} 1, if\ Feat(x) \leq \Theta, \\ -1, if\ Feat(x) > \Theta, \end{cases} \tag{2}$$

where Θ - weak classifier's threshold. The cascade of weak classifiers is a linear combination of weak classifiers (Viola & Jones, 2004):

$$H(x) = \sum_t^T \eta_t \times h_t(x), \tag{3}$$

where T - weak classifiers' number, η_t - t-weak classifier's weight. The AdaBoost algorithm (Freund & Schapire, 1997) is used for training of the cascade of weak classifiers and the selection of the most important Haar-like features. The second level uses the Convolutional Neural Network (Lecun et al., 1998) which is more robust to variations of the input image, compared to other known classifiers. The output value of a neuron with bipolar sigmoid transfer function and with the coordinates (m, n) of p-plane and l-layer is (Kurylyak et al., 2009):

$$y_{m,n}^{i,p} = \frac{2}{1 + exp(-WSUM_{m,n}^{i,p}(x))}, \tag{4}$$

where x is input face candidate's image, $WSUM$ - neuron's weighted sum calculated by (Kurylyak et al., 2009):

$$WSUM_{m,n}^{i,p}(x) = \left(\sum_{k=0}^{K-1} \sum_{r=0}^{R-1} \sum_{c=0}^{C-1} y_{2m+r,2n+c}^{l-1,k}(x) \times w_{r,c}^{l,p,k} \right) - b^{l,p}. \tag{5}$$

Here K is input planes' number (as well as convolutional kernels), R and C are convolutional kernel's height and width, $w_{r,c}^{l,p,k}$ is synaptic weight with coordinates (r, c) in the convolutional kernel between k-plane of the $(l-1)$-layer and p-plane of the l-layer, $b^{l,p}$ is neurons' bias of the p-plane and l-layer. The CNN uses a sparse structure instead of a fully-connected one; also its number of layers is decreased. In order to increase the neural network's processing speed, convolution and subsampling operations are performed in each plane simultaneously (Simard et al., 2003) Fig. 2.

2.3 Facial features extraction with subspace projection

2.3.1 Principal Component Analysis

Principal Component Analysis (PCA) is a very popular method for dimensionality reduction in machine learning and statistics communities. It can be considered as method learning the basis vectors spanning the linear subspace, called *principal subspace*, containing all data points embedded in that subspace. This basis is determined by the non-zero eigenvectors of data covariance matrix, and the dimension of the subspace is their number. Usually this number is much less than the dimension of the ambient space and dimensionality reduction is performed by projection the data points onto that subspace.

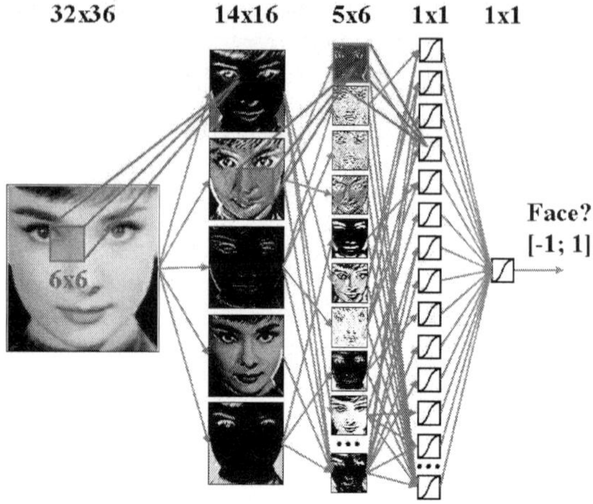

Fig. 2. Convolutional neural network's structure for the combined cascade of neural network classifiers

PCA can be formulated in two ways (Bishop, 2007). The firs one is called *minimum-error formulation* and the target is to minimize the mean squared error between the data and its projections onto the subspace. In the second approach, called *maximum variance formulation*, the goal is to maximize the variance of the projected data. In this work we approach PCA by the former one. Let $\{\mathbf{x}_i\}_{i=1}^{M} \subset \mathbb{R}^N$ be a data set of measurements of physical phenomenon, drawn from an unknown probability distribution. We seek a subspace of dimension $D \ll N$ such that the variance of the projected data onto it is maximized. First step of PCA involves computation of data covariance matrix by:

$$\mathbf{C} = \sum_{i=1}^{M} (\mathbf{x}_i - \bar{\mathbf{x}})(\mathbf{x}_i - \bar{\mathbf{x}})^{\mathrm{T}}, \tag{6}$$

where $\bar{\mathbf{x}} = \sum_{i=1}^{M} \mathbf{x}_i$ is the mean vector. Then the variance of the projected data onto single direction $\mathbf{v} \in \mathbb{R}^N$ is:

$$\frac{1}{N} \sum_{i=1}^{N} (\mathbf{v}^{\mathrm{T}}\mathbf{x}_i - \mathbf{v}^{\mathrm{T}}\bar{\mathbf{x}})^2 = \mathbf{v}^{\mathrm{T}}\mathbf{C}\mathbf{v}. \tag{7}$$

Maximizing the variance, means maximizing the quadratic term $\mathbf{v}^{\mathrm{T}}\mathbf{C}\mathbf{v}$. But when $\|\mathbf{v}\| \to \infty$ it follows that $|\mathbf{v}^{\mathrm{T}}\mathbf{C}\mathbf{v}| \to \infty$, hence it is desired to constrain the problem in appropriate way. This is achieved by the normalized condition $\mathbf{v}^{\mathrm{T}}\mathbf{v} = 1$. By introducing Lagrange multipliers we reformulate the problem from constrained to unconstrained maximization one:

$$\mathbf{v}^{\mathrm{T}}\mathbf{C}\mathbf{v} + \lambda \left(1 - \mathbf{v}^{\mathrm{T}}\mathbf{v}\right) \to \max, \tag{8}$$

where λ is a Lagrange multiplier. Taking the partial derivatives of 8 with respect to $\mathbf{v}$ and setting to zero, we end up with the eigenvalue problem $\mathbf{C}\mathbf{v} = \lambda\mathbf{v}$. Thus the direction defining maximum variance of projected data is the leading eigenvector called *principal component*. Additional principal component can be found by calculating new direction which maximizes

8 and it is orthogonal to the previous ones considered. By arranging the eigenvalues on the diagonal of the diagonal matrix Λ and the eigenvectors as e columns of the matrix $\mathbf{V}$ we can write the eigenvalue problem as $\mathbf{CV} = \Lambda\mathbf{V}$. By selecting only D eigenvectors of $\mathbf{V}$ corresponding to the leading eigenvalues and arranging them in matrix $\tilde{\mathbf{V}}$ the subspace projection is defined by $\tilde{\mathbf{x}}_i = \mathbf{x}_i^T\tilde{\mathbf{V}}$ for i=1...M.

2.3.2 Spectral regression framework for dimensionality reduction

Dimensionality reduction can be well interpreted in a *graph embedding* fashion. Such interpretation is very intuitive and also opens the possibility of developing a new approaches to the problem of pattern recognition. Advantage of graph embedding is the unifying power to the most of the dimensionality reduction methods such as Linear Discriminant Analysis (LDA), Locally Linear Embeddings (LLE), Locality Preserving Projections (LPP), ISOMAP as proposed in (Yan et al., 2007). Consider a data set $\{\mathbf{x}_i\}_{i=1}^{M} \subset \mathbb{R}^N$ represented as points in Euclidean space. Each data point can be viewed as e vertex of adjacency graph $\Gamma = \{X, W\}$ with edges W defined under some rule. Depending on the rule, graph edges can be distance, or a measure of similarity, between data points (vertices of Γ). The *graph Laplacian* is defined by $L = D - W$, where $Dii = \sum_{j \neq j} W_{ij}$. The goal of graph embedding is to find a low dimensional representation of each vertex, while preserving similarities between the vertex pairs. This is achieved by minimizing (Chung, 1997):

$$\sum_{ij}(y_i - y_j)^2 W_{ij} = 2\mathbf{y}^T L\mathbf{y}, \tag{9}$$

where $\mathbf{y} = [y_1, y_2, \ldots, y_M]^T$ is a map of graph vertices (data points) on the real line. If we select a linear map of the form $y_i = \mathbf{a}^T\mathbf{x}_i$, the problem 9 can be reduced to:

$$\mathbf{a}^* = \arg\min_{\mathbf{a}} \frac{\mathbf{a}^T XWX^T\mathbf{a}}{\mathbf{a}^T XDX^T\mathbf{a}}, \tag{10}$$

where $X = [\mathbf{x}_1, \mathbf{x}_2, \ldots, \mathbf{x}_M]$ is the data matrix. This approach is called *Linear extension of graph embedding*, and the optimal solution of 10 can be found by the *generalized eigenvalue problem*:

$$XWX^T\mathbf{a} = \lambda XDX^T\mathbf{a}. \tag{11}$$

With different choices of W, 11 can be formulated as LDA, LPP etc. Solving 11 can be computationally intensive in cases where W is a dense matrix of very high dimensions. To overcome this issue in (Cai et al., 2007) proposed a method, called Spectral Regression, which casts solving the generalized eigenvalue problem 11 into regression framework. The advantages of this approach are: (1) Solving a regression problem less computationally intensive than eigenvalue problem; (2) Regression can be solved with regularization term controlling the complexity and avoid overfitting; (3) By selecting regression terms, various properties can be achieved, such as sparsity. Furthermore, if we chose a function in a Reproducing Kernel Hilbert Space $y_i = \sum_{j=1}^{M} \alpha_i K(\mathbf{x}_j, \mathbf{x}_i)$, where $K(.,.)$ is a Mercer kernel, SR can be extended in kernel mode.

Algorithmically SR is performed as follows (Cai, 2009):

1. **Construct the weights matrix W**:
   - Set $W_{ij} = 1/l_k$ if $\mathbf{x}_i$ and $\mathbf{x}_j$ both belong to the k-th class where l_k is the number of samples in it;

- Set $W_{ij} = \delta.s(i,j)$ if $\mathbf{x}_i$ is among the p-nearest neighbor of $\mathbf{x}_j$ or vice-versa. The parameter δ is used to adjust the weight between supervised and unsupervised neighbor information. $s(i,j)$ is a function evaluating the similarity between $\mathbf{x}_i$ and $\mathbf{x}_j$. This function can be the heat the kernel function $s(i,j) = exp(-\frac{\|\mathbf{x}_i - \mathbf{x}_j\|}{2\sigma})$ or simple minded $s(i,j) = 1$, where $\sigma \in \mathbb{R}$;
- Set $W_{ij} = 0$ otherwise.

2. **Responses Generation**: Solve the eigenvalue problem $\mathbf{W}\mathbf{y} = \lambda\mathbf{D}\mathbf{y}$ and select the K largest eigenvectors $\{\mathbf{y}_k\}_{k=1}^{K}$, where $\mathbf{D}$ is diagonal matrix with elements on the diagonal $D_{jj} = \sum_i W_{ij}$.
3. **Regression**: Ridge regression solves the quadratic regression problem with Euclidean norm penalty:

$$\mathbf{a}_k = \arg\min_{\mathbf{a}} \left(\sum_{i=1}^{M} (\mathbf{a}^{\mathrm{T}}\mathbf{x}_i - y_i^k)^2 + \alpha \|\mathbf{a}\|^2 \right) \tag{12}$$

In Lasso regression, the regression problem is solved with L_1-norm penalty:

$$\mathbf{a}_k = \arg\min_{\mathbf{a}} \left(\sum_{i=1}^{M} (\mathbf{a}^{\mathrm{T}}\mathbf{x}_i - y_i^k)^2 + \alpha \sum_{j=1}^{N} \left| a_j \right| \right) \tag{13}$$

In both 12 and 13 the solutions are represented by $\{\mathbf{a}_k\}_{k=1}^{K} \subset \mathbb{R}^N$.

4. **Subspace Projection**: Perform dimensionality reduction by projecting on a lower dimensional space $\mathbf{x} \to \mathbf{z} = \mathbf{A}^{\mathrm{T}}\mathbf{x}$, where $\mathbf{A} = [\mathbf{a}_1, \mathbf{a}_2, \ldots, \mathbf{a}_{K},]$ is matrix of the solution vectors from the regression step.

For the purpose of this work, we select two different regularization terms in the regression step. The first one, called Ridge, is the standard quadratic regression 12. The dimensionality of the output space is this case is $c - 1$ and $K = c$. The second mode, called Lasso, uses L_1-norm for the regularization term which induces sparsity on $\{\mathbf{a}_k\}_{k=1}^{K}$. We test the proposed framework for face recognition with both modes of SR.

2.4 Classification of facial images

Support Vector Machine is a supervised learning algorithm used for classification in two classes. The aim of SVM is to find a N-dimensional hyperplane that optimally separates the data. Optimally in this case means that the margin, between nearest data points and the hyperplane, will be maximized. Unfortunately in real problems data is rarely separable by a hyperplane but can be separated by a non-linear surface. SVM can be transformed to a non-linear classifier by applying the kernel trick (Vapnik, 1998). This way data is mapped implicitly in a higher dimensional space where it can be separated by a hyperplane. Let $\{\mathbf{x}_i\}_{i=1}^{M} \subset \mathbb{R}^N$ be a data set with class labels vector $\mathbf{y} = [y_1, y_2, \ldots, y_M]$ such that $y_i \in \{-1, 1\}$, $i = 1, 2, \ldots, M$. Learning the parameters of the support vectors is performed by solving the constrained optimization problem:

$$\begin{aligned} \boldsymbol{\alpha}^* &= \arg\min_{\boldsymbol{\alpha}} \frac{1}{2}\boldsymbol{\alpha}^{\mathrm{T}}\mathbf{Q}\boldsymbol{\alpha} - \mathbf{e}^{\mathrm{T}}\boldsymbol{\alpha} \\ s.t.\ & \mathbf{y}^{\mathrm{T}}\boldsymbol{\alpha} = 0, \\ & 0 \le \alpha_i \le C,\ i = 1,2,\ldots,M, \end{aligned} \tag{14}$$

where $C > 0$ is an upper bound, $\mathbf{e}$ is a vector with unit elements, $\mathbf{Q}$ is $M \times M$ *positive semi-definite* matrix defined by $Q_{ij} = y_i y_j K(\mathbf{x}_i, \mathbf{x}_j)$ and $K(.,.)$ is a Mercer kernel (Vapnik, 1998). The decision function for unknown sample $\mathbf{x}$ is given by:

$$y^* = \mathrm{sgn}\left(\sum_{i=1}^{M} \alpha_i^* y_i K(\mathbf{x}_i, \mathbf{x}) + b\right). \tag{15}$$

2.5 Experimental testing and results

2.5.1 Data set

We apply the proposed framework on a part of the face image database from the Computational Vision at the California Institute of Technology, USA (Caltech-CV-Group, 1999). The original database contains JPEG images of faces of 19 persons with different lighting/expressions/backgrounds/ and male or female. We select prepare a subset of the database with 10 images per person. Next we split this subset in two groups with even images per class, i.e. 5 images per person for the first and second groups.

2.5.2 Experimental setup and testing

For each group of the subset we perform training followed by testing with the remaining group and for each run we calculate the recognition rate. The final recognition rate is calculated by averaging the rates of each run. With this protocol we test our framework with 'Ridge' and 'Lasso' regression setting of SR, where the former is tested with different value of sparsity. The dimensionality of the subspace for 'Ridge' and 'Lasso' is 18 and 30 respectively. For the first one it is controlled by the number of classes and for the second one it is determined by experiments. In Table 1 are displayed the results of our experiments.

Method	Accuracy,%
SR Ridge	94.21
SR Lasso (30)	38.94
SR Lasso (50)	73.15
SR Lasso (80)	**95.26**

Table 1. Recognition rate for FR framework setups

Controlling the sparsity parameter over 80 is not possible because it is bounded above by the cardinality of the training set which is 95 in our case. With value 80 of sparsity parameter the proposed framework achieves best recognition rate as results reveal. For this reason, we select Lasso mode of SR for combining with ECG biometric modality.

3. ECG personal identification

3.1 State of art

The ECG is widely used as diagnostic tool in the cardiology because of its clinical significance and considering its noninvasive nature. It is a registration of the electrical activity of the human heart over time. The genesis of this electrical activity is discussed in detail in (Malmivuo & Plonsey, 1995) . The registration of ECG is performed by measuring the generated electrical voltage between pair of leads attached on the human body according to defined standards. Usually more than 2 electrodes are used and the standard for clinical electrocardiography is an ECG acquisition taken from 12 channels, thus "viewing" the electrical activity from 12 different angles. According to electrodes placement, the leads can be divided into two groups: limb leads (RA, LA, LL) and augmented leads (V1-6) (Fig. 3). Each of these groups represents the electric field of the heart in a given plane (frontal or horizontal respectively).

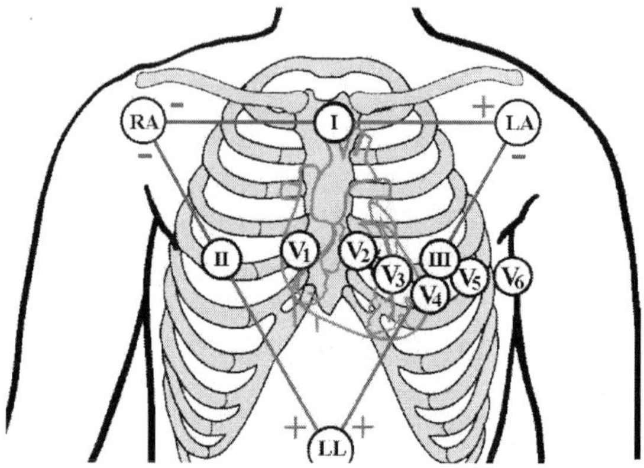

Fig. 3. Electrodes placement in standard 12 channel electrocardiography

Obviously the anatomical differences in heart muscle among individuals can be "seen" better in ECG when using all 12 channels. This type of acquisition is very impractical for real world application, so it is important to extract appropriate features when using only easy accessible leads, for example left and right arm.

A typical "healthy" ECG waveform has three distinct regions called waves (Malmivuo & Plonsey, 1995) (Fig. 4), however this morphology strongly depends on used leads, patient's condition, etc. A cardiac cycle starts with P wave, which represents the depolarization (electrical discharge) of the heart's atria. The QRS complex is a transient signal deflection related to the depolarization of the ventricles. The cardiac cycle ends with repolarization (electrical recharge) of the ventricles, which is seen in ECG as T wave. The remaining regions are referred as baseline.

The researches on the use of ECG as modality for personal identification have started since the beginning of the 21-st century. The first published articles covering this field are concentrated mainly in proving the personal discriminative characteristics of ECG as well as theirs relatively time invariance. In (Biel et al., 2001) are studied the time domain characteristics of such signals taken from 20 subjects (Fig. 4). In the cited article time domain features were reduced down to 7 according to achieved results from Principal Component Analysis (PCA). It was proven by experiments that it is possible to identify a person using ECG taken from only one lead.

In (Israel et al., 2005) can be seen a similar approach for ECG features extraction. The time domain characteristics are referred in this article as analytic features. The signal is preprocessed in terms of bandpass filtration. After this, the peaks of P wave, R wave and T wave are found as local maxima determination is a given region. The minimum radius curvature is incorporated to find the onsets and offsets of such waves. The relevant features are selected using Wilks' Lambda method.

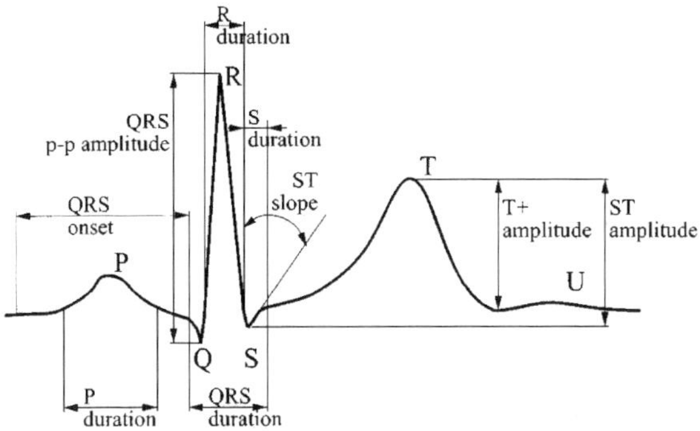

Fig. 4. Graphical representation of time domain ECG characteristics used as features for ECG personal identification (Biel et al., 2001)

Use of analytic ECG features has some disadvantages. Firstly, it requires sophisticated methods for automated ECG segmentation. Secondly, the amplitude parameters of the ECG depend not only on the unique anatomic characteristics of the human heart but also on electrodes contact, electrodes placement, etc. Finally, when the signal is highly influenced by noise, automated determination of analytic features with acceptable accuracy could fail.
As described above, the ECG segmentation is a primary and unavoidable process for any kind automated technique for features extraction. In order to avoid the need of precise determination of ECG wave boundaries, in (Plataniotis et al., 2006) is described an original approach. Firstly, the signal is divided into subsets in way that the subsets contain at least two complete cardiac cycles. Secondly, the normalized Autocorrelation Function (ACF) is calculated for a subset. Finally, the Discrete Cosine Transform (DCT) is applied on the ACF. About 40 of the most significant DCT coefficients serve as features for the classification process. The motivation behind using the ACF is the capability of non random patterns detection in the signal.

3.2 Methods for automated ECG segmentation

ECG segmentation is a key process in automated ECG identification systems. In healthy persons the ECG has strong cyclic recurrence. This is exploited in numerous approaches for full automated segmentation using Hidden Markov Models (HMM) (Hughes et al., 2004), (Boumbarov et al., 2009).
When dealing with morphological features, the full segmentation is not necessary to be performed. Instead of that some simple, fast and still effective ways can be used to extract the subsets consisting all waves and complexes in a complete cardiac cycle. The approach in (Velchev, 2010) proposes an identification of R peaks based on morphological filtration and histogram analysis. The morphological filtration is an interaction between the signal $x(t)$ and a predefined simple shape $g(t)$ called structuring element. The domains $x(t)$ and $g(t)$ are respectively X and G . The morphological filtration is based on two simple operations: dilatation $D(x,g)(t)$ and erosion $E(x,g)(t)$. They are defined as:

$$D(x,g)(y) = \max_{t-u \in X, u \in G} (x(t-u) + g(u)) \tag{16}$$

and

$$E(x,g)(y) = \min_{t-u \in X, u \in G} (x(t-u) - g(u)) \,. \tag{17}$$

The morphological operations *Close* and *Open* are defined as follows:

$$Close(x,g) = E(D(x,g),g) \tag{18}$$

and

$$Open(x,g) = D(E(x,g),g). \tag{19}$$

The operation $Open(x,g)$ smoothers the convex peaks of the signal, while Close(x, g) smoothers the concave peaks.
The structure of morphological filter for QRS complexes exaggeration is according to the following relation:

$$x'(t) = x(t) - \frac{1}{2}(Open(Close(x,g),g) + Close(Open(x,g)g)), \tag{20}$$

where $x'(t)$ is the filtered signal. The chosen structuring element is "disk". Determining its radius is a highly empirical procedure. However, it is clear that the diameter of the "disk" should be less than wf_s ,where w is the smallest width of the P and T waves and f_s is the sampling frequency. In Fig. 5 are shown two examples achieved according to 20 and using the "disk" as structuring element. As can be seen it is possible to choose the parameter of the structuring element in such way, the QRS complexes are remaining as much close to the original while the P and T waves are considerably suppressed.
The next procedure is to find an appropriate threshold for coarse detection of the R peaks. A histogram is calculated for each detected local maxima of the signal. This histogram for most of the cases will be bimodal. The optimal threshold is calculated according to Otsu's method (Otsu, 1979).

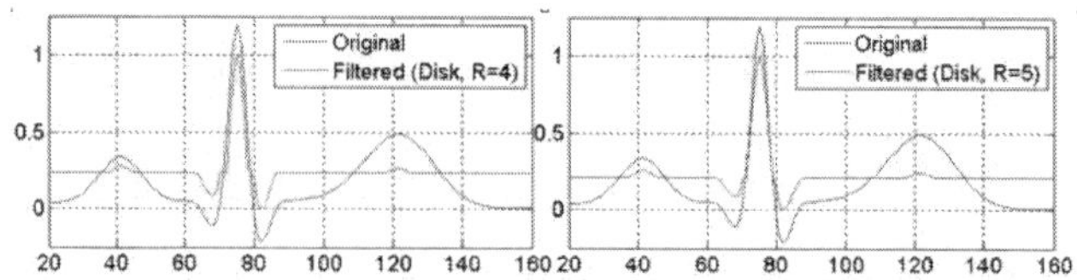

Fig. 5. Results from morphological filtration of ECG signal with structuring element 'disk" with radius 4 (left) and 5 (right). The sampling frequency is 256Hz

The last procedure in the segmentation process is to find a sample in the baseline between T and next P wave. This sample has to belong to the smoothest part of the region and its determination doesn't have to be extremely precise. Let $\mathbf{z} = [z1, z2, \ldots, z_{N-1}]$ be a vector which elements are the numbers of the wanted samples and N is the number of the identified R peaks. An element of **z** is calculated according to:

$$z_i = \arg\min_{\tau} \left(\sqrt{\left(CW_x^{\psi_{gauss1}}(s,\tau)\right)^2 + \left(CW_x^{\psi_{gauss2}}(s,\tau)\right)^2} \right), \quad \tau \in \left[r_i + (\Theta_2 - \Theta_1)(r_{i+1} - r_i)\right], \tag{21}$$

where $CW_x^{\psi_{gauss1}}$ and $CW_x^{\psi_{gauss2}}$ are the continuous wavelet transforms of x with wavelet function first and second derivative of the Gaussian, s is the scale of the wavelet transform, τ is the translation, r_i is the position of the i-th detected R peak and $\Theta_2 > \Theta_1$ are time values measured form R peak position between which the wanted sample is expected to be found. The values of s should be chosen large enough in terms for noise robustness and small enough in terms to achieve an accepted accuracy of identification. In Fig. 6 is shown an example result using the described approach.

As a final result from these procedures the ECG signal is segmented into complete cardiac cycles containing all important information in the middle of the resulting segments. Additionally the proposed approach doesn't require any supervised training.

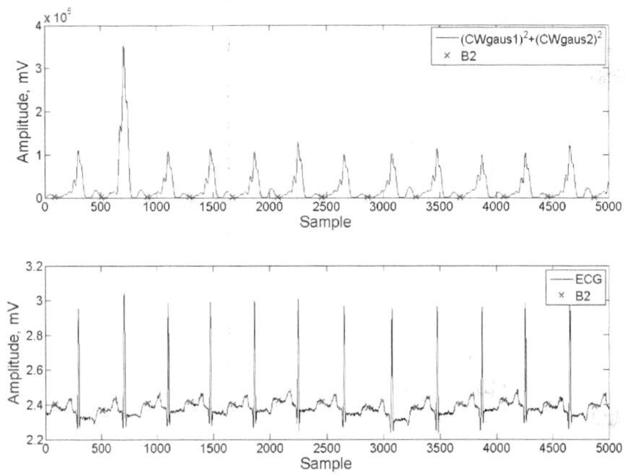

Fig. 6. An example of detection of samples belonging to the baseline between T and next P wave using combination of wavelet transforms with first and second derivative of the Gaussian as wavelet functions

3.3 ECG features extraction using linear and nonlinear projections in subspaces

As mentioned above, using the time domain characteristics as unique ECG features for personal identification has many significant drawbacks. An alternative approach is extraction of morphological features from ECG. These features could be extracted from a whole cardiac cycle in ECG, thus the need of full segmentation is eliminated. In this sense these features can be considered as holistic. They consist simultaneously amplitude and temporal characteristics of the ECG waves as well as their shape.

In this section two approaches for holistic features extraction are described. The first is based on linear projections in subspaces: PCA and Linear Discriminant Analysis (LDA). The second uses nonlinear versions of PCA and LDA: Kernel Principal Component Analysis

(KPCA) and Generalized Discriminant Analysis (GDA). A block diagram for an ECG personal identification system based on described features is shown in Fig. 7.

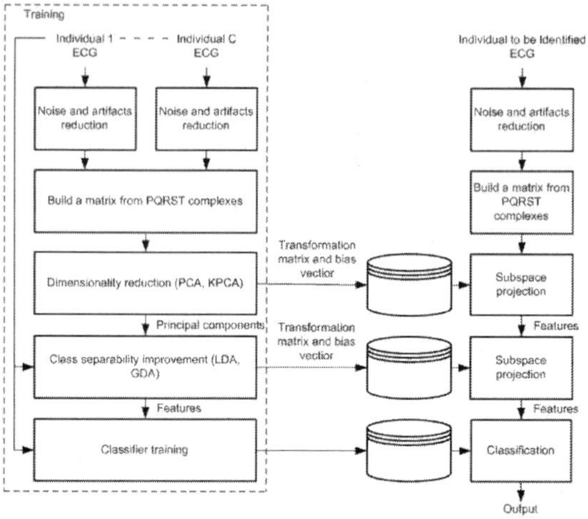

Fig. 7. Block diagram for ECG personal identification system based on features extracted using linear or nonlinear projections in subspaces

3.3.1 ECG features extraction using linear projections in subspaces

PCA is a statistical technique for dimensionality reduction of high dimensional correlated data down to a very few coefficients called principal components (Castells et al., 2007). This technique is optimal in terms of retaining as small as possible the Mean Squared Error (MSE) between the original signal and the restored signal form reduced set of principal components. Let the ECG signal is automatically segmented into PQRST complexes. These complexes are aligned and arranged as rows in a matrix. In general they differ in their length. The number of columns of the matrix is chosen larger than the maximal expected length of the PQRST complexes. The PQRST complexes are arranged in the matrix in such way the R peaks are in the middle of rows (Fig. 8). The elements in each row before and after the copied PQRST complex are set as same as the first and last sample of the complex respectively.

The training procedure requires a common training matrix $\tilde{\mathbf{X}} = [\tilde{x}_{ij}]_{N \times M}$ built from PQRST complexes taken from all individuals. The normalized training matrix $\tilde{\mathbf{Z}} = [\tilde{z}_{ij}]_{N \times M}$ is calculated according to:

$$\tilde{\mathbf{Z}} = [\tilde{z}_{ij}]_{NxM} = \frac{1}{\mathbf{u}\tilde{\boldsymbol{\sigma}}} \left(\tilde{\mathbf{X}} - \mathbf{u}\tilde{\boldsymbol{\mu}} \right), \tag{22}$$

where $\mathbf{u}$ is a column vector of ones with number of elements N, $\tilde{\boldsymbol{\mu}} = [\tilde{\mu}_1, \tilde{\mu}_2, \ldots, \tilde{\mu}_M]$ is the mean vector (bias vector) and $\tilde{\boldsymbol{\sigma}} = [\tilde{\sigma}_1, \tilde{\sigma}_2, \ldots, \tilde{\sigma}_M]$ is the vector of the standard deviation of $\tilde{\mathbf{X}}$. The covariance matrix $\boldsymbol{\Sigma}_{\tilde{Z}} = [\tilde{\sigma}_{\tilde{Z},ij}]_{M \times M}$ of $\tilde{\mathbf{Z}}$ is calculated according to:

$$\boldsymbol{\Sigma}_{\tilde{Z}} = E\left\{ \tilde{\mathbf{Z}}\tilde{\mathbf{Z}}^{\mathrm{T}} \right\} = \frac{1}{N-1} \tilde{\mathbf{Z}}\tilde{\mathbf{Z}}^{\mathrm{T}}, \tag{23}$$

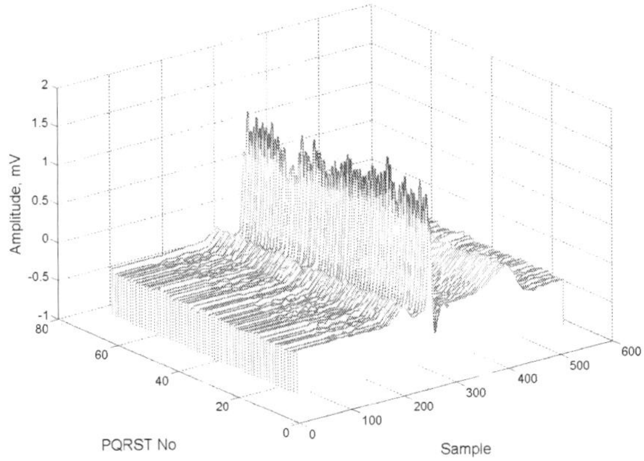

Fig. 8. Aligned PQRST complexes form ECG signal

where $E\{\}$ denotes expectation.
The next procedure is called eigendecomposition of $\mathbf{\Sigma}_{\tilde{Z}}$ according to the relation:

$$\mathbf{\Sigma}_{\tilde{Z}}\mathbf{V} = \mathbf{V\Lambda}, \tag{24}$$

where $\mathbf{\Lambda} = [\tilde{\lambda}_{ij}]_{M\times M}$ is a diagonal matrix and the values in its main diagonal are the eigenvalues of $\mathbf{\Sigma}_{\tilde{Z}}$.The columns of $\mathbf{V} = [\tilde{v}_{ij}]_{M\times M}$ are the eigenvectors of $\mathbf{\Sigma}_{\tilde{Z}}$.
The eigenvalues in $\mathbf{\Lambda}$ and the columns of $\mathbf{V}$ have to be rearranged in descending order of the eigenvalues. The number of principal components is equal to the original length of the input observations M so they have to be reduced. The reduced number of principal components L is calculated according to:

$$L = \min_{i}\left(mse(\hat{\tilde{\mathbf{Z}}})_i \leq mse_{tr}\right), \tag{25}$$

where $\mathbf{mse}(\hat{\tilde{\mathbf{Z}}}) = \left[mse(\hat{\tilde{\mathbf{Z}}})_1, mse(\hat{\tilde{\mathbf{Z}}})_2, \ldots, mse(\hat{\tilde{\mathbf{Z}}})_M\right]$ is the vector from MSE values between original matrix $\tilde{\mathbf{Z}}$ and the matrix $\hat{\tilde{\mathbf{Z}}}$, which is restored back from reduced set of principal components and mse_{tr} is a preliminary determined value. The elements of $\mathbf{mse}\left(\hat{\tilde{\mathbf{Z}}}\right)$ are calculated according to (Sanei & Chambers, 2007):

$$\mathbf{mse}\left(\hat{\tilde{\mathbf{Z}}}\right)_i = \sum_{i=1}^{M}\lambda_{jj} - \sum_{k=1}^{i}\lambda_{kk}, \; i = 1, \ldots, M. \tag{26}$$

The transformation matrix $\mathbf{W} = [w_{ij}]_{M\times L}$ is built as:

$$w_{ij} = v_{ij}, \; i = 1, \ldots, M. \tag{27}$$

In the process of authentication the features $\mathbf{Y}$ of the authenticated individual are calculated as (Franc & Hlavac, 2004):

$$\mathbf{y} = \left(\mathbf{W}^{\mathrm{T}}\left(\frac{1}{\mathbf{u}\tilde{\sigma}}(\mathbf{X} - \mathbf{u}\tilde{\mu})\right)\right), \tag{28}$$

where $\mathbf{u}$ is a column vector of ones with number of elements N . Typically it the set of PQRST complexes there exist an amount of non specific information. This is due to the presence of noise and artifacts or can be caused by an arrhythmia. These non specific PQRST complexes should be excluded from the analysis. A convenient way is to use the Hotelling statistics $\hat{\mathbf{T}}^2 = \left[\hat{\mathbf{t}}_1^2, \hat{\mathbf{t}}_2^2, \ldots, \hat{\mathbf{t}}_N^2\right]$. A given element indicates how far is the training sample from the centroid of the formed cluster (Hotelling, 1931):

$$\hat{t}_i^2 = \sum_{j=1}^{L} \frac{\hat{y}_{ij}^2}{\lambda_j}, \; i = 1, \ldots, N. \tag{29}$$

The criterion for excluding a given PQRST complex from the analysis is $\hat{t}_i^2 > \hat{t}_{tr}^2$, where the threshold value $\hat{t}_{tr}^2$ is:

$$\hat{t}_{tr}^2 = \frac{1}{N}\sum_{i=1}^{N} \hat{t}_i^2 + \sqrt{\frac{1}{N-1}\sum_{j=1}^{N}\left(\hat{t}_j^2 - \frac{1}{N}\sum_{i=1}^{N}\hat{t}_i^2\right)} \tag{30}$$

3.3.2 ECG features extraction using linear projections in subspaces

In Fig. 9 is given an example of features distribution of five individuals using two different techniques - PCA and KPCA. As can be seen the extracted features using PCA aren't linearly separable. Despite the complications of the process, the results from KPCA are much better.

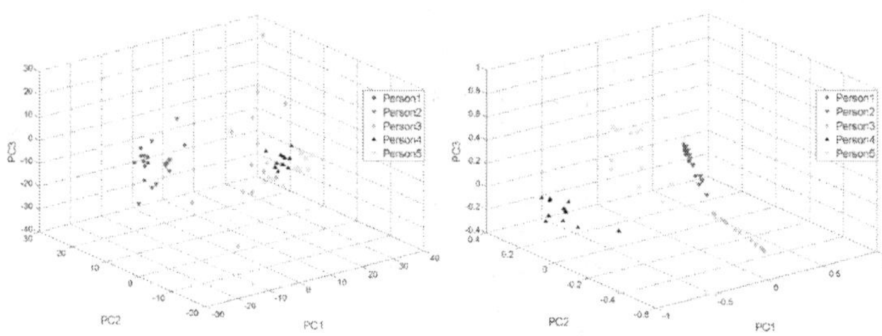

(a) (b)

Fig. 9. Distribution of the first three principal components from ECG signals taken from 5 individuals using PCA - (a) and KPCA - (b)

According to (Schölkopf et al., 1998) KPCA could be considered as performing PCA in other (usually higher dimensional) space $\mathcal{F}$. The mapping of the original input vector (PQRST complex) in the new space is expressed as $\mathbf{x} \rightarrow \mathbf{\Phi}(\mathbf{x})$. Let $\mathbf{\Phi}(\mathbf{x}_1), \ldots, \mathbf{\Phi}(\mathbf{x}_M)$ be the projections of the input vectors, where M is the number of observations. An input vector is expressed as $\mathbf{x}_j = \left[x_{j,1}, x_{j,2}, \ldots, x_{j,N}\right]$, where N is the original dimensionality. The key process behind KPCA is to perform an eigendecomposition of the covariance matrix $\mathbf{\Sigma}_{\mathbf{\Phi}(\mathbf{X})}$:

$$\lambda \mathbf{v} = \mathbf{\Sigma}_{\mathbf{\Phi}(\mathbf{X})}\mathbf{v}, \tag{31}$$

where $\mathbf{v}$ is an eigenvector of $\mathbf{\Sigma}_{\mathbf{\Phi}(\mathbf{X})}\mathbf{v}$, λ is the corresponding eigenvalue and $\mathbf{\Sigma}_{\mathbf{\Phi}(\mathbf{X})} = \frac{1}{M}\sum_{j=1}^{M}\mathbf{\Phi}(\mathbf{x}_j)\mathbf{\Phi}(\mathbf{x}_j)^{\mathrm{T}}$. Taking into account the definition of $\mathbf{\Sigma}_{\mathbf{\Phi}(\mathbf{X})}$, $\mathbf{v}$ is a linear combination of the vectors $\mathbf{\Phi}(\mathbf{x}_j)$, $j = 1, \ldots, M$:

$$\mathbf{v} = \sum_{j=1}^{M} w_j \mathbf{\Phi}(\mathbf{x}_j). \tag{32}$$

Using 31 and 32 we obtain:

$$\lambda \sum_{j=1}^{M} w_j \mathbf{\Phi}(\mathbf{x}_j) = \frac{1}{M} \sum_{j,k=1}^{M} w_j \mathbf{\Phi}(\mathbf{x}_k) \left(\mathbf{\Phi}(\mathbf{x}_k)^{\mathrm{T}} \mathbf{\Phi}(\mathbf{x}_j) \right) \tag{33}$$

which is equivalent of system of M equations:

$$\begin{aligned} &\lambda \sum_{j=1}^{M} w_j \left(\mathbf{\Phi}(\mathbf{x}_k)^{\mathrm{T}} \mathbf{\Phi}(\mathbf{x}_j) \right) = \\ &= \frac{1}{M} \sum_{j,k=1}^{M} w_j \left(\mathbf{\Phi}(\mathbf{x}_k)^{\mathrm{T}} \mathbf{\Phi}(\mathbf{x}_j) \right) \left(\mathbf{\Phi}(\mathbf{x}_k)^{\mathrm{T}} \mathbf{\Phi}(\mathbf{x}_j) \right), \forall l = 1, \ldots, M. \end{aligned} \tag{34}$$

In the last expression the inner products in the new space give the possibility not to deal directly with $\mathbf{\Phi}$, but to use the kernel matrix $\mathbf{K}[k_{ij}]_{M \times M}$:

$$k_{ij} = k(\mathbf{x}_i, \mathbf{x}_j) = \left\langle \mathbf{\Phi}(\mathbf{x}_i), \mathbf{\Phi}(\mathbf{x}_j) \right\rangle, \, i, j = 1 \ldots M. \tag{35}$$

where the operator $\langle \cdot, \cdot \rangle$ stands for inner product and k is the kernel function. Using kernel matrix 34 is transformed into:

$$M\lambda\mathbf{K}\mathbf{w} = \mathbf{K}^2\mathbf{w}, \tag{36}$$

where $\mathbf{w} = [w_1, w_2, \ldots, w_M]^{\mathrm{T}}$. The last is equal to (Schölkopf et al., 1998):

$$M\lambda\mathbf{K}\mathbf{w} = \mathbf{K}\mathbf{w}, \tag{37}$$

Determining $\mathbf{w}$ for each principal component is calculating the eigenvectors of $\mathbf{K}$. Let $\lambda_1 \leq \lambda_1 \leq, \ldots, \lambda_M$ is the full set of arranged eigenvalues, $\mathbf{w}^1, \mathbf{w}^2, \ldots, \mathbf{w}^M$ is the set of eigenvectors and λ_p is the first nonzero eigenvalue. According to (Schölkopf et al., 1998) $\mathbf{w}^1, \mathbf{w}^2, \ldots, \mathbf{w}^M$ have to be normalized in the way $\left\langle \mathbf{v}^k, \mathbf{v}^k \right\rangle = 1, \forall k = p, \ldots, M$. Using 37 the normalization is expressed as:

$$\begin{aligned} &\sum_{i,j=1}^{M} w_i^k, w_j^k \left\langle \mathbf{\Phi}(\mathbf{x}_i), \mathbf{\Phi}(\mathbf{x}_j) \right\rangle = \sum_{i,j=1}^{M} w_i^k, w_j^k k_{ij} \\ &= \left\langle \mathbf{w}^k, \mathbf{K}\mathbf{w}^k \right\rangle = \lambda_k \left\langle \mathbf{w}^k, \mathbf{w}^k \right\rangle = 1, \, \forall k = p, \ldots, M \end{aligned} \tag{38}$$

Obtaining the projection of x onto its principal components subspace would require calculating the projections of the eigenvectors $\mathbf{v}^k, k = p, \ldots, M$ onto $\mathcal{F}$:

$$\left\langle \mathbf{v}^k, \mathbf{\Phi}(\mathbf{x}) \right\rangle = \sum_{j=1}^{M} w_j^k \left\langle \mathbf{\Phi}(\mathbf{x}_j), \mathbf{\Phi}(\mathbf{x}) \right\rangle \tag{39}$$

A traditional approach for improving the class separability is the Linear Discriminant Analysis (LDA) (Theodoridis & Koutroumbas, 2006), however if the features are not linearly separable LDA usually fails. It is possible to use a more generalized approach called Generalized Discriminant Analysis (GDA) (Baudat & Anouar, 2000).
Let the input matrix **X** be composed from columns arranged following their class membership. The covariance matrix of the centers of the classes $\mathbf{\Sigma}_{\bar{\mathbf{\Phi}}(\mathbf{X})}$ is:

$$\mathbf{\Sigma}_{\bar{\mathbf{\Phi}}(\mathbf{X})} = \frac{1}{M} \sum_{c=1}^{C} n_c \bar{\mathbf{\Phi}}(\mathbf{x}_c)(\bar{\mathbf{\Phi}}(\mathbf{x}_c))^{\mathrm{T}}, \tag{40}$$

where n_c is the number of observations belonging to the class c. C is the number of the classes, and $\bar{\mathbf{\Phi}}(\mathbf{x}_c) = E\{\mathbf{\Phi}(\mathbf{x}_c)\}$. For the covariance matrix $\mathbf{\Sigma}_{\mathbf{\Phi}(\mathbf{X})}$ the following is valid (Baudat & Anouar, 2000):

$$\mathbf{\Sigma}_{\mathbf{\Phi}(\mathbf{X})} = \sum_{c=1}^{C} \sum_{k=1}^{n_c} \bar{\mathbf{\Phi}}(\mathbf{x}_{c,k})(\bar{\mathbf{\Phi}}(\mathbf{x}_{c,k}))^{\mathrm{T}}, \tag{41}$$

where $\mathbf{x}_{c,k}$ is the k-th observation from the class c. The goal in GDA is maximization of between-class variance and minimization of within-class variance. This is done by the following eigendecomposition:

$$\lambda \mathbf{\Sigma}_{\mathbf{\Phi}(\mathbf{X})} \mathbf{v} = \mathbf{\Sigma}_{\bar{\mathbf{\Phi}}(\mathbf{X})} \mathbf{v}, \tag{42}$$

where $\mathbf{v}$ is an eigenvector, and λ is the corresponding eigenvalue. The maximum eigenvalue maximizes the ratio (Baudat & Anouar, 2000):

$$\lambda = \frac{\mathbf{v}^{\mathrm{T}} \mathbf{\Sigma}_{\bar{\mathbf{\Phi}}(\mathbf{X})} \mathbf{v}}{\mathbf{v}^{\mathrm{T}} \mathbf{\Sigma}_{\mathbf{\Phi}(\mathbf{X})} \mathbf{v}}, \tag{43}$$

This criterion 43 in the space $\mathcal{F}$ is:

$$\lambda = \frac{\mathbf{w}^{\mathrm{T}} \mathbf{KPKw}}{\mathbf{w}^{\mathrm{T}} \mathbf{KKw}}, \tag{44}$$

where $\mathbf{K}$ is the kernel matrix, $\mathbf{w} = [w_1, w_2, \ldots, w_M]^{\mathrm{T}}$ is a vector from coefficients for which the equation 32 is valid, and $\mathbf{P}[p_{ij}]_{M \times M} = \mathbf{P}_1 \oplus \mathbf{P}_2 \oplus \ldots \oplus \mathbf{P}_c \oplus \ldots \oplus \mathbf{P}_C$ is a block matrix. The elements of $\mathbf{P}_c[p_{c,ij}]_{n_c \times n_c}$ are:

$$p_{c,ij} = \frac{1}{n_c}, \; i = 1, \ldots n_c; j = 1, \ldots n_c. \tag{45}$$

The solution of 45 can be found in (Baudat & Anouar, 2000).

3.4 Results and discussion

3.4.1 Datasets for experiments

The ECG signals for the experiments are collected from 28 individuals using own ECG registration hardware. The sampling rate is 512Hz and the resolution is 12bit. The system was trained using subsets from these signals. The testing was performed two weeks later in order to prove the time invariance of the features.

3.4.2 Experimental results

All experiments with kernel versions of PCA and LDA were made using the Gaussian kernel function $k(\mathbf{x},\mathbf{y}) = exp\left(-\frac{\|\mathbf{x}-\mathbf{y}\|^2}{2\sigma^2}\right)$ with $\sigma^2 = 1$. In Table 2 are presented the achieved values of the accuracy of the ECG personal identification.

Method	Accuracy,%
PCA	78.4
PCA-LDA	79.2
KPCA	91.8
PCA-GDA	95.7
Time domain features (Biel et al., 2001)	95.0

Table 2. Accuracy of the ECG identification

As can be seen the results using holistic features extracted with linear projections in subspaces are relatively poor. The GDA outperforms all approaches but the significant disadvantage of this method is the computation complexity. In addition the maximal dimensionality of the features is limited up to the number of identified individuals minus one. For combining with facial biometric modality we select KPCA approach for feature extraction. Despite its lower performance we prefer it because there is an algorithm, called GreedyKPCA, in which the kernel matrix does not have to be stored.

4. Combining ECG personal identification and face recognition

4.1 Classifier combination approaches

There are different approaches for combining classifiers, depending on their output. If only class labels are available as output, voting schemes can be used for final decision. If *a posteriori* probabilities are available however, different linear combinations rules can be applied. We will follow the former approach.

The strategies for combination utilize the fact that the classifier output reflects its confidence, and not the final decision. The confidence of a single classifier is represented by (Duin, 2002): $P_i(\mathbf{x}) = Prob(\omega_i \mid \mathbf{x})$, where ω_i is the i-th class and $i = 1, 2, \ldots, M$ is the class number. In the case of multiple classifiers, the confidence need to be defined for each of the classifiers $j = 1, 2, \ldots, C$. The output of each classifier can be viewed as a feature vector $\mathbf{z}_j$. Then, following the Bayesian rule for optimal classification a sample $\mathbf{x}$ is assigned to class ω_i if:

$$P(\omega_i \mid \mathbf{z}_1, \mathbf{z}_2, \ldots, \mathbf{z}_C) = \max_n P(\omega_n \mid \mathbf{z}_1, \mathbf{z}_2, \ldots, \mathbf{z}_C) \quad (46)$$

Using the Bayesian rule we can express 46 by the likelihood functions (Kittler et al., 1998):

$$P(\omega_n \mid \mathbf{z}_1, \mathbf{z}_2, \ldots, \mathbf{z}_C) = \frac{P(\mathbf{z}_1, \mathbf{z}_2, \ldots, \mathbf{z}_C \mid \omega_n) P(\omega_i)}{p(\mathbf{z}_1, \mathbf{z}_2, \ldots, \mathbf{z}_C)}, \quad (47)$$

where $p(\mathbf{z}_1, \mathbf{z}_2, \ldots, \mathbf{z}_C)$ is the unconditional joint probability density. It can be expressed through the conditional distributions and for this reason it could be used for the calculation of the final decision (Kittler et al., 1998). Also, using the numerator term only allows for various classifier combination rules to be obtained (Kittler et al., 1998):

- **Min rule:** According to this rule a sample $\mathbf{x}$ is assigned to class ω_i if

$$\min_j P(\omega_i \mid \mathbf{z}_j) = \max_n \min_j P(\omega_n \mid \mathbf{z}_j),\ j = 1,2,\dots C,\ n = 1,2,\dots,M. \tag{48}$$

This rule selects a classifier with least objection against a certain class.

- **Max rule:** In this rule a classifier with a most confidence is selected

$$\max_j P(\omega_i \mid \mathbf{z}_j) = \max_n \max_j P(\omega_n \mid \mathbf{z}_j),\ j = 1,2,\dots C,\ n = 1,2,\dots,M. \tag{49}$$

This rule selects a classifier with least objection against a certain class.

- **Product rule:** This rule assigns $\mathbf{x}$ to class ω_i if

$$p^{-(C-1)}(\omega_i)\prod_{j=1}^{C} P(\omega_i \mid \mathbf{z}_j) = \max_n p^{-(C-1)}(\omega_n)\prod_{j=1}^{C} P(\omega_n \mid \mathbf{z}_j),\ n = 1,2,\dots,M. \tag{50}$$

- **Sum rule:** This rule can be derived from the product rule and assigns $\mathbf{x}$ to class ω_i if

$$(1-C)P(\omega_i) + \sum_{j=1}^{C} P(\omega_i \mid \mathbf{z}_j) = \max_n \left[(1-C)P(\omega_n) + \sum_{j=1}^{C} P(\omega_n \mid \mathbf{z}_j)\right],\ n = 1,2,\dots,M. \tag{51}$$

4.2 Experimental results

We approach the combination of Face and ECG biometric modalities by combining both classifier's output probabilities using the rules specified in the previous section. For ECG identification we use the output probabilities of Radial Basis Neural Network classifier and for Face Recognition framework we use *LIBSVM library* (Chang & Lin, 2001) for calculating the output probabilities of SVM classifier. For both modalities we select 19 persons for identification with 5 samples for training and 5 samples for testing. Hence, the output of both classifiers is a 19 elements feature vector of probabilities with totally 10 samples per person, per modality. In our experiments, we test all rules for classifier combination considered and the results are displayed in Table 3. Also, we compare our work with the best results achieved by *Combining Attributes* in (Israel et al., 2003), .

Combination Rule	Accuracy,%
Max Rule	93.15
Min Rule	99.0
Sum Rule	99.1
Product Rule	**99.5**
Combining Attributes (Israel et al., 2003)	99.0

Table 3. Accuracy of the ECG identification

Experimental results reveal that combining probabilities output of ECG identification and Face Recognition framework with the *Product Rule* achieved best results.

5. Conclusion

In this work an approach for personal identification based on biometric modality fusion was presented. The presented combination of classifiers is characterized by its high accuracy and it is particularly suitable for precise biometric identification in intelligent video surveillance systems.

6. Acknowledgement

This work was supported by National Ministry of Education and Science of Bulgaria under contract DO02-41/2008 "Human Biometric Identification in Video Surveillance Systems", Ukrainian-Bulgarian R&D joint project References

7. References

Abate, A. F., Nappi, M., Riccio, D. & Sabatino, G. (2007). 2d and 3d face recognition: A survey, *Pattern Recogn. Lett.* 28: 1885–1906.

Baudat, G. & Anouar, F. (2000). Generalized discriminant analysis using a kernel approach, *Neural Comput.* 12: 2385–2404.

Biel, L., Pettersson, O., Philipson, L. & Wide, P. (2001). ECG Analysis – A new approach in Human Identification, *IEEE Transactions on Instrumentation and Measurement* 50(3): 808–812.

Bishop, C. M. (2007). *Pattern Recognition and Machine Learning (Information Science and Statistics)*, 1 edn, Springer.

Bolle, R. & Pankanti, S. (1998). *Biometrics, Personal Identification in Networked Society: Personal Identification in Networked Society*, Kluwer Academic Publishers, Norwell, MA, USA.

Boumbarov, O., Velchev, Y. & Sokolov, S. (2009). Ecg personal identification in subspaces using radial basis neural networks, *Intelligent Data Acquisition and Advanced Computing Systems: Technology and Applications, 2009. IDAACS 2009. IEEE International Workshop on*, pp. 446 –451.

Cai, D. (2009). *Spectral Regression: A Regression Framework for Efficient Regularized Subspace Learning*, PhD thesis, Department of Computer Science, University of Illinois at Urbana-Champaign.

Cai, D., He, X. & Han, J. (2007). Spectral regression for efficient regularized subspace learning, *Proc. Int. Conf. Computer Vision (ICCV'07)*.

Caltech-CV-Group (1999). Faces database.
URL: *http://www.vision.caltech.edu/html-files/archive.html*

Chang, C.-C. & Lin, C.-J. (2001). *LIBSVM: a library for support vector machines.* Software available at `http://www.csie.ntu.edu.tw/~cjlin/libsvm`.

Chang, K. I., Bowyer, K. W. & Flynn, P. J. (2003). Face recognition using 2d and 3d facial data, *ACM Workshop on Multimodal User Authentication*, pp. 25–32.

Chang, K. I., Bowyer, K. W. & Flynn, P. J. (2005). An evaluation of multimodal 2d+3d face biometrics, *IEEE Trans. Pattern Anal. Mach. Intell.* 27: 619–624.

Chung, F. (1997). *Spectral Graph Theory*, American Mathematical Society.

Duin, R. P. W. (2002). The combining classifier: to train or not to train?, *Pattern Recognition, 2002. Proceedings. 16th International Conference on*, Vol. 2, pp. 765–770.

Franc, V. & Hlavac, V. (2004). Statistical Pattern Recognition Toolbox for Matlab, *Prague, Czech: Center for Machine Perception, Czech Technical University* .
URL: *http://cmp.felk.cvut.cz/cmp/software/stprtool/index.html*

Freund, Y. & Schapire, R. E. (1997). A decision-theoretic generalization of on-line learning and an application to boosting, *J. Comput. Syst. Sci.* 55: 119–139.

Hotelling, H. (1931). The generalization of student's ratio, *The Annals of Mathematical Statistics* (3): 360–378.

Hughes, N. P., Peter, N. & Philosophy, H. D. (2004). Markov models for automated ecg interval analysis, *in Proceedings NIPS 16*, MIT Press, pp. 611–618.

Israel, S. A., Irvine, J. M., Cheng, A., Wiederhold, M. D. & Wiederhold, B. K. (2005). Ecg to identify individuals, *Pattern Recogn.* 38: 133–142.

Israel, S. A., Scruggs, W. T., Worek, W. J. & Irvine, J. M. (2003). Fusing face and ecg for personal identification, *Applied Image Pattern Recognition Workshop,* 0: 226.

Kittler, J., Society, I. C., Hatef, M., Duin, R. P. W. & Matas, J. (1998). On combining classifiers, *IEEE Transactions on Pattern Analysis and Machine Intelligence* 20: 226–239.

Kurylyak, Y., Paliy, I., Sachenko, A., Chohra, A. & Madani, K. (2009). Face detection on grayscale and color images using combined cascade of classifiers, *Computing* 8: 61–71.

Lecun, Y., Bottou, L., Bengio, Y. & Haffner, P. (1998). Gradient-based learning applied to document recognition, *Proceedings of the IEEE* 86(11): 2278 –2324.

Malmivuo, J. & Plonsey, R. (1995). *Bioelectromagnetism : Principles and Applications of Bioelectric and Biomagnetic Fields,* 1 edn, Oxford University Press, USA.
URL: *http://www.worldcat.org/isbn/0195058232*

Otsu, N. (1979). A threshold selection method from gray-level histograms, *IEEE Transactions on Systems, Man and Cybernetics* 9(1): 62–66.

Plataniotis, K., Hatzinakos, D. & Lee, J. (2006). Ecg biometric recognition without fiducial detection, *Biometric Consortium Conference, 2006 Biometrics Symposium: Special Session on Research at the*, pp. 1 –6.

Sanei, S. & Chambers, J. A. (2007). *EEG Signal Processing*, Wiley-Interscience.
URL: *http://www.worldcat.org/isbn/0470025816*

Schölkopf, B., Smola, A. & Müller, K.-R. (1998). Nonlinear component analysis as a kernel eigenvalue problem, *Neural Comput.* 10: 1299–1319.

Simard, P. Y., Steinkraus, D. & Platt, J. C. (2003). Best practices for convolutional neural networks applied to visual document analysis, *Proceedings of the Seventh International Conference on Document Analysis and Recognition - Volume 2,* ICDAR '03, IEEE Computer Society, pp. 958–.

Theodoridis, S. & Koutroumbas, K. (2006). *Pattern Recognition, Third Edition*, Academic Press.
URL: *http://www.worldcat.org/isbn/0123695317*

Vapnik, V. N. (1998). *Statistical Learning Theory*, Wiley-Interscience.
URL: *http://www.worldcat.org/isbn/0471030031*

Velchev, Y. (2010). Using morphological filtering and histogram analysis, *Computer and Communications Engineering* 4(1): 51–55.

Viola, P. & Jones, M. J. (2004). Robust real-time face detection, *Int. J. Comput. Vision* 57: 137–154.

Yan, S., Xu, D., Zhang, B., Zhang, H.-J., Yang, Q. & Lin, S. (2007). Graph embedding and extensions: A general framework for dimensionality reduction, *Pattern Analysis and Machine Intelligence, IEEE Transactions on* 29(1): 40 –51.

5

Biometrical Fusion – Input Statistical Distribution

Luis Puente, María Jesús Poza, Belén Ruíz and Diego Carrero
Universidad Carlos III de Madrid
Spain

1. Introduction

Biometric systems are based on the use of certain distinctive human traits, be they behavioral, physicial, biological, physiological, psychological or any combination of them. As reflected in the literature, some of the most frequently used biometric modalities include fingerprint, face, hand geometry, iris, retina, signature, palm print, voice, ear, hand vein, body odor and DNA. While these traits may be used in an isolated manner by biometric recognition systems, experience has shown that results from biometric systems analyzing a single trait are often insufficiently reliable, precise and stable to meet specific performance demands (Ross et al. 2006). In order to move system performance closer to the level expected by the general public, therefore, novel biometric recognition systems have been designed to take advantaje from taking multiple traits into account.

Biometric fusion represents an attempt to take fuller advantage of the varied and diverse data obtainable from individuals. Just as is the case with human recognition activities in which decisions based on the opinions of multiple observers are superior to those made by only one, automatic recognition may also be expected to improve in both precision and accuracy when final decisions are made according to data obtained from multiple sources.

its discussion of data fusion in biometric systems, the present chapter will analyze distinct types of fusion, as well as particular aspects related to the normalization process directly preceding data fusion.

2. Biometrics

Biometric recognition involves the determination of the identity of an individual according to his/her **personal qualities** in opposition to the classical identification systems which depend on the users' **knowledge** of a particular type of information (e.g., passwords) or **possession** of a particular type of object (e.g., ID cards).

In biometrics, 'recognition' may be used to refer to two distinct tasks. In the first one, that is called verification, an individual claims to be certain user who has been previously registered(enrolled) into the system. It is also possible that the individual does not indicate his/her identity but there exist some additional information that allow to suppose it. In this case system operation is reduced to confirm or reject that a biometric sample belongs to the claimed identity. In the second task, identification, it is not available such prior information about individual's identity, the system must determine which among all of the enrollees the subject is, if any. In the present chapter, only verification will be discussed insofar as identification may be understood as the verification of multiple entities.

In both cases (verification and identification), a **sample** of a predetermined biometric **trait** (e.g., face, voice or fingerprint) is **captured** (e.g., via photo, recording or impression) from a subject under scrutiny (**donor**), this is done using an adequate **sensor** for the task (e.g., camera, microphone or reader/scanner). A sample is called **genuine** when its donor identity and the identity of the **claimed user** are the same and it is called **impostor** when they are not. Following its capture, the sample is processed (**feature extraction**) in order to obtain the values of certain predefined aspects of the sample. This set of values constitute the **feature vector**. The feature vector is then **matched** against the **biometric model** corresponding to the individual whose identity has being claimed[1]. This model has been created at the time of that user enrols into the system. As a result of the matching, an evaluation of the degree of similarity between the biometric sample and the model is obtained, it is called **score**[2]. With this information, the **decision is taken** either to **accept** the subject as a genuine user or to **reject** the subject as an impostor (see Fig. 1).

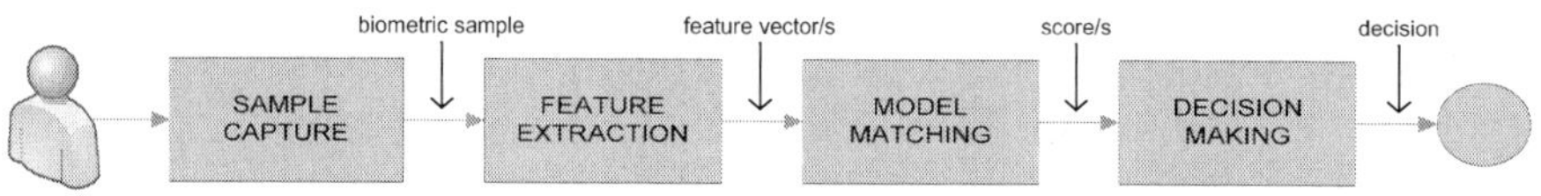

Fig. 1. Monobiometric[3] process.

In the majority of biometric recognition systems currently in use only a single biometric trait is captured in order to confirm or reject the claimed users's identity. Such systems are known as monobiometric. Nevertheless, at they heart is a pattern recognizer, which arrives at a final decision with the results obtained from a single sample processed according to a single algorithm.

Fig. 1 presents a simple representation of the biometric recognition process in monobiometric systems. After a subject presents the biometric trait which the system's sensor is designed to process, in the first stage (i.e., capturing sample), a biometric sample is obtained by the sensor and processed by the system to eliminate noise, emphasize certain features of interest and, in general, prepare the sample for the following stage of the process. In the next step (i.e., feature extraction), characteristic parameters of the sample are quantified and a feature vector that groups them is obtained. Following quantification, the system proceeds to match the feature vector (i.e., model matching) against others captured during the training phase that correspond to the individual whose identity is being claimed.

[1]The common structure of all biometric recognition systems is performed in two phases: (1) an initial training phase in which one or various biometric models are generated for each subject, and a later one called recognition phase, in which biometric samples are captured and matched against the models.

[2]The present chapter interprets scores as representing similarity. While, in practice, scores may also indicate difference, no generality is lost here by interpreting scores in this way since a linear transformation of the type ($s' = K\text{-}s$) can always be established.

[3]Term derived from the Greek monos (one) + bios (life) + metron (measure) and preferred by the authors of the present chapter over the term "unibiometric", also found in the literature but involving a mix of Greek and Latin morphological units. The same comment should be made about polybiometric and multibiometric terms.

These latter vectors are often represented in biometric systems with models that summarizes their variability. As a result of the matching process, a score is obtained quantifying the similarity between the sample and the model. In the final stage (i.e., decision making) and as a result of the score generated, the biometric system makes a decision to accept the sample genuine or to reject it as impostor.

3. Biometric fusion

In polybiometric systems, various data sources are used and combined in order to arrive at the final decision about the donor's identity. These systems are composed of a set of monobiometric parallel subprocesses that operate the data obtained from the distinct sources in order to finally combine them (i.e., fusing data). This fused data is then processed by system through a single subprocess until th final decision can be made regarding the truth of the claimed identity

In the construction of polybiometric recognitions systems, certain parameters must be set in response to the following questions:

- What are the distinct sources of biometric data being analyzed?
- At what point in the biometric process will the data be fused or, said another way, what intermediate data will be used for this fusion?
- What algorithm is most adequate for bringing about a particular type of fusion?

The following sections of the present chapter will analyze these different aspects of system design.

4. Data sources

In order to respond to the first question of the previous paragraph regarding multiple data sources, the following specific questions must also be considered (Ross 2007).

- How many sensors are to be utilized? In the construction of multi-sensor systems, different sensors (each with distinct performances) are used in order to capture multiple samples of a single biometric trait. In one example of this sort of polybiometric system, simultaneous photographs are captured of a subject's face using both infrared and visible light cameras.

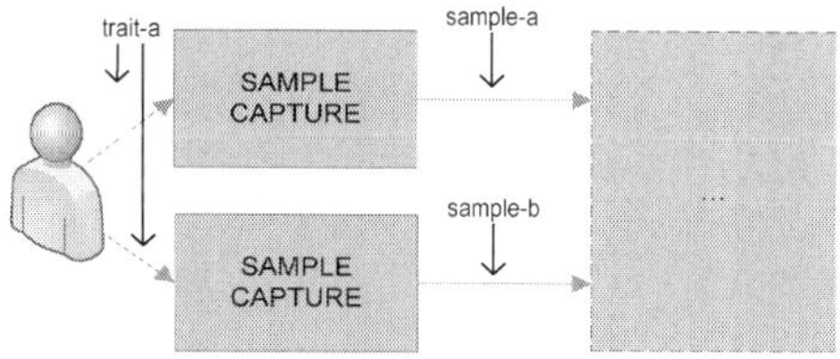

Fig. 2. Multi-sensor system.

- How many instances of the same biometric trait are to be captured? Human beings can present multiple versions of particular biometric traits (e.g., fingerprints for different fingers, hand geometry and veins for each hand and irises for each eye). As a result and with a schema similar to that of multi-sensor systems, multi-instance systems are designed to capture various instances of the same biometric trait.

- How many times is an instance of a particular trait to be captured? Using a single sensor and a single instance of a particular trait, it is nevertheless possible to obtain distinct samples of that instance under different conditions (e.g., video images taken of a trait instance from different angles or voice recordings taken at different moments and with different speech content). These multi-sample systems may also be represented by a schema similar to that of multi-sensor systems.
- How many different biometric traits are to be captured? Biometric recognition systems may be designed to analyze a single biometric trait (i.e., unimodal systems) or various traits (i.e., multimodal systems). The particularities of the latter type of system are represented by the schema below.

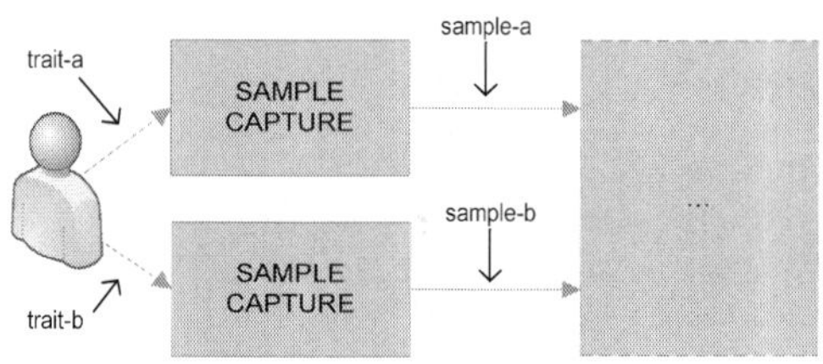

Fig. 3. Multimodal systems.

- How many distinct feature extraction algorithms are to be utilized in the processing of the biometric samples? Multi-algorithm systems are designed to use various algorithms for the feature extraction from biometric samples. In this case, the use of different extraction algorithms may allow the system to emphasize different biometric features of interest (e.g., spectral or prosodic features of a voice sample) and produce different feature vectors for each.

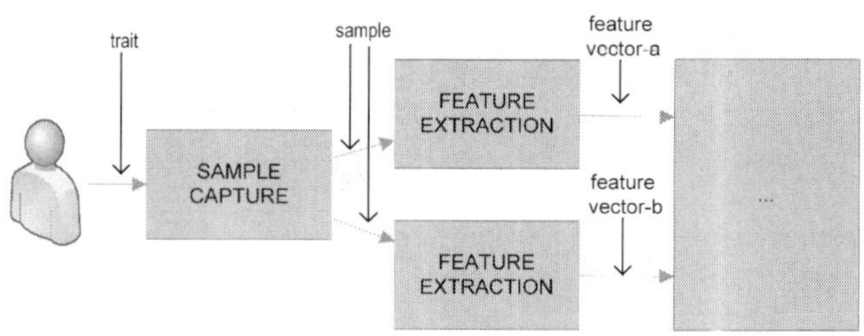

Fig. 4. Multi-algorithm systems.

- Against how many types of patterns and using how many methods are the feature vectors to be matched? Multi-matching systems are biometric recognition systems that allow match the feature vectors against various types of models or/and. using multiple techniques.
- Finally, it is also possible to construct hybrid systems systems of an even greater complexity that incorporate more than one type of the multiple data source discussed above.

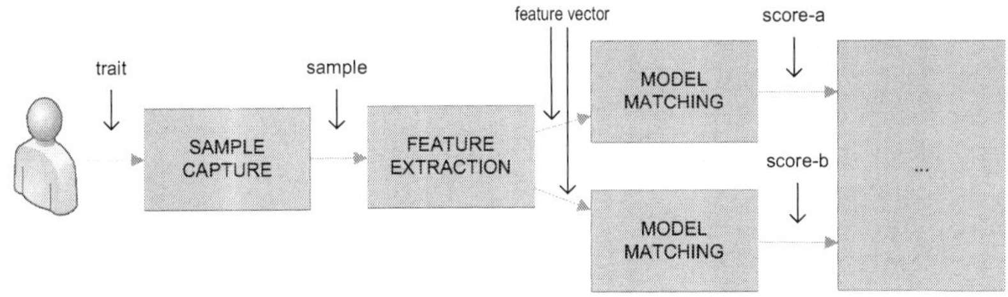

Fig. 5. Multi-matching schema.

5. Fusion level

As discussed earlier, biometric fusion is composed by a set of monobiometric subprocesses that work in parallel to operate the data obtained from distinct sources. Once this different data has been operated and fused, it is then handled by the system through a single subprocess until the point where the donor's identity final decision can be made. This process is represented in Fig. 6 below.

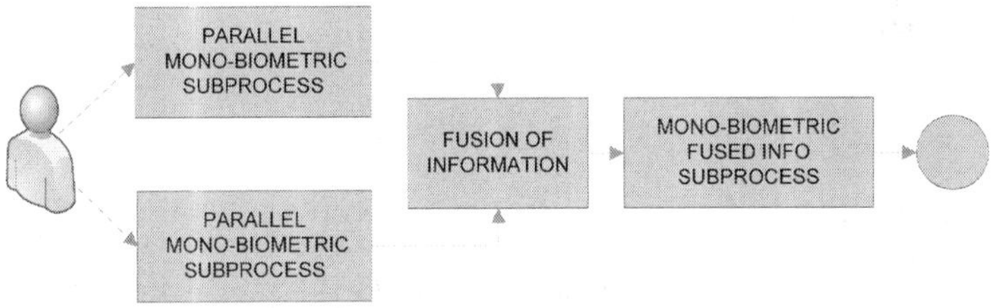

Fig. 6. Biometric fusion process.

Having considered the biometric fusion schema, it is time to return to the questions articulated earlier in the chapter and analyze now at what level of the process the fusion should be carried out or, in other words, what type of data the system should fuse. The possible responses in the literature to these points allow to establish diverse characterizations of data fusion systems defined as fusion levels(Ross 2007) (Joshi et al. 2009) (Kumar et al. 2010).

The first point at which data fusion may be carried out is at the sample level, that means immediately following sample capture by system sensors. This type of fusion is possible in multi-sensor, multi-instance and multi sample systems and may be obtained by following a particular sample fusion method. The form that this method takes in each case depends on the type of biometric trait being utilized. While fusion may range from a simple concatenation of the digitalized sample data sequence to more complex operations between multiple sequences, but it is almost always carried out for the same reason: to eliminate as many negative effects as possible associated with the noise encrusted in the data samples during capture. Once the fused sample has been generated, it may be used by the system for feature extraction.

The second point at which data may be combined is immediately following the feature extraction. At the feature level, vectors derived from the different sources are combined, yielding a single, fused vector.

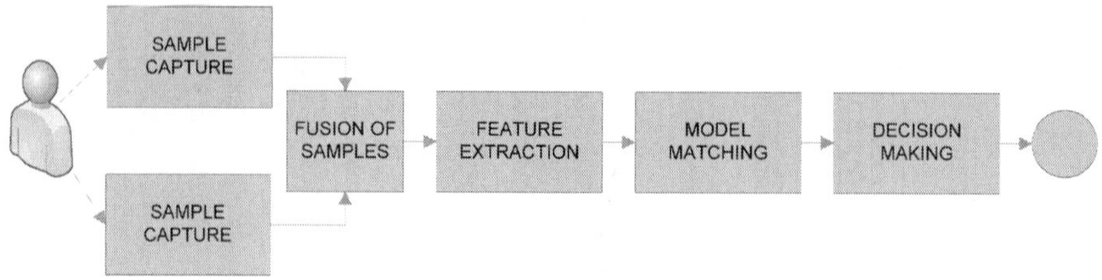

Fig. 7. Sample level fusion.

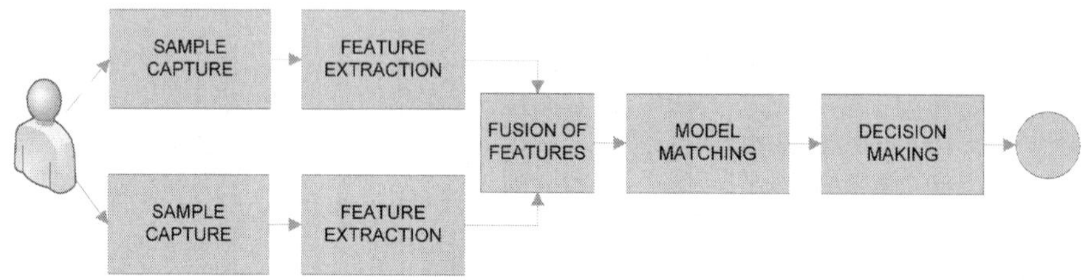

Fig. 8. Feature level fusion.

Another alternative is the fusion of scores obtained following the matching of different sample data against corresponding models. The new score resulting from this fusion is then used by the system to reach the final decision. This sort of fusion is normally carried out according to mathematic classification algorithms ranging in type from decision trees to techniques from the field of artificial intelligence, the latter of which offering learning capabilities and requiring prior training. The present chapter focuses particularly on this latter type of fusion which will be developed in much greater detail in sections below.

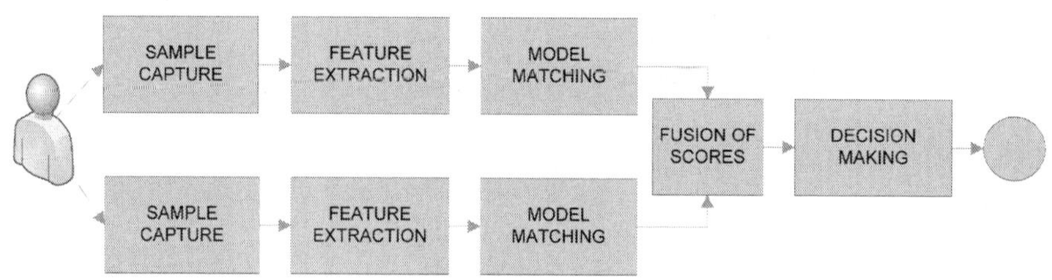

Fig. 9. Score level fusion.

Fusion may also be carried out on the final decisions obtained for each monobiometric process through the use of some kind of Boolean function. The most frequent algorithms used in this type of fusion are AND, OR and VOTING. With the first type, the final, combined decision is *GENUINE* if and only if each monobiometric process decision is also *GENUINE*. For the second type, the final, combined decision is *IMPOSTOR* if and only if each monobiometric process decision is also *IMPOSTOR*. Finally, for the third type combined decision is that of the majority of monobiometric process decisions which may or may not have been previously weighted.

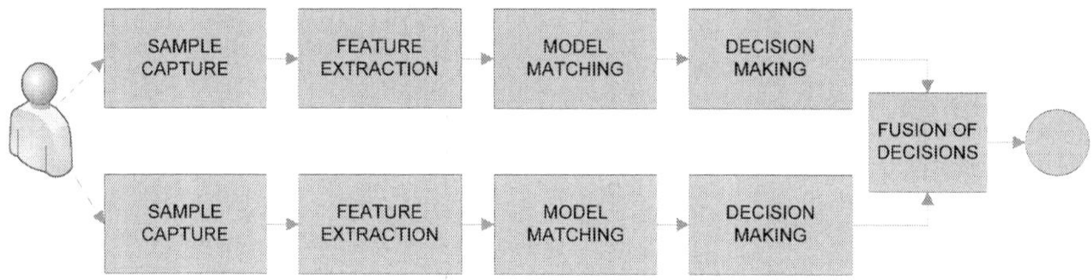

Fig. 10. Decision level fusion.

Finally, dynamic classifier selection schema uses scores generated at the data matching level in order to determine what classifier offers the highest degree of confidence. The system then arrives at a final decision through the application of solely the selected classifier. This is represented in Fig. 11 below.

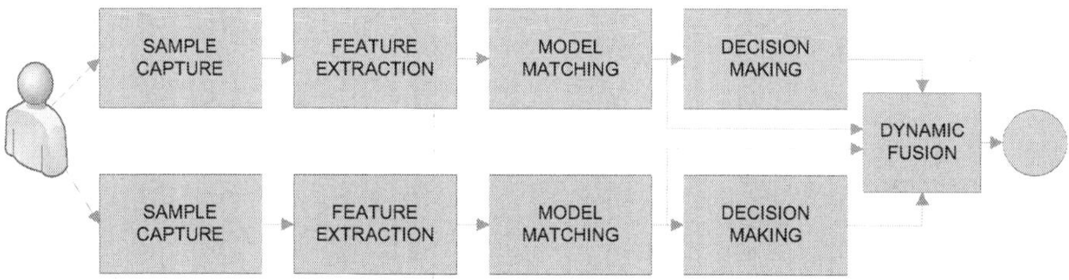

Fig. 11. Dynamic classifier selection.

6. Biometric performances

For the recognition of a individual by a classical recognition system, the data collected (e.g., passwords or ID cards information) from the subject must be identical to the previously recorded data into the system. In biometric recognition systems, however, almost never the data captured from a subject (nor the feature vectors obtained from them) are identical to the previous ones (Ross et al. 2006). The reasons for these variations are manifold and include the following:

- Imperfections in the capture process that create alterations (e.g., noise) in the data;
- Physical changes in the capture environment (e.g., changes in lighting and degradation of the sensors used); and
- Inevitable changes over time in individual's biometric traits.
- As a result of the unrepeatibility of biometric samples, the process of biometric recognition can not be deterministic and it must be based on the stochastic behaviour of samples. In this way, rather than flatly asserting correspondence between a biometric sample and the corresponding model, biometric systems only permit the assertion that this correspondence has a certain probability of being true.

The differences observed among the distinct biometric samples taken of a single trait from a single donor are known as intra-class variations. On the other hand, inter-class variation refers to the difference existing between the samples captured by the system from one subject and those of others. The level of system confidence in the correctness of its final decision is determined according to these two types of variation. The lesser the intra-class variation and the greater the inter-class variation are, the greater the probability that the final decision is correct.

In the matching model step, the system assigns a score to the sample feature vector reflecting the system's level of confidence in the correspondence between sample and claimed identity. If this score (*s*) lies above a certain threshold (*th*) (i.e., if: s ≥ th), the system will decide that the sample is genuine. However, if the score lies below the threshold the system will decide that the sample is an impostor one.

Insofar as score, as understood here, is a random variable, the probability that any particular score corresponds to a genuine sample can be defined by its probability density function (pdf) $f_g(s)$. Similarly, the probability that the score corresponds to an impostor sample can be defined by a pdf $f_i(s)$. As a result, the terms 'false match rate' (FMR) or 'false acceptance rate' (FAR) may be defined as the probability that an impostor sample be taken by the biometric system as genuine. Similarly, the terms 'false not match rate' (FNMR) or 'false rejection rate' (FRR) may be defined as the probability that a genuine sample be taken for as an impostor one.

When the decision score threshold is established in a system (see Fig. 12), the level of system performance is therefore established, because FAR and FRR directly depend on its value. Wether threshold values increases, FAR will also increase while FRR will decrease [Stan et al. 2009]. The optimal value of *th* can be obtained by minimizing the cost function established for the concrete system use. This cost function defines the balance between the damage that can be done by a false acceptance (e.g., a subject is granted access by the system to a protected space in which he or she was not authorized to enter) and that done by a false rejection (e.g., a subject with authorization to enter a space is nevertheless denied entry by the system).

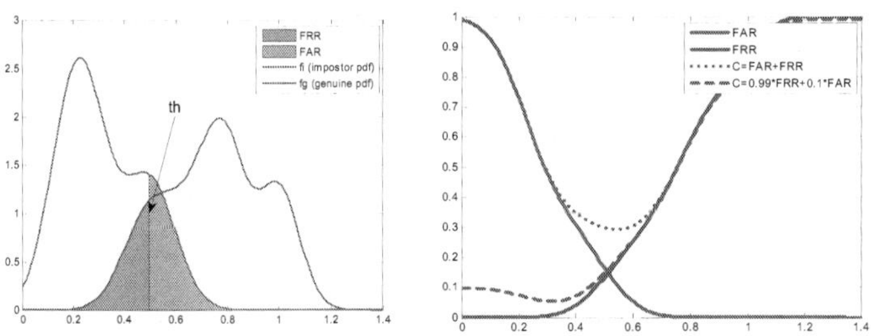

Fig. 12. Error rates and pdfs (left). Error rates and cost functions (right).

$$FRR(s) = \int_{-\infty}^{th} f_g(s)ds, \quad FAR(s) = \int_{th}^{\infty} f_i(s)ds \tag{1}$$

The National Institute of Standards and Technology (NIST) proposes as a cost function the one shown in formula 2, which is a weighted sum of both error rates. C_{FR} and C_{FA} correspond to the estimated costs of a false rejection and false acceptance, respectively, and P_g and P_i indicate the probabilities that a sample is genuine or impostor. Is obviously true that $Pi+Pg=1$ (Przybocki et al. 2006):

$$C = C_{FR} \cdot FRR \cdot P_g + C_{FA} \cdot FAR \cdot P_i \tag{2}$$

In NIST recognition system evaluations, the costs of a false acceptance and a false rejection are quantified, respectively, at 10 and 1, whereas the probabilities that a sample is genuine or impostor are considered to be 99% and 1%, respectively. With these parameters and normalizing the resulting expression, the following equation is obtained (formula 3):

$$C = 0.99 \cdot FRR + 0.1 \cdot FAR \tag{3}$$

For reasons of expediency, however, the present chapter utilizes other criteria that nevertheless enjoy wide use in the field. According to these criteria, $C_{FA} = C_{FR}$ and $P_g = P_i$, such that the resulting cost function may be defined as the following (formula 5):

$$C = FRR + FAR \tag{4}$$

Another value used in the characterization of biometric systems is the equal error rate (EER) which, as shown below, indicates the point at which the error rates are equal:

$$EER(s) = FRR(s) \quad | \quad FRR(s) = FAR(s) \tag{5}$$

As a final concept to consider here, the receiver operating characteristic curve (ROC curve) is a two-dimensional measure of classification performance and is regularly used in the comparison of two biometric systems. The ROC curve represents the evolution of the true acceptance rate (TAR) with respect to that of the FAR (Martin 1997):

$$TAR = f(FAR) \quad where\ TAR = 1 - FRR \tag{6}$$

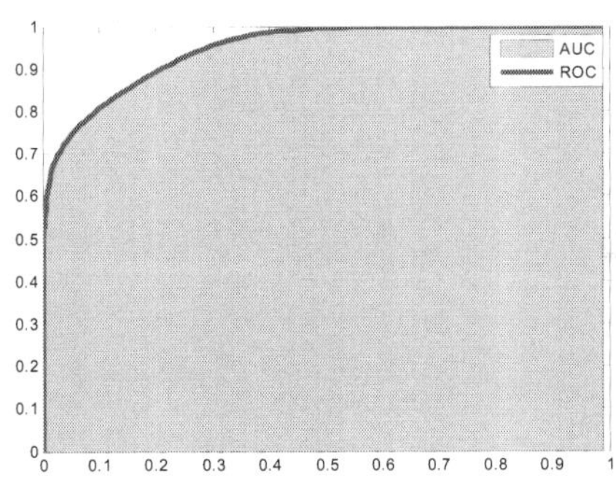

Fig. 13. ROC curve and area under the curve (AUC).

Through the analysis of the ROC curve, the evaluation of a recognition system may be carried out by considering the costs associated with errors even where the latter have not been previously established. In particular, using the area under the convex ROC curve (AUC), system performance may be expressed as a single numeric value and evaluated: the system considered the best being that with the greatest AUC (Villegas et al. 2009)(Marzban 2004).

$$AUC = \int_0^1 ROC(FAR) \tag{7}$$

7. Single scores distribution

Let the simplest case of match score distribution be supposed where, for a single source, scores are distributed according to the following criteria:

$$s \in \Re, f_g = pdf(s \mid gen.) = N(1,1), \; f_g = pdf(s \mid imp.) = N(-1,1) \tag{8}$$

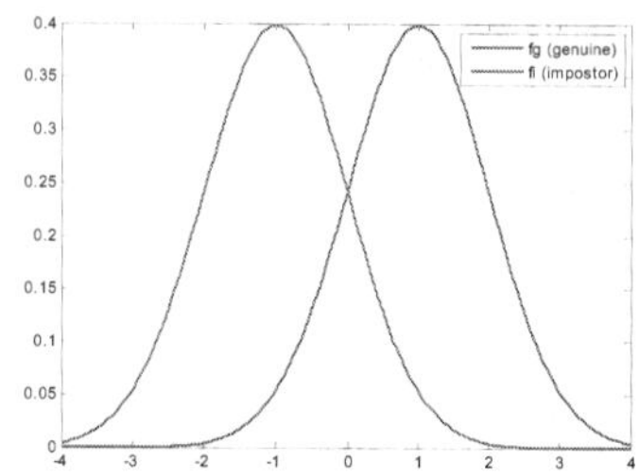

Fig. 14. Gaussian distribution of scores.

Given the symmetry of the functions, it can be held that the threshold value minimizing the cost function can be located at *th=0,* point at which FAR and FRR are equal, defining also the EER as shown in formula 9:

$$EER = \int_{-\infty}^{0} f_g(s)\, ds = \int_{0}^{\infty} f_i(s)\, ds \tag{9}$$

From an estimation, the value *EER=15.85%* is obtained. It is clear, then, that the farther apart the centroids or the smaller the deviations of the distribution functions are, the smaller the error rates.

8. Multiple score fusion

Let it be supposed that match score fusion is to be applied to the results of two processes having generated independent scores (s_1 and s_2) and with distribution functions identical to those described in the previous section of the present chapter. Thus, a match score vector is formed with Gaussian distribution functions for both genuine and impostor subject samples. This vector will have two components, each of which integrating the results from each of the monobiometric classifiers.

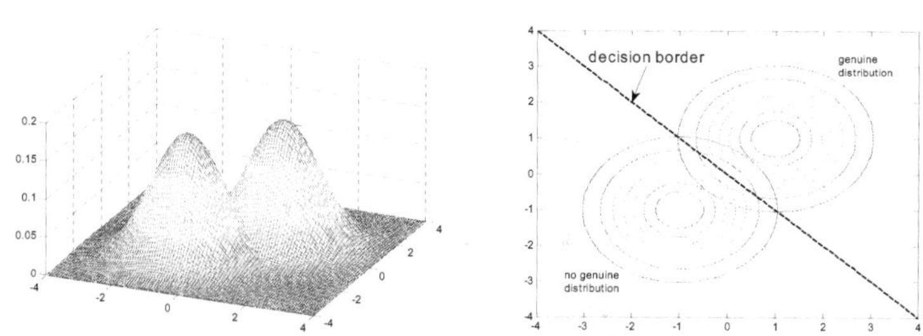

Fig. 15. Representation of two-dimensional Gaussian distribution scores.

$$v = (s_1, s_2) \mid s_1, s_2 \in \Re$$
$$f_{g1} = pdf(s_1 \mid genuine) = N(1,1), \quad f_{g2} = pdf(s_2 \mid genuine) = N(1,1) \tag{10}$$
$$f_{f1} = pdf(s_1 \mid impostor) = N(-1,1), f_{f2} = pdf(s_2 \mid impostor) = N(-1,1)$$

In Fig.15, the distribution functions are presented together for both genuine and impostor subject score vectors. Right image represents the contour lines of the distribution functions. Observing it, it seems intuitive that, just as was done in the previous section of the present chapter and applying the criteria for symmetry discussed therein, the best decision strategy is that which takes as a genuine subject score vector any vector found above and to the right of the dotted line which, in this particular case, corresponds to $s1+s2 \geq 0$. This converts the threshold, for the one-dimension scores, to a boundary line decision in this two-dimension space (an hiperplane if n dimensions space).

$$f_g(v) = f_{g1}(s_1) \times f_{g2}(s_2), f_g(v) = f_{f1}(s_1) \times f_{f1}(s_2) \tag{11}$$

Following this, the resulting estimation of the EER is shown in formula 12. In the specific case proposed here, the resulting EER is found to be 7.56% indicating an important improvement owing to the fact that the centroids of the distribution functions have been separated here by a factor of $\sqrt{2}$.

$$EER = \iint_G f_i(v)\, ds_1\, ds_2, \quad G: s_1 + s_2 \geq 0$$
$$EER = \iint_I f_g(v)\, ds_1\, ds_2, \quad I: s_1 + s_2 < 0 \tag{12}$$

9. Using gaussian mixture model classifiers

Gaussian mixture model (GMM) classifiers are used in order to create a model of statistical behaviour represented by the weighted sum of the gaussian distributions estimated for the class of genuine training score vectors and another similar model to represent the class of impostor vectors. Using the two models, the vectors are classified using the quotient of the probabilities of belonging to each of the two classes. If this quotient is greater than a given threshold (established during the system training phase), the vector is classified as genuine. If the quotient is below the given threshold, the vector is classified as an impostor. Such a procedure is quite similar to that discussed in the previous section of the chapter.
In a situation such as that described in the paragraph above, the following points indicate the expectations for a training process and test using GMMs:

- These models (f_g', f_i') of sums of Gaussian functions should maintain a certain similarity to the generative sample distribution ;
- The established threshold may be equivalent to the theoretical decision boundary between genuine and impostor score vectors; and
- Test results clearly approach the theoretic FAR and FRR.

In order to test the fitness of these premises, 1000 two-dimensional random vectors (Vg) following the distribution function of the genuine vectors and another 1000 vectors (Vi) following the distribution function of the impostor vectors have been taken as training data. With these vectors, GMMs were created to approximate the distribution functions[4].

[4]For the training and tests of the GMMs performed here a version of EM algorithm has been used. http://www.mathworks.com/matlabcentral/fileexchange/8636-emgm

$$f_g \approx f_g^{'} = \sum_{k=1}^{n_g} w_{gk} \cdot N\left(\mu_{gk}, \sigma_{gk}\right), f_i \approx f_i^{'} = \sum_{k=1}^{n_i} w_{ik} \cdot N\left(\mu_{ik}, \sigma_{ik}\right) \quad (13)$$

The models obtained in the training phase for 10 Gaussian models (10G) derived from the simulated data training are presented below in Table 1:

Genuine s_1

K	W	μ	σ	CV[5].
1	0.1374	1.6513	0.1366	-0.0159
2	0.0930	1.5475	0.3065	-0.0689
3	0.0954	-0.2555	0.3610	0.0162
4	0.1504	0.7806	0.1926	0.0199
5	0.0885	2.1518	0.3166	0.0184
6	0.0767	1.1427	0.3610	-0.0333
7	0.0776	-0.1481	0.4024	0.0222
8	0.1362	0.7321	0.1797	0.0017
9	0.0783	0.0929	0.4629	-0.0726
10	0.0666	2.6161	0.3393	0.0316

Genuine s_2

K	W	μ	σ	CV.
1	0.1374	0.9382	0.1875	-0.0159
2	0.0930	-0.2543	0.3520	-0.0689
3	0.0954	0.8522	0.2825	0.0162
4	0.1504	1.5416	0.1641	0.0199
5	0.0885	1.8540	0.3072	0.0184
6	0.0767	2.5772	0.4198	-0.0333
7	0.0776	2.0269	0.3084	0.0222
8	0.1362	0.4505	0.1629	0.0017
9	0.0783	-0.4208	0.3757	-0.0726
10	0.0666	0.5487	0.4818	0.0316

Impostor s_1

	W	μ	σ	CV.
1	0.0806	-1.8355	0.2769	-0.0462
2	0.0811	-0.6579	0.3288	0.0030
3	0.1356	-0.3908	0.1417	0.0027
4	0.0966	-2.3122	0.3943	-0.0498
5	0.1183	-1.4191	0.1553	-0.0279
6	0.0994	0.2029	0.3683	0.0397
7	0.0823	-0.8118	0.3312	0.0032
8	0.0811	-2.1787	0.3847	-0.0317
9	0.0668	0.5181	0.3286	0.0824
10	0.1583	-1.0723	0.2032	0.0163

Impostor s_2

	W	μ	σ	CV.
1	0.0806	0.2294	0.3898	-0.0462
2	0.0811	0.4612	0.3891	0.0030
3	0.1356	-1.3111	0.1581	0.0027
4	0.0966	-0.7795	0.2842	-0.0498
5	0.1183	-1.4853	0.1977	-0.0279
6	0.0994	-0.4434	0.2838	0.0397
7	0.0823	-2.4402	0.3102	0.0032
8	0.0811	-2.0999	0.3894	-0.0317
9	0.0668	-1.7791	0.4830	0.0824
10	0.1583	-0.4784	0.1519	0.0163

Table 1. Gaussian Mixture Model (GMM)

In Fig.16 (left), genuine and impostor models are presented for the score s_1 of the score vector. With red lines indicating the impostor model and black lines indicating the genuine sample model, each of the 10 individual Gaussian distributions with which the GMM classifier approximated the distribution of the training data are represented by the thin lines on the graph. The weighted sums of these Gaussian functions (see Formula 13) are represented by the thick lines on the graph. The result has an appearance similar to two Gaussian distributions around +1 and -1. Fig. 18 (right) shows the contour lines of the two-dimensional models.

For a value of *th = 0.9045* (calculated to minimize) it was found that FAR = 8.22% and FRR = 7.46%.

[5]Covariance S1-S2 o S2-S1

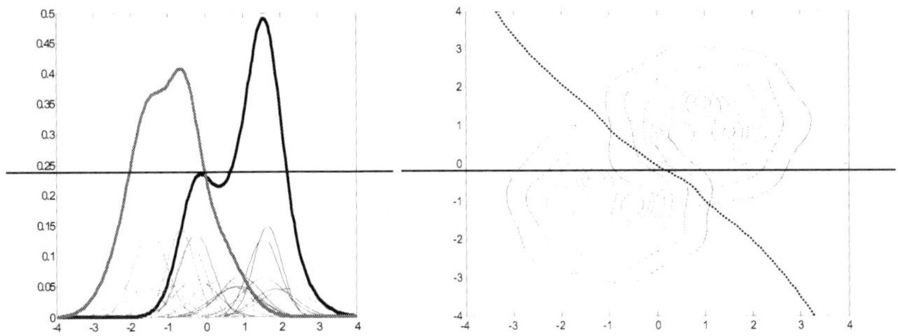

Fig. 16. GMM with 10G. Single axe (left). Contours for two-dimension model (right).

$$decision(v) = genuine \Leftrightarrow \frac{f_g'(v)}{f_i'(v)} \geq th \quad (14)$$

In Fig.16. the decision boundary line, at which the quotient of pdfs is equal to the threshold and which separates genuine and impostor decisions, presented as a dotted line. This line is quite near to the proposed boundary. Then the formula 14 represents a transformation from a two-dimension criterion to a one-dimension threshold, which, of course, is easier to manage.

If the same exercise is repeated for a model with 3 Gaussians (3G) and for another with only 1 Gaussian (1G), the following results are obtained:

N Gaussian	FAR	FRR	MER[6]	th	AUC
10	8.22%	7.46%	7.84%	0.9045	97.12%
3	7.97%	7.52 %	7.75%	0.9627	97.05%
1	7.99%	7.56%	7.77%	0.9907	97.10%

Table 2. Gaussian Mixture Model (GMM).

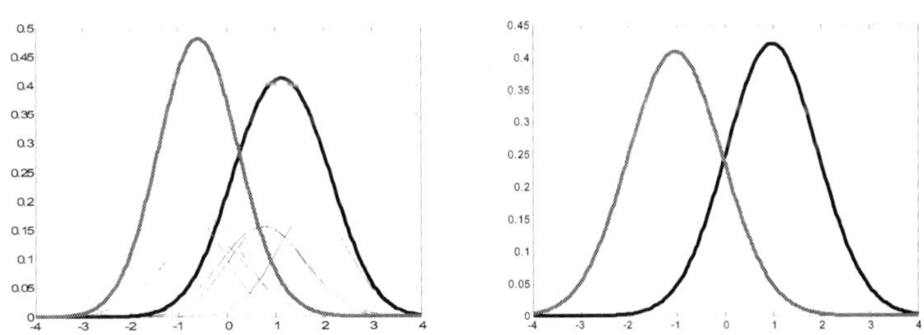

Fig. 17. Single axe GMM with 3G (left) and 1G (right).

[6]MER = (FAR*FRR)/2

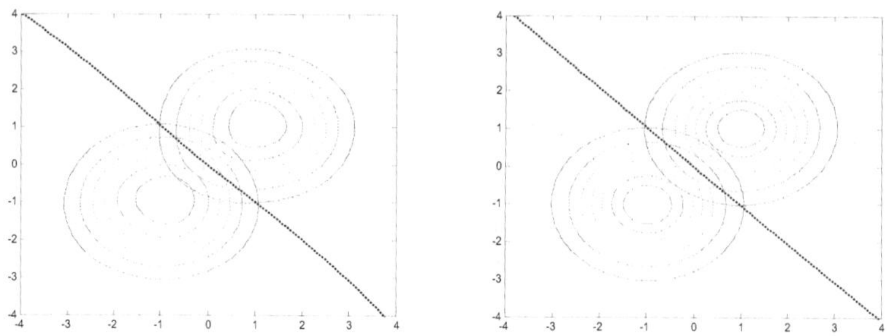

Fig. 18. Contour lines for two-dimension GMM with 3G (left) and 1G (right).

Changing the threshold value (see Fig.19), distinct decision boundaries and their corresponding error rates may be obtained. With these values, a ROC curve may be drawn and its AUC estimated.

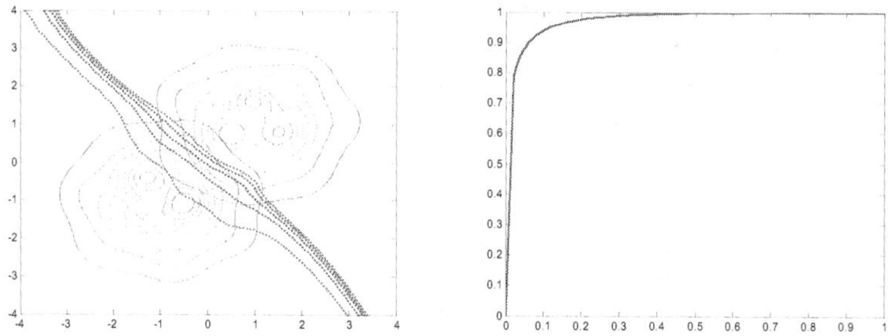

Fig. 19. Various decision boundaries (left) and ROC Curve (right) for GMM 10G.

10. Using support vector machine classifiers

A support vector machine (SVM) is a classifier that estimates a decision boundary between two vector sets (genuine and impostor ones) such that maximizes the classification margin. In the training phase of a SVM, a model is constructed that defines this boundary in terms of a subset of data known as support vectors (*SV*), a set of weights (*w*) and an offset (*b*).

$$decision(v) = genuine \Leftrightarrow -b + \sum_{k} w_k \left(SV_j \cdot v' \right) \geq 0 \quad (^7) \tag{15}$$

The equation above defines the distance of a vector (v) to the boundary, where positive distances indicate genuine samples and negative distances indicate impostor samples[8]. For other kind of boundary lines is possible to select between different kernel functions. Then the general decision function is shown in formula 16, where *K(sv,v)* represents the adequate kernel function. The kernel implied in formula 15 is called "linear kernel".

---

[7]*v*' indicates the transpose v vector.

[8]For the examples presented in this chapter, SVM-Light software has been used.

$$decision(v) = genuine \Leftrightarrow -b + \sum_k w_k \cdot K\left(SV_j, v'\right) \geq 0 \qquad (16)$$

Given the data distribution and the fact that the expected separation boundary is a straight line, it may be assumed that the linear kernel is the most adequate kernel function here.
Fig.20 shows the distribution of genuine samples (in blue) and impostor samples (in red). Points indicated with circles correspond to the support vectors generated in the training phase. The central black line crossing the figure diagonally represents the set of points along the boundary line, which is also quite close to the theoretical boundary.

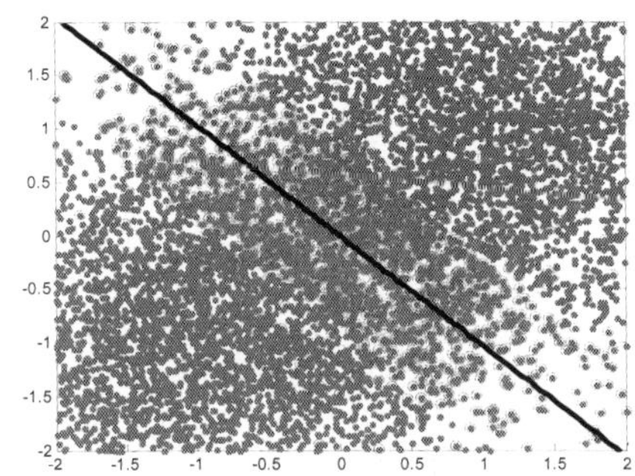

Fig. 20. SVM with linear kernel.

The results of the test data classification demonstrate the performance indicated below for

Kernel	FAR	FRR	MER	nSV[9]	AUC
Linear	8.01%	7.56%	7.78%	1956	95.06%

Table 3. Results of SVM test

The classifier establishes a transformation of the vector space into a real value whose module is the distance from the boundary, calculated such that the system be optimized to establish the decision threshold at the distance of 0. Just as in the case of GMMs, system behavior can be analyzed using the ROC curve and, more specifically, the AUC through the adjustment of this threshold value (see Table 3).

11. Using neural network classifiers

An artificial neural network (ANN) simulates an interconnected group of artificial neurons using a computational model. In this context, a neuron is a computational element that operates n-inputs in order to obtain just one output following a transfer function like the one shown at formula 16. Where s_k is the k-esime neuron input, w_k is the weigth of k-esime input w_0 represent the offset and finaly represents a function (typically sigmoid or tanh) that performs the transference between neurons.

[9]nSV: Number of support vectors

$$z = \text{func}(w_0 + \sum_{k=1}^{N}(w_k \cdot s_k)) \quad (16)$$

A typical ANN groups its neurons in a few layers, so that, the certain layer neuron outputs are only connected to the next layer neuron inputs.
The neural network training step gets as a result the weight for every neuron input that minimizes the error rates.
Then the simplest network is one which has only one neuron with two input and one output (2-1-1). This way, the transfer function has no effect on the system and at the end decision function becomes a linear combination of the inputs and therefore the training estimates a linear separator similar to the one seem before for SVM with linear kernel.
Applying neural networks to above described data, is possible to obtain the following results:[10]

Struct	FAR	FRR	MER	AUC
2-1-1	7.94%	7.62 %	7.78%	95.06%

Table 4. Results of 3 ANN test

12. Beta distributions

One common way in which monobiometric systems present their scores is through likelihood estimates (the probability that the sample is genuine). In such cases, the score rangeis limited to 0-1 (0-100%). Ideally, instances of genuine subject scores would be grouped together around 1 or a point close to 1, while impostor subject scores would be grouped together around zero or near it. Both would demonstrate beta distributions. An example of this ideal situation is plotted in Fig.23 with the pdf for genuine samples follows *Beta(5,1)* and the pdf for impostor samples follows *Beta(1,5)*. Because the symmetrical properties of these functions, the equilibrium point can be clearly located at $s = 0.5$ with an solving the integral in formula 1 the $EER = 3.12\%$.
As it was done for the Gaussians , identical distribution functions are established for both dimensions of the two-dimension score space, then a theoretical value of EER= 0.396% would be obtained. Also is possible the same routine and evaluate system performances for GMM, SVM and NN classifiers[11].
Fig. 21 shows the pdf's used in this example and de model obtained for them, while table 5 display test results.

Classifier.	FAR	FRR	MER
GMM 10G	0.55%	0.31%	0.43%
GMM 3G	0.56%	0.28%	0.42%
GMM 1G	0.54%	0.28%	0.41%
SVM Linear	0.48%	0.34%	0.38%
NN (2-1-1)	0.43%	0.35%	0.39%

Table 5. Test results.

[10]For the examples with ANN, Neural Network Toolbox™ have been used. http://www.mathworks.com/products/neuralnet/

[11]For the examples, the same number of genuine and imposter vectors were randomly generated as the previous sections

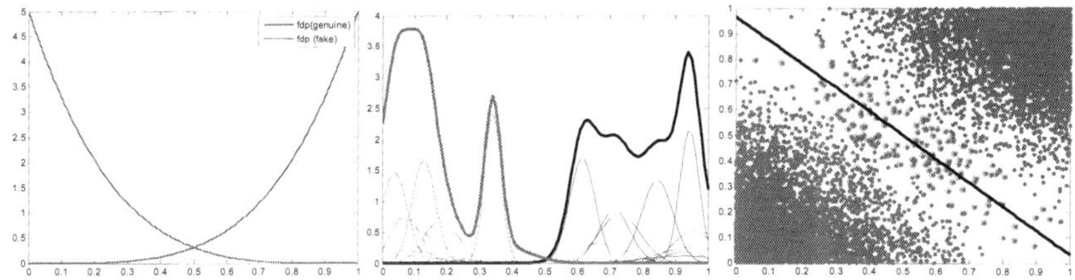

Fig. 21. Single axe pdfs (left), model with 10G (centre), vector plots and SVM linear model.

13. More realistic distributions

Unfortunately, the distribution functions for real scores are not as clear-cut as those presented in Fig. 21. Scores for impostor subject samples, for example, are not grouped around 0, but rather approach 1. Similarly, genuine subject sample scores often tend to diverge from 1. Distributions similar to those in Fig. 22 are relatively common. To illustrate this, *pdf.genuine = Beta (9,2)* and *pdf.impostor = Beta (6,5)* have been chosen for the first score (s_1).

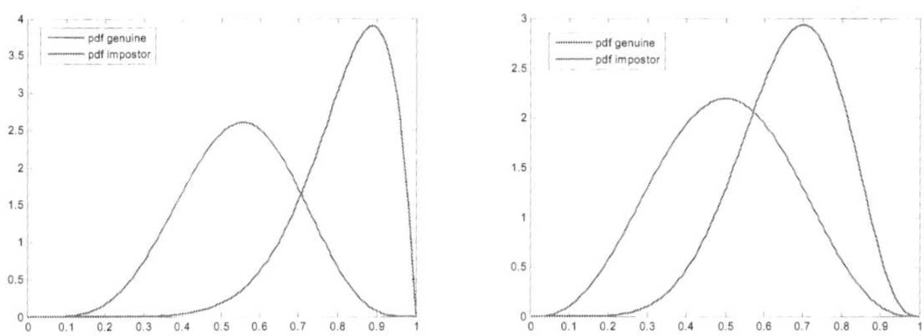

Fig. 22. Single axe score1 pdfs (left) and score2 pdfs (right)..

These particular distributions don't display any symmetrical property then the equilibrium point estimated loking for *FAR = FRR* and as a result the threshold value of *th = 0.7* with an *EER of 15.0%* has been obtained. Eve more, the optimal threshold value does not coincide here with the ERR. Then in order to minimize the cost function threshold must adopt a value of *0.707* yielding the error rates of *FAR = 16.01%, FRR = 13.89%* and *MER = 14.95%*.
In order to further simulate real conditions, score 2 has been supposed here to display a different behavior, to wit, *pdf.genuine = Beta (8,4)* and *pdf.impostor = Beta (4,4)*, as it is displayed in Fig 22
As can be seen, the equilibrium point is found here at a value of *EER = 29.31%* and *th = 0.5896* and the minimum of the cost function at *FAR =34.84%, FRR = 23.13%, MER = 28.99%* and *th = 0.5706%*.
If these two distributions are combined and a two-dimensional score space is established, the resulting pdfs can be represented as the one in Fig.23. It plots these two-dimensional density distributions where de genuine one is found near the point (1,1) while the impostor one is located farther from it.

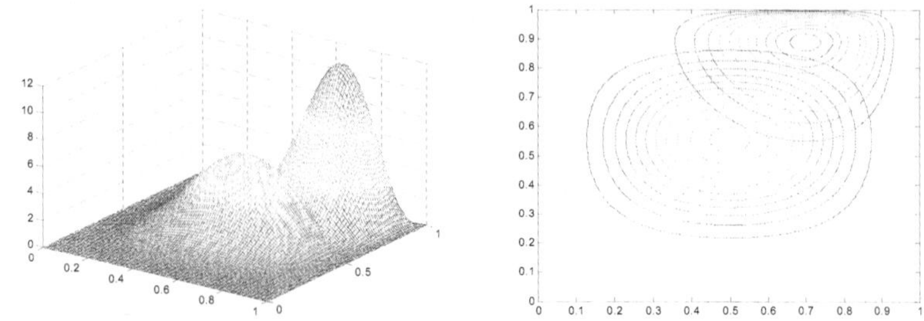

Fig. 23. Combined density distribution, 3D view (left), contour lines (right).

Applying the GMM trainer with 10 Gaussian functions to these distributions, the images in Fig. 24 are obtained representing the set of the 10 Gaussians making up the genuine model; the set of 10 Gaussians making up the impostor model and representing the impostor and genuine models as the weighted sum of each of their Gaussian functions.

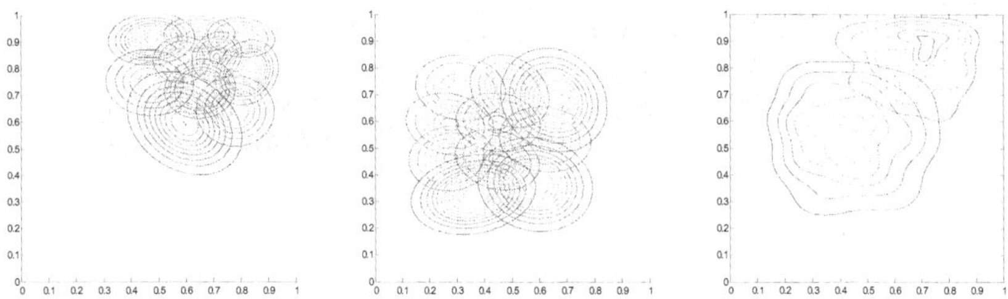

Fig. 24. Contour lines for GMM 10G models. Individual genuine Gaussians (left), individual impostor Gaussians (centre) , both models right (right).

Equivalent representations can be obtained using a GMM with 3 Gaussians

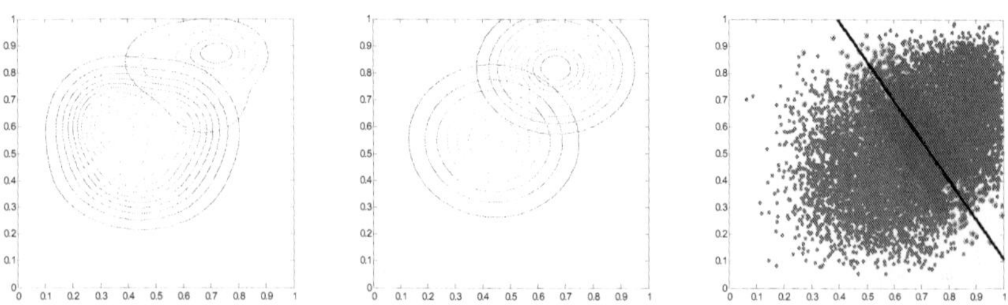

Fig. 25. Contour lines for GMM 3G models (left),and GMM 1G models (centre), vectors and SVM boundary (right).

As in previous sections tests conducted with GMM, SVM and ANN classifiers yield the following results:

Classifier	FAR	FRR	MER
GMM 10G	10.19%	9.64%	9.91%
GMM 3G	10.96%	8.94%	9.96%
GMM 1G	10.92%	9.19 %	10.06%
SVM Linear	10.44 %	9.29 %	9.86 %
ANN (2-1-1)	10.49%	9.285 %	9.87 %

Table 6. Results from classifiers test

14. Match score normalization

As described in earlier sections of the present chapter, the data sources in a system of match score fusion are the result of different monobiometric recognition subprocesses working in parallel. For this reason, the scores yielded are often not homogeneous.

In the most trivial case, the source of this lack is different meaning of the scores, they may represent the degree of similarity between the sample and the model or the degree of disimilarity or directly represent the degree of subsystem confidence in the decision made.

Other sources of non-homogeneous scores are include the different numeric scales or the different value ranges according to which results are delivered, as well as the various ways in which the non-linearity of biometric features is presented. Finally, the different statistical behavior of scores must also be taken into account when performing the fusion. For these reasons, the score normalization, transferring them to a common domain, is essential prior to their fusion. In this way, score normalization must be seen as a vital phase in the design of a combination schema for score level fusion.

Score normalization may be understood as the change in scale, location and linearity of scores obtained by distinct monobiometric recognition subprocesses. In a good normalization schema, estimates of transformation parameters must not be overly sensitive to the presence of outliers (robustness) and must also obtain close to optimal results (efficiency) (Nandakumar et al. 2005)(Jain et al 2005)(Huber 1981)

There are multiple techniques that can be used for score normalization. Techniques such as min-max, z-score, median and MAD, double sigmoid and double linear transformations have been evaluated in diverse publications (Snelick et al. 2003) (Puente et al. 2010).

15. Min-max normalization

Perhaps the simplest of currently existing score normalization techniques is min-max normalization. In min-max normalization, the goal is to reduce dynamic score ranges to a known one (tipically: 0-1) while, at the same time, retaining the form of the original distributions.

For the use this technique, it is necessary that maximum and minimum values (*max, min*) were provided by the matcher prior to normalize. Alternatively, these may be evaluated as the maximum and minimum data of the values used during system training. In this way, a linear transformation is carried out where 0 is assigned to the minimum value 1 to the maximum value[12]. This transformation function is shown in the following formula 16:

[12]Where match scores indicate the difference between a sample and reference, 1 should be assigned to the minimum value and 0 to the maximum.

$$s_k^{'} = \frac{s_k - \min}{\max - \min} \tag{16}$$

Applying this transformation to the observations generated in a previous section of the current chapter (see '13. More realistic distributions'), the following results are obtained:

Classifier	FAR	FRR	MER
GMM 10G	17.62%	17.67 %	17.64%
GMM 3G	17.27%	18.47 %	17.87%
GMM 1G	19.49%	16.36%	17.93%
SVM Linear	18.22%	17.31 %	17.77 %
ANN (2-1-1)	16.70%	18.80%	17.75%

Table 7. Result after min-max normalization

16. Z-score normalization

Due to its conceptual simplicity, one of the most frequently used transformations is z-score normalization. In z-score normalization, the statistical behavior of the match scores is homogenized through their transformation into other scores with a mean of 0 and a standard deviation of 1.

$$s_k^{'} = \frac{s_k - \mu}{\sigma}, where\ \mu = \frac{1}{N}\sum_{i=1}^{N} s_i\ and\ \sigma^2 = \frac{1}{N-1}\sum_{1=1}^{N}(s_i - \mu) \tag{17}$$

Clearly, it is necessary that the mean and standard deviation of the original match scores be known prior to normalization or, as in min-max normalization, they should be estimated from training data.

Z-score distributions do not retain the forms of the input distributions, save in cases of scores with a Gaussian distribution, and this technique does not guarantee a common numerical range for the normalized scores.

Test results from z-score normalization are shown below:

Classifier	FAR	FRR	MER
GMM 10G	10.13%	9.76 %	9.94%
GMM 3G	10.96%	8.96%	9.96%
GMM 1G	10.92%	9.19%	10.06%
SVM Linear	10.38%	9.33%	9.86%
ANN (2-1-1)	9.43%	10.36%	9.90%

Table 8. Result after z-score normalization

17. Median and MAD

The median and MAD (median absolute deviation) normalization technique uses the statistical robustness resulting from the median of a random distribution to make it less sensitive to the presence of outliers. Nevertheless, median and MAD is generally less effective than z-score normalization, does not preserve the original distribution and does not guarantee a common range of normalized match scores.

$$s_k^{'} = \frac{s_k - mediam}{MAD}, \; where \quad MAD = mediam(|mediam - s_i|) \tag{18}$$

Classifier	FAR	FRR	MER
GMM 10G	10.15%	9.64 %	9.89%
GMM 3G	10.40%	9.51%	9.96%
GMM 1G	10.49%	9.44%	9.97%
SVM Linear	9.94%	9.77%	9.85%
ANN (2-1-1)	9.53%	10.28%	9.90%

Table 9. Test results wiht mediam-MAD notmalization

18. Double sigmoid normalization

In one particular study from the literature, a double sigmoid transformation is proposed as a normalization scheme (Cappelli et al. 2000):

$$s_k^{'} = \begin{cases} \dfrac{1}{1 + e^{-2\left(\frac{s_k - d}{r_1}\right)}} & s_k \leq d \\ \dfrac{1}{1 + e^{-2\left(\frac{s_k - d}{r_2}\right)}} & s_k > d \end{cases} \tag{19}$$

According to the double sigmoid normalization technique, match scores are converted to the interval [0,1]. While the conversion is not linear, scores located on the overlap are nevertheless mapped onto a linear distribution (Fahmy et al. 2008).

Classifier	FAR	FRR	MER
GMM 10G	11.40%	10.04%	10.72%
GMM 3G	10.32%	9.94%	10.13%
GMM 1G	11.19%	10.20%	10.70%
SVM Linear	11.10%	9.66%	10.38%
ANN (2-1-1)	10.70%	9.65%	10.06%

Table 10. Test results for double sigmoid normalization.

19. Double linear normalization

Scores yielded by monobiometric classifiers are interpreted as pair of a decision and confidence. The decision, thus, is made according to the location side of the score is located respect to the threshold while confidence is de distance between them. Thus, the greater the distance to the threshold, the greater will be the weight assigned to the score for the final decision.

Generally, this distance does not enjoy a homogeneous distribution for scores of genuine and impostor observations. As a result, in distributions such as that presented in Fig.23, scores of impostor samples tend to have a greater likelihood than those of genuine scores.
In order to compensate that kind of heterogeneity, a transformation has been proposed in (Puente et al. 2010) to make distributions more uniform around the decision threshold:

$$s_k' = \begin{cases} \dfrac{0.5}{th - \min}(s_k - \min) & s_k \le d \\ \dfrac{0.5}{\max - th}(s_k - th) + 0.5 & s_k > d \end{cases} \tag{19}$$

GMM	FAR	FRR	MER
GMM 10G	11.32%	9.61%	9.92%
GMM 3G	11.03%	8.90%	9.96%
GMM 1G	10.92%	9.19%	10.06%
SVM Linear	10.44%	9.29%	9.86%
ANN (2-1-1)	9.98%	9.78%	9.88%

Table 11. Test results for double linear normalization.

20. Conclusions

The principal conclusion that can be drawn from the present chapter is undoubtedly the great advantage provided by score fusion relative to monobiometric systems. In combining data from diverse sources, error rates (EER, FAR and FRR) can be greatly reduced and system stability greatly increased through a higher AUC.
This improvement has been observed with each of the classifiers discussed in the present chapter. Nevertheless and in consideration of comparative studies of normalization techniques and fusion algorithms, it can be noted that the specific improvement produced depends on the algorithms used and the specific case at hand. It is not possible, therefore, to state *a priori* which techniques will be optimal in any given case. Rather, it is necessary to first test different techniques in order to pinpoint the normalization and fusion methods to be used.
One final conclusion that stands out is that improvements in error rates are directly linked to the number of biometric features being combined. From this, it may be deduced that the greater the number of features being fused, the larger the improvement will be in the error rates.

21. References

Huber, P. J. (1981). *Robust Statistics* (John Wiley & Sons, 1981).
Martin, A.; Doddington, G. ; Kamm, T. ; Ordowski, M. & Przybocki M. (1997). *The DET Curve in Assessment of Detection Task Performance*, Eurospeech, 1997, pp. 1895–1898.

Cappelli, R. ; Maio, D. & Maltoni, D. (2000). *Combining fingerprint classifiers*, in: Proceedings of First International Workshop on Multiple Classifier Systems, 2000, pp. 351–361.

Ghosh, J. ; *Multiclassifier Systems: Back to the Future*, Proceedings of the Third International Workshop on Multiple Classifier Systems, p.1-15, June 24-26, 2002.

Kumar, H. G. & Imran M. (2010). *Research Avenues in Multimodal Biometrics*. IJCA, Special Issue on RTIPPR(1):1–8, 2010. Published By Foundation of Computer Science.

Ross, S. ; Nandakumar, K. & Jain, A. K. (2006). *Handbook of Multibiometrics*, Springer Publishers, 1st edition, 2006. ISBN: 0-3872-2296-0.

Ross, A. (2007). *An Introduction to Multibiometrics*, Proc. of the 15th European Signal Processing Conference (EUSIPCO), (Poznan, Poland), September 2007.

Ross, A. & Poh, N. (2009). *Multibiometric Systems: Overview, Case Studies and Open Issues*, in Handbook of Remote Biometrics for Surveillance and Security, M. Tistarelli, S. Z. Li and R. Chellappa (Eds.), Springer, 2009. ISBN: 978-1-84882-384-6.

Tejas, J. ; Somnath, D. & Debasis, S. (2009). *Multimodal biometrics: state of the art in fusion techniques*, International Journal of Biometrics, v.1 n.4, p.393-417, July 2009

Li ; S.Z. (2009). *Encyclopedia of Biometrics*. Springer Science + Business Media, LLC. 2009. ISBN 978-0-387-732002-8.

Villegas, M. & Paredes R. (2009). *Score Fusion by Maximizing the Area under the ROC Curve*. IbPRIA '09 Proceedings of the 4th Iberian Conference on Pattern Recognition and Image Analysis. Springer-Verlag Berlin, Heidelberg 2009. ISBN: 978-3-642-02171-8. DOI 10.1007/978-3-642-02172-5_61.

Marzban, C. (2004). *A comment on the roc curve and the area under it as performance measures*. Technical report, The Applied Physics Laboratory and the Department of Statistics, University of Washington (2004)

Doddington, G. ; Liggett, W. ; Martin, A. ; Przybocki, M. & Reynolds, D. (1998). *Sheeps, Goats, Lambs and Wolves: A Statistical Analysis of Speaker Performance* in the NIST 1998 Speaker Recognition Evaluation, Proc. ICSLD 98, Nov. 1998.

Nandakumar, K., Jain, A.K. & Ross, A.A. (2005). *Score normalization in multimodal biometric systems*. Pattern Recognition 38 (1212), 2270–2285 (2005)

Jain, K. ; Nandakumar K. & Ross, A. (2005). *Score Normalization in Multimodal Biometric Systems*, Pattern Recognition, Vol. 38, No. 12, pp. 2270-2285, December 2005. Winner of the Pattern Recognition Society Best Paper Award (2005).

Ross, A. ; Rattani, A. & Tistarelli, M. (2009). *Exploiting the Doddington Zoo Effect in Biometric Fusion*, Proc. of 3rd IEEE International Conference on Biometrics: Theory, Applications and Systems (BTAS), (Washington DC, USA), September 2009.

Snelick, R. ; Indovina, M. ; Yen, J. & Mink, A. (2003). *Multimodal Biometrics: Issues in Design and Testing*, in Proceedings of Fifth International Conference on Multimodal Interfaces, (Vancouver, 2003), pp. 68–72.

Puente, L. ; Poza, M.J. ; Ruíz, B., García, A. (2010). *Score normalization for Multimodal Recognition Systems*. Journal of Information Assurance and Security, volume 5, 2010, pp 409-417.

Fahmy, M.S.; Atyia, A.F. & Elfouly, R.S. (2008). *Biometric Fusion Using Enhanced SVM Classification*, Intelligent Information Hiding and Multimedia Signal Processing,

2008. IIHMSP '08 International Conference on, vol., no., pp.1043-1048, 15-17 Aug. 2008

Przybocki , M. A. ; Martin, A. F. & Le, A. N. (2006). *NIST speaker recognition evaluation chronicles - Part 2",* Proc. Odyssey 2006: Speaker Lang. Recognition Workshop, pp. 1 2006.

Part 2

Novel Biometric Applications

6

Normalization of Infrared Facial Images under Variant Ambient Temperatures

Yu Lu, Jucheng Yang, Shiqian Wu, Zhijun Fang and Zhihua Xie
School of Information Technology,
Jiangxi University of Finance and Economics
China

1. Introduction

Face recognition, being straightforward, passive and non-invasive comparing with other biometrics such as fingerprint recognition (Yang et al., 2006; Yang & Park, 2008a; Yang & Park, 2008b), has a nature place in biometric technology and computer vision. Currently, most researches on face recognition focus on visual images. The reason is obvious: such sensors are cheap and visual images are widely used. The key problem in visual face recognition is to cope with different appearances due to the large variations both in intrinsic (pose, expression, hairstyle etc) and extrinsic conditions (illumination, imaging system etc). It is difficult to find the unique characteristics for each face, and it is accordingly not easy to develop a reliable system for face recognition by using visual images.

Infrared face recognition, being light-independent and not vulnerable to facial skin, expressions and posture, can avoid or eliminate the drawbacks of face recognition in visible light. Some methods (Buddharaju, et al, 2004, Chen, et al, 2005, Kong, et al, 2005, Wu, et al, 2005A) based on thermal images are proposed for infrared face recognition in last decade. It is highlighted that the infrared images which are the character of the human skins can be affected by ambient temperature, psychological, as well as physiological conditions. Therefore, the recognition systems based on thermal images have the problem that achieving high performance when the test and train images are captured in the same ambient temperature, while the performance is poor if the test and train samples are collected under different temperature (time-lapse data).

To improve the performance on time-lapse data, it is important to normalize the training images and test images. Linear gray transform and histogram equalization are two common methods for image normalization. However, these approaches change the grayscales which represents the skin temperature so that the thermal images have no physical significance. Therefore, both methods are not suitable for normalization of infrared facial images.

In this chapter, we dedicate to provide a novel study on normalization of infrared facial images, especially resulting from variant ambient temperatures. Three normalization methods are proposed to eliminate the effect of variant ambient temperatures. The experimental results show that the proposed methods can increase the robustness of infrared face recognition system and greatly improve its performance on time-lapse data.

The organization of the chapter is as below. In section 2, effect of ambient temperatures on thermal images is analyzed. Three normalization methods are presented in Section 3. An improved Normalized Cross Correlation (NCC) for similarity measurement is proposed in Section 4. A variety of experiments on normalized images are performed in Section 5, and the conclusions are drawn in Section 6.

2. Effect of ambient temperatures on facial thermal patterns

Prokoski et al. (1992) indicated that humans are homoiotherm and capable of maintaining constant temperature under different surroundings. However, it should be highlighted that the so-called "homoiotherm" only refers to the approximately constant temperature in deep body (i.e., the core temperature), whereas the skin temperature distribution fluctuates with ambient temperature, changes from time to time, as shown in Houdas & Ring(1982), Guyton & Hall (1996), Jones & Plassmann (2000). The infrared camera can only capture the apparent temperature instead of deep temperature. It is pointed by Housdas & Ring (1982) that the variations in facial thermograms result from not only external conditions, such as environmental temperature, imaging conditions, but also various internal conditions, such as physiological and psychological conditions. Socolinsky & Selinger (2004A, 2004B) also studied such variations. It is necessary to learn how the thermal patterns vary in different conditions.

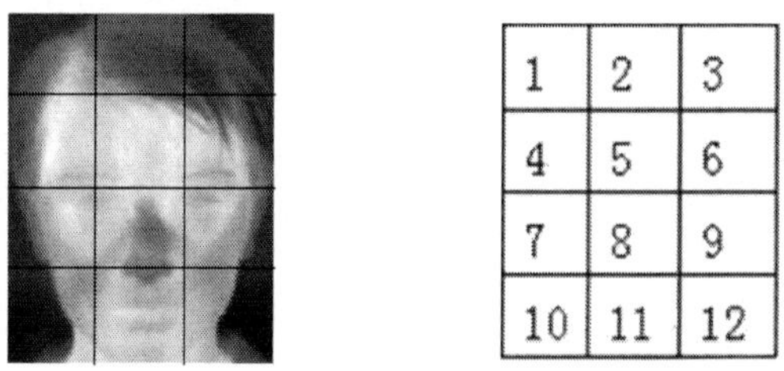

(a) (b)

Fig. 1. An image is divided into blocks.(a)original image (b)block representation

Wu et al. (2005A, 2007) have illustrated that variations in ambient temperatures significantly change the thermal characteristics of faces, and accordingly affect the performance of recognition. It is indicated by Professor Wilder et al. (1996) that the effect of ambient temperatures on thermal images was essentially equivalent to that of external light on visible images. We only consider the effect of ambient temperatures on thermal images in the following analysis.

To study the characteristics of thermal images under different ambient temperatures T_e, a sequence of images are captured and then divided into un-overlap blocks as shown in Figure 1. We can then obtain the mean values from each blocked image.

Table 1 shows the temperature mean in each block of different people in the identical ambient temperature. It is seen that different people have different thermal distributions. Table 2 indicates the means of skin temperatures in each blocked images of the same person. As is shown, the temperatures in each blocked images increase if the ambient temperatures increase. It reveals that the skin temperature variations and the ambient temperatures variations have the same tendency for the same person. In other hand, the skin temperature variations of different people have the same tendency when ambient temperature changes.

Block	Image 1	Image 2	Image 3	Image 4
1	30.62℃	31.46℃	30.26℃	31.32℃
2	31.85℃	32.21℃	31.15℃	32.59℃
3	34.46℃	34.73℃	34.32℃	34.05℃
4	33.58℃	33.24℃	33.35℃	33.17℃
5	35.02℃	34.53℃	35.04℃	34.96℃
6	33.43℃	33.74℃	33.50℃	33.49℃
7	33.34℃	32.83℃	33.58℃	32.97℃
8	35.02℃	34.56℃	34.61℃	33.63℃
9	33.62℃	33.35℃	33.45℃	32.83℃
10	32.14℃	32.83℃	32.89℃	32.75℃
11	34.47℃	34.11℃	34.39℃	34.40℃
12	32.41℃	32.67℃	33.06℃	33.01℃

Table 1. Skin temperature mean in each block of different people when ambient temperature is $T_e = 25.9$

Block/ T_e	$T_{e1} = 24$ ℃	$T_{e2} = 25.9$ ℃	$T_{e3} = 28$ ℃	$T_{e4} = 28.5$ ℃
1	29.74℃	31.71℃	32.96℃	33.12℃
2	30.91℃	31.88℃	33.37℃	33.58℃
3	30.22℃	30.66℃	33.53℃	33.65℃
4	31.98℃	32.97℃	34.04℃	34.18℃
5	34.51℃	34.52℃	35.38℃	35.52℃
6	32.35℃	33.03℃	34.39℃	35.69℃
7	33.06℃	33.44℃	34.81℃	34.96℃
8	34.25℃	34.25℃	35.41℃	35.63℃
9	33.21℃	33.29℃	35.06℃	35.29℃
10	32.62℃	32.67℃	34.17℃	34.45℃
11	34.25℃	34.42℃	35.49℃	35.85℃
12	30.94℃	31.93℃	34.09℃	34.48℃

Table 2. Skin temperature mean in each block of same person in different ambient temperatures, where T_e is the ambient temperature

What's more, with the increasing of ambient temperature, the skin temperature variations of each blocked images are different. In order to study the relation between skin temperature variations of each blocked images and ambient temperatures, the maximum and minimum values in thermal images are used to analyze.

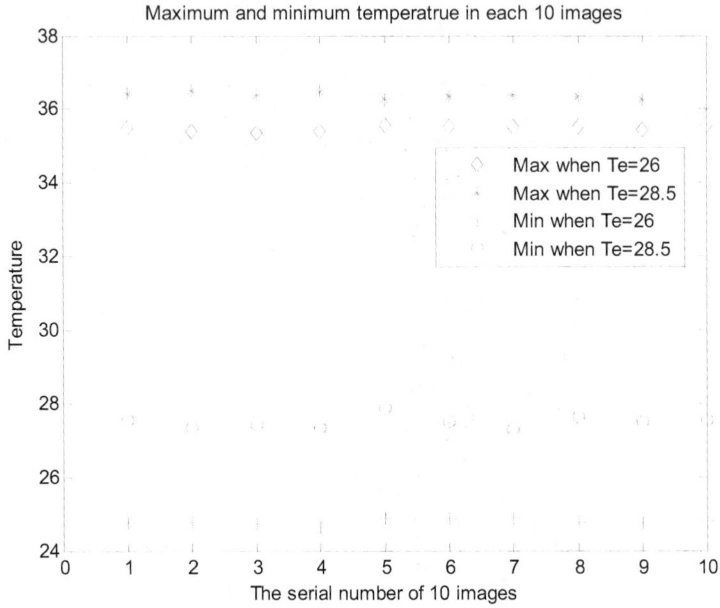

Fig. 2. Maximum and minimum values in thermal facial images captured under different ambient temperatures

As shown in Figure 2, ten thermal images are collected in ambient temperatures 26°C and 28.5°C respectively. The maximal and minimal temperature in thermal images, are relatively stable. They all increase when ambient temperatures rise from 26°C to 28.5°C. It's worth noting that the variations of maximums are lower than those of minimums, which implies the temperature variations are different in blocks.

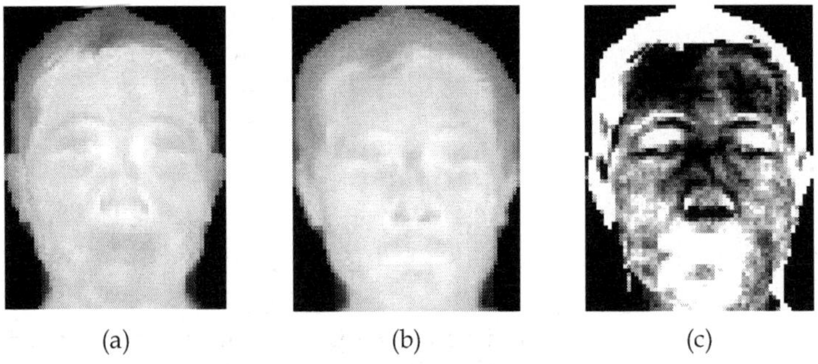

Fig. 3. Thermal images in different environments and their subtraction

In Figure 3, the input image (a) and reference image (b) (the thermal images of same person) were captured when ambient temperatures are 26°C and 28.5°C, which are geometrically normalized to size 80X60. The background areas are set to 0.000001°C. Image (c) is the subtraction of image (b) and image (a). According to the characteristics of thermal images, the point which is brighter (white) has higher gray value, while the gray value of dark point is lower. Through observation of the images captured under different ambient temperatures, reference image, and their subtracted images, we find:

1. The temperatures of forehead and nose parts are higher than those of the facial edge, hair. Moreover, the higher the skin temperature, the smaller its temperature variations while ambient temperature changing;
2. In a thermal image, the difference of skin temperature among different parts is about 10°C;
3. Due to the effect of breathing pattern, the temperature fluctuation in mouth part is larger than any other parts.

3. Normalization of infrared facial images

Wilder et al. (1996) illustrated that the effect of ambient temperatures on thermal images is essentially equivalent to that of external light on visible images. Therefore, analogous to the methods solving illumination in visible face recognition, the methods to eliminate ambient temperatures variations can be divided into the following three kinds: the methods based on transformation, invariant features and temperature model.

In order to eliminate the effect of ambient temperatures on infrared imaging and improve the robustness and recognition performance, we convert the image captured in unknown ambient temperature into the reference ambient temperature. This is the so-called normalization of infrared images, which is the key issue addressed in this section.

The traditional infrared image normalization methods are classified as linear gray transform methods (Chen, et al, 2005, Zhang, et al, 2005) and equilibrium methods (Song, et al, 2008). The former methods expand the range of gray levels to a dynamic range, with some linear methods. These methods aim to strengthen the detail characteristics of images. The second equilibrium methods are also dedicated to improve the information which people are interested in. These methods are not suitable for application to infrared image normalization, because these methods change the distribution of facial temperatures. Kakuta et al. (2002) have proposed a method for infrared image normalization which utilizes the skin temperature subtraction to convert infrared images, as shown in Figure 4.

This method can convert IR images obtained under various thermal environments into those under reference environments, in which the IR images contain foot, chest, hand and forearm. Every point in image gets the same offset. However, it should be noted that the offset of each point in face is different, as indicated in Section 2. Such processing will lead to wrong temperature variation. Therefore, we should use different normalized methods to deal with the infrared facial images. Three methods will be introduced below separately.

3.1 Normalization based on transformation (TN)

Normalization of infrared facial images based on transformation aims to reduce the effect of ambient temperatures using transformation. The proceedure using the block and least squares method is explained in details in Wu, et al, 2010. Firstly, the images collected under different ambient temperatures were divided into some block images, the values of the change of ambient temperatures and the change of corresponding temperature in face were obtained, which were used to be fitted into a function through the least squares method. Then, each

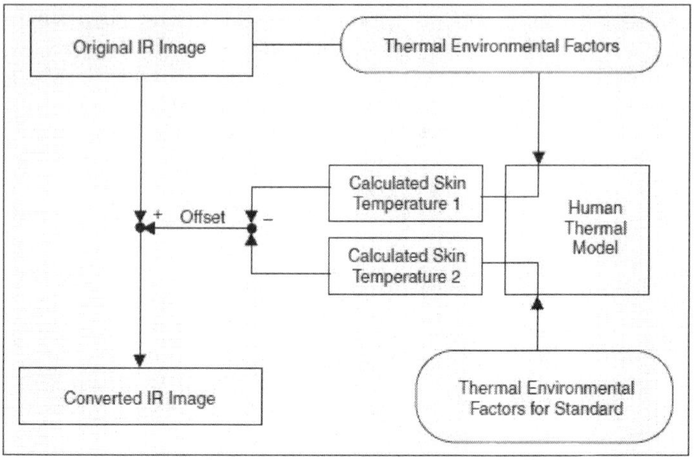

Fig. 4. Conversion of infrared images (Kakuka, et al., 2002)

block image was normalized by the relevant function and the whole image could be normalized into the reference condition. Specific processes are as follows.

1. Obtaining q images captured in every i different ambient temperatures $T_{e1}, T_{e2}, ..., T_{ei}$ and blocking them;
2. For each image containing $m \times n$ blocks, the means $T_1, T_2, ..., T_q$ of q reference images and their mean T_{mean} are calculated. With this calculation, the means $T_{i1}, T_{i2}, ..., T_{iq}$ of q reference images in i different ambient temperatures and their mean T_{imean} are obtained;
3. $|T_e - T_{ei}|$ and $|T_{mean} - T_{imean}|$ correspond to the environmental temperature variation and skin temperature variation respectively. The function $y_j = f_j(t)$, which represents the relation between the variations of ambient temperatures and skin temperatures, is obtained using the block and least squares method. Finally, the function $y_j = f_j(t)$ is used to execute temperature normalization.

$$g_j' = g_j - f_j(t) \tag{1}$$

   In formula (1), t is the variation of ambient temperatures and $f_j(t)$ is variation in facial temperature. g_j represents the j th blocked image in image g under different ambient temperatures, while g_j' is the image after temperature normalization.
4. Each blocked image can be normalized by the same process. With these executions, the thermal image captured in unknown ambient temperature can converted into the reference environmental temperature.

3.2 Normalization based on invariable features (IFN)

Principal Component Analysis (PCA) is one of the most successful feature extraction methods, which frequently applied to various image processing tasks. The magnitudes of eigenvalues correspond to the variances accounted for the corresponding component. For example, discarding principal components with small eigenvalues is a common technique

for eliminating small random noise. It has been theoretically justified from the point of spherical harmonics of view that eliminating three components with largest eigenvalues has been used to handle illumination variations (Ramamoorthi, 2002). In addition, the traditional PCA algorithm operates directly on the whole image and obtains the global features, which are vulnerable to external ambient temperatures, psychological and physiological factors. Such global procedure cannot obtain enough information because the local part of the face is quite different. It is necessary to extract more detailed local features. Xie et al. (2009) proposed a weighted block-PCA and FLD infrared face recognition method based on blood perfusion images, which highlighted the contribution of local features to the recognition and had a good performance when the test samples and training samples were captured in the same environmental temperatures.
Temperature normalization using block-PCA is proposed in this section (Lu, et al, 2010). In order to take full advantage of both the global information and the local characteristics of facial images, the images are partitioned into blocks. Then PCA is performed on each block and the component corresponding to the biggest eigenvalue of each block is discarded to eliminate the ambient effect.

3.2.1 PCA feature extraction

The main idea of PCA algorithm is using a small number of characteristics of features to describe the samples, reduce feature space dimension while the required identifying information are reserved as much as possible. These steps are briefly descripted below.

3.2.1.1 PCA algorithm

Assume that a set $X = \{x_1, x_2, \cdots, x_N\} \in \mathbb{R}^{u \times N}$ of N training samples is given, where x_i is column vector which connects the rows of image matrix x_i, u is the number of pixels.
The average face of the training set is as follow:

$$\bar{M} = \frac{1}{N} \times \sum_{i=1}^{N} x_i \tag{2}$$

The distance from each face to the average face is as follow:

$$M_i = x_i - \bar{M}, \quad i = 1, 2, \cdots N \tag{3}$$

We assume the matrix $M = (M_1, M_2, \cdots, M_N)$ and the corresponding covariance matrix is $S = M \times M^T$. The next task is to select the v largest eigenvalues with corresponding eigenvectors and form a projection space W_{opt}. Then the original image vector of dimension u is projected to the space of dimension v, where $u>v$. The eigenvector after projection is as follow:

$$y_k = W_{opt}^T \times x_k, \quad k = 1, 2, \cdots, N \tag{4}$$

There are usually two methods to determine the number of principal components, which is the number v of eigenvalues of the covariance matrix.
The first method is determined by the pre-specified information compression ratio ($\eta \le 1$):

$$\eta = \frac{\lambda_1 + \lambda_2 + \lambda_3 + \cdots + \lambda_v}{\lambda_1 + \lambda_2 + \lambda_3 + \cdots + \lambda_N} \tag{5}$$

Where $\lambda_1 > \lambda_2 > \lambda_3 > \cdots > \lambda_v > \cdots > \lambda_N$ is the eigenvalues of S.
The second method that directly reserves the $N-1$ largest eigenvalues is a common method, where $v = N-1$.

3.2.1.2 Improved PCA algorithm

Although the eigenvalues and their corresponding eigenvectors produced by PCA take into account all the differences between the images, it cannot distinguish these differences caused by the human face or by the external factors. If the test samples and training samples are not collected at the same time, the ambient temperatures, psychological and physiological factors will affect the infrared image greatly when the images are collected. Some variations affected by these factors will arise in the principal components with the corresponding eigenvectors. In addition, each principal component represents a kind of image feature. Some of these features pertain to the subjects depicted in the thermal images, and others represent irrelevant sources of image variations. Thus, it is natural to handle these principal components to reduce the effects caused by the ambient temperatures, psychological and physiological factors on infrared images.
After determining the number of principal component, eliminating three components with largest eigenvalues has been used to handle illumination variations (Ramamoorthi, 2002). Although the effect of ambient temperatures on thermal images is essentially equivalent to that of external light on visible images, it cannot ensure the three components with the largest eigenvalues in thermal images must contain the useless information for identification. The principal components may contain feature information which distinguishing the images from different classes. Therefore, a method calculates the standard deviation of each principal component is proposed to estimate the effects of ambient temperatures, psychological and physiological factors on each principal component. The procedure describes as follows:
After obtaining the projection matrix, ten test samples, as an example, of the same person are extracted and projected into the projection matrix. The images with high dimension are mapped to the space with low dimension. Then the standard deviation of each principal component belongs to the test samples are calculated and shown as follows:
$\sigma = [53.1,\ 21.1,\ 31,\ 30,\ 35.2,\ 44.2,\ 83.1,\ 97.9,\ 49.1,\ 17.1,\ \cdots]$
We can see from the standard deviations of principal components that the standard deviations of the first, 7th and 8th principal component are very large in comparison with other standard deviations of the ten principal components, which means the image information in these three principal component change much more strenuous with the variations of ambient temperatures, psychological and physiological factors. These factors account for the differences between test samples and training samples collected in different time. The standard deviations of the second and the third principal component are smaller than any other standard deviations of the ten principal components. The useful feature information for identification will lose if these two principal components are discarded. For a human face, the energies of first three or five components are 90% and 94% respectively. The remaining principal components contain less information comparing with the preceding principal components. Therefore, instead of discarding three components with largest eigenvalues, the component with the biggest eigenvalue is discarded in this section to reduce the effects of the ambient temperatures, psychological and physiological in infrared images.

The v largest eigenvalues and their corresponding eigenvectors are obtained by the covariance matrix S firstly. Then the eigenvector corresponding to the biggest eigenvalue is removed and the reminders of the v-1 eigenvalues with the eigenvectors constitute a new projection space W_{new}. In this way, the original u-dimensional images vector project onto the (v-1)-dimensional space have better separability. The eigenvector after projection is as follows:

$$y'_k = W_{new}^T \times x_k, \quad k = 1, 2, \cdots, N \tag{6}$$

3.2.2 Feature extraction by block-PCA

Traditional PCA algorithm is performed on the original images directly, which obtained are the global features. For IR images, only partial areas of the temperature information on the face change obviously when ambient temperature changes. The global features extracted tend to strengthen the part changes obviously, and neglect some other information representing the image classes.

Apart from the effect of uniform illumination on visible recognition system, the infrared images affecting by ambient temperatures are an overall process; all the human faces would change with the variation of the ambient temperatures. In fact, different parts of IR images have different variations with the changes of ambient temperatures. It does not primarily eliminate the effect of environment on the local thermal images by removing the largest principal component of the whole images.

Therefore, the idea of block-PCA is proposed to extract local features. It is assumed that a set $X = \{x_1, x_2, \cdots, x_N\}$ of N training samples is given. Each image is divided into $m \times n$ blocks and the size of each sub-block is $p \times q$.

$$x = \begin{bmatrix} x_{11}, x_{12}, \cdots, x_{1n} \\ x_{21}, x_{22}, \cdots, x_{1n} \\ \vdots \;, \vdots \;, \vdots \;, \;\vdots \\ x_{m1}, x_{m2}, \cdots, x_{mn} \end{bmatrix} \tag{7}$$

Using the method of information compression ratio, the eigenvectors of each block image is obtained and then the eigenvector with the largest eigenvalues of each block image is discarded. The remainders of eigenvectors of each block constitute a new eigenvector to be used for feature extraction.

The new combined eigenvector not only reflects the global features of images, but also reflects the global features. Moreover, it can reduce the effects of ambient temperatures, physiological and psychological factors, which enhance the robustness of the recognition system.

3.3 Normalization based on temperature model (TMN)

3.3.1 0-1 normalization

As aforementioned in section 2, the higher the skin temperature, the smaller its temperature variation while ambient temperature changes. Therefore, every point in thermal image may have a weight coefficient to represent its own temperature variation weight. The high temperature corresponds to small weight coefficient while the low temperature point has a

high weight coefficient. Therefore, a weighted linear normalization method of infrared image is proposed in this section (Wu, et al 2010). The weight coefficients of each point in the face are first obtained through the improved 0-1 standardization. Then each coefficient is used to temperature normalization.
Assume a sample $x = (x_1, x_2, ..., x_m)$, where x_m is one of the skin temperature in face.

$$x_k \to f(x_k) = (x_{\max} - x_k)/(x_{\max} - x_{\min}) = \varpi_k \tag{8}$$

In which ϖ_k is corresponding to weight coefficient while ambient temperature changes, $x_{\max} = \max(x) = \max(x_1, x_2, ..., x_m)$, $x_{\min} = \min(x) = \min(x_1, x_2, ..., x_m)$.
As shown in the formula (8), the point x_k whose skin temperature is high has a high gray value, which results in small weight coefficient, and vise reverse.

3.3.2 Temperature normalization based on maximin model

Suppose two thermal images $f(x, y)$ and $g(x, y)$ to be captured in different ambient temperatures, in which $f(x, y)$ is collected in reference temperature T_{e1} while $g(x, y)$ is gathered in unknown temperature T_{e2}. State parameters are extracted from $g(x, y)$ and the ambient temperature T_{e2} is obtained. Then the difference of ambient temperature of the two thermal images is calculated. $\Delta T = T_{e1} - T_{e2}$, each point in test image has the following calculation:

$$\begin{cases} x_k = x_k - \dfrac{x_{\max} - x_k}{x_{\max} - x_{\min}} * |\Delta T|, \text{ if } T_{e2} > T_{e1} \\ x_k = x_k + \dfrac{x_{\max} - x_k}{x_{\max} - x_{\min}} * |\Delta T|, \ else \end{cases} \tag{9}$$

The result obtained by the maximin model is anastomotic with the analysis shown in Section 2.

4. Improved Normalized Cross Correlation for similarity measurement

The purpose of temperature normalization is to convert the images captured under different environmental temperature into the images collected in reference temperature. The images after temperature normalization have the similar ambient temperature and temperature distribution. Therefore, the similarity of test image and reference image can be used to test the performances of the proposed normalization methods.
Normalized Cross Correlation (NCC) is defined as follow:

$$NCC = \frac{\sum_{i=1}^{m}\sum_{j=1}^{n} S(i,j) \times T(i,j)}{\sqrt{\sum_{i=1}^{m}\sum_{j=1}^{n} S(i,j)^2} \times \sqrt{\sum_{i=1}^{m}\sum_{j=1}^{n} T(i,j)^2}} \tag{10}$$

where S is the reference image, T is the test image. Both S and T have same size m×n. Obviously, higher value of NCC indicates more similar of the two images. NCC equals to 1 when the two images are completely same.

It is noted that the NCC method is pixelwised, which requires that the images have the exact location for measurement. As faces are 3D moving subjects, which poses 3D rotation and deformation, it is impossible to achieve exact one-to-one correspondence computation even the two faces images are well segmented. Namely, the point $I(x,y)$ corresponding to nose in test image is not the same location of nose in reference image. Therefore, the traditional NCC method is not accurate for face measurement. An improved NCC method is proposed to test the normalized performance.
Assuming S, T are normalized reference image and normalized test image. Each image is blocked into p×q and M is the mean of block image. The improved NCC is defined as follow:

$$NCC_{new} = \frac{\sum_{i=1}^{p}\sum_{j=1}^{q}(MS(p,q)-T_{e1})\times(MT(p,q)-T_{e1})}{\sqrt{\sum_{i=1}^{p}\sum_{j=1}^{q}(MS(p,q)-T_{e1})^2}\times\sqrt{\sum_{i=1}^{p}\sum_{j=1}^{q}(MT(p,q)-T_{e1})^2}} \tag{11}$$

$$NCC_{nor} = \frac{\sum_{i=1}^{p}\sum_{j=1}^{q}(MS(p,q)-T_{e1})\times(MT_{nor}(p,q)-T_{e1})}{\sqrt{\sum_{i=1}^{p}\sum_{j=1}^{q}(MS(p,q)-T_{e1})^2}\times\sqrt{\sum_{i=1}^{p}\sum_{j=1}^{q}(MT_{nor}(p,q)-T_{e1})^2}} \tag{12}$$

In which $MS(p,q)$ is the mean temperature value of one block image, T_{e1} is the ambient temperature when reference image S is captured. If $NCC_{nor} - NCC_{new} > 0$, It reveals that the test image has more similar temperature distribution with reference image. On the other hand, it also shows the temperature normalized method can effectively reduce the effect of external temperature on thermal facial recognition system.

5. Experimental results and discussions

5.1 Infrared database

Currently, there is no international standard infrared facial database stored in temperature. We captured the infrared images using the ThermoVision A40 made by FLIR Systems Inc. The training database comprises 500 thermal images of 50 individuals which were carefully collected under the similar conditions: environment under air-conditioned control with temperature around 25.6~26.3°C. Each person stood at a distance of about 1 meter in front of the camera. The ten templates are: 2 in frontal-view, 2 in up-view, 2 in down-view, 2 in left-view, and 2 in right-view. As glass is opaque to long-wavelength infrared, subjects are required to remove their eyeglasses in database collection. The original resolution of each image is 240×320. The size turns to be 80×60 after face detection and geometric normalization, as illustrate in Figure 5 and Figure 6.
Time-lapse data were collected from one month to six months. There are totally 85 people involved, some are included in training subjects and some are not. This time, the subjects are allowed to wear eyeglasses. The data were acquired either in air-conditioned room from morning (24.5~24.8 °C), afternoon (25.4~26.3 °C) to night (25.4~25.8 °C) or outdoor, where the ambient temperature was 29.3~29.5 °C. It is noted that these data were collected on the condition that the subjects had no sweat. The total test images is 1780.

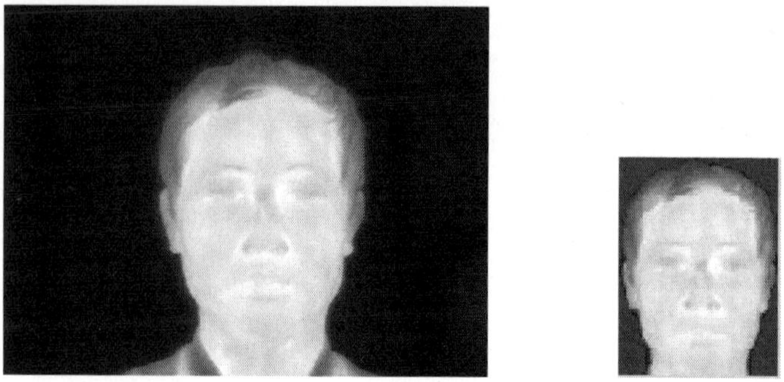

(a) Original image (b) Geometrically normalized image

Fig. 5. Original image and its geometrically normalized image. (a) Original image. (b) Geometrically normalized image

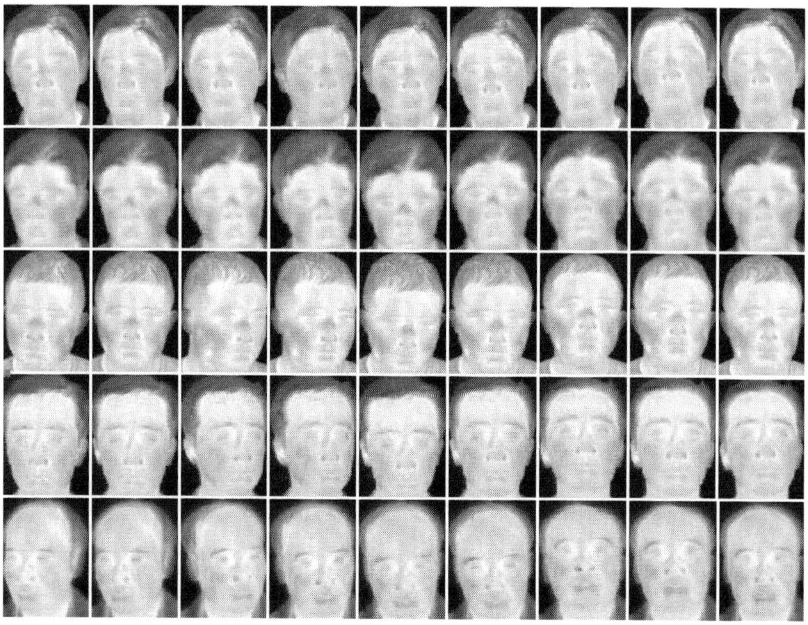

Fig. 6. Some samples after geometrical normalization

5.2 NCC experimental results

In Figure 7, image (c) represents the normalized image. It is shown that the test image and its normalized image look similar that it is difficult to judge the effect of the proposed normalization method.

We can see from Figure 8 that the NCC after TMN method is higher than the original NCC without normalization.

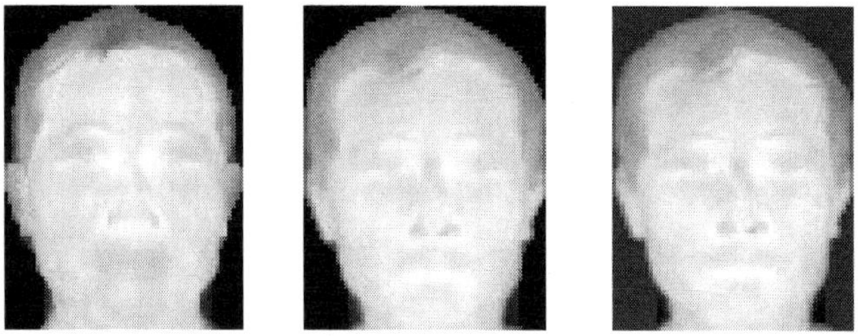

(a) (b) (c)

Fig. 7. (a) training image (b) test image and (c) normalized image using TN method

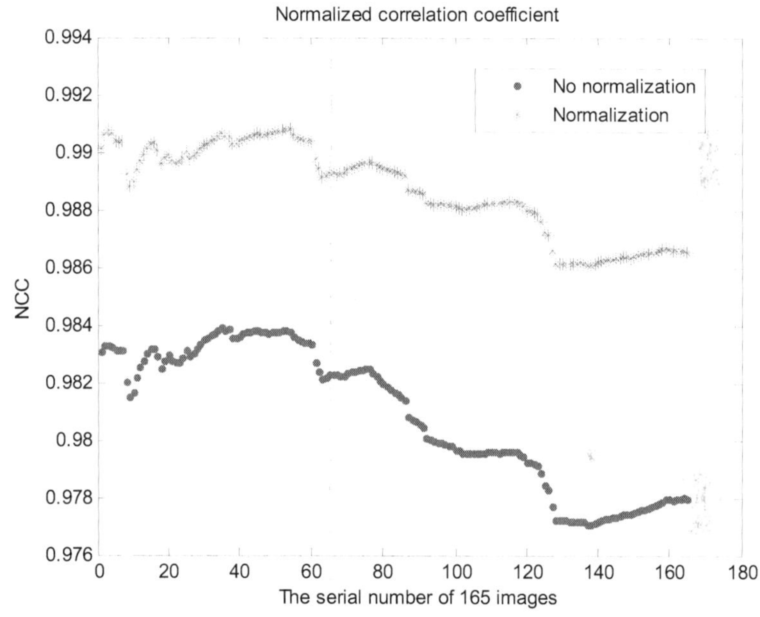

Fig. 8. NCC of training image with the original test images and their normalized images using TMN method

Further, we verify the performance of normalization via recognition rate. In our expreiments, ,linear least squares model and polynomial least squares modedl are applied to verify the influence of different matching curves (Wu, S. Q. et al, 2010a). After temperature normalization, the thermal images are converted into simplified blood perfusion domain to get stable biological features (Wu et al, 2007).

$$\omega = \frac{\varepsilon\sigma(T^4 - T_e^4)}{\alpha c_b(T_a - T)} \tag{13}$$

Where T is the temperatures in images, T_a, T_e, are the core temperature and ambient temperature respectively. ω is the blood perfusion, ε, σ, α and c_b are four constants.
To extract features, PCA and discrete wavelet transform (DWT) (Wu, S. Q. et al, 2009) are chosen to test the effect of normalization. 3NN and Euclidean distance are selected in classification and recognition. Experimental results are shown in the Table 3.

5.3 Experimental results using TN method

It is shown in Table 3 that the normalization using block size 80×60 did not improve the performance, because the skin temperature variations vary with different facial parts. However, the normalized methods using small blocks have better performance, whatever which method are chosen. Moreover, the normalized method using linear least squares achieves better performance than that using polynomial least squares. It is highlighted that smaller block size results in higher performance. But the computation increases along with the increasing of the number of block images.

Feature extraction method	Block size	Pre-normalization	Linear normalization	Polynomial normalization
PCA	80×60	56.97%	56.97%	56.97%
PCA	40×20	56.97%	59.39%	60.61%
PCA	20×20	56.97%	63.64%	70.30%
DWT	80×60	49.70%	49.70%	49.70%
DWT	40×20	49.70%	53.94%	56.46%
DWT	20×20	49.70%	55.76%	63.64%

Table 3. Recognition rate in comparing pre- and post-normalization

We also using 2D linear discriminant analysis (2DLDA) (Wu, S. Q. et al, 2009), DWT+PCA (Wu, S. Q. et al, 2009), and PCA+LDA (Xie, Z. H et al, 2009) for feature extraction. The recognition rates with or without normalization are shown in Table 4.

Feature extraction method	Block size	Pre-normalization	Polynomial normalization
2DLDA	20×20	61.82%	71.52%
DWT+PCA	20×20	50.30%	57.58%
PCA+LDA	20×20	55.76%	61.82%

Table 4. Recognition rate with/without normalization

As shown is Table 4, no matter which feature extraction method is chosen, the thermal recognition system after temperature normalization has better performances.

5.4 Experimental results using IFN method

Fig.9 shows how the recognition rates vary with number of eigenvectors removed, and Table 5 demonstrates the recognition rates when discarding different components.

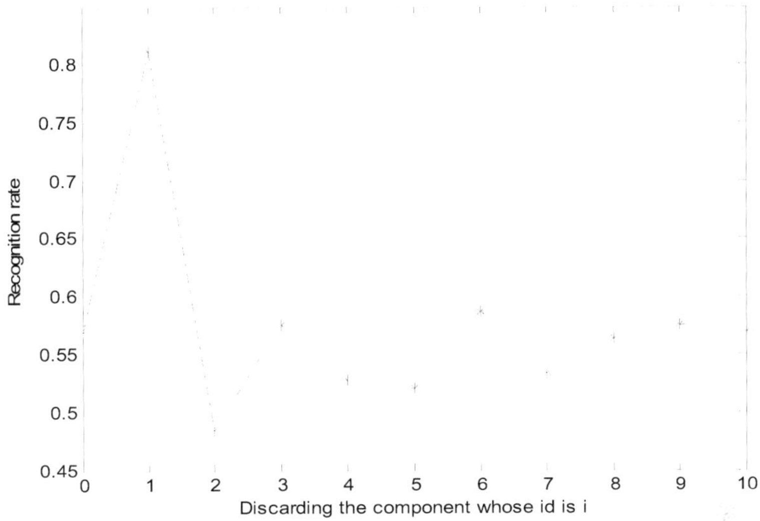

Fig. 9. Recognition rate when discarding some components

As shown in Table 5 that the recognition performance has remarkable improvement by getting rid of the principal components with the largest eigenvalues. Moreover, the performance by discarding three principal components with the largest eigenvalues is better than that discarding two principal components with the largest eigenvalues, which verifies that the second principal component is useful for identification and it cannot be discarded. These results agree with the previous work in Section 3.2.

s	v	Recognition rate
0	499	56.97%(94/165)
1	498	81.21%(134/165)
2	497	74.55%(123/165)
3	496	80.61%(133/165)
4	495	80.61%(133/165)
5	494	77.58%(128/165)

Table 5. Recognition varies with the number of discarded components with biggest components, where v is the total number of principal components and s is the component number to discard

As can be seen from Figure 10, the recognition rate with $\eta_{\min} = 0.9$ is only 3.64 percentages lower than that with $\eta_{\max} = 0.99$. Hence, the block-PCA experiments are operated in $\eta_{\min} = 0.9$ and $\eta_{\max} = 0.99$. Sub-block size is chosen to be 40×30，20×30，20×20 and 10×10. Since the positions in each block-image are different and their temperature changes are different with the variations of the ambient temperatures, the method that determined the number of eigenvectors in this chapter is information compression ratio. Accordingly, the total number of eigenvalues after the PCA operation on each block-image is different. After the block-PCA performed, and the component with the biggest eigenvalue of each block-image is discarded, the reminders of the eigenvectors combine into a new eigenvector by the tactic of the block-images and are used for recognition.

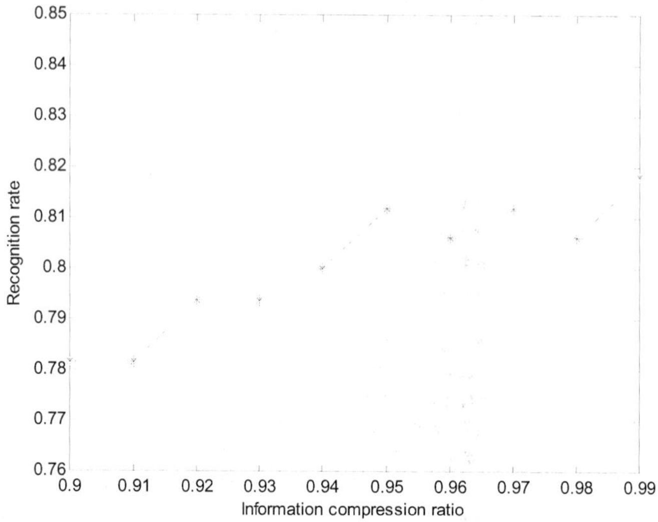

Fig. 10. Recognition rate by deleting the biggest component when the information compression rate $\eta = 0.99$

η	$p \times q$	Recognition rate
0.99	80×60	81.82%(135/165)
0.99	40×30	83.03%(137/165)
0.99	20×30	86.06%(142/165)
0.99	20×20	85.45%(141/165)
0.99	10×10	84.24%(139/165)

Table 6. Recognition rate with different block size in $\eta_{\max} = 0.99$

Table 6 and Table 7 show the recognition rate in different block size, where η is the information compression ratio, $p \times q$ is the size of the sub-block.

η	$p \times q$	Recognition rate
0.9	80×60	78.18%(129/165)
0.9	40×30	81.82%(135/165)
0.9	20×30	87.27%(144/165)
0.9	20×20	86.06%(142/165)
0.9	10×10	84.24%(139/165)

Table 7. Recognition rate with different block size in $\eta_{\min} = 0.9$

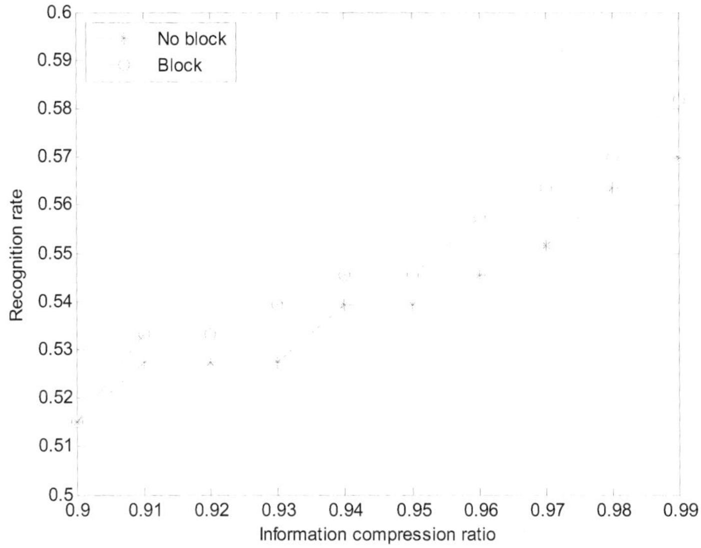

Fig. 11. Recognition rate of normalization based on block and normalization based on non-block.

It is shown in Table 6 and Table 7 that the recognition performance is best when the size of the sub-block is 20×30 in the case of $\eta_{\min} = 0.9$ and $\eta_{\max} = 0.99$. This is because the own temperature change of each sub-block is different with the change of the ambient temperatures. The different parts in the face can locate well in each sub-block when the size of sub-block is 20×30.

In order to show the contribution of the block on recognition rate, the results using block and without using block are shown in Figure 11. No matter how much is the information compression rate, the performance when use the block is better than those without using block. It is seen from all the experimental results that the recognition rates using block-PCA and discarding the principal components are significantly higher than those using PCA directly. Therefore, the method using block-PCA proposed here can fully utilize the local characteristics of the images and thus improve the robustness of the infrared face recognition system.

5.5 Experimental results using TMN method

As shown in Table 8, the normalized method without weighting coefficients does not improve the performance. This is because the skin temperature variations of some points in thermal images are lower than the ambient temperature variation. Such processing leads to great change of skin temperatures, especially in the parts with high temperatures. By using the normalized method with weighted coefficients, the performances are all improved greatly, no matter which kind of features are extracted. It illustrates the normalization method using maximin model can improve the robustness and performance of the system.

Feature extraction method	Pre-normalization	Normalization without weigh coefficient	Normalization with weigh coefficient
PCA	56.97%(94/165)	55.15%(91/165)	89.09%(147/165)
DWT	49.70%(82/165)	49.70%(82/165)	86.06%(142/165)
2DLDA	61.82%(102/165)	52.73%(87/165)	78.79%(130/165)
DWT+PCA	50.30%(83/165)	47.88%(79/165)	81.21%(134/165)
PCA+LDA	55.76%(92/165)	50.91%(84/165)	76.36%(126/165)

Table 8. Recognition rate using TMN method with different feature extraction methods

6. Conclusion

This chapter dedicates to eliminate the effect of ambient temperatures on thermal images. The thermal images of a face are severely affected by a variety of factors, such as environmental temperature, eyeglasses, hairstyle and so on. To alleviate the ambient temperatures variations, three methods are proposed to normalize thermal images. The normalization based on transformation uses the function obtained by the block and least squares method. Features which do not change with ambient temperatures are extracted. The third method aims to normalize image through the maximin model. The extensive experiments demonstrated that the recognition performances with temperature normalization are substantially better than that with no temperature normalization process.

It should be highlighted that psychological (e.g., happy, angry, and sad etc) and physiological (e.g., fever) conditions also affect the thermal patterns of faces. How to analysis and eliminate these variations will be our future work.

7. Acknowledgement

This work was partially supported by the National Natural Science Foundation of China (No. 61063035), and it is also supported by the Merit-based Science and Technology Activities Foundation for Returned Scholars, Ministry of Human Resources of China.

8. References

Buddharaju, P.; Pavlidis, I. & Kakadiaris, I. A. (2004). Face recognition in the thermal infrared spectrum, *Proceedings of IEEE International Conference on Computer Vision and Pattern Recognition Wokrshop*, pp. 133, Washington DC, USA, 2004

Chen, X.; Flynn, P. J. & Bowyer, K. W. (2005). IR and visible light face recognition, *Computer Vision and Image Understanding*, Vol. 99, No. 3, pp. 332-358, 2005

Chen, X. X.; Lee, V. & Don, D. (2005). A simple and effective radiometric correction method to improve landscape change detection across sensors and across time, *Remote Sensing of Environment*, Vol. 98, No. 1, pp. 63-79, 2005

Guyton, A. C. & Hall, J. E. (1996). Textbook of Medical Physiology, *9th ed., Philadelphia: W.B.Saunders Company*, 1996

Houdas, Y. & Ring, E. F. J. (1982). Human Body Temperature: Its Measurement and Regulation. *New York: Plenum Press*, OSTI ID: 6601231, 1982

Jones, B. F. & Plassmann, P. (2002). Digital infrared thermal imaging of human skin, *IEEE Enginerring in Medicine & Biology Magazine*, Vol. 21, No. 6, pp.41-48, 2002

Kakuta, N; Yokoyama, S, & Mabuchi, K. (2002). Human thermal models for evaluating infrared images. *IEEE Medicine & Biology Society*, pp.65-72, ISBN 0739-5175, 2002

Kong, S. G.; Heo. J.; Abidi, B. R.; Paik, J. & Abidi, M. A. (2005). Recent advances in visual and infrared face recognition - a review, *Computer Vision and Image Understanding*, Vol. 97, No. 1, pp, 103-105, 2005

Lu, Y.; Li, F.; Xie, Z. H. et al. (2010). Time-lapse data oriented infrared face recognition method using block-PCA, *Proceedings of 2010 International Conference on Multimedia Technology*, pp. 410-414, Ningbo, China, October, 2010

Prokoski, F. J.; Riedel, B. & Coffin, J. S. (1992). Identification of individuals by means of facial thermography, *Proceedings of IEE Int. Conf. Security Technology, Crime Coutermeasures*, pp. 120-125, Atlanta, USA, Oct. 1992

Ramamoorthi R. (2002). Analytic PCA construction for theoretical analysis of lighting variability, including attached shadows, in a single image of a convex Lambertian object. *IEEE Transactions on Pattern Analysis and Machine Intelligence*, Vol. 24, pp. 1322-1333, 2002.

Socolinsky, D. A. & Selinger, A. (2004A). Thermal face recognition in an operational scenario, *Proceedings of IEEE Conference on Computer Vision and Pattern Recognition*, pp. 1012-1019, Washington DC, USA, 2004

Socolinsky, D. A. & Selinger, A. (2004B). Thermal face recognition over time, *Proceedings of Int. Conf. Pattern Recognition*, pp. 187-190, Cambridge, UK, 2004

Song, Y. F; Shao, X. P. & Xu, J. (2008). New enhancement algorithm for infrared image based on double plateaus histogram, *Infrared and Laser Engineering*, 2008

Wilder, J.; Phillips, P. J.; Jiang, C. & Wiener, S. (1996). Comparison of visible and infrared imagery for face recognition, *Proceedings of the 2nd Int. Conf. Automatic Face and Gesture Recognition*, pp. 182-187, Killington, Vermont, USA, 1996

Wu, S. Q.; Song, W.; Jiang, L. J. et al. (2005A). Infrared face recognition by using blood perfusion data, *Proceedings of Audio- and Video-based Biometric Person Authentication*, pp. 320-328, Rye Brook, NY, USA, 2005

Wu, S. Q.; Gu, Z. H.; China, K. A. & Ong, S. H. (2007). Infrared facial recognition using modified blood perfusion, *Proceedings 6th Int. Conf. Inform., Comm. & Sign. Proc*, pp. 1-5, Singapore, Dec, 2007

Wu, S. Q.; Lu, Y.; Fang, Z. J. et al. (2010a). Infrared image normalization using block and least-squares method, *Proceeding of Chinese Conference on Pattern Recognition*, pp. 873-876, Chongqing, China, October, 2010

Wu, S. Q.; Lu, Y.; Fang, Z. J. & Xie, Z. H. (2010b). A weighted linear normalization method of infrared image, *Journal of Wuhan University of Technology*, Vol. 32, No. 20, pp. 1-5, 2010.

Wu, S. Q.; Liang, W.; Yang, J. C. & Yuan, J. S. (2009). Infrared face recognition based on modified blood perfusion model and 2DLDA in DWT domain, *The 6th International Symposium on Multispectral Image Processing and Pattern Recognition*, Yichang, China, 2009.

Xie, Z. H.; Wu, S. Q; FANG, Z. J etc. (2009). Weighted block-PCA and FLD infrared face recognition method based on blood perfusion images. *Journal of Chinese Computer Systems*, Vol. 30, No.10, pp. 2069-2072, 2009

Yang, J.C.; Yoon, S.; Park, D.S. (2006). Applying learning vector quantization neural network for fingerprint matching, *Lecture Notes in Artificial Intelligence (LNAI 4304)* (Springer, Berlin) , pp. 500-509, 2006

Yang, J.C.; Park, D. S. (2008a). A fingerprint verification algorithm using tessellated invariant moment features, Neurocomputing, Vol. 71, pp. 1939-1946, 2008

Yang, J.C.; Park, D. S. (2008b). Fingerprint verification based on invariant moment features and nonlinear BPNN, *International Journal of Control, Automation, and Systems,* Vol.6, No.6, pp. 800-808, 2008

Zhang, X. J; Sun, X. L. (2005). A research on the piecewise linear transformation in adaptive IR image enhancement, *IT AGE,vol.3,pp.13-16,2005.*

7

Use of Spectral Biometrics for Aliveness Detection

Davar Pishva
Ritsumeikan Asia Pacific University, ICT Institute
Beppu City,
Japan

1. Introduction

Numerous technologies are available for automatic verification of a person's identity. The authentication process usually involves verification of what a person knows (e.g., passwords, pass phrases, PINs), has (e.g., tokens, smart cards), is (e.g., fingerprint, hand geometry, facial features, retinal print, iris pattern), or generates (e.g., signature, voice). Use of something known by a person and use of something held by a person are two simple identification/verification solutions widely used today. Biometrics (also known as biometry) is defined as "the identification of an individual based on biological traits, such as fingerprints, iris patterns, and facial features" (McFedries, 2007), and relies on what a person is or can generate.

Using something one knows requires only a good memory, but can on the other hand be easily overheard, seen, or even guessed. An item that one holds can be stolen and used or copied later. Using biometrics might at first seem to overcome these problems since fingerprints, iris patterns, etc. are part of one's body and thus not easily misplaced, stolen, forged, or shared. Indeed, biometrics technology is becoming a preferred standard for identification and authentication in ATMs, credit card transactions, electronic transactions, e-passports, airports, international borders, nuclear facilities and other highly restricted areas. Presently Europe leads the way but, the highest growth potential is forecasted to be in Asia as many Asian countries have already started adopting the technology. Its market size is estimated to be US$7.1 billion by 2012 (Bailey, 2008). Ironically however, this widespread acceptance of biometrics technology has been attracting the attention of attackers and has provoked interest in exploration of spoofing mechanisms against biometric systems. For example, the thousands of fingerprints that one leaves everywhere in one's daily life can be recovered and molded into artificial fingers for fooling biometrics devices based on fingerprint detection. In an experiment conducted by Matsumoto et al., eleven optical and silicon fingerprint sensors accepted artificial fingers in at least sixty percent of attempts (Matsumoto et al., 2002). Furthermore, with a commercially available high resolution digital camera, the iris pattern of a person's eye can be readily extracted from the person's facial picture and molded into contact lenses to be used to fool machines employing iris pattern recognition. An experiment conducted on two commercial iris recognition devices also showed that one of these devices could be fooled 50% of the time and the other 100% of the time (Matsumoto et al., 2002, 2004).

Although susceptibility of most biometric system to spoofing have been experimented on fingerprint and iris recognition devices as these technologies are used in a variety of commercial products, other biometrics devices can also be spoofed, and to give examples, a dummy hand can be used on a hand geometry system, a high resolution picture can be used on a face recognition system, etc.

In view of this, international biometrics standard organizations are quite concerned about the vulnerabilities of biometrics system and reliabilities of corresponding countermeasures (Tilton, 2006). As a matter of fact, biometrics security, including spoofing, dominated agenda of UK Biometric Working Group (BWG) in their annual report during 2003/2004. The group, which helps the British government implement biometric systems, was instrumental in setting up the European Biometrics Forum (EBF) and creating BIOVISION (a one-year European initiative funded by the EC with the principal aim of developing a "Roadmap" for European biometrics for the next 10 years). BWG, which also serves as JCT1/SC37 (a formal standard body) and liaisons to SC27 (the subcommittee on information security), considers aliveness testing as an appropriate countermeasure against spoofing of biometric authentication (UK Biometric, 2003, 2004).

In an aliveness detection scheme, biometric authentication is augmented by a means for detecting that an object being presented to an authentication system is not an artificial dummy but is a part of a living person. For example, a fingerprint identification means may be augmented by a means that detects the blood pulse from a fingertip so that the fingertip presented for authentication can be judged to be that of a living person. However, even this method can be fooled, for example, by covering a living person's fingertip, which will provide a pulse, with a thin, plastic-molded artificial fingertip that can provide an authentic fingerprint pattern.

Although there are more reliable aliveness detection methods such as perspiration detection (Derakshani et al., 2003), skin color (Brownlee, 2001), medical-based measurement (Lapsley et al., 1998, Osten et al., 1998), rate of warming patents (O'Gorman & Schuckers, 2001), or challenges/responses methods (Fukuzumi, 2001), these are cumbersome in terms of device size, performance, cost, power requirements, operating environment, and human interaction requirements. Conversely, compact spectroscopy-based technologies which have been proposed for biometric identity determination (Rowe et al., 2007) can only work under a controlled measurement environment, as there are spectral alterations due to consumption of alcohol, exposure to warm/cold temperature, or other situation that could alter an individual's complexion, blood circulation, etc. The author has shown that although spectroscopy can be used to capture even differences in fingerprint pattern (Pishva, 2007, 2008, 2010) relying solely on spectroscopy for biometric identification can only worsen the biometrics false reject ratio as intra-individual spectral variation under a non-controlled measurement environment can be more than the spectral differences that exist due to fingerprint pattern differences.

2. Objective and spectroscopic method

As can be understood from the abovementioned examples, many spoofing techniques against biometrics authentication systems make use of an artificial or nonhuman material, such as a plastic fingertip, contact lens, copy medium, etc., to provide a false biometric signature. In view of this, the author considered that biometrics authentication systems can be significantly reinforced against spoofing by incorporating a means that enables judgment not simply of aliveness but judgment that an object being presented for authentication is a portion of a living human being that is free of any intervening artificial or prosthetic material.

An object of this work is therefore to provide a method and a system that enhances existing biometrics technology with a spectroscopic method in order to prevent spoofing. It goes beyond the simple approach of aliveness detection and proposes the implementation of verification of 'spectral signatures' or 'spectral factors' that are unique to human beings or a predetermined class or group of human beings in addition to currently employed methodologies in a multi-factor manner to reduce the likelihood of an imposter getting authenticated. Another aim of the work is to provide methods and systems that augment two widely used biometrics systems (fingerprint and iris recognition devices) with spectral biometrics capabilities in a practical manner and without creating much overhead or inconveniencing the users.

2.1 Use of spectroscopic techniques

Spectroscopy refers to a method of examining matter and its properties by analyzing light, sound, or particles that are emitted, absorbed or scattered by the matter under investigation (Wikipedia, 2011). A multiple biometrics system employing spectroscopy can make spoofing very difficult and time consuming, if not impossible. This is because a spectroscopic approach using various wavelengths allows us to examine various parameters of skin, underlying tissue, blood, fat, melanin pigment in eyes, etc. that vary from person to person, and makes spoofing a very difficult task of imitating multiple physiological characteristics.

2.2 Skin morphology

Skin is a complex biological structure made of different layers with distinct morphologies and optical properties. Conventionally, it is described by dividing it into two major layers. The inner layer, or the dermis, is between 1 to 4 mm thick and consists mainly of connective tissue composed of collagen fibers. Other dermal structures include nerves, blood vessels, lymph vessels, muscles, and gland units. The outer layer, the epidermis, is typically 40-μm thick, but it can be much thicker on load-bearing areas such as palms and soles.

2.3 Skin reflectance

When we look at light reflected from the skin, we usually see two distinct reflection components: a specular or interface reflection component L_s and a diffuse or body reflection component L_b (Shafer, 1985). The specular or interface reflection occurs at the surface and in only one direction, such that the incident light beam and the surface normal are coplanar, and angles between incident and reflected light are equal with respect to the surface normal. As shown in Fig. 1, not the entire incident light is reflected at the surface and some penetrate into the skin. The refracted light beam travels through the skin, hitting various physiological particles from time to time. Within the body, the light rays repeatedly get reflected and refracted at boundaries that have different refractive indices. Some of the scattered light ultimately return to the surface and exit from the skin in various directions, forming the diffuse reflection component L_b. This component carries information about the person's skin color and his/her unique biological "spectral signature".

Using the 2-layer model, Ohtsuki and Healey (Ohtsuki & Healey, 1998) determined the surface reflectance, which takes place at the epidermis surface, to be about 5% of the incident light, independent of the lighting wavelength and the human race (Anderson & Parrish, 1981). The rest of the incident light (95%) enters the skin and becomes absorbed and scattered within the two skin layers. The absorption is mainly due to such ingredients in the

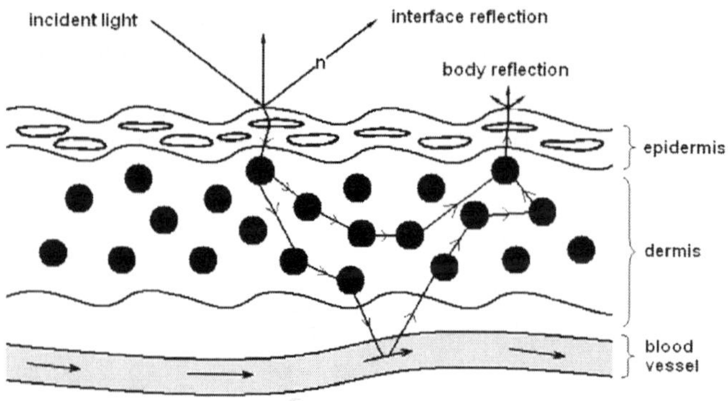

Fig. 1. Principle of body reflectance.

blood as hemoglobin, bilirubin, and beta-carotene. Fig. 2(a) shows a spectral reflectance curve of a typical Caucasian skin (Anderson & Parrish, 1981, Melanoma, 2006). Fig. 2(b) shows an exploded form of this spectrum into its distinct components, namely: epidermis and hemoglobin (there are also spectra of water and collagen substances, but these do not play a significant role in the indicated wavelength range). As can be observed, the melanin in the epidermis absorbs the most part of blue light at ~ 470 nm; and hemoglobin absorbs green light at ~ 525 nm and red light at ~ 640 nm. Also, though not shown, near infrared light at 850nm is used to identify papillary dermis (Melanoma, 2006).

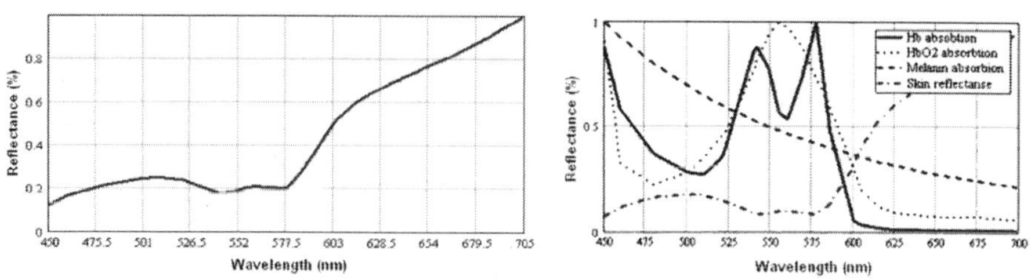

(a) Typical Caucasian skin spectrum (b) Its distinct components

Fig. 2. A typical Caucasian skin spectrum and its distinct components (Melanoma, 2006).

3. Proposed methodology and technical solution

In order to achieve the above mentioned objectives, this work proposes to augment a base authentication technique, such as optical fingerprint matching, in which a non-spectrometric biometric signature, such as a fingerprint image, is acquired from a biometric signature source, such as a fingertip, with a means of extracting spectral information from the same biometric signature source in a practical manner that does not affect the size, performance, cost, power requirements, operating environment, and human interaction requirements of the base authentication technique.

3.1 Multi-factor authentication approach

A multi-factor authentication method relies on either multiple biometrics or, biometrics in conjunction with smart cards and PINs in order to reduce the likelihood of an imposter being authenticated. One aspect according to this work provides: a multifactor authentication method including the steps of: acquiring a primary signature of a primary signature source of a subject to be authenticated; acquiring a secondary signature source of the subject; using the primary signature to determine the unique identity of the primary signature source; and using the using the secondary signature to verify that the subject to be authenticated belongs to a predetermined class of objects.

Here, the primary source is a non-spectrometric biometric signature of a biometric signature source of the subject to be authenticated; and the secondary source is a spectral information of the biometric (primary) signature source; wherein the non-spectrometric biometric signature is used for determining the unique identity of the biometric signature source; and the spectral information for verifying that the subject to be authenticated is an authentic living human being.

Here, the multifactor authentication method may further include the steps of: registering a non-spectrometric biometric signature of a biometric signature source of a subject to be authenticated; and registering spectral information of the biometric signature source; and in the step of using the non-spectrometric biometric signature to determine the unique identity of the biometric signature source, the acquired non-spectrometric biometric signature may be compared with the registered non-spectrometric biometric signature to determine the unique identity of the biometric signature source, and in the step of using the spectral information to verify that the subject to be authenticated belongs to the predetermined class of objects, the acquired spectral information may be compared with the registered spectral information to verify that the subject to be authenticated belongs to a predetermined class of objects.

Here, a 'non-spectrometric biometric signature' refers to an image, pattern, set of geometrical parameters, or other form of biological trait data obtained by an existing biometrics technology. Thus for example, the subject to be authenticated may be a person, and with this example, the predetermined class of objects may be 'living human beings with predetermined spectral characteristics,' the biometric signature source may be a fingertip, the non-spectrometric biometric signature may be a fingerprint image of the fingertip, and the spectral information of the biometric signature source may be a diffuse reflectance spectrum of the fingertip. That is, with this example, first, a fingerprint image of a person's fingertip is registered and a diffuse reflectance spectrum of the person's same fingertip is registered. Thereafter, a fingerprint image of a fingertip of a person, who is to be authenticated, is acquired, and a diffuse reflectance spectrum of this person's same fingertip is acquired. The acquired fingerprint image is then compared with the registered fingerprint image to determine the unique identity of the person, in other words, to determine that the fingerprint is that of the person to be authenticated, that is, the person whose fingerprint had been registered in advance and not that of anybody else, and the acquired diffuse reflectance spectrum of the fingertip is compared with the registered reflectance spectrum to verify that the person is actually a living human body with the predetermined spectral characteristics.

3.2 Reliability of the approach

Here, because the non-spectrometric biometric signature, such as a fingerprint image, of the biometric signature source, such as the fingertip, is augmented by the spectral information of the biometric signature source, such as the diffuse reflectance spectrum of the fingertip, so that while the non-spectrometric biometric signature (e.g. fingerprint image) ensures the unique identity of the object or the person to be authenticated, the spectral information (e.g. diffuse reflectance spectrum) ensures that the non-spectrometric biometric signature (e.g. fingerprint image) is a genuine signature of the predetermined class of objects (e.g. living human beings), spoofing, for example, that uses the non-spectrometric biometric signature (e.g. fingerprint image) formed on an object (e.g. copy medium, plastic finger, etc.) not belonging to the predetermined class of objects (e.g. living human beings) can be prevented. That is, the spectral information of an object reflects the optical complexity of that object, and the more complex an object is, the more complex the spectral information. In particular, skin or other portion of a living human is a complex biological structure made of different layers with distinct morphologies and optical properties. Thus for example, a diffuse reflectance spectrum obtained from a fingertip includes spectral components of such substances as melanin, hemoglobin, and other constituents of skin, muscle, blood, etc., with which the proportions present, etc. differ among individual persons. The spectral information obtained from a fingertip or other portion of a living human is thus extremely complex and cannot be replicated readily by the use of artificial dummies and prosthetic devices, and especially because in the present approach, the non-spectrometric biometric signature of the same portion is acquired for identification, spoofing is made a practically insurmountable task.

In the above example of spoofing using a fingertip image printed on a copy medium, because any copy medium is an artificial object, such as paper, plastic, etc., or in the least, a non-living object, such as non-living skin, it cannot provide the same spectral information as that of a portion of a living human being. If an imposter attaches a fingertip cover, which is molded to provide the image of an authentic fingerprint image, to his/her own fingertip, the detected spectral information may contain spectral information of the imposter's fingertip, which is spectral information of a living human being. However, as long as the fingertip cover that is attached is an artificial object, or in the least, a non-living object, the detected spectral information will contain spectral information that differs from that of a living human being and thus as a whole, the detected spectral information will not be the same as that of a living human being.

In the present approach, the spectral information is used to verify that the subject to be authenticated belongs to a predetermined class of objects. The predetermined class of objects is preferably broad enough to provide allowance for intra-object variations and yet narrow enough to preclude spoofing. In the above example, 'living human beings with predetermined spectral characteristics' is the predetermined class of objects, and this allows for intra-personal variations due to such external conditions as injury and exposure to high or low temperatures, chemicals, ultraviolet rays, or such internal conditions as changes in blood flow due to consumption of medicine, alcohol, etc., and at the same time precludes the use of artificial and non-living-human objects for spoofing.

3.3 Implementation steps and means

Here, the steps of acquiring the non-spectrometric biometric signature of the biometric signature source of the subject to be authenticated and acquiring the spectral information of the biometric signature source may be carried out simultaneously. This significantly shortens the time required for authentication.

In the step of comparing the acquired spectral information with the registered spectral information to verify that the subject to be authenticated belongs to the predetermined class of objects, cluster analysis may be performed on the acquired spectral information and the registered spectral information to determine a similarity value of the acquired spectral information and the registered spectral information, and the subject to be authenticated may be verified as belonging to the predetermined class of objects when the determined similarity value is within a predetermined range.

Another aspect according to this approach provides: a multifactor authentication system including: a means for acquiring a non-spectrometric biometric signature of a biometric signature source of a subject to be authenticated; a means for acquiring spectral information of the biometric signature source; and a means that uses the non-spectrometric biometric signature to determine the unique identity of the biometric signature source and uses the spectral information to verify that the subject to be authenticated belongs to a predetermined class of objects.

Here, the multifactor authentication system may further include: a means for storing an acquired non-spectrometric biometric signature as a registered non-spectrometric biometric signature and storing an acquired spectral information as registered spectral information; and the means that uses the non-spectrometric biometric signature to determine the unique identity of the biometric signature source and uses the spectral information to verify that the subject to be authenticated belongs to a predetermined class of objects may compare a newly acquired non-spectrometric biometric signature with the stored, registered non-spectrometric biometric signature to determine the unique identity of the biometric signature source and compare newly acquired spectral information with the stored, registered spectral information to verify that the subject to be authenticated belongs to a predetermined class of objects.

In the above-described example where the subject to be authenticated is a person, the predetermined class of objects is 'living human beings with predetermined spectral characteristics,' the biometric signature source is a fingertip, the non-spectrometric biometric signature is a fingerprint image of the fingertip, and the spectral information of the biometric signature source is a diffuse reflectance spectrum of the fingertip, the means for acquiring the non-spectrometric biometric signature may be a CCD or CMOS detecting system, with which an image of the fingerprint is formed on a detecting surface of a CCD or CMOS sensor, the means for acquiring the spectral information may be a photodiode array (PDA) detecting system, with which diffusely reflected light from the fingertip is spectrally dispersed onto a PDA, and a computer or other information processing means may be used as the means that uses the fingerprint image (non-spectrometric biometric signature) to determine the unique identity of the fingertip (biometric signature source) and uses the spectral information to verify that the person (subject to be authenticated) is a 'living human being with predetermined spectral characteristics' (belongs to the predetermined class of objects).

Here, a half-mirror or a beam splitter may be used to simultaneously acquire the non-spectrometric biometric signature (e.g. fingerprint image) and the spectral information (e.g. diffuse reflectance spectrum), and an extended portion of the CCD/ CMOS detector may be configured as PDAs for simultaneously capturing numerous identical spectra to be integrated into a single spectrum having a sufficient S/N ratio for spectral analysis. The system can thereby be made compact and high in the speed of authentication.

4. Spectroscopic investigation

To thoroughly investigate the applicability, effectiveness and usability of the spectroscopic method as an enhancement technique for preventing spoofing in existing biometrics technology, a number of spectra from real fingers, real fingers covered with a fingertip molds that provide fingerprint pattern of authentic persons and artificial fingers made of different materials that contain authentic fingerprint patterns, were measured and analyzed. In the analysis phase, reflectance values of numerous physiological components ('spectral factors') were extracted from the measured spectra and Euclidean distances (Wikipedia, 2011) among the corresponding extracted factors of the spectra were computed to verify authenticity of an identified individual.

A through explanation of the investigation is given in (Pishva, 2008) and the author will simply highlight the main findings here. Furthermore, even though the spectroscopic investigation was only carried out on fingerprint system, the approach is general and can very well be applied to other biometrics systems such as iris pattern, hand geometry, etc.)

4.1 Measurement

Initially a total of 150 reflectance spectra (350nm ~ 1050nm) from 10 fingers of 5 Japanese men having a similar complexion was measured at three different times in order to investigate intra-individual and inter-individual spectral variations. The experimental conditions during the three measurements were set so that there were no possible spectral alterations due to consumption of alcohol, exposure to warm/cold temperature, or other situation that could alter an individual's complexion, blood circulation, etc., as it was done for the sake of a preliminary examination.

In the second stage, under a similar condition, the five peoples' fingers were covered with fingertip covers made of transparent plastic and rubber materials in order to provide fingerprint pattern of an authentic person. A set of similar measurements were also taken from artificial fingers that were made of craft paper, wooden and plastic materials.

In the final stage, 750 spectra data were also measured from different fingers of a man (the above mentioned P1), a woman (W1) and a child (C1) under different conditions to study the effects of finger size, finger texture, finger orientation and as well as stableness of the 'spectral factors'. Some spectra were measured when fingers placed flat, while others when rotated to counterclockwise or clockwise directions by about 45°. Some spectra were measured right after having lunch while others late in the evening. Some spectra were also measured after consumption of alcoholic drink (i.e. sometimes after drinking two bottles of beer).

4.1.1 Spectral patterns of real fingers

Fig. 3(a) shows the three spectra that were captured from the right index finger of an individual at three different times. As can be observed, the spectra look identical, understandably because, it comes from an object of the same finger pattern, skin color and physiological structure. Fig. 3(b) shows spectra of the right index fingers of five different persons. As can be observed, around a general pattern, there are significant variations in the spectra as there are numerous physiological differences among individuals. Thus, it looks feasible to use such pattern and variations at different wavelengths to monitor and check aliveness and authenticity of the person during a biometrics' verification process.

4.1.2 Spectral patterns of bogus fingers

Fig. 4 (a) shows spectral patterns of artificial fingers made of craft paper, wooden and rubber materials (non-living objects) which are supposedly contain fingerprint of the person whose finger spectral pattern is shown in Fig. 3(a). As can be clearly observed, spectral patterns of craft paper, wooden and rubber fingers are quite different from each other and

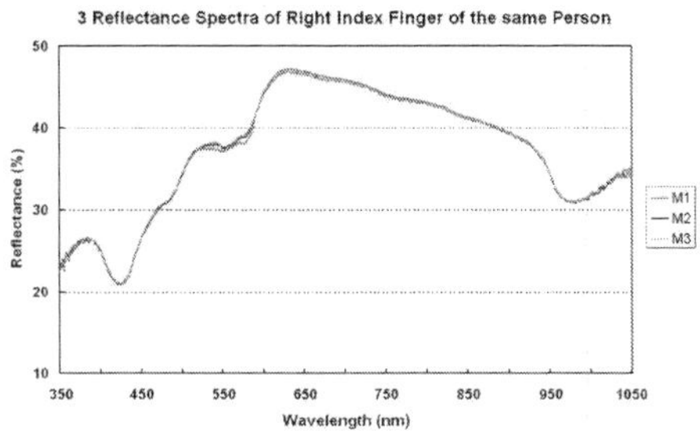

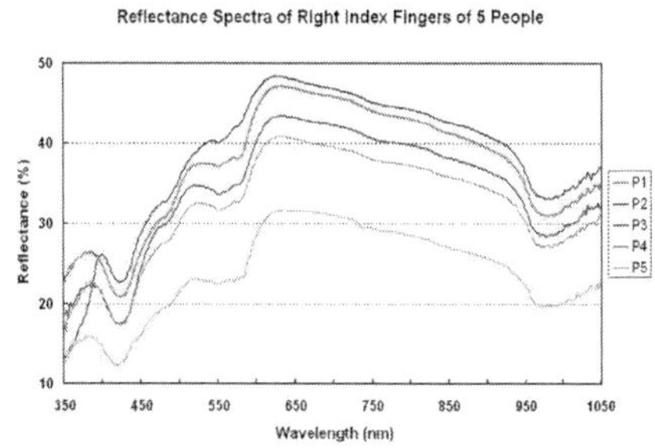

(a) The same person's finger spectra (b) Variation among different persons

Fig. 3. An individual's finger spectra and spectral variation among different individuals.

they are also very much different from that of the authentic person's finger spectra. It is obvious that shape of the spectra from artificial finger highly depends on the base materials used in making them rather than the fingerprint pattern that is molded on them. Fig. 4(b) shows spectral patterns of the right index finger of the person whose finger spectral pattern is shown in Fig. 3(a), when covered with his fingertip prints made of transparent plastic and rubber materials. As can be observed, each spectrum, though different, have a general shape as the transparent fingertips are worn by the same person. It should also be noted that the spectra are different from those of the artificial finger spectra shown in Fig. 4(a) and the authentic one indicated in Fig. 3(a). This justifies an earlier claim that attachment of artificial fingertip cover on a real finger alters the reflectance spectra and is a proof of the robustness of the multi-factor spectral biometrics approach.

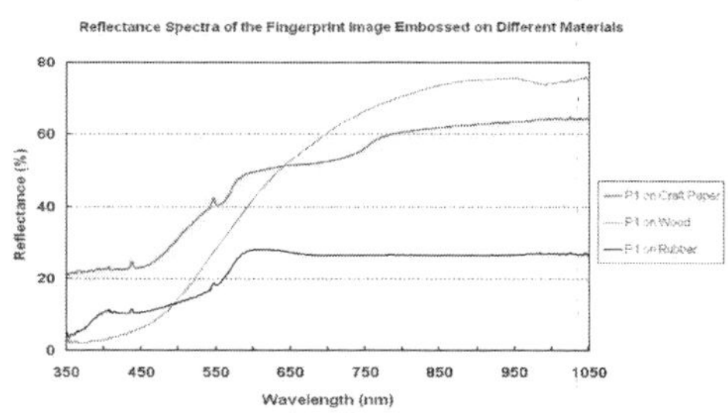

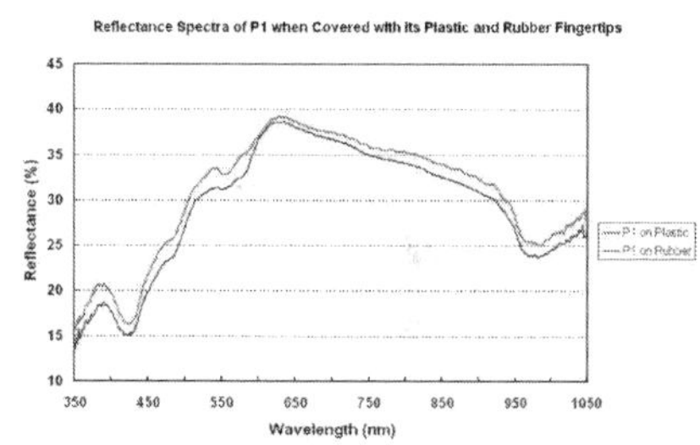

(a) Spectral patterns of artificial fingers (b) Spectral patterns with fingertips

Fig. 4. Spectral Patterns of Artificial Fingers and fingers covered with fingertips.

4.2 Data analysis

Cluster analysis with MINITAB statistical software (Minitab, 2009) was used for data analysis after extracting 'spectral factors' from the measured spectra. In this work spectral factors' refers to reflectance values at certain wavelengths or regions on a spectrum which correspond to specific physiological components. For example, in the finger reflectance spectrum of Fig. 5, shaded areas correspond to certain physiological components (i.e., 350 to 470 nm indicate melanin reflectance, vicinity of 525 nm, 640 nm and 850 nm indicate hemoglobin, vicinity of 650 and 750 nm are for arterial blood peak, and 750 and 925 nm are for venous blood peak (Melanoma, 2006, Pishva, 2007). Each 'spectral factor' can be extracted by computing area of the corresponding shaded region.

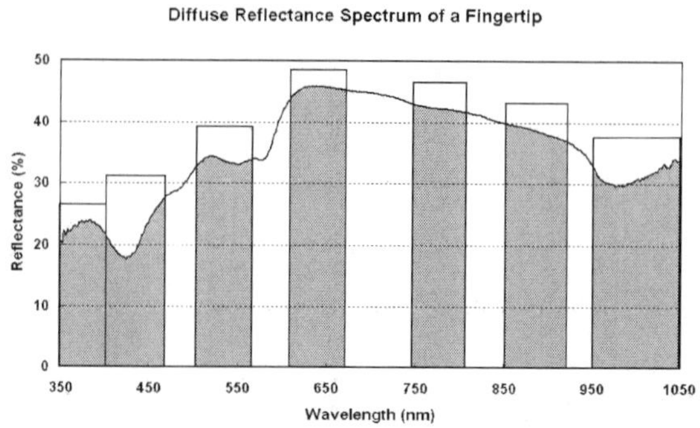

Fig. 5. 'Spectral Factors' Regions.

4.2.1 Results at verification phase – presentation of real fingers

Table 1 shows the 'Similarity' values that are obtained when P1-1 is used as the registered template set and the extracted 'spectral factor' sets of the five persons are presented, one set at a time, to the analysis routine as the newly acquired 'spectral factors' for verification.

Template Value	Newly Acquired 'Spectral Factors'	'Similarity' Level
P1-1	P1-1	100.00
P1-1	P1-2	99.47
P1-1	P1-3	99.49
P1-1	P2-1	96.03
P1-1	P2-2	96.39
P1-1	P2-3	96.29
P1-1	P3-1	92.20
P1-1	P3-2	92.01
P1-1	P3-3	92.02
P1-1	P4-1	86.86
P1-1	P4-2	86.87
P1-1	P4-3	86.01
P1-1	P5-1	65.49
P1-1	P5-2	65.49
P1-1	P5-3	64.71

Table 1. 'Similarity' values of real finger spectra.

As can be seen, whereas 'Similarity' values higher than 99% are obtained when the 'spectral factors' of the same finger of the same person are presented in the verification phase, the

'Similarity' value drops significantly (96% to 64%) when a different person's 'spectral factors' are presented.

4.2.2 Results at verification phase – presentation of bogus fingers

In the previous section, it was shown that a 'Similarity' value higher than 99% is obtained when the same person's 'spectral factors' are presented to the analysis routine. As such, some researchers have even proposed that skin spectroscopy alone can be used for biometrics identity determination of an individual (Rowe et al., 2007). This work, however, proposes spectral biometrics as an enhancement technique and not as an identification method.

Table 2 shows the 'Similarity' values determined in the verification phase when P1-1 is used as the registered template, and the 'spectral factors,' extracted from the index finger of P1 at different times under various conditions, the fingers of five persons covered with transparent plastic and rubber fingertip molds having the fingerprint pattern of the authentic person (P1) and artificial fingers made of craft paper, wooden and plastic materials that contain the authentic fingerprint pattern of P1, are presented one set at a time as the newly acquired 'spectral factors.'

Template Value	Newly Acquired 'Spectral Factors'	'Similarity' Level
P1-1	P1-2 (authentic, controlled)	99.47
P1-1	P1-m (authentic, after meal)	96.23
P1-1	P1-n (authentic after alcohol)	95.46
P1-1	P1onP1-PlasticCvr-1	82.00
P1-1	P1onP1-RubberCvr-1	78.28
P1-1	P1onP2-PlasticCvr-1	91.20
P1-1	P1onP2-RubberCvr-1	93.13
P1-1	P1onP3-PlasticCvr-1	85.05
P1-1	P1onP3-RubberCvr-1	84.09
P1-1	P1onP4-PlasticCvr-1	88.22
P1-1	P1onP4-RubberCvr-1	90.09
P1-1	P1onP5-PlasticCvr-1	81.21
P1-1	P1onP5-RubberCvr-1	79.44
P1-1	P1onCraftPaperFngr-1	62.85
P1-1	P1onWoodenFngr-1	48.94
P1-1	P1onPlasticFngr-1	59.27

Table 2. 'Similarity' values during verification process.

As can be observed from Table 2, 'Similarity' values obtained from artificial fingers and fingers containing fingertip covers are quite different from that of the real authentic finger. However, even for the real authentic finger, 'Similarity' values obtained under a controlled measurement environment is much better than those obtained under relaxed conditions (e.g., after having meal or being under the influence of alcohol). In fact some 'Similarity' values obtained under a more relaxed condition are comparable to those values that were obtained from another person (e.g., person P2 in Table 1). This clearly indicates that although the use of spectral biometrics as the sole means for identity determination may be difficult, spectral biometrics can be used as an enhancement technique, as proposed in this work, since it discriminates authentic fingers from artificial fingers and fingers containing fingertip covers.

4.3 Optimal boundary conditions

In order to obtain a consistently reliable result, determination of optimal boundary conditions including establishment of optimal settings and effect of spectrum resolution was carried out. It was found out that the stability of the 'spectral factors' for consistently generating a 'Similarity' value higher than 95%, largely depended on the size of the measurement spot rather than on the physiological or environmental factors, or spectrum resolution. Sampling the center of a 1 cm^2 measurement spot, a condition which can easily be provided by thumb fingers, provided a uniform reflectance condition at the measurement point (Pishva, 2008).

5. System configuration and application scope

As mentioned earlier, this work proposes spectral biometrics methods as enhancement techniques for preventing spoofing in existing biometrics technologies. The idea is to double check the authenticity of an identified subject in order to ensure that a live person with a matching biological 'spectral signature' is being authenticated. As such, implementation and configuration of multi-factor spectral biometrics would depend on the configuration of the base biometrics authentication technology used in the system.

Moreover, when complementing existing biometrics technology with a spectral biometrics method, factors such as the device size, performance, cost, power requirements, operating environment, and human interaction requirements must also be considered. Taking these into account, this section shows how the two widely used biometrics systems (fingerprint and iris recognition devices) can be augmented with spectral biometrics without creating much overhead or inconvenience to users.

5.1 Preferred configuration for fingerprint authentication system

Preferred embodiments of this approach shall now be described. Fig. 6 is a schematic diagram of a basic arrangement of a spectral biometrics enhanced authentication system according to a first embodiment of this approach, which is a fingerprint authentication device that authenticates a person's identity based on his/her fingerprint and biospectral characteristics of his/her finger.

As shown in Fig. 6, this fingerprint authentication device 1 includes a measurement unit 2, a controller 120, a memory (storage device) 130, and a monitor 140. The measurement unit 2 includes an optical system 10 and a CCD (charge coupled device; image sensor) 100. The optical system 10 includes an I2 lamp (light source) 15, a sheet prism (prism means) 20, a first lens 40, a second lens 60, a mirror 70, and a diffraction grating 80. As shown in Fig. 7, the CCD 100 is an image sensor with pixels arranged in 1280 rows and 1024 columns and has an image acquisition portion 102 (first portion of a detecting surface of the CCD sensor),

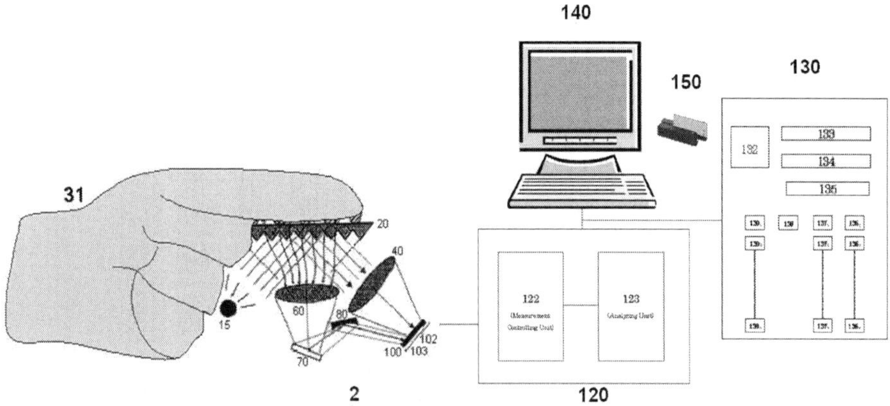

Fig. 6. A Schematic diagram of a basic arrangement according to a first embodiment.

which is a region of 960×960 pixels at an upper portion of the CCD 100 that excludes the pixels of 32 edge rows at the top side and 32 columns at each of the left and right sides of the CCD 100 as boundary pixels, and a spectrum acquisition portion 103 (second portion of a detecting surface of the CCD sensor), which is a region of 160×960 pixels at a lower portion of the CCD 100 that excludes the pixels of 32 edge rows at the bottom side and 32 columns at each of the left and right sides of the CCD 100 as boundary pixels. 96 rows of pixels between the image acquisition portion and the spectrum acquisition portion 103 are also handled as boundary pixels. The controller 120 is electrically connected to the CCD 100, the memory 130, and the monitor 140 and controls operations of these components by issuing appropriate instruction signals. The memory 130 has a measured image (matrix) storage area 132, a reference spectrum (vector) storage area 133, a measured spectrum (vector) storage area 134, a reduced measured spectrum (vector) storage area 135, registered image

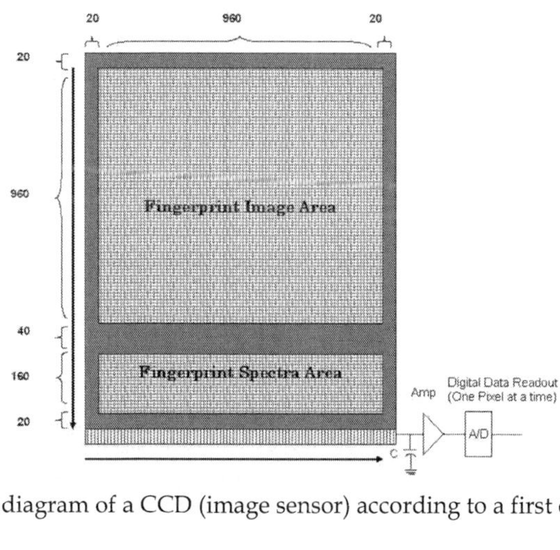

Fig. 7. A schematic diagram of a CCD (image sensor) according to a first embodiment.

pattern (template) storage areas 136_1 to 136_n (where n is an integer greater than 1), registered spectral template data storage areas 137_1 to 137_n, an identity storage area 138, and registered identity storage areas 139_1 to 139_n. The controller 120 is also electrically connected to a card reader 150 that serves as an identity inputting means.

A manner in which a fingerprint image is acquired as a non-spectrometric biometric signature of a fingertip (biometric signature source) of a person (subject to be authenticated) and a diffuse reflectance spectrum of the fingertip is acquired as spectral information of the fingertip (biometric signature source) by this fingerprint authentication device 1 shall now be described.

5.1.1 Incident light and its reflected components

As shown in Fig. 6, with this fingerprint authentication device 1, light from the I2 lamp 15 is made incident via a sheet prism 20 onto a finger 31_1, which belongs to a person 30_1 to be authenticated and is being pressed against an upper surface of the sheet prism 20. A portion of the light made incident on the finger 31_1 is reflected as a specular reflection component Ls from the surface of the finger 31_1, and a first lens 40 forms an image of this specular reflection component Ls on the image acquisition portion 102 of the CCD 100.

Another portion of the light made incident on the finger 31_1 penetrates into the skin, is refracted, reflected, absorbed, or re-emitted as fluorescence or phosphorescence, etc. by internal tissue, blood, and other various physiological components inside and below the skin, and some of this light ultimately returns to the surface and exits from the skin in various directions, thus forming a diffuse reflection component Lb. Because this light component results from light that has traveled inside the skin, it carries information concerning the person's skin color and his/her unique biological 'spectral signature' (Fig 4). After exiting from the skin, the diffuse reflection component Lb passes through the sheet prism 20 and is converged, via the second lens 60 and the mirror 70, onto the diffraction grating 80, which spectrally disperses and makes the diffuse reflection component Lb incident on the spectrum acquisition portion 103 of the CCD 100 in a manner such that a fingertip diffuse reflection spectrum of a range of 350nm to 1050nm is acquired from each row of the spectrum acquisition portion 103.

5.1.2 Reflected components and their detection process

Light made incident on the CCD 100 is photoelectrically converted into electrical charges at the respective pixels. In accordance to an instruction signal from a measurement controlling unit 122 of the controller 120, these charges are electronically shifted into a horizontal shift register 104, one row at a time, and thereafter, the contents of the horizontal shift register 104 are shifted, one pixel at a time, into a capacitor 105. The charges in the capacitor 105 are then provided as an analog voltage to an amplifier 106, which performs amplification to an appropriate analog voltage level (e.g., 0 to 10 volts). The amplified voltage output by the amplifier is then converted to a digital value by an analog-to-digital (A/D) converter 107. The digital values output by the A/D converter 107 are then input as data into the memory 130 according to instruction signals from the measurement controlling unit 122 of the controller 120. The digital values obtained by reading the charges from the image acquisition portion 102 of the CCD 100 are thus stored as data in the measured image storage area 132 in accordance to an instruction signal from the controller 120, and the digital values obtained by reading the charges from the spectrum acquisition portion 103 of

the CCD 100 are binned as data in the measured spectrum storage area 134 in accordance to an instruction signal from the measurement controlling unit 122 of the controller 120.
In this readout process, the data of the boundary pixels (i.e. the pixels of the 32 edge rows at the top and bottom sides, the 32 columns at each of the left and right sides, and the 96 rows between the image acquisition portion 102 and the spectrum acquisition portion 103 of the CCD 100) are ignored as data that may not be reliable in comparison to data of other portions or as data that may be hybrid data of the image and the spectrum.

5.1.3 Authentication process

An authentication process using the fingerprint authentication device 1 shall now be described with reference to the flowcharts of Fig 8. This authentication process is constituted of an enrollment process (Fig. 8a), in which a person's fingerprint image and fingertip diffuse reflectance spectrum are registered along with the person's identity, and a verification process (Fig. 8b), which is performed each time a person needs to be verified.

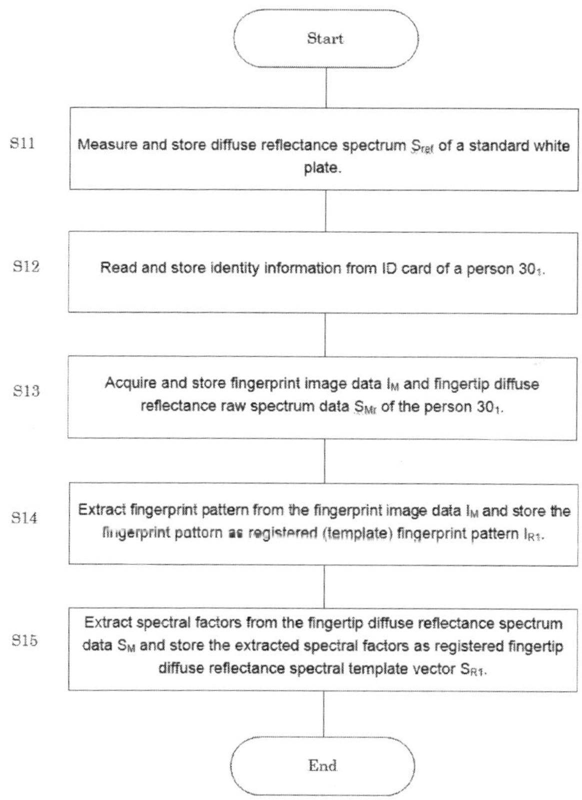

(Fig 8.a) Flowchart of an enrolment process

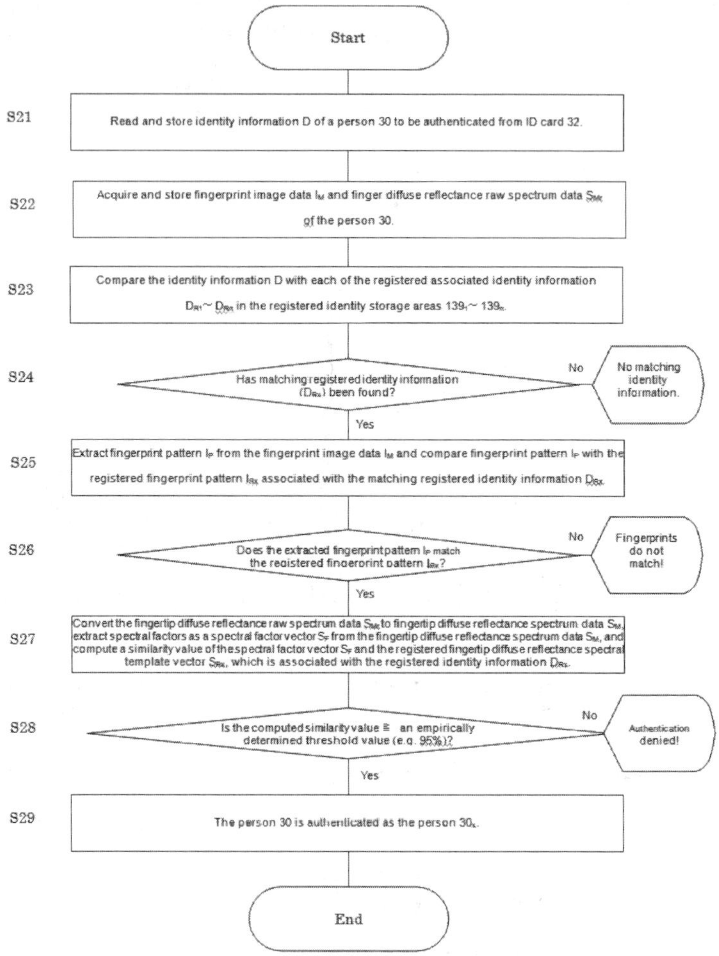

(Fig 8.b) Flowchart of a verification process

Fig. 8. A flowchart of an enrolment and a verification process in an authentication process according to an embodiment of this work.

5.1.3.1 Enrollment process

Firstly, in the enrollment process shown in Fig. 8(a), a diffuse reflectance spectrum S_{ref} of a standard white plate (not shown) is set on the upper surface of the sheet prism 20 and its diffuse reflectance spectrum S_{ref} is measured. The spectrum data S_{ref} that are obtained by this measurement and stored in the measured spectrum storage area 134 of the memory 130 are then transferred and re-stored in the reference spectrum storage area 133 of the memory 130 (step S11).

Identity information, such as the name, etc. of the person 30_1, are then read from an ID card 32_1, belonging to the person 30_1, by means of the card reader 150 and stored as registered identity information D_{R1} in the registered identity storage area 139_1 (step S12).

Fingerprint image data I_M and fingertip diffuse reflectance raw spectrum data S_{Mr} of the person 30_1 are then captured and measured as described above and stored in the measured image storage area 132 and the measured spectrum storage area 134, respectively, of the memory 130 (step S13).

The controller 120 then issues an instruction signal to the memory 130 to make the fingerprint image data I_M, stored in the measured image storage area 132, be transmitted to an analyzing unit 123, where a fingerprint pattern is extracted from the fingerprint image data I_M. Methods of extracting a fingerprint pattern from such fingerprint image data are well-known and described, for example, in 'Handbook of Fingerprint Recognition,' (Maltoni et al., 2003), and a detailed description thereof shall not be provided here. The controller 120 then stores the extracted fingerprint pattern, for example, as a registered (template) fingerprint pattern IR_1 in the registered image pattern (template) storage area 136_1 so that this fingerprint pattern is associated with the registered associated identity information D_{R1} in the registered identity storage area 139_1 (step S14).

Next, the controller 120 issues an instruction signal to the memory 130 to make the fingertip diffuse reflectance raw spectrum data S_{Mr}, stored in the measured spectrum storage area 134, and the reference reflectance spectrum data S_{ref}, stored in the reference spectrum storage area 133, be transmitted to the analyzing unit 123. In the analyzing unit 123, the fingertip diffuse reflectance raw spectrum data S_{Mr} are converted to fingertip diffuse reflectance spectrum data S_M by using the values of reference reflectance spectrum data S_{ref} as 100% reflectance. Spectral factors are then extracted from the fingertip diffuse reflectance spectrum data S_M. In the present example, the fingertip diffuse reflectance spectrum data S_M is integrated in the respective ranges of 350 to 400nm, 401 to 470nm, 500 to 560nm, 600 to 660nm, 730 to 790nm, 830 to 900nm, and 925 to 1025nm to obtain seven integration values (Fig. 5). Here, the ranges of 350 to 400nm and 401 to 470nm correspond to peaks due to melanin, the ranges of 500 to 560nm, 600 to 660nm, and 830 to 900nm correspond to peaks due to hemoglobin, the range of 730 to 790nm corresponds to arterial blood, and the range of 925 to 1025nm corresponds to venous blood. The resulting seven values are then stored as a registered fingertip diffuse reflectance spectral template vector S_{R1} in the registered spectral template data storage area 137_1, and this spectral template vector is thereby associated with the registered associated identity information D_{R1} in the registered identity storage area 139_1 (step S15).

This enrollment process is not performed each time a person needs to be authenticated but is performed just once or once every predetermined interval (months, years, etc.). Also, for persons besides the person 30_1, the procedure from step S12 to step S15 of this enrollment process may be performed at any time to register a registered fingerprint image IR and a registered fingertip diffuse reflectance spectrum S_R in association with an associated identity information D_R for each of an arbitrary number n of persons.

5.1.3.2 Verification process

In the verification process (Fig. 8b), first, the identity information of a person 30 to be authenticated are read from an ID card 32, belonging to the person 30, by means of the card reader 150 and stored as identity information D in the identity storage area 138 (step S21).

Fingerprint image data I_M and fingertip diffuse reflectance raw spectrum data S_{Mr} of the person 30 are then captured and measured as described above and stored in the measured

image storage area 132 and the measured spectrum storage area 134, respectively, of the memory 130 (step S22).
The controller 120 then issues an instruction signal to the memory 130 to make the identity information D, stored in the identity storage area 138, be transmitted to the analyzing unit 123. At the analyzing unit 123, the identity information D is compared with each of the registered associated identity information D_{R1} # D_{Rn} in the registered identity storage areas 139_1 # 139_n to find matching registered identity information (step S23, S24). If matching registered identity information is found, step S25 is entered. On the other hand, if matching registered identity information is not found, step S41 is entered, in which a message, such as 'No matching identity information,' is displayed on the monitor 140, and then the process is ended without authentication of the person 30.

5.1.3.3 Process at control and analysis unit

For the present description, it shall be deemed that the identity information D matches the registered associated identity information D_{Rx} of a person 30_x (where x is a value in the range of 1 to n). In this case, upon entering step S25, the controller 120 issues an instruction signal to the memory 130 to make the fingerprint image data I_M, stored in the measured image storage area 132, be transmitted to the analyzing unit 123, where a fingerprint pattern I_P is extracted from the fingerprint image data I_M. At the analyzing unit 123, the extracted fingerprint pattern I_P is compared with the registered fingerprint pattern I_{Rx} in the registered image pattern storage area 136x, which is the fingerprint pattern associated with the registered identity information D_{Rx}, to judge whether the extracted fingerprint pattern I_P matches the registered fingerprint pattern I_{Rx} (step S26). Methods of comparing fingerprint patterns from such fingerprint image data are well-known and described, for example, in the abovementioned 'Handbook of Fingerprint Recognition,' (Maltoni et al., 2003), and a detailed description thereof shall not be provided here.
If by the above analysis of step S26, the extracted fingerprint pattern I_P is found to match the registered fingerprint pattern I_{Rx}, step S27 is entered. On the other hand, if the fingerprint patterns do not match, step S42 is entered, in which a message, such as 'Fingerprints do not match!' is displayed on the monitor 140, and then the process is ended without authentication of the person 30.
For the present description, it shall be deemed that the extracted fingerprint pattern I_P matches the registered fingerprint pattern I_{Rx}. In this case, upon entering step S27, the controller 120 issues an instruction signal to the memory 130 to make the fingertip diffuse reflectance raw spectrum data S_{Mr}, which are of the person 30 and are stored in the measured spectrum storage area 134, and the reference reflectance spectrum data S_{ref}, which are stored in the reference spectrum storage area 133, be transmitted to the analyzing unit 123. In the analyzing unit 123, the fingertip diffuse reflectance raw spectrum data S_{Mr} are converted to fingertip diffuse reflectance spectrum data S_M of the person 30 by using the values of reference reflectance spectrum data S_{ref} as 100% reflectance. Seven spectral factors are then extracted as a spectral factor vector S_F from the fingertip diffuse reflectance spectrum data S_M in the same manner as described above. A similarity value of the spectral factor vector S_F thus acquired and the registered fingertip diffuse reflectance spectral template vector S_{Rx} in the registered spectral template data storage area 137_x, which is associated with the registered identity information D_{Rx}, is then computed by cluster analysis using single linkage Euclidean distance. The computation of the similarity value is performed, for example, using a cluster analysis software, such as Minitab Statistical Software© (made by Minitab Inc., 2009), and using a seven-valued vector R_0, having zero

entries for all seven spectral factors, as a dissimilarity reference vector corresponding to a similarity value of 39.11%. Because the computation of the similarity value by cluster analysis using single linkage Euclidean distance is a well-known art (see for example, Ragnemalm, PhD Thesis 1993), a detailed description thereof shall be omitted here.

The computed similarity value is then compared with, for example, an empirically determined threshold value of 95% (S28). If the computed similarity value is greater than or equal to this threshold value, the process ends upon authentication of person 30 as the person 30_x (step S29). On the other hand, if the computed similarity value is less than the threshold value, step S44 is entered, in which a message, such as 'Authentication denied!' is displayed on the monitor 140 and then the process is ended without authentication of the person 30.

5.1.4 Reliability of the approach

As can be understood from the above description of the embodiment, with the present approach, because a non-spectrometric biometric signature (fingerprint image) of a biometric signature source (fingertip) is augmented by spectral information of the biometric signature source (diffuse reflectance spectrum of the fingertip) in a manner such that the non-spectrometric biometric signature (fingerprint image) is used to ensure the unique identity of the object (person) to be authenticated and the spectral information (diffuse reflectance spectrum) is used to ensure that the non-spectrometric biometric signature (fingerprint image) is a genuine signature of the predetermined class of objects (living human beings with fingerprint diffuse spectral characteristics within a predetermined similarity range of predetermined characteristics), spoofing, for example, that uses a non-spectrometric biometric signature (fingerprint image) formed on an object (e.g. copy medium, plastic finger, etc.) not belonging to the predetermined class of objects (living human beings with fingerprint diffuse spectral characteristics within a predetermined similarity range) can be prevented. That is, the spectral information of an object reflects the optical complexity of that object, and the more complex an object is, the more complex the spectral information. In particular, skin or other portion of a living human is a complex biological structure made of different layers with distinct morphologies and optical properties. Thus for example, a diffuse reflectance spectrum obtained from a fingertip includes spectral components of such substances as melanin, hemoglobin, and other constituents of skin, muscle, blood, etc., with which the proportions present, etc. differ among individual persons (see, for example, Fig. 3b). The spectral information obtained from a fingertip or other portion of a living human is thus extremely complex and cannot be replicated readily by the use of artificial dummies and prosthetic devices, and especially because in this approach, the non-spectrometric biometric signature of the same portion is acquired for identification, spoofing is made a practically insurmountable task.

5.2 Preferred configuration for iris authentication system

To illustrate a further scope of application of this approach, a second embodiment according to this approach shall now be described. Fig. 9 is a schematic diagram of a basic arrangement of a spectral biometrics enhanced authentication system according to the second embodiment of this approach, which is an iris authentication device 200 that authenticates a person's identity based on his/her iris pattern and biospectral characteristics of his/her iris.

5.2.1 Incident light and its reflected components

As shown in Fig. 9, this iris authentication device 200 uses the same CCD 100, having the image acquisition portion 102 and the spectrum acquisition portion 103, as that used in the

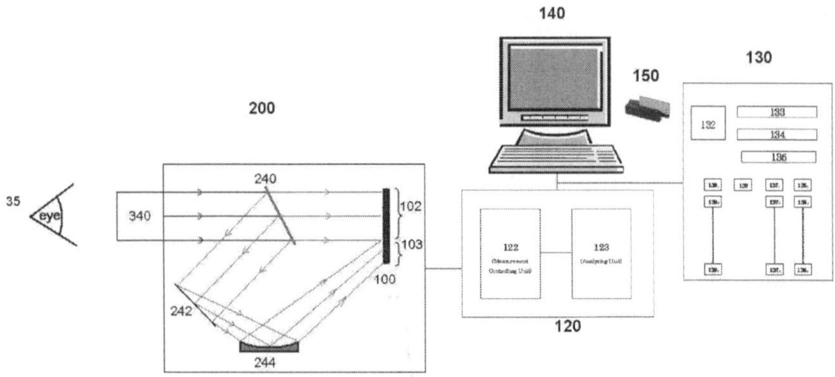

Fig. 9. A schematic diagram of a basic arrangement of a spectral biometrics enhanced authentication system according to a second embodiment of this work.

first embodiment. With this authentication device 200, an image of an iris 35 of a person 30 to be authenticated is formed on the image acquisition portion 102 by a lens 340. A portion (10% to 20%) of the light propagating from the iris 35 to the CCD 100 is reflected by a half-mirror 240 and then reflected by a mirror 242 onto a diffraction grating 244, which spectrally disperses and makes the component, reflected by the half-mirror 240, incident on the spectrum acquisition portion 103 of the CCD 100 in a manner such that a reflection spectrum of the iris within a range of 350nm to 1050nm can be acquired from each row of the spectrum acquisition portion 103.

5.2.2 Data acquisition and authentication process

The iris image, acquired by the image acquisition portion 102 of the CCD 100, and the iris reflection spectrum, acquired by the spectrum acquisition portion 103, are then handled in the same manner as the fingerprint image and fingertip diffuse reflectance spectrum, respectively, of the first embodiment to obtain an iris pattern and an iris spectral information vector, which are then handled in the same manner as the fingerprint image pattern and the fingertip spectral information vector of the first embodiment to perform the authentication process.

5.2.3 Reliability of the approach

As with the fingertip diffuse reflectance spectrum, the iris reflection spectrum contains information on internal tissue, blood, and other various physiological components of the eye (iris) and thus provides information concerning a person's unique biological spectral signature that cannot be spoofed readily.

5.3 More general system configuration approach

The present approach is not limited to the embodiments described above, and various modifications can be made within the scope of the approach. For example, although the CCD 100, having pixels arranged in 1280 rows and 1024 columns, was used as the image sensor in the embodiment described above, a CCD of any other size may be used or a CMOS device may be used instead as the image sensor. Also together with a CCD, CMOS sensor, or other image sensor; a PDA (photodiode array) or a sensor having just the same number of photodetecting elements as the number of spectral information to be determined (seven in the case of the above-described embodiments) may be configured to perform image acquisition and spectrum acquisition, respectively. Other variations such as employment of different light source (e.g., D2 lamp, laser, etc.) for acquisition, use of a more flexible ID reading mechanism for database access, or use of different biometrics signature (e.g., hand geometry, facial features, retinal print, etc.) may also be employed instead. Furthermore, during the analysis process a different wavelength range, wavelength number, or even a different pattern recognition method; such as neural networks, fuzzy logic, or linear programming may be employed instead.

6. Conclusion

This chapter showed that it is quite feasible to use spectral biometrics as a complementary method for preventing spoofing of existing biometrics technologies. The proposed method ensures that the identity obtained through the primary biometrics signature comes from a living, authentic person. It also showed how spectral biometrics can be implemented in two widely-used biometrics systems in a practical manner without introducing much overhead to the base biometrics technology or inconvenience to users.

7. References

Anderson, R., Hu, J. & Parrish, J. (1981). Optical radiation transfer in the human skin and applications in vivo remittance spectroscopy. In: *Bioengineering and the Skin, MTP Press Limited*, pp. 253-265.

Bailey, E. (2008). Europe Lead the Way with Biometric Passports, But Highest Growth Potential is in Asia, In: *Report Buyer*, Date of access: 2011, Available from: <http://www.sbwire.com/press-releases/new-report-predicts-that-global-biometrics-market-will-reach-71-billion-by-2012-18766.htm>

Brownlee, K., (2001). Method and apparatus for distinguishing a human finger from a reproduction of a fingerprint, In: *US Patent 6,292,576*, Digital Persona Inc.

Derakshani, R., Schuckers, S., Hornak, L. & Gorman, L. (2003). Determination of vitality from a non-invasive biomedical measurement for use in fingerprint scanners, In: *Pattern Recognition*.

Fukuzumi, S. (2001). Organism identifying method and device, In: *US patent 6,314,195*.

Ingemar R. (1993). The Euclidean distance transform, In: *PhD Thesis, Linköping University, E.E.Dept., Dissertation #304*, Linköping studies in science and technology.

Jenkins, F. & White, H. (1976). Fundamentals of Optics, Macmillan, New York.

Lapsley, P., Lee, J. & Pare, D. (1998). SmartTouch LLC Anti-fraud biometric scanner that accurately detects blood flow, In: *US Patent 5,737,439*.

Maltoni, D., Jain, A. & Prabhakar, S. (2005). Handbook of Fingerprint Recognition, Springer, 1st ed.

Maltoni, D., Maio, D., Jain, A. & Prabhakar, S. (2003). Handbook of Fingerprint Recognition. Springer Verlag, New York, NY, USA.

Matsumoto, T., Hirabayashi, M. & Sato, S. (2004). A Vulnerability of Irsi Matching (Part 3), In: *Proceedings of the 2004 Symposium on Cryptography and Information Security*, the Institute of Electronics, Information and Communication Engineers, pp. 701-706.

Matsumoto, T., Matsumoto, H., Yamada, K. & Hoshino, S. (2002). Impacts of Artificial 'Gummy' Fingers on Fingerprint System, In: *Optical Society and Counterfeit Deterrence Techniques IV*, Proceedings of SPIE, 4677, pp. 275-289.

McFedries, P. (2007). Biometrics, In: *The Word Spy*, Date of access: 2011, Available from: <http://www.wordspy.com/words/biometrics.asp>

Melanoma, (2006). Skin Reflectance Spectra, In: *Melanoma*, Date of access: 2011, Available from: <http://melanoma.blogsome.com/2006/03/24/skin-reflectance-spectra>

MINITAB Inc., (n.d.). In: *Minitab Statistical Software*, Date of access: 2007, Available from: <www.minitab.com/contacts>

O'Gorman, L. & Schuckers, S. (2001). Spoof detection for biometric sensing systems, In: *WO 01/24700*, Veridicom, Inc.

Ohtsuki T. & Healey, G. (1998). Using color and geometric models for extracting facial features, In: *Journal of Imaging Science and Technology*, 42(6), pp. 554-561.

Osten, D., Carim H., Arneson, M. & Blan, B. (1998). Biometric, personal authentication system, In: *US Patent 5,719,950*, Minnesota Mining and Manufacturing Company.

Pishva, D. (2007). Multi-factor Authentication Using Spectral Biometrics, In: *Journal of Japan Society for Fuzzy Theory and Intelligent Informatics - Special Issue on Security and Trust*, Vol.19, No.3, pp. 256-263.

Pishva, D. (2007). Spectroscopic Approach for a Liveness Detection in Biometrics Authentication, In: *41st Annual IEEE International Carnahan Conferences on Security Technology*, pp.133-137.

Pishva, D. (2008). Spectroscopic Method and System for Multi-factor Biometric Authentication, In: *International Patent Application Number PCT/JP2007/060168, International Publication No: WO/2008/139631 A1.*

Pishva, D. (2008). Spectroscopically Enhanced Method and System for Multi-factor Biometric Authentication, In: *IEICE Trans. Inf. & Syst.*, Vol. E91-D, No. 5, pp. 1369-1379.

Pishva D. (2010). Spectroscopic Method and System for Multi-factor Biometric Authentication, In: *International Patent Application Number PCT/JP2007/060168, Australian Patent Number: 2007352940.*

Rowe, R., Corcoran, S., Nixon, K. (n.d.). Biometric Identity Determination using Skin Spectroscopy, In: *Lumidigm, Inc.*, Date of access: 2007, Available from: <www.lumidigm.com>

Shafer, S. (1985). Using color to separate reflection components, In: *Color Research and Application*, Vol. 10, no. 4, pp. 210-218.

Tilton, C. (2006). Biometric Standards – An Overview, In: *Daon*, Date of access: 2007, Available from: <http://www.daon.com/downloads/standards/Biometric%20Standards%20White%20Paper%20Jan%2006.pdf>

UK Biometric Working Group Annual Report for 2003/2004, (2003, 2004). Date of access: 2011, Available from: <http://www.cesg.gov.uk/policy_technologies/biometrics/media/annual_report_03-04.pdf>

van der Putte, T. & Keuning, J. (2000). Biometrical fingerprint recognition: don't get your fingers burned, In: *Proceedings of IFIP TC8/WG8.8 Fourth Working Conference on Smart Card Research and Advanced Applications*, Kluwer Academic Publishers, pp. 289-303.

Wikipedia, the free encyclopaedia (n.d.). Euclidean distance, In: *Wikipedia*, Date of access: 2011, Available from: <http://en.wikipedia.org/wiki/Euclidean_distance>

Wikipedia, the free encyclopaedia (n.d.). Spectroscopy, In: *Wikipedia*, Date of access: 2011, Available from: <http://en.wikipedia.org/wiki/Spectroscopy>

Wildes, R., Asmuth, J., Green, G., Hsu, S., Kolczynski, R., Matey, J. & McBride, S. (1996). A machine vision system for iris recognition, In: *Machine Vision and Application*, Vol. 9, pp. 1-8.

8

A Contactless Biometric System Using Palm Print and Palm Vein Features

Goh Kah Ong Michael, Tee Connie and Andrew Beng Jin Teoh
[1]*Multimedia University,*
[2]*Yonsei University,*
[1]*Malaysia*
[2]*Korea*

1. Introduction

Recently, biometrics has emerged as a reliable technology to provide greater level of security to personal authentication system. Among the various biometric characteristics that can be used to recognize a person, the human hand is the oldest, and perhaps the most successful form of biometric technology (Hand-based biometrics, 2003). The features that can be extracted from the hand include hand geometry, fingerprint, palm print, knuckle print, and vein. These hand properties are stable and reliable. Once a person has reached adulthood, the hand structure and configuration remain relatively stable throughout the person's life (Yörük et al., 2006). Apart from that, the hand-scan technology is generally perceived as nonintrusive as compared to iris- or retina-scan systems (Jain et al., 2004). The users do not need to be cognizant of the way in which they interact with the system. These advantages have greatly facilitated the deployment of hand features in biometric applications.

At present, most of the hand acquisition devices are based on touch-based design. The users are required to touch the device or hold on to some peripheral or guidance peg for their hand images to be captured. There are a number of problems associated with this touch-based design. Firstly, people are concerned about the hygiene issue in which they have to place their hands on the same sensor where countless others have also placed theirs. This problem is particularly exacerbated during the outbreak of epidemics or pandemics like SARS and Influenza A (H1N1) which can be spread by touching germs leftover on surfaces. Secondly, latent hand prints which remain on the sensor's surface could be copied for illegitimate use. Researchers have demonstrated systematic methods to use latent fingerprints to create casts and moulds of the spoof fingers (Putte & Keuning, 2000). Thirdly, the device surface will be contaminated easily if not used right, especially in harsh, dirty, and outdoor environments. Lastly, some nations may resist placing their hands after a user of the opposite sex has touched the sensor.

This chapter presents a contactless hand-based biometric system to acquire the palm print and palm vein features. Palm prints refer to the smoothly flowing pattern formed by alternating creases and troughs on the palmar surface of the hand. Three types of line patterns are clearly visible on the palm. These line patterns are known as the principal lines, wrinkles, and ridges. Principal lines are the longest, strongest and widest lines on the palm. The principal lines characterize the most distinguishable features on the palm. Most people have three principal lines, which are named as the heart line, head line, and life line (Fig. 1).

Wrinkles are regarded as the thinner and more irregular line patterns. The wrinkles, especially the pronounced wrinkles around the principal lines, can also contribute for the discriminability of the palm print. On the other hand, ridges are the fine line texture distributed throughout the palmar surface. The ridge feature is less useful for discriminating individual as they cannot be perceived under poor imaging source.

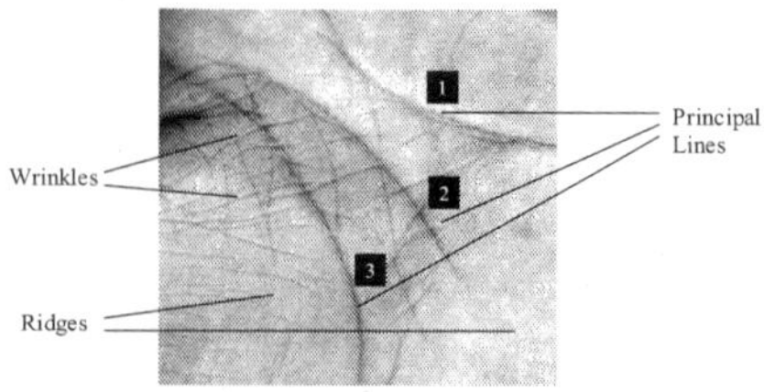

Fig. 1. The Line Patterns on the Palm Print. The Three Principal Lines on a Palm: 1-heart line, 2-head line and 3-life line (Zhang et al., 2003)

On the other hand, hand vein refers to the vascular pattern or blood vein patterns recorded from underneath the human skin. The subcutaneous blood vein flows through the human hand, covering the wrist, palm, and fingers. Every person has unique structure and position of veins, and this does not change significantly from the age of ten (Vein recognition in Europe, 2004). As the blood vessels are believed to be "hard-wired" into the body at birth, even twins have unique vein pattern. In fact, the vascular patterns on the left and right hands are also different. As the complex vein structure resides in the human body, it is not possible (except using anatomically surgery) to copy or duplicate the vein pattern. Besides, external conditions like greasy and dirty, wear and tear, dry and wet hand surface do not affect the vein structure. The properties of stability, uniqueness, and spoof-resilient make hand vein a potentially good biometrics for personal authentication. Fig. 2 depicts the vein patterns captured from the palmar surface, and back of the hand.

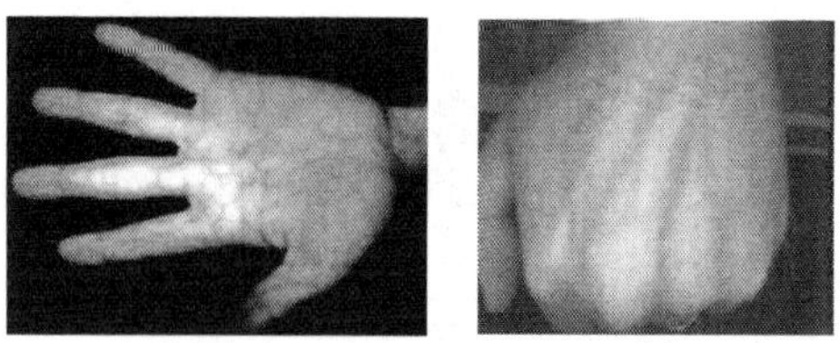

(a) (b)

Fig. 2. (a) Vein Image on the Palm and Fingers (PalmSecure™, 2009). (b) Vein Image at the Hand Dorsum (Hitachi and Fujitsu win vein orders in diverse markets, 2007).

2. Literature review

2.1 Palm print biometrics

2.1.1 Image acquisition

Most of the palm print systems utilized CCD scanners to acquire the palm print images (Zhang et al., 2003; Han, 2004; Kong & Zhang, 2004). A team of researchers from Hong Kong Polytechnic University pioneered CCD-based palm print scanner (Zhang et al., 2003). The palm print scanner was designed to work with predefined controlled environment. The proposed device captured high quality palm print images and aligned palms accurately with the aid of guidance pegs.

Although CCD-based palm print scanners could capture high quality images, they require careful device setup. This design involves appropriate selection and configuration of the lens, camera, and light sources. In view of this, some researchers proposed to use digital cameras and video cameras as this setting requires less effort for system design (Doublet et al., 2007). Most of the systems that deployed digital cameras and video cameras posed less stringent constraint on the users. They did not use pegs for hand placement and they did not require special lighting control. This was believed to increase user acceptance and reduce maintenance effort of the system. Nevertheless, they might cause problem as the image quality may be low due to uncontrolled illumination variation and distortion due to hand movement.

Apart from CCD scanners and digital camera/video camera, there was also research which employed digital scanner (Qin et al., 2006). Nonetheless, digital scanner is not suitable for real-time applications because of the long scanning time. Besides, the images may be deformed due to the pressing effect of the hand on the platform surface. Fig. 3 shows the palm print images collected using CCD scanner, digital scanner, and video camera.

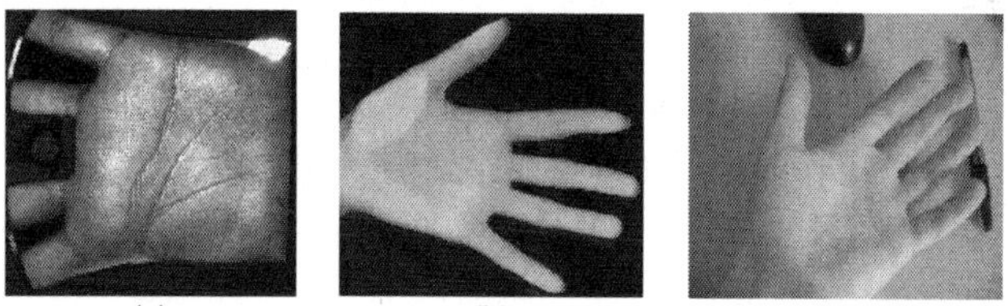

(a) (b) (c)

Fig. 3. Palm print images captured with (a) CCD scanner (Zhang et al., 2003), (b) digital scanner (Qin et al., 2006), and (c) video camera (Doublet et al., 2007).

2.1.2 Feature extraction

A number of approaches have been proposed to extract the various palm print features. The works reported in the literature can be broadly classified into three categories, namely line-based, appearance-based, and texture-based (Zhang & Liu, 2009). Some earlier research in palm print followed the line-based direction. The line-based approach studies the structural information of the palm print. Line patterns like principle lines, wrinkles, ridges, and creases are extracted for recognition (Funada, et al., 1998; Duta et al., 2002; Chen et al., 2001). The later researches used more flexible approach to extract the palm lines by using edge detection methods like Sobel operator (Wu et al., 2004a; Boles & Chu, 1997; Leung et al., 2007), morphological operator (RafaelDiaz et al., 2004), edge map (Kung et al., 1995), and modified radon transform (Huang et al., 2008). There were also researchers who implemented their own edge detection algorithms to extract the line patterns (Liu & Zhang, 2005; Wu et al., 2004b; Huang et al., 2008).

On the other hand, the appearance-based approach is more straightforward as it treats the palm print image as a whole. Common methods used for the appearance-base approach include principal component analysis (PCA) (Lu et al., 2003; Kumar & Negi, 2007), linear discriminant analysis (LDA) (Wu et al., 2003), and independent component analysis (ICA)

(Connie et al., 2005). There were also researchers who developed their own algorithms to analyze the appearance of the palm print (Zuo et al., 2005; Feng et al., 2006; Yang et al., 2007; Deng et al., 2008).

Alternatively, the texture-based approach treats the palm print as a texture image. Therefore, statistical methods like Law's convolution masks, Gabar filter, and Fourier Transform could be used to compute the texture energy of the palm print. Among the methods tested, 2-D Gabor filter has been shown to provide engaging result (You et al., 2004; Wu et al., 2004b; Kong et al., 2006). Ordinal measure has also appeared as another powerful method to extract the texture feature (Sun et al., 2005). It detects elongated and line-like image regions which are orthogonal in orientation. The extracted feature is known as ordinal feature. Some researchers had also explored the use of texture descriptors like local binary pattern to model the palm print texture (Wang et al., 2006). In addition to this, there were other techniques which studied the palm print texture in the frequency domain by using Fourier transform (Li et al., 2002) and discrete cosine transform (Kumar & Zhang, 2006).

Apart from the approaches described above, there were research which took a step forward to transform the palm print feature into binary codes representation (Kong & Zhang, 2004b; Kong & Zhang, 2006; Zhang et al., 2003; Sun et al., 2005; Kong & Zhang, 2002). The coding methods are suitable for classification involving large-scale database. The coding algorithms for palm print are inspired by the IrisCode technique (Daugman, 1993). PalmCode (Kong & Zhang, 2002; Zhang, Kong, You, & Wong, 2003) was the first coding based technique reported for palm print research. Later on, more variations had evolved from PalmCode which included Fusion Code (first and second versions) (Kong & Zhang, 2004b), and Competitive Code (Kong & Zhang, 2004; Kong, Zhang, & Kamel, 2006b). In addition, there were also other coding approaches like Ordinal code (Sun et al., 2005), orientation code (Wu et al., 2005), and line orientation code (Jia et al., 2008).

2.1.1 Matching

Depending on the types of features extracted, a variety of matching techniques were used to compare two palm print images. In general, these techniques can be divided into two categories: geometry-based matching, and feature-based matching (Teoh, 2009). The geometry-based matching techniques sought to compare the geometrical primitives like points (Duta, Jain, & Mardia, 2002; You, Li, & Zhang, 2002) and lines features (Huang, Jia, & Zhang, 2008; Zhang & Shu, 1999) on the palm. When the point features were located using methods like interesting point detector (You, Li, & Zhang, 2002), distance metric such as Hausdorff distance could be used to calculate the similarly between two feature sets. When the palm print pattern was characterized by line-based feature, Euclidean distance could be applied to compute the similarity, or rather dissimilarity, between two line segments represented in the Z^2 coordinate system. Line-based matching on the whole is perceived as more informative than point-based matching because the palm print pattern could be better characterized using the rich line features as compared to isolated datum point. Besides, researchers conjectured that simple line features like the principal lines have sufficiently strong discriminative ability (Huang, Jia, & Zhang, 2008).

Feature-based matching works well for the appearance-based and texture-based approaches. For research which studied the subspace methods like PCA, LDA, and ICA, most of the authors adopted Euclidean distances to compute the matching scores (Lu, Zhang, & Wang, 2003; Wu, Zhang, & Wang, 2003; Lu, Wang, & Zhang, 2004). For the other studies, a variety of distance matrices like city-block distance and chi square distances were deployed (Wu, Wang, & Zhang, 2004b; Wu, Wang, & Zhang, 2002; Wang, Gong, Zhang, Li, & Zhuang, 2006). Feature-based matching has a great advantage over geometry-based matching when low-resolution images are used. This is due to the reason that geometry-based matching usually requires higher resolution images to acquire precise location and orientation of the geometrical features.

Aside from the two primary matching approaches, more complicated machine learning techniques like neural networks (Han, Cheng, Lin, & Fan, 2003), Support Vector Machine (Zhou, Peng, & Yang, 2006), and Hidden Markov models (Wu, Wang, & Zhang, 2004c) were also tested. In most of the time, a number of the matching approaches can be combined to yield better accuracy. You et al. (2004) showed that the integration can be performed in a hierarchical manner for the boost in performance and speed.

When the palm print features were transformed into binary bit-string for representation, Hamming distance was utilized to count the bit difference between two feature sets (Zhang, Kong, You, & Wong, 2003; Kong & Zhang, 2004b; Sun, Tan, Wang, & Li, 2005). There was an exception to this case where the angular distance was employed for the competitive coding scheme (Kong & Zhang, 2004).

2.2 Hand vein biometrics

2.2.1 Image acquisition

In visible light, the vein structure of the hand is not always easily discernible. Due to biological composition of the human tissues, the vein pattern can be observed under infrared light. In the entire electromagnetic spectrum, infrared refers to a specific region with wavelength typically spanning from 0.75µm to 1000µm. This region can be further divided into four sub-bands, namely near infrared (NIR) in the range of 0.75µm to 2µm, middle infrared in the range of 2µm to 6µm, far infrared (FIR) in the range of 6µm to 14µm, and extreme infrared in the range of 14µm to 1000µm. In the literature, the NIR (Cross & Smith, 1995; Miura, Nagasaka, & Miyatake, 2004; Wang, Yau, Suwandya, & Sung, 2008; Toh, Eng, Choo, Cha, Yau, & Low, 2005) and FIR (Wang, Leedhamb, & Cho, 2008; Lin & Fan, 2004) sources were used to capture the hand vein images.

FIR imaging technology forms images based on the infrared radiation emitted from the human body. Medical researchers have found that human veins have higher temperature than the surrounding tissues (Mehnert, Cross, & Smith, 1993). Therefore, the vein patterns can be clearly displayed via thermal imaging (Fig. 4(a) and (b)). No external light is required for FIR imaging. Thus, FIR does not suffer from illumination problems like many other imaging techniques. However, this technology can be easily affected by external conditions like ambient temperature and humidity. In addition, perspiration can also affect the image quality (Wang, Leedham, & Cho, 2007).

On the other hand, the NIR technology functions based on two special attributes, (i) the infrared light can penetrate into the hand tissue to a depth of about 3mm, and (ii) the reduced haemoglobin in the venous blood absorbs more incident infrared radiation than the surrounding tissues (Cross & Smith, 1995). As such, the vein patterns near the skin surface are discernible as they appear darker than the surrounding area. As shown in Fig. 4(c), NIR can capture the major vein patterns as effectively as the FIR imaging technique. More importantly, it can detect finer veins lying near the skin surface. This increases the potential discriminative ability of the vein pattern. Apart from that, the NIR has better ability to withstand the external environment and the subject's body temperature. Besides, the colour of the skin does not have any impact of the vein patterns (Wang, Leedhamb, & Cho, 2008).

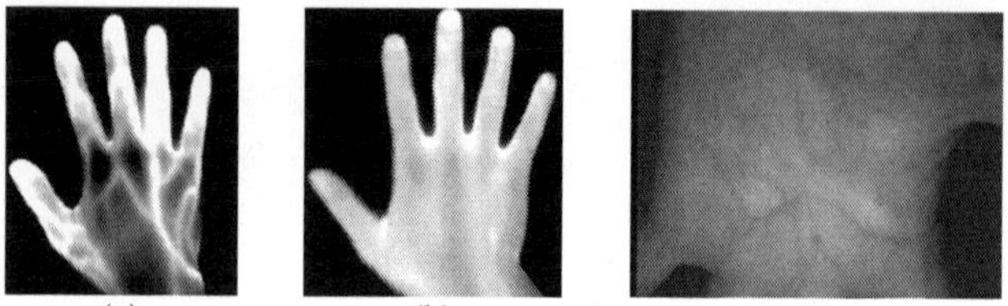

(a) (b) (c)

Fig. 4. (a) FIR image in normal office environment, (b) FIR image in outdoor environment (Wang, Leedhamb, & Cho, 2008). (c) NIR taken for vein image (Wang, Leedhamb, & Cho, 2008

Infrared sensitive CCD cameras like Takena System NC300AIR (Miura, Nagasaka, & Miyatake, 2004), JAI CV-M50 IR (Toh, Eng, Choo, Cha, Yau, & Low, 2005; Wang, Yau, Suwandya, & Sung, 2008), and Hitachi KP-F2A (Wang, Leedham, & Cho, 2007) were used to capture images of veins near to body surface. Near infrared LEDs with wavelength from 850nm (Wu & Ye, 2009; Wang, Leedham, & Cho, 2007; Kumar & Prathyusha, 2009) to 880nm (Cross & Smith, 1995) were used as the light source. To cutoff the visible light, IR filter with different cutoff wavelengths, λ, were devised. Some researches deployed IR filter with λ ~ 800nm (Wang, Leedham, & Cho, 2007; Wu & Ye, 2009) and some used higher cutoff wavelengths at 900nm (Cross & Smith, 1995).

2.2.2 Feature extraction

The feature extraction methods for vein recognition can be broadly categorized into: (i) structural-, and (ii) global-based approaches. The structural method studies the line and feature points (like minutiae) of the vein (Cross & Smith, 1995; Miura, Nagasaka, & Miyatake, 2004; Wang, Zhang, Yuan, & Zhuang, 2006; Kumar & Prathyusha, 2009). Thresholding and thinning techniques (Cross & Smith, 1995; Wang, Zhang, Yuan, & Zhuang, 2006), morphological operator (Toh, Eng, Choo, Cha, Yau, & Low, 2005), as well as skeletonization and smoothing methods (Wang, Leedhamb, & Cho, 2008) were used to extract the vein structure. These geometrical/topological features were used to represent the vein pattern.
On the contrary, the global-based method characterizes the vein image in its entirety. Lin and Fan (2004) performed multi-resolution analysis to analyze the palm-dorsa vein patterns. Wang et al. (2006) carried out multi feature extraction based on vein geometry, K-L conversion transform, and invariable moment and fused the results of these methods. The other global-based approaches adopted curvelet (Zhang, Ma, & Han, 2006) and Radon Transform (Wu & Ye, 2009) to extract the vein feature.

2.2.3 Matching

Most of the works in the literature deployed the correlation technique (or its other variations) to evaluate the similarity between the enrolled and test images (Cross & Smith, 1995; Kono, Ueki, & Umemur, 2002; Miura, Nagasaka, & Miyatake, 2004; Toh, Eng, Choo, Cha, Yau, & Low, 2005). Other simple distance matching measure like Hausdorff distance was also adopted (Wang, Leedhamb, & Cho, 2008). More sophisticated methods like back propagation neural networks (Zhang, Ma, & Han, 2006) and Radial Basis Function (RBF) Neural Network and Probabilistic Neural Network (Wu & Ye, 2009) were also used as the classifiers in vein recognition research.

3. Proposed solution

In this research, we endeavour to develop an online acquisition device which can capture hand images in a contactless environment. In specific, we want to acquire the different hand modalities, namely palm print and palm vein images from the hand simultaneously without incurring additional sensor cost or adding user complication. The users do not need to touch or hold on to any peripheral for their hand images to be acquired. When their hand images are captured, the regions of interest (ROI) of the palm will be tracked and extracted. ROIs contain the important information of the hand that is used for recognition. The ROIs are pre-processed so that the print and vein textures become distinguishable from the background. After that, distinguishing features in the ROIs are extracted using a proposed technique called directional coding. The hand features are mainly made up of line-like texture. The directional coding technique encodes the discriminative information of the hand based on the orientation of the line primitives. The extracted palm print and palm vein features are then fused at score level to yield better recognition accuracy. We have also included an image quality assessment scheme to evaluate the image quality. We distribute more weight to better quality image when fusion is performed. The framework of our proposed system is shown in Fig. 5.

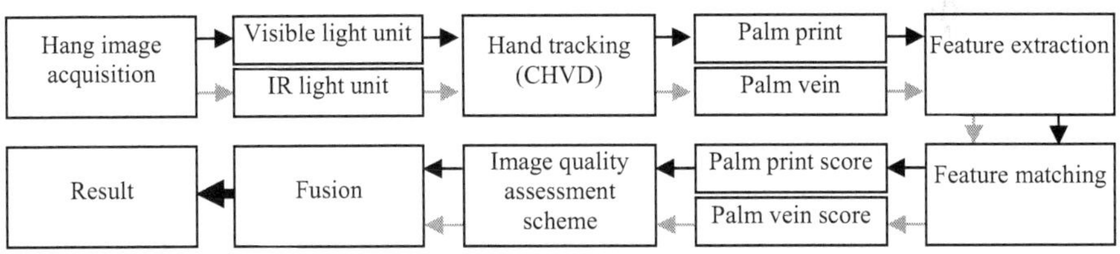

Fig. 5. Framework of the proposed system.

3.1 Design and implementation of acquisition device

Image acquisition is a very important component because it generates the images to be used and evaluated in this study. We aim to develop a real-time acquisition device which can capture hand images in a contactless environment. The design and implementation of an efficient real-time hand acquisition device must contend with a number of challenges. Firstly, the acquisition device must be able to provide sufficient contrasted images so that the hand features are discernable and can be used for processing. The hardware setup plays a crucial role in providing high quality images. Arrangement of the imaging sensor and design of the lighting units also have great impact on the quality of the images acquired. Therefore, the capturing device should be calibrated carefully to obtain high contrasted images. Secondly, a single acquisition device should be used to capture multiple image sources (e.g. visible and infrared images). It is not efficient and economical for a multimodal biometric system to install multiple capturing devices, for example, using a normal camera to acquire visible image and using another specialized equipment to obtain IR image. Therefore, an acquisition device with low development cost is expected for a multimodal biometric system from the system application view. Thirdly, speed is a major concern in an online application. The capturing time of the acquisition device should be fast enough to make it unnoticeable to the user that multiple biometric features are being acquired by the system for processing. In other words, a real-time acquisition system should be able capture all of the biometric features in the shortest time possible.

In this research, we design an acquisition device that aims to fulfil the requirements above. The hardware setup of the capturing device is shown in Fig. 6. Two low-cost imaging units are mounted side by side on the device. The first imaging unit is used to capture visible light images while the second for obtaining infrared images. Both units are commercially available off-the-shelf webcams. Warm-white light bulbs are placed around the imaging units to irradiate the hand under visible light. The bulbs emit yellowish light source that enhances the lines and ridges of the palm. To acquire IR image, we do not use any specialized IR camera. Instead, we modify the ordinary webcam to be an IR-sensitive camera. The webcam used for infrared imaging is fitted with an infrared filter. The filter blocks the visible (non-IR) light and allows only the IR light to reach the sensor. In this study, we find that IR filter which passes infrared rays above 900nm gives the best quality images. A number of infrared LEDs are arranged on the board to serve as the infrared cold source to illuminate the vein pattern. We have experimented with different types of infrared LEDs and those emitting light in the range of 880nm to 920nm provide relatively good contrast of the vein pattern. A diffuser paper is used to attenuate the IR source so that the radiation can be distributed more uniformly around the imaging unit.

During image acquisition, we request the user to position his/her hand above the sensor with the palm facing the sensor (Fig. 6(a)). The user has to slightly stretch his/her fingers apart. There is no guidance peripheral to restraint the user's hand. The user can place his/her hand naturally above the sensor. We do not restrict the user to place his/her hand at a particular position above the sensor nor limit them to pose his/her at a certain direction. Instead, we allow the user to move his/her hand while the images are being acquired. Besides, the user can also rotate his/her hand while the images are being taken. The optimal viewing region for the acquisition sensor is 25 cm from the surface of the imaging unit. We allow a tolerable focus range of 25 cm ± 4 cm to permit more flexibility for the users to interact with the device (Fig. 6(c) and Fig. 7).

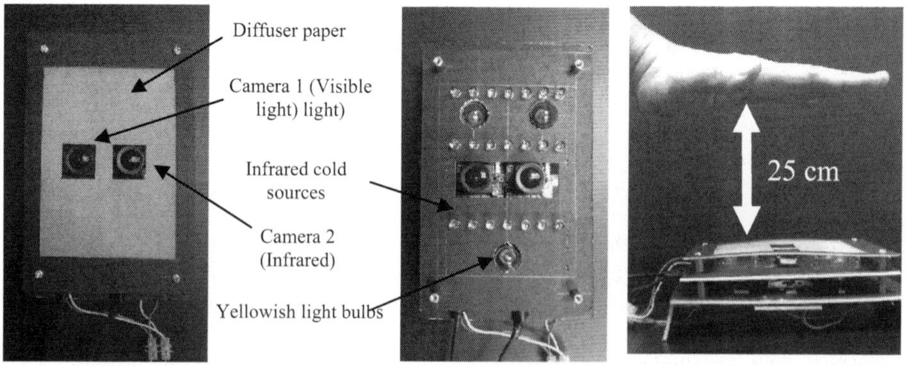

(a) (b) (c)

Fig. 6. (a) Image acquisiton device (covered). (b) Image acquisiton device (uncovered). (c) Acquiring the hand images.

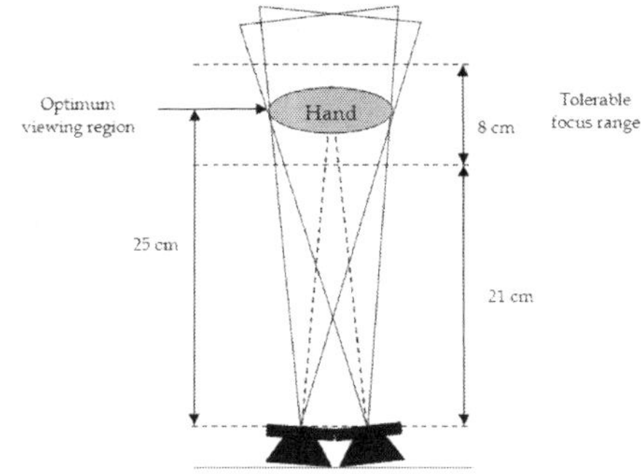

Fig. 7. Tolerable focus range for the image acquisition device.

In this study, a standard PC with Intel Core 2 Quad processor (2.4 GHz) and 3072 MB RAM was used. The program was developed using Visual Studio .NET 2008. The application depicted in Fig. 8 shows a live video sequence of the hand image recorded by the sensor. Both of the visible light and IR images of the hand can be captured simultaneously. The interface provides direct feedback to the user that he/she is placing his/her hand properly inside the working volume. After the hand was detected in the working volume, the ROIs of the palm and fingers were captured and stored as bitmap format from the video sequence. The hand image was detected in real-time video sequence at 30 fps. The image resolution was 640 x 480 pixels, with color output type in 256 RGB (8 bits-per-channel). The delay interval between capturing the current and the next ROI was 2 seconds.

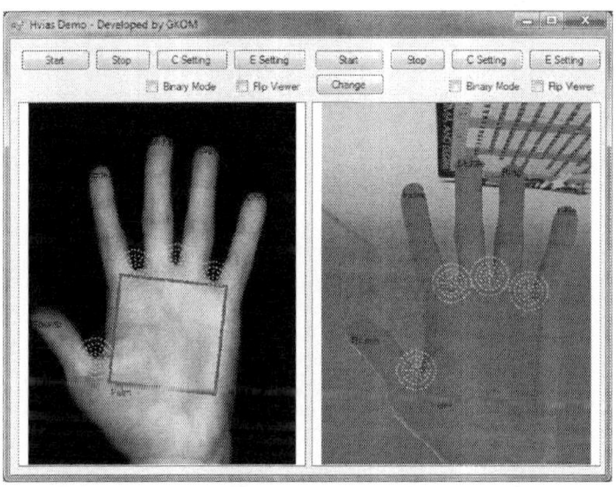

Fig. 8. Software interface depicting the image acquisition process.

We used the setup described above in an office environment to evaluate the performance of the proposed multimodal hand-based biometric system. We have recorded the hand images from 136 individuals. 64 of them are females, 42 of them are less than 30 years old. The users come from different ethnic groups such as Chinese, Malays, Indians, and Arabians. Most of them are students and lecturers from Multimedia University. Ten samples were captured for each user. The samples were acquired in two different occasions separated at an interval of two months.

3.2 Pre-processing

We adopt the hand tracking algorithm proposed in our previous work (Goh et al., 2008) to detect and locate the region of interest (ROI) of the palm. After obtaining the ROIs, we enhance the contrast and sharpness of the images so that the dominant palm vein features can be highlighted and become distinguishable from the background. Gamma correction is first applied to obtain better image contrast (Gonzalez, & Woods, 2002). To bring out the detail of the ridge pattern, we have investigated a number of well-known image enhancement methods like Laplacian filters, Laplacian of Gaussian, and unsharp masking method. Although these techniques work well for sharpening the images, the noise elements tend to be over-enhanced. For this reason, we propose a local-ridge-enhancement (LRE) technique to obtain a sharp image without overly amplifying the noise. This method discovers which part of the image contains important lines and ridge patterns, and amplifies only these areas.

The proposed LRE method uses a "ridge detection mask" to find the palm vein structures in the image. LRE first applies a low-pass filter, $g(x,y)$, on the original image, $I(x,y)$, shown in Fig. 9a to obtain a blur version of the image, $M(x,y)$,

$$M(x,y) = g(x,y) * I(x,y) \tag{1}$$

In this research, Gaussian filter with σ=60 is used for this purpose. After that, we use a high-pass filter, $h(x,y)$, to locate the ridge edges from the blur image,

$$M^{'}(x,y) = h(x,y) * M(x,y) \tag{2}$$

Note that since the trivial/weak ridge patterns have already been "distilled" in the blur image, only the edges of the principal/strong ridges show up in $M^{'}(x,y)$. In this work, the Laplacian filter is used as the high-pass filter.

At this stage, $M^{'}(x,y)$ exhibit the edges of the primary ridge structure (Fig. 9c). We binarize $M^{'}(x,y)$ by using a threshold value, τ. Some morphological operators like opening and closing can be used to eliminate unwanted noise regions. The resultant image is the "mask" marking the location of the strong ridge pattern.

We "overlay" $M^{'}(x,y)$ on the original image to amplify the ridge region,

$$I^{'}(x,y) = \begin{cases} c \cdot I(x,y) & \text{if } M'(x,) = 1 \\ I(x,y) & \text{otherwise} \end{cases} \tag{3}$$

where $I^{'}(x,y)$ is the enhanced image and c is the coefficient to determine the level of intensity used to highlight the ridge area. The lower the value of c, the more the ridge pattern will be amplified (the darker the area will be). In this work, the value of c is empirically set to 0.9. Fig. 9f shows the result of the enhanced image. We wish to point out that more variations can be added to determine different values for c in order to highlight the different ridge areas according to their strength levels. For example, gray-level slicing can be used to assign larger weight, c, to stronger ridge pattern, and vice versa. We do not perform this additional step due to the consideration for computation overhead (computation time is a critical factor for an online application). Fig. 10 depicts some sample image enhancement results for the palm print and palm vein images.

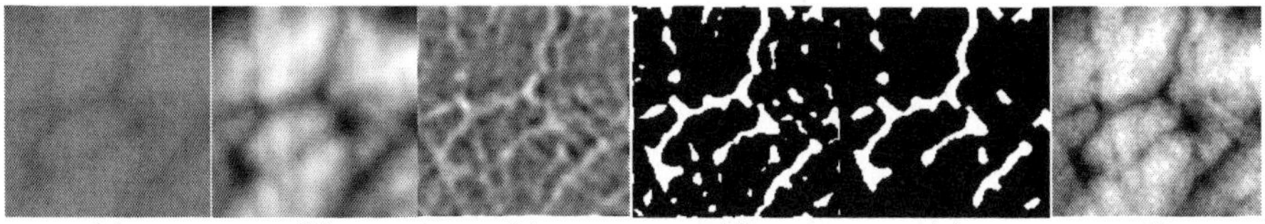

(a) (b) (c) (d) (e) (f)

Fig. 9. Processes involved in the proposed LRE method. (a) Original image. (b) Response of applying low-pass filter. (c) Response of applying high-pass filter on the response of low-pas filter. (d) Image binarization. (e) Applying morphological operations. (f) Result of LRE.

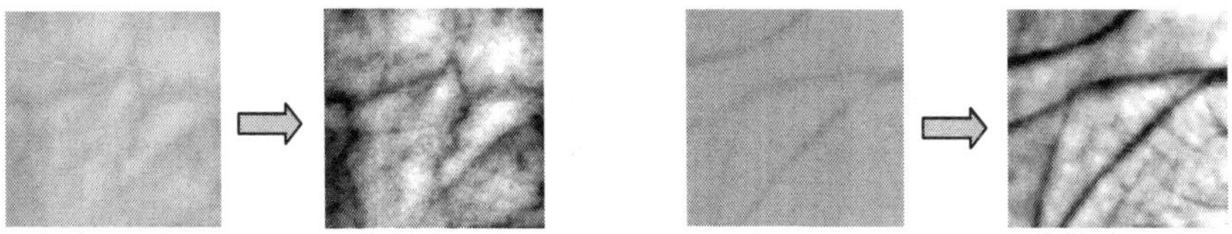

Palm vein LRE Result Palm print LRE Result

Fig. 10. Result of applying the proposed LRE method.

3.3 Feature extraction

We propose a new scheme named Directional Coding method to extract the palm print and palm vein features. These hand features contain similar textures which are primarily made up of line primitives. For example, palm prints are made up of strong principal lines and some thin wrinkles, whilst palm vein contains vascular network which also resembles line-like characteristic. Therefore, we can deploy a single method to extract the discriminative line information from the different hand features. The aims is to encode the line pattern based on the proximal orientation of the lines. We first apply Wavelet Transform to decompose the palm print images into lower resolution representation. The Sobel operator is then used to detect the palm print edges in horizontal, vertical, +45°, and -45° orientations. After that, the output sample, $\Phi(x,y)$, is determined using the formula,

$$\Phi(x,y) = \delta\left(\arg\max_f(\omega_R(x,y))\right) \tag{4}$$

where $\omega_R(x,y)$ denotes the responses of the Sobel mask in the four directions (horizontal, vertical, +45°, and -45°), and $\delta \in \{1,2,3,4\}$ indicates the index used to code the orientation of . The index, δ, can be in any form, but we use decimal representation to characterize the four orientations for the sake of simplicity. The output, $\Phi(x,y)$, is then converted to the corresponding binary reflected Gray code. The bit string assignment enables more effective matching process as the computation only deals with plain binary bit string rather than real or floating point numbers. Besides, another benefit of converting bit string to Gray code representation is that Gray code exhibits less bit transition. This is a desired property since we require the biometric feature to have high similarity within the data (for the same subject). Thus, Gray code representation provides less bit difference and more similarity in the data pattern. Fig. 11(b) to (e) shows the gradient responses of the palm print in the four directions. Fig. 11(f) is the result of taking the maximum gradient values obtained from the four responses. This image depicts the strongest directional response of the palm print and it closely resembles the original palm print pattern shown in Fig. 11(a). The example of directional coding applied on palm vein image is illustrated in Fig. 12.

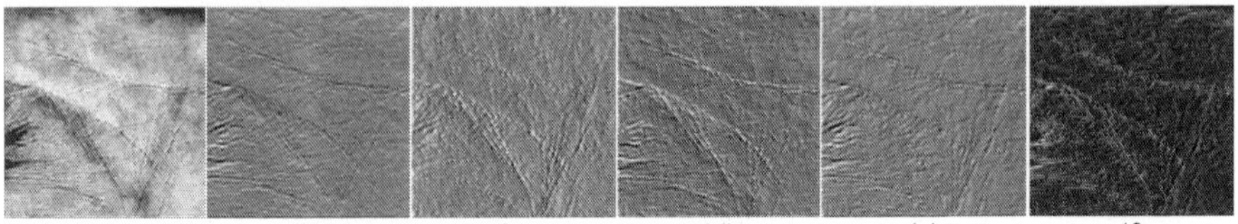

(a) (b) (c) (d) (e) (f)

Fig. 11. Example of Directional Code applied on palm print image.

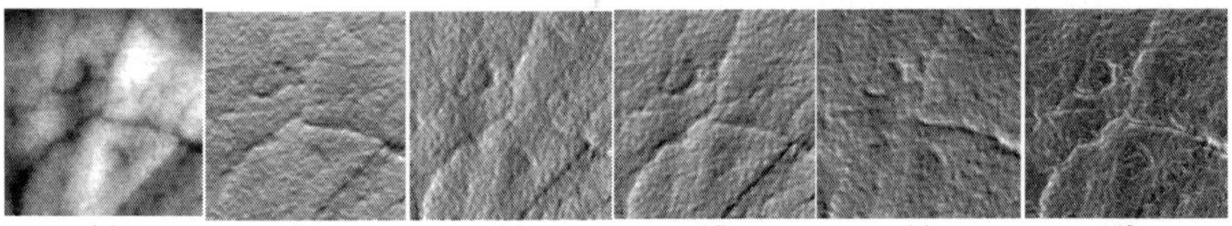

(a) (b) (c) (d) (e) (f)

Fig. 12. Example of Directional Code applied on palm vein image.

3.4 Feature matching

Hamming distance is deployed to count the fraction of bits that differ between two code strings for the Directional Coding method. Hamming distance is defined as,

$$d_{ham}(G,P) = XOR(G,P) \tag{5}$$

3.5 Fusion approach

In this research, the sum-based fusion rule is used to consolidate the matching scores produced by the different hand biometrics modalities. Sum rule is defined as,

$$\overline{S} = \sum_{i=1}^{k} s_i \tag{6}$$

where s denotes the scores generated from the different experts and k signifies the number of experts in the system. The reason of applying sum rule is because studies have shown that sum rule provides good results as compared to other decision level fusion techniques like likelihood ratio-based fusion (He, et al., 2010), neural networks (Ross & Jain, 2003) and decision trees (Wang, Tan, & Jain, 2003). Another reason we do not apply sophisticated fusion technique in our work is because our dataset has been reasonably cleansed by the image pre-processing and feature extraction stages (as will be shown in the experiment section).

Sum rule is a linear-based fusion method. To conduct more thorough evaluation, we wish to examine the use of non-linear classification tool. Support Vector Machine (SVM) is adopted for this purpose. SVM is a type of machine learning technique which is based on Structural

Risk Minimization (SRM) principal. It has good generalization characteristics by minimizing the boundary based on the generalization error, and it has been proven to be successful classifier on several classical pattern recognition problems (Burges, 1998). In this research, the Radial Basis Kernel (RBF) function is explored. RBF kernel is defined as (Saunders, 1998; Vapnik, 1998),

$$K(x,x_i) = \exp\left(-\frac{(x-x_i)^2}{2\sigma^2}\right) \tag{7}$$

where $\sigma > 0$ is a constant that defines the kernel width.

3.6 Incorporating image quality assessment in fusion scheme

We propose a novel method to incorporate image quality in our fusion scheme to obtain better performance. We first examine the quality of the images captured by the imaging device. We distribute more weight to better quality image when fusion is performed. The assignment of larger weight to better quality image is useful when we fuse the images under visible (e.g. palm print) and infrared light (e.g. palm vein). Sometimes, the vein images may not appear clear due to the medical condition of the skin (like thick fatty tissue obstructing the subcutaneous blood vessels), thus, it is not appropriate to assign equal weight between these poor quality images and those having clear patterns.

We design an evaluation method to assess the richness of texture in the images. We quantify/measure the image quality by using the measures derived using Gray Level Co-occurrence Matrix (GLCM) (Haralick, Shanmugam, & Dinstein, 1973). We have discovered several GLCM measures which can describe image quality appropriately. These measures were modelled using fuzzy logic to produce the final image quality metric that can be used in the fusion scheme.

3.6.1 Brief overview of GLCM

GLCM is a popular texture analysis tool which has been successfully applied in a number of applications like medical analysis (Tahir, Bouridane, & Kurugollu, 2005), geological imaging (Soh & Tsatsoulis, 1999), remote sensing (Ishak, Mustafa, & Hussain, 2008), and radiography (Chen, Tai, & Zhao, 2008). Given an $M \times N$ image with gray level values range from 0 to L-1, the GLCM for this image, $P(i,j,d,\theta)$, refers to the matrix recording the joint probability function, where i and j are the elements in the GLCM defined by a distance d in θ direction. More formally, the $(i, j)^{th}$ element in the GLCM for an image can be expressed as,

$$P(i,j,d,\theta) = \#\begin{Bmatrix} [(x_1,y_1),(x_2,y_2)] \mid f(x_1,y_1)=i, f(x_2,y_2)=j, \\ dis((x_1,y_1),(x_2,y_2))=d, \angle((x_1,y_1),(x_2,y_2))=\theta \end{Bmatrix} \tag{8}$$

where $dis()$ refers to a distance metric, and $\angle$ is the angle between two pixels in the image. # denotes the number of times the pixel intensity $f(x, y)$ appears in the relationship characterized by d and θ. To obtain the normalized GLCM, we can divide each entry by the number of pixels in the image,

$$P_{norm}(i,j,d,\theta) = \frac{P(i,j,d,\theta)}{M \times N} \tag{9}$$

Based on the GLCM, a number of textural features could be calculated. Among the commonly used features are shown in Table 1.

No.	Feature	Equation
1	Angular second moment (ASM)/Energy	$ASM = \sum_{i,j} P(i,j,d,\theta)^2$
2	Contrast	$con = \sum_{i,j} \lvert i-j \rvert^2 P(i,j,d,\theta)$
3	Correlation	$corr = \sum_{i,j} \frac{(i-\mu_i)(j-\mu_j)}{\sigma_i \sigma_j} P(i,j,d,\theta)$
4	Homogeneity	$hom = \sum_{i,j} \frac{P(i,j,d,\theta)}{1+\lvert i-j \rvert}$
5	Entropy	$ent = -\sum_{i,j} P(i,j,d,\theta) \log(P(i,j,d,\theta))$

Table 1. Some common GLCM textual features.

These measures are useful to describe the texture of an image. For example, ASM tells how orderly an image is, and homogeneity measures how closely the elements are distributed in the image.

3.6.2 Selecting image quality metrics

Based on the different texture features derived from GLCM, the fuzzy inference system can be used to aggregate these parameters and derive a final image quality score. Among the different GLCM metrics, we observe that contrast, variance, and correlation could characterize image quality well. Contrast is the chief indicator for image quality. An image with high contrast portrays dark and visible line texture. Variance and correlation are also good indicators of image quality. Better quality images tend to have higher values for contrast and variance, and lower value for correlation. Table 2 shows the values for contrast, variance, and correlation for the palm print and palm vein images.

When we observe the images, we find that images constituting similar amount of textural information yield similar measurements for contrast, variance, and correlation. Both the palm print and palm vein images for the first person, for instance, contain plenty of textural information. Thus, their GLCM features, especially the contrast value, do not vary much. However, as the texture is clearly more visible in the palm print image than the palm vein image for the second person, it is not surprising to find that the palm print image contains much higher contrast value than the vein image in this respect.

3.6.3 Modeling fuzzy inference system

The three image quality metrics namely contrast, variance and correlation are fed as input to the fuzzy inference system. Each of the input sets are modelled by three functions as depicted in Fig. 13(a)-(c).

The membership functions are formed by Gaussian functions or a combination of Gaussian functions given by,

Subject	Palm print	Quality metrics	Palm vein	Quality metrics
1	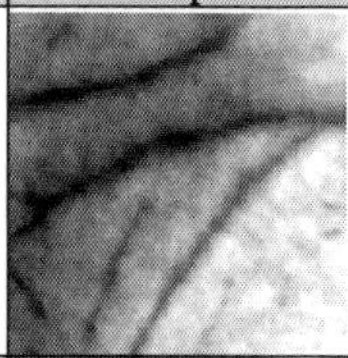	Contrast : 5.82 Variance : 12.91 Correlation : 3.21 **Defuzzified Output : 0.74**	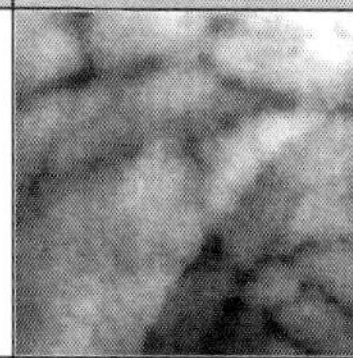	Contrast : 5.82 Variance : 12.91 Correlation : 3.21 **Defuzzified Output : 0.74**
2	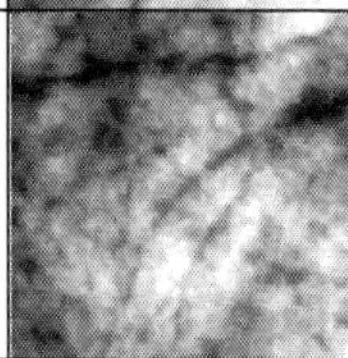	Contrast : 7.71 Variance : 7.92 Correlation : 2.63 **Defuzzified Output : 0.81**	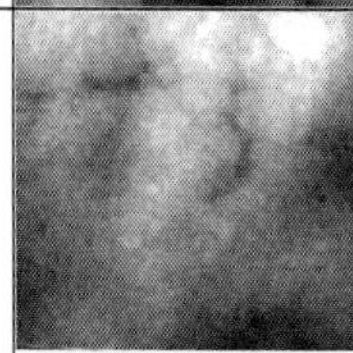	Contrast : 2.97 Variance : 6.44 Correlation : 3.45 **Defuzzified Output : 0.49**
3	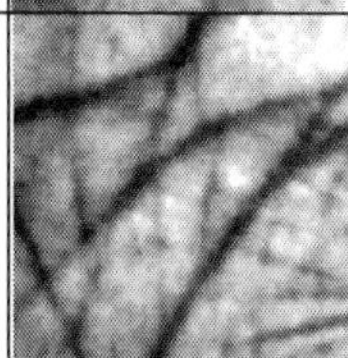	Contrast : 12.13 Variance : 8.28 Correlation : 1.90 **Defuzzified Output : 0.81**	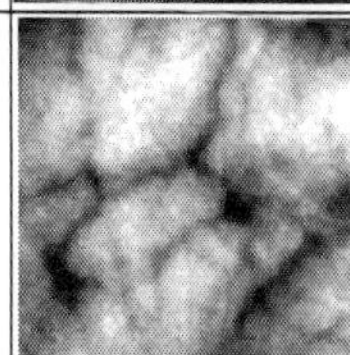	Contrast : 8.47 Variance : 8.44 Correlation : 2.54 **Defuzzified Output : 0.80**
4	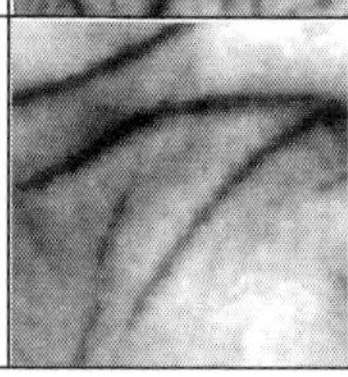	Contrast : 7.05 Variance : 8.57 Correlation : 2.80 **Defuzzified Output : 0.80**	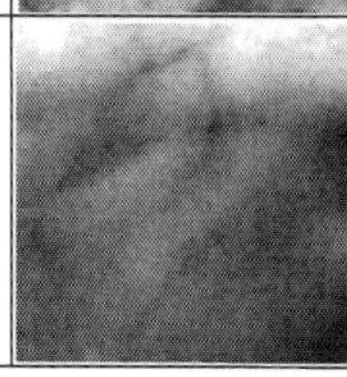	Contrast : 2.04 Variance : 4.10 Correlation : 3.58 **Defuzzified Output : 0.26**

Table 2. The GLCM measures and image quality metrics for the sample palm print and palm vein images.

$$f(x,\sigma,c) = e^{-(x-c)^2/2\sigma^2} \tag{10}$$

where c indicates the centre of the peak and σ controls the width of the distribution. The parameters for each of the membership functions are determined by taking the best performing values using the development set.

The conditions for the image quality measures are expressed by the fuzzy IF-THEN rules. The principal controller for determining the output for image quality is the contrast value. The image quality is good if the contrast value is large, and vice versa. The other two inputs, variance and correlation, serve as regulators to aggregate the output value when the contrast value is fair/medium. Thirteen rules are used to characterize fuzzy rules. The main properties for these rules are,

- If all the inputs are favourable (high contrast, high variance, and low correlation), the output is set to high.
- If the inputs are fair, the output value is determined primarily by the contrast value.
- If all the inputs are unfavourable (low contrast, low variance, and high correlation), the output is set to low.

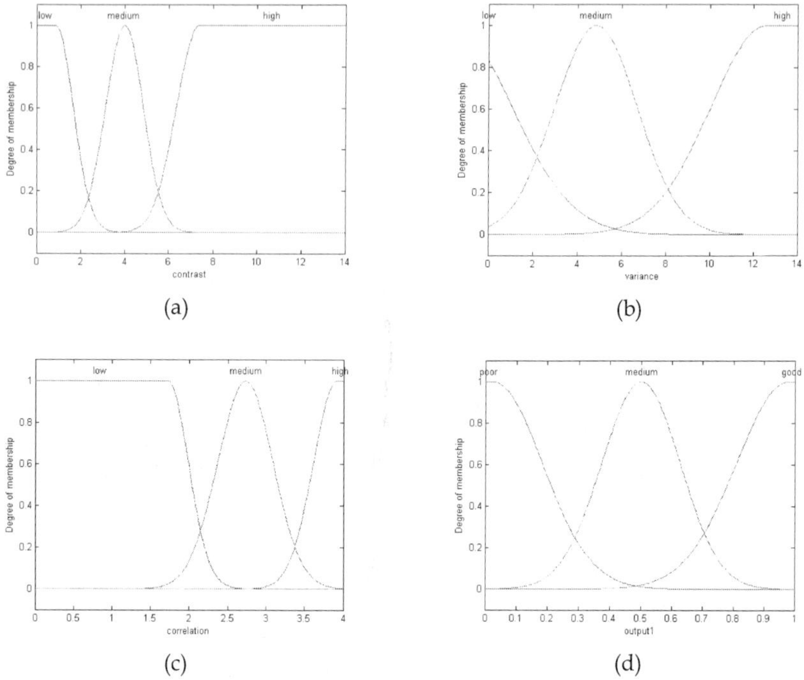

Fig. 2. Three membership functions defined for the input variables, (a) the contrast parameters, (b) the variance parameter and (c) the correlation parameters. (d) Output membership functions.

We use the Mamdami reasoning method to interpret the fuzzy set rules. This technique is adopted because it is intuitive and works well with human input. The output membership functions are given as O={*Poor, Medium, Good*}. The output functions are shown in Fig. 13(d) and they comprise of similar distribution functions as the input sets (combination of Gaussian functions). The defuzzified output score are recorded in Table 2. The output values adequately reflect the quality of the input images (the higher the value, the better the image quality).

3.6.4 Using image quality score in fusion scheme

The defuzzified output values are used as the weighting score for the biometric features in the fusion scheme. Let say we form a vector $\left(d_1, d_2, \ldots, d_j\right)$ from the individual outputs of the biometrics classifiers, the defuzzified output can be incorporated into the vector as $\left(\omega_1 d_1, \omega_2 d_2, \ldots, \omega_j d_j\right)$, where j stands for the number of biometric samples, and ω refers to the defuzzified output value for the biometric samples. Note that ω is the normalized value in which $\omega_1 + \omega_2 + \ldots + \omega_j = 1$. The weighted vector can then be input to the fusion scheme to perform authentication.

4. Results and discussion

4.1 Performance of uni-modal hand biometrics

An experiment was carried out to assess the effectiveness of the proposed Directional Coding method applied on the individual palm print and palm vein modalities. The results for both the left and right hands were recorded for the sake of thorough analysis of the hand features. The values for EER were taken at the point where FAR was equalled, or nearly equalled, to FRR. In the experiment, we also examined the performance of the system when FAR was set to 0.01% and 0.001%. The reason for doing this is because FAR is considered as one of the most significant parameter settings in a biometric system. It measures the likelihood of unauthorized accesses to the system. In some security critical applications, even one failure to detect fraudulent break-in could cause disruptive consequence to the system. Therefore, it is of paramount importance to evaluate the system at very low FAR.

The performances of the individual hand modalities are presented in Table 3. We observe that palm vein performs slightly better by yielding GAR approximately equals 97% when FAR was set to 0.001%. Thus, we find that there is a need to combine these modalities in order to obtain promising result. We also discover that the results for both of the hands do not vary significantly. This implies that the users can use either hand to access the biometric system. This is an advantage in security and flexibility as the user can choose to use either hand for the system. If one of the user's hand information is tampered or the hand is physically injured, he/she can still access the system by using the other hand. Apart from that, allowing the user to use both hands reduces the chance of being falsely rejected. This gives the user more chances of presentation and thereby reduces the inconvenience of being denied access.

We had also included an experiment to verify the usefulness of the proposed local ridge enhancement (LRE) pre-processing technique to enhance the hand features. The result of applying and without applying the pre-processing procedure is depicted in Fig. 14. The pre-processing step had indeed helped to improve the overall performance by 6%.

Hand Side	Biometrics	EER%	GAR% when FAR = 0.01%	GAR% when FAR = 0.001%
Right	Palm print (PP)	2.02	95.46	94.65
Right	Palm vein (PV)	0.71	98.34	97.95
Left	Palm print (PP)	1.97	95.77	94.05
Left	Palm vein (PV)	0.80	98.26	97.49

Table 3. Performance of individual biometric experts.

4.2 Performance of multimodal hand biometrics

4.2.1 Analysis of biometric features

Correlation analysis of individual experts is important to determine their discriminatory power, data separability, and information complementary ability. A common way to identify the correlation which exists between the experts is to analyze the errors made by them. The fusion result can be very effective if the errors made by the classifiers are highly de-correlated. In other words, the lower the correlation value, the more effective the fusion will become. This is due to the reason that more new information will be introduced when the dependency between the errors decreases (Verlinde, 1999). One way to visualize the

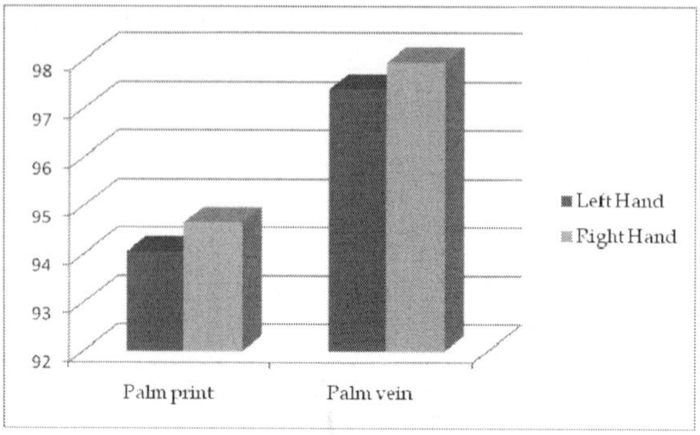

Fig. 14. Improvement gain by applying the proposed LRE pre-processing technique for left and right hands.

correlation between two classifiers is to plot the distribution graph of the genuine and imposter populations. In the correlation observation shown in Fig. 15, the distribution of the genuine and imposter populations take the form of two nearly independent clusters. This indicates that the correlation between the individual palm print and palm vein modality is low. In other words, we found that both biometrics are independent and are suitable to be used for fusion.

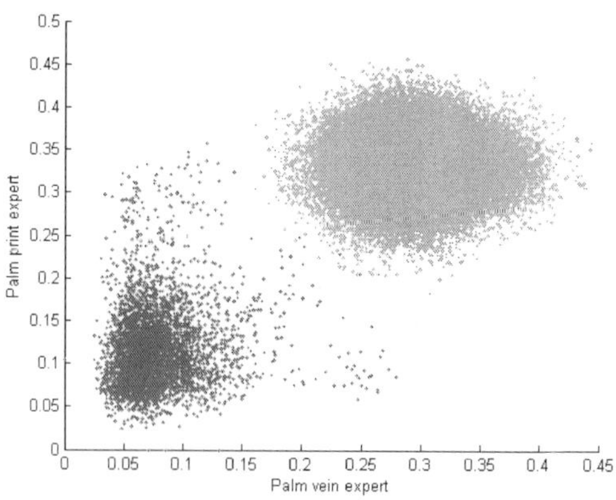

Fig. 15. Visual representation of the correlation of between palm print and palm vein experts.

4.2.2 Fusion using sum-rule

In this experiment, we combine the palm print and palm vein experts using the sum-based fusion rule. Table 4 records the results when we fused the two different hand modalities. We observe that, in general, the fusion approach takes advantage of the proficiency of the individual hand modalities. The fusions of palm print and palm vein yielded an overall increase of 3.4% in accuracy as compared to the single hand modalities.

Hand Side	Fused Biometrics	EER%	GAR% when FAR = 0.01%	GAR% when FAR = 0.001%
Right	PP + PV	0.040	99.84	99.73
Left	PP + PV	0.090	99.75	99.56

Table 4. Performance of using sum-based fusion rule.

4.2.3 Fusion using support vector machine

In this portion of study, we examine the use of SVM for our fusion approach. In the previous experiment, we use sum-rule (linear-based) method to fuse the different experts. Although sum-rule can yield satisfying result especially in the fusion of three or more modalities, the fusion result can be further improved by deploying a non-linear classification tool. The fusion result of using SVM is presented in Table 5.

Hand Side	Fused Biometrics	EER%	GAR% when FAR = 0.01%	GAR% when FAR = 0.001%
Right	PP + PV	0.020	99.90	99.82
Left	PP + PV	0.040	99.86	99.64

Table 5. Performance of using SVM.

As a whole, SVM has helped to reduce the error rates of the fusion of the experts. This improvement is due to the fact that SVM is able to learn a non-linear decision plane which could separate our datasets more efficiently. Fig. 16 shows the decision boundary learnt by SVM in classifying the genuine and imposters score distributions.

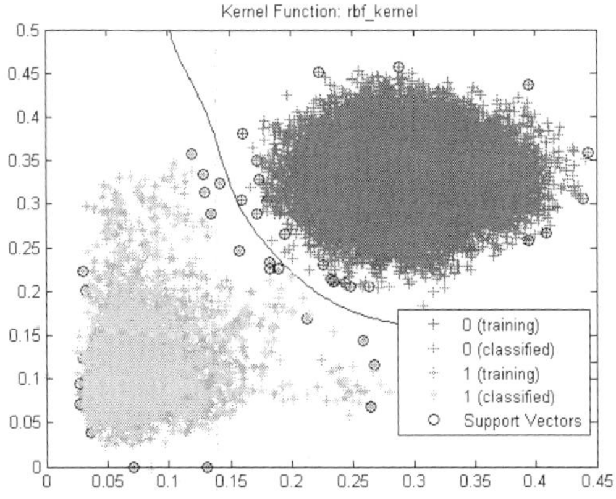

Fig. 16. Decision boundaries learnt by SVM.

4.3 Fuzzy-weighted quality-based fusion

In order to testify the proposed fuzzy-weighted (FW) image quality-based fusion scheme is useful, we carried out an experiment to evaluate the technique. Fig. 17 depicts the comparison of applying the proposed fuzzy-weighted method over the standalone sum-rule and SVM fusion approaches.

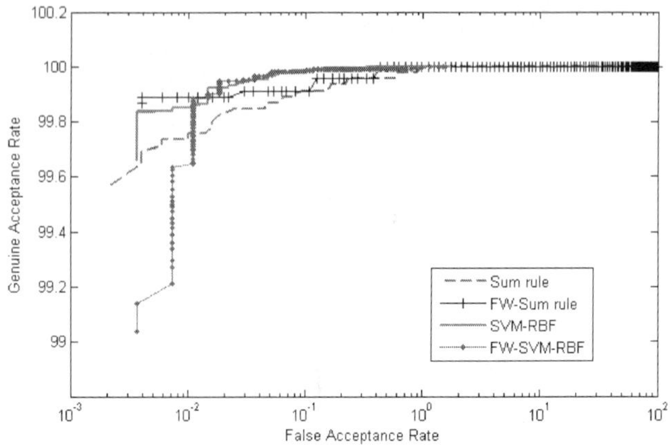

Fig. 17. Improvement gained of fuzzy-weighted fusion scheme for palm print and palm vein.

We observe that the performance of the fusion methods could be improved by incorporating the image quality assessment scheme. The gain in improvement is particularly evident when the fuzzy-weighted quality assessment method is applied on sum-rule. This result shows that the proposed quality-based fusion scheme offers an attractive alternative to increase the accuracy of the fusion approach.

5. Conclusions

This chapter presents a low resolution contactless palm print and palm vein recognition system. The proposed system offers several advantages like low-cost, accuracy, flexibility, and user-friendliness. We describe the hand acquisition device design and implementation without the use of expensive infrared sensor. We also introduce the LRE method to obtain good contrast palm print and vein images. To obtain useful representation of the palm print and vein modalities, a new technique called directional coding is proposed. This method represents the biometric features in bit string format which enable speedy matching and convenient storage. In addition, we examined the performance of the proposed fuzzy-weighted image quality checking scheme. We found that performance of the system could be improved by incorporating image quality measures when the modalities were fused. Our approach produced promising result to be implemented in a practical biometric application.

6. References

Boles, W. & Chu, S. (1997). Personal identification using images of the human palms. *Proceedings of IEEE Region 10 Annual Conference, Speech and Image Technologies for Computing and Telecommunications*, Vol. 1, pp. 295–298

Chen, J., Zhang, C. & Rong, G. (2001). Palmprint recognition using creases. *Proceedings of International Conference of Image Processing*, pp. 234-237

Connie, T., Jin, A., Ong, M. & Ling, D. (2005). An automated palmprint recognition system. *Image and Vision Computing*, Vol. 23, No.5, pp. 501–515

Cross, J. & Smith, C. (1995). Thermographic imaging of the subcutaneous vascular network of theback of the hand for biometric identification. *Proceedings of IEEE 29th International Carnahan Conference on Security Technology*, pp. 20-35

Daugman, J. (1993). High confidence visual recognition of persons by a test of statistical independence. *IEEE Transactions on Pattern Analysis and Machine Intelligence*, Vol. 15, No.11, pp. 1148–1161

Deng, W., Hu, J., Guo, J., Zhang, H. & Zhang, C. (2008). Comment on globally maximizing locally minimizing: unsupervised discriminant projection with applications to face and palm biometrics. *IEEE Transactions on Pattern Analysis and Machine Intelligence*, Vol.30, No.8, pp. 1503–1504

Doublet, J., Revenu, M. & Lepetit, O. (2007). Robust grayscale distribution estimation for contactless palmprint recognition. *First IEEE International Conference on Biometrics: Theory, Applications, and Systems*, pp. 1-6

Duta, N., Jain, A. & Mardia, K. (2002). Matching of palmprint. *Pattern Recognition Letters*, Vol. 23, pp. 477-485

Feng, G., Hu, D., Zhang, D. & Zhou, Z. (2006). An alternative formulation of kernel LPP with application to image recognition. *Neurocomputing*, Vol. 67, No.13-15, pp. 1733–1738

Funada, J., Ohta, N., Mizoguchi, M., Temma, T., Nakanishi, T., Murai, K., et al. (1998). Feature extraction method for palmprint considering elimination of creases. *Proceedings of the 14th International Conference of Pattern Recognition*, Vol. 2, pp. 1849-1854

Goh, M., Connie, T., Teoh, A. (2008). Touch-less Palm Print Biometric System. *The 3rd International Conference on Computer Vision Theory and Applications*, pp. 423 - 430.

Gonzalez, R.C., & Woods, R.E. (2002). Digital Image Processing (Second Edition), *Prentice-Hall Inc.*.

Han, C. C. (2004). A hand-based personal authentication using a course-to-fine strategy. *Image and Vision Computing*, Vol.22, No.11, pp.909-918

Han, C., Cheng, H., Lin, C. & Fan, K. (2003). Personal authentication using palm-print features. *Pattern Recognition*, Vol. 36, No.2, pp. 371–381

Hand-based Biometrics. (2003). *Biometric Technology Today*, Vol.11, No.7, pp. 9-11

Hitachi and Fujitsu win vein orders in diverse markets. (2007, March). *Biometric Technology Today*, pp. 4

Huang, D., Jia, W. & Zhang, D. (2008). Palmprint verification based on principal lines. *Pattern Recognition*, Vol.41, No.4, pp. 1316–1328

Jain, A. K., Ross, A. & Prabhakar, S. (2004). An Introduction to biometric recognition. *IEEE Transactions on Circuits System and Video Technology*, Vol.14, No.1, pp. 4-20

Jia, W., Huang, D., & Zhang, D. (2008). Palmprint verification based on robust line orientation code. *Pattern Recognition*, Vol.41, No.5, pp. 1504–1513

Kong, A. & Zhang, D. (2004). Competitive coding scheme for palmprint verification. *Proceedings of International Conference on Pattern Recognition*, Vol. 1, pp. 520–523

Kong, A. & Zhang, D. (2006). Palmprint identification using feature-level fusion. *Pattern Recognition*, Vol.39, No.3, pp. 478–487

Kong, A., Zhang, D. & Kamel, M. (2006a). Palmprint identification using feature-level fusion. *Pattern Recognition*, Vol. 39, pp. 478–487

Kong, A., Zhang, D. & Kamel, M. (2006b). A study of brute-force break-ins of a palmprint verification system. *IEEE Transactions on Systems, Man and Cybernetics, Part B*, Vol. 36, No.5, pp. 1201–1205

Kong, W. & Zhang, D. (2002). Palmprint texture analysis based on low-resolution images for personal authentication. *Proceedings of 16th International Conference on Pattern Recognition*, Vol. 3, pp. 807–810

Kumar, A. & Zhang, D. (2006). Personal recognition using hand-shape and texture. *IEEE Transactions on Image Processing*, Vol. 15, pp. 2454–2461

Kumar, K. V. & Negi, A. (2007). A novel approach to eigenpalm features using feature-partitioning framework. *Conference on Machine Vision Applications*, pp. 29-32

Kung, S., Lin, S. & Fang, M. (1995). A neural network approach to face/palm recognition. *Proceedings of IEEE Workshop on Neural Networks for Signal Processing*, pp. 323–332

Leung, M., Fong, A. & Cheung, H. (2007). Palmprint verification for controlling access to shared computing resources. *IEEE Pervasive Computing*, Vol. 6, No.4, pp.40–47

Li, W., Zhang, D., & Xu, Z. (2002). Palmprint identification by Fourier transform. *International Journal of Pattern Recognition and Artificial Intelligence*, Vol.16, No.4, pp. 417–432

Lin, C.-L., & Fan, K.-C. (2004). Biometric verification using thermal images of palm-dorsa vein patterns. *IEEE Transactions on Circuits and Systems for Video Technology*, Vol. 14 , No. 2, pp. 199 - 213

Liu, L. & Zhang, D. (2005). Palm-line detection. *IEEE International Conference on Image Processing*, Vol. 3, pp. 269-272

Lu, G., Wang, K. & Zhang, D. (2004). Wavelet based independent component analysis for palmprint identification. *Proceedings of International Conference on Machine Learning and Cybernetics*, Vol. 6, pp. 3547–3550

Lu, G., Zhang, D. & Wang, K. (2003). Palmprint recognition using eigen palms features. *Pattern Recognition Letters*, Vol.24, No.9, pp. 1463–1467

Miura, N., Nagasaka, A. & Miyatake, T. (2004). Feature extraction of finger-vein patterns based on repeated line tracking and its application to personal identification. *Machine Vision and Applications*, Vol. 15, pp. 194-203

PalmSecure™. (2009). In: *Fujitsu*, 10.12.2010, Available from http://www.fujitsu.com/us/services/biometrics/palm-vein/

Putte, T. & Keuning, J. (2000). Biometrical fingerprint recognition: don't get your fingers burned. *Proceedings of the Fourth Working Conference on Smart Card Research and Advanced Applications*, pp. 289-303

Qin, A. K., Suganthan, P. N., Tay, C. H. & Pa, H. S. (2006). Personal Identification System based on Multiple Palmprint Features. *9th International Conference on Control, Automation, Robotics and Vision*, pp. 1 – 6

RafaelDiaz, M., Travieso, C., Alonso, J. & Ferrer, M. (2004). Biometric system based in the feature of hand palm. *Proceedings of 38th Annual International Carnahan Conference on Security Technology*, pp. 136–139

Saunders, C. (1998). Support Vector Machine User Manual. *RHUL, Technical Report*

Sun, Z., Tan, T., Wang, Y. & Li, S. (2005). Ordinal palmprint representation for personal identification. *Proceeding of Computer Vision and Pattern Recognition*, Vol. 1, pp. 279–284

Teoh, A. (2009). Palmprint Matching. In S. Z. Li, *Encyclopedia of Biometrics*, pp. 1049-1055. Springer

Toh, K.-A., Eng, H.-L., Choo, Y.-S., Cha, Y.-L., Yau, W.-Y., & Low, K.-S. (2005). Identity verification through palm vein and crease texture. *International Conference on Biometrics*, pp. 546-553

Vapnik V. (1998). Statistical Learning Theory, *Wiley-Interscience publication*

Vein recognition in Europe. (2004). *Biometric Technology Today*, Vol.12, No.9, pp. 6

Wang, J. G., Yau, W. Y., Suwandya, A. & Sung, E. (2008). Person recognition by fusing palmprint and palm vein images based on "Laplacianpalm" representation. *Pattern Recognition*, Vol.41, pp. 1514-1527

Wang, L., Leedham, G., & Cho, S. (2007). Infrared imaging of hand vein patterns for biometric purposes. *IET Computer Vision*, Vol. 1, No. 3-4, pp. 113-122

Wang, X., Gong, H., Zhang, H., Li, B. & Zhuang, Z. (2006). Palmprint identification using boosting local binary pattern. *Proceedings of International Conference on Pattern Recognition*, pp. 503–506

Wu, J.-D. & Ye, S.-H. (2009). Driver identification using finger-vein patterns with Radon transform and neural network. *Expert Systems with Applications*, No. 36, pp. 5793–5799

Wu, X., Wang, K. & Zhang, D. (2002). Line feature extraction and matching in palmprint. *Proceeding of the Second International Conference on Image and Graphics*, pp. 583–590

Wu, X., Wang, K. & Zhang, D. (2004a). A novel approach of palm-line extraction. *Proceeding of the Third International Conference on Image and Graphics*, pp. 230–233

Wu, X., Wang, K., & Zhang, D. (2004b). Palmprint recognition using directional energy feature. *Proceedings of International Conference on Pattern Recognition*, Vol. 4, pp. 475–478

Wu, X., Wang, K. & Zhang, D. (2004c). HMMs based palmprint identification. *Lecture Notes in Computer Science*, Vol. 3072, pp. 775–781

Wu, X., Wang, K. & Zhang, D. (2005). Palmprint authentication based on orientation code matching. *Proceeding of Fifth International Conference on Audio- and Video-based Biometric Person Authentication*, pp. 555–562

Wu, X., Zhang, D. & Wang, K. (2003). Fisherpalms based palmprint recognition. *Pattern Recognition Letters*, Vol. 24, No.15, pp. 2829–2838

Yang, J., Zhang, D., Yang, J. & Niu, B. (2007). Globally maximizing locally minimizing: unsupervised discriminant projection with applications to face and palm biometrics. *IEEE Transactions on Pattern Analysisand Machine Intelligence*, Vol.29, No.4, pp. 650–664

Yörük, E.; Dutağaci, H. & Sankur, B. (2006). Hand biometrics. *Image and Vision Computing*, Vol.24, No.5, pp. 483-497

You, J., Kong, W., Zhang, D. & Cheung, K. (2004). On hierarchical palmprint coding with multiple features for personal identification in large databases. *IEEE Transactions on Circuits and Systems for Video Technology*, Vol.14, No.2, pp. 234–243

You, J., Li, W. & Zhang, D. (2002). Hierarchical palmprint identification via multiple feature extraction. *Pattern Recognition*, Vol.35, No.4, pp. 847–859

Zhang, D. & Liu, L. L. (2009). Palmprint Features, In *Encyclopedia of Biometrics*, S. Z. Li, pp. 1043-1049, Springer

Zhang, D. & Shu, W. (1999). Two novel characteristics in palmprint verification: datum point invariance and line feature matching. *Pattern Recognition*, Vol.32, No.4, pp. 691–702

Zhang, D., Kong, W., You, J. & Wong, M. (2003). On-line palmprint identification. *IEEE Transaction on PAMI*, Vol.25, No.9, pp. 1041-1050

Zhang, Z., Ma, S. & Han, X. (2006). Multiscale feature extraction of finger-vein patterns based on curvelets and local interconnection structure neural network. *The 18th International Conference on Pattern Recognition (ICPR'06)*, pp. 145 - 148

Zhou, X., Peng, Y. & Yang, M. (2006). Palmprint Recognition Using Wavelet and Support Vector Machines. *Lecture Notes in Computer Science*, Vol. 4099, pp. 385–393

Zuo, W., Wang, K. & Zhang, D. (2005). Bi-directional PCA with assembled matrix distance metric. *Proceeding of IEEE International Conference on Image Processing*, Vol. 2, pp. 958–961

9

Liveness Detection in Biometrics

Martin Drahanský
Brno University of Technology, Faculty of Information Technology
Czech Republic

1. Introduction

The biometric systems, oriented in this chapter especially on fingerprints, have been introduced in the previous chapters. The functionality of such systems is influenced not only by the used technology, but also by the surrounding environment (including skin or other diseases). Biased or damaged biometric samples could be rejected after revealing their poor quality, or may be enhanced, what leads to the situation that samples, which would be normally rejected, are accepted after the enhancement process. But this process could present also a risk, because the poor quality of a sample could be caused not only by the sensor technology or the environment, but also by using an artificial biometric attribute (imitation of a finger(print)). Such risk is not limited just to the deceptional technique, but if we are not able to recognize whether an acquired biometric sample originates from a genuine living user or an impostor, we would then scan an artificial fake and try to enhance its quality using an enhancement algorithm. After a successful completion of such enhancement, such fake fingerprint would be compared with a template and if a match is found, the user is accepted, notwithstanding the fact that he can be an impostor! Therefore the need of careful liveness detection, i.e. the recognition whether an acquired biometric sample comes from a genuine living user or not, is crucial.

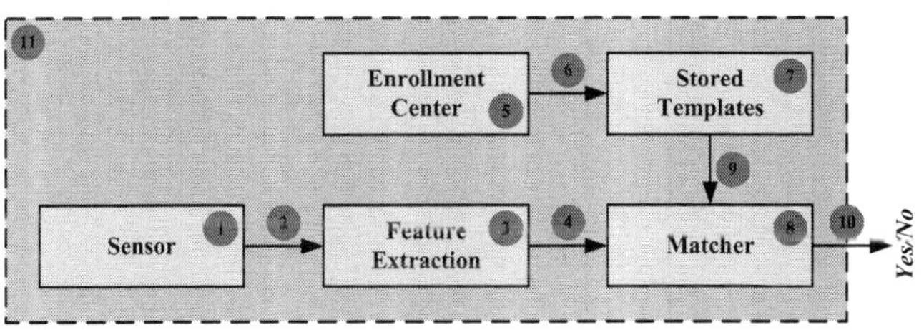

Fig. 1. Basic components of a biometric system.

Each component of a biometric system presents a potentially vulnerable part of such system. The typical ways of deceiving a biometric system are as follows (Fig. 1) (Dessimoz et al., 2006; Jain, 2005; Ambalakat, 2005; Galbally et al., 2007):

1. *Placing fake biometrics on the sensor.* A real biometric representation is placed on the device with the aim to achieve the authentication, but if such representation has been obtained in an unauthorized manner, such as making a fake gummy finger, an iris printout or a face mask, then it is considered as a deceiving activity.

2. *Resubmitting previously stored digitized biometric signals (replay attack)*. A digitized biometric signal, which has been previously enrolled and stored in the database, is replayed to the system, thus circumventing the acquisition device.
3. *Overriding the feature extraction process*. A pre-selected template is produced in the feature extraction module using a Trojan horse.
4. *Tampering with the biometric feature representation*. During the transmission between the feature extraction and matching modules, a fraudulent feature set replaces the template acquired and processed by the device.
5. *Attacking the enrollment center*. The enrollment module is also vulnerable to spoof attacks such as those described in the previous points 1 to 4.
6. *Attacking the channel between the enrollment center and the database*. During the transmission, a fraudulent template replaces the template produced during the enrollment.
7. *Tampering with stored templates*. A template, previously stored in the database (distributed or not), can be modified and used afterward as corrupted template.
8. *Corrupting the matcher*. A pre-selected score is produced in the matching extraction module using a Trojan horse.
9. *Attacking the channel between the stored templates and the matcher*. During the transmission between the database and the matching module, a fraudulent template replaces the template previously stored.
10. *Overriding the final decision*. The result of the decision module can be modified and then used for the replacement of the output obtained previously.
11. *Attacking the application*. The software application can also be a point of attack and all possible security systems should be used to reduce the vulnerability at this level.

From the above list of possible attacks we can deduce that most security risks or threats are quite common and could be therefore resolved by traditional cryptographic tools (i.e. encryption, digital signatures, PKI (Public Key Infrastructure) authentication of communicating devices, access control, hash functions etc.) or by having vulnerable parts at a secure location, in tamper-resistant enclosure or under constant human supervision (Kluz, 2005).

When a legitimate user has already registered his finger in a fingerprint system, there are still several ways how to deceive the system. In order to deceive the fingerprint system, an attacker may put the following objects on the fingerprint scanner (Matsumoto et al., 2005; Ambalakat, 2005; Roberts, 2006):

- *Registered (enrolled) finger*. The highest risk is that a legitimate user is forced, e.g. by an armed criminal, to put his/her live finger on the scanner under duress. Another risk is that a legitimate user is compelled to fall asleep with a sleeping drug in order to make free use of his/her live finger. There are some deterrent techniques against similar crimes, e.g. to combine the standard fingerprint authentication with another method such as a synchronized use of PINs or identification cards; this can be helpful to deter such crimes.
- *Unregistered finger (an impostor's finger)*. An attack against authentication systems by an impostor with his/her own biometrics is referred to as a non-effort forgery. Commonly, the accuracy of authentication of fingerprint systems is evaluated by the false rejection rate (FRR) and false acceptance rate (FAR) as mentioned in the previous chapters. FAR is an important indicator for the security against such method (because a not enrolled

finger is used for authentication). Moreover, fingerprints are usually categorized into specific classes (Collins, 2001). If an attacker knows what class the enrolled finger is, then a not enrolled finger with the same class (i.e. similar pattern) can be used for the authentication at the scanner. In this case, however, the probability of acceptance may be different when compared with the ordinary FAR.

- *Severed fingertip of enrolled finger.* A horrible attack may be performed with the finger severed from the hand of a legitimate user. Even if it is the finger severed from the user's half-decomposed corpse, the attacker may use, for criminal purposes, a scientific crime detection technique to clarify (and/or enhance) its fingerprint.
- *Genetic clone of enrolled finger.* In general, it can be stated that identical twins do not have the same fingerprint, and the same would be true for clones (Matsumoto et al., 2005). The reason is that fingerprints are not entirely determined genetically but rather by the pattern of nerve growth in the skin. As a result, such pattern is not exactly the same even for identical twins. However, it can be also stated that fingerprints are different in identical twins, but only slightly different. If the genetic clone's fingerprint is similar to the enrolled finger, an attacker may try to deceive fingerprint systems by using it.
- *Artificial clone of enrolled finger.* More likely attacks against fingerprint systems may use an artificial finger. An artificial finger can be produced from a printed fingerprint made by a copy machine or a DTP technique in the same way as forged documents. If an attacker can make then a mold of the enrolled finger by directly modeling it, he can finally also make an artificial finger from a suitable material. He may also make a mold of the enrolled finger by making a 3D model based on its residual fingerprint. However, if an attacker can make an artificial finger which can deceive a fingerprint system, one of the countermeasures against such attack is obviously based on the detection of liveness.
- *Others.* In some fingerprint systems, an error in authentication may be caused by making noise or flashing a light against the fingerprint scanner, or by heating up, cooling down, humidifying, impacting on, or vibrating the scanner outside its environmental tolerances. Some attackers may use such error to deceive the system. This method is well known as a "fault based attack" (e.g. denial of service), and may be carried out by using one of the above mentioned techniques. Furthermore, a fingerprint image may be made protruding as an embossment on the scanner surface, if we spray some special material on such surface.

Many similar attacks are documented in the literature, including all the above mentioned types. In this chapter, however, we will focus only on finger(print) fakes. One example of the attack on fingerprint technology has been presented in (LN, 2008). Hackers in the club-magazine "Die Datenschleuder" (4,000 copies in one edition) have printed a fingerprint of the thumb from the right hand of the German minister of the interior - Dr. Wolfgang Schäuble, and invited readers to make a fake finger(print) of the minister and to try to pretend that their identity is those of the minister. This could be considered as a bad joke, as a fingerprint also serves as a conclusive proof of a person's identity. A hacker has acquired this fingerprint from a glass after some podium discussion. Nevertheless, biometric travel documents (issued in Germany starting from 2007, issued in the Czech Republic from 2009), enforced not only by Dr. Schäuble, should be protected just against this situation. The implementation of fingerprints into the travel documents was prescribed by a direction of the European Union in 2004.

It is clear from (Matsumoto et al., 2005) that the production of a fake finger(print) is very simple (Drahanský, 2010). Our own experiments have shown that to acquire some images (e.g. from glass, CD, film or even paper) is not very difficult and, in addition, such image could be enhanced and post-processed, what leads to a high-quality fingerprint. The following production process of a fake finger(print) is simple and can be accomplished in several hours. After that, it is possible to claim the identity as an impostor user and common (nearly all) fingerprint recognition systems confirm this false identity supported by such fake finger.

Therefore, the application of liveness detection methods is a very important task, and should be implemented (not only) in all systems with higher security requirements, such as border passport control systems, bank systems etc. The biometric systems without the liveness detection could be fooled very easily and the consequences might be fatal.

The security of a biometric system should never be based on the fact that biometric measurements are secret, because biometric data can be easily disclosed. Unlike typical cryptographic measures where a standard challenge–response protocol can be used, the security of a biometric system relies on the difficulty of replicating biometric samples (Kluz, 2005). This quality known as the liveness ensures that the measured characteristics come from a live human being and are captured at the time of verification. We should realize that any testing of liveness is worthless unless the capture device and communication links are secure. Due to the fact that a biometric system uses physiological or behavioral biometric information, it is impossible to prove formally that a capture device provides only genuine measurements. Consequently, it cannot be proven that a biometric system as a whole is fool-proof (Kluz, 2005). Each solution of this problem has its own advantages and disadvantages; it is more suitable for a certain particular type of the biometric system and environment than for other. Some solutions are software-based; other require a hardware support. Methods which combine both approaches can also be used.

2. Liveness detection

Securing automated and unsupervised fingerprint recognition systems used for the access control is one of the most critical and most challenging tasks in real world scenarios. Basic threats for a fingerprint recognition system are repudiation, coercion, contamination and circumvention (Drahanský et al., 2006; Drahanský, 2007). A variety of methods can be used to get an unauthorized access to a system based on the automated fingerprint recognition. If we neglect attacks on the algorithm, data transport and hardware (all these attacks demand good IT knowledge), one of the simplest possibilities is to produce an artificial fingerprint using soft silicon, gummy and plastic material or similar substances (Matsumoto et al., 2005; Tan et al., 2008). A fingerprint of a person enrolled in a database is easy to acquire, even without the user's cooperation. Latent fingerprints on daily-use products or on sensors of the access control system itself may be used as templates.

To discourage potential attackers from presenting a fake finger (i.e. an imitation of the fingertip and the papillary lines) or, even worse, to hurt a person to gain access, the system must be augmented by a liveness detection component (Drahanský et al., 2006; Drahanský, 2007). To prevent false acceptance we have to recognize if the finger on the plate of the fingerprint sensor (also referred to as fingerprint scanner) is alive or not.

2.1 Perspiration

A non-invasive biomedical measurement for determination of the liveness for use in fingerprint scanners was developed by the Biomedical Signal Analysis Laboratory at Clarkson University/West Virginia University (Schuckers et al., 2003). This software-based method processes the information already acquired by a capture device and the principle of this technique is the detection of perspiration as an indication of liveness (see Fig. 2).

It is worth noting that the outmost layer of the human skin houses around 600 sweat glands per square inch (Schuckers et al., 2003). These sweat glands diffuse the sweat (a dilute sodium chloride solution) on to the surface of the skin through pores. The position of skin pores does not change over time and their pore-to-pore distance is approximately 0.5 mm over fingertips.

Fig. 2. Example of live fingerprint images acquired some time apart (Schuckers et al., 2003).

The perspiration method is based on a high difference in the dielectric constant and electrical conductivity between the drier lipids that constitute the outer layer of the skin and the moister sweaty areas near the perspiring pores. The dielectric constant of sweat is around 30 times higher than the lipid, so the electrical model of the skin thanks to perspiration can be created.

The sweat creation and ascent from sweat pores during the scanning with 4× zoom factor could be seen in Fig. 3.

Fig. 3. Ascent of sweat from sweat pores on a fingertip (4× zoomed).

2.2 Spectroscopic characteristics

The technology discussed in this section was developed by the Lumidigm company (Rowe, 2005; Kluz, 2005) from Albuquerque and is based on the optical properties of human skin. This hardware method may be regarded not only as a liveness detection mechanism but also as an individual biometric system with an inherent liveness capability.

Living human skin has certain unique optical characteristics due to its chemical composition, which predominately affects optical absorbance properties, as well as its multilayered structure, which has a significant effect on the resulting scattering properties (Rowe, 2005; Rowe, 2008). By collecting images generated from different illumination wavelengths passed into the skin, different subsurface skin features may be measured and used to ensure that the material is living human skin. When such a multispectral sensor is combined with a conventional fingerprint reader, the resulting sensing system can provide a high level of certainty that the fingerprint originates from a living finger.

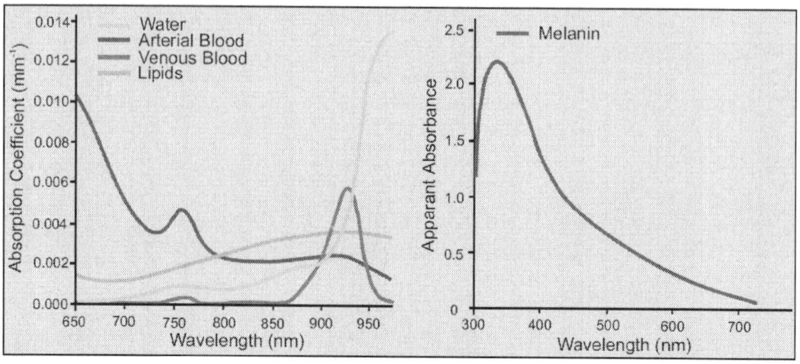

Fig. 4. Spectrographic properties of different components of living tissue (suitable for detection of spoofing attacks on iris recognition) (Toth, 2005).

The principle of this technique lies in passing light of different wavelengths through a sample and measuring the light returned, which is affected by the structural and chemical properties of the sample. Different wavelengths have to be used to measure the sample satisfactorily, because diverse wavelengths penetrate to different depths into the sample and are differently absorbed and scattered (Kluz, 2005). For example, when we put a flashlight against the tip of a finger only the red wavelengths can be seen on the opposite side of the finger. This is because shorter (mostly blue) wavelengths are absorbed and scattered quickly in the tissue, unlike longer (red and very near infrared) ones, which penetrate deep into the tissue. The measurements can be transformed into a graph (Fig. 4) that shows the change in all measured wavelengths after interacting with a sample and is known as a spectrum. Next, the proper analysis of tissue spectra, based on multivariate mathematical methods has to be done to provide correct results.

Figure 5 shows the layout of an optical fingerprint sensor that combines a conventional frustrated total internal reflection (FTIR) fingerprint reader with a multispectral imager.

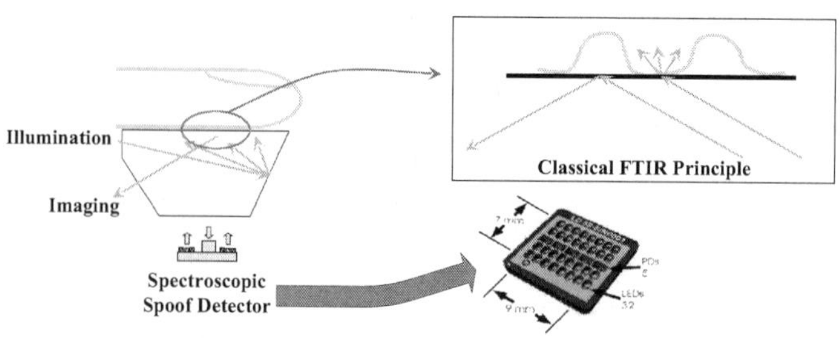

Fig. 5. FTIR and multispectral imager (Rowe, 2008).

The key components of a multispectral imager (Rowe, 2008; Rowe, 2005) suitable for imaging fingers are shown in Fig. 5. The light sources are LEDs of various wavelengths spanning the visible and short-wave infrared region. Crossed linear polarizers may be included in the system to reduce the contribution of light that undergoes a simple specular reflection to the image, such as light that is reflected from the surface of the skin. The crossed polarizers ensure that the majority of light seen by the imaging array has passed through a portion of skin and undergone a sufficient number of scattering events to have randomized the polarization. The imaging array is a common silicon CMOS or CCD detector.

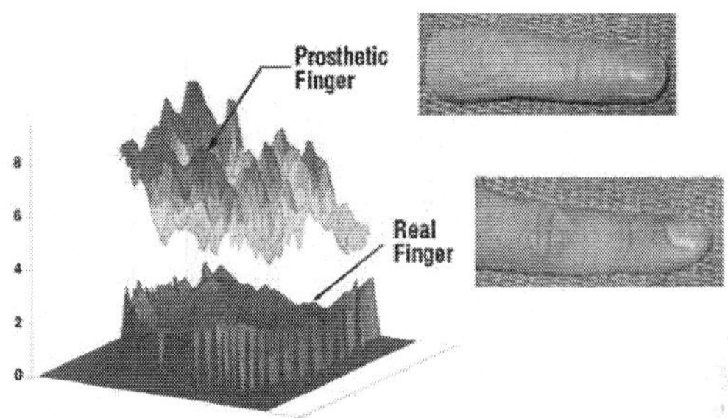

Fig. 6. Multispectral image data can clearly discriminate between a living finger and an ultra-realistic spoof. The graphs on the left side show how similar the spectral content of each image is to that expected for a genuine finger (Rowe, 2005; Toth, 2005).

A highly realistic artificial finger made by Alatheia Prosthetics (Rowe, 2005) was one of a number of different spoof samples used to test a multispectral imager's ability to discriminate between real fingers and spoofs. Figure 6 shows the results of a multivariate spectral discrimination performed to compare the consistency of the spectral content of a multispectral image of a real finger with both a second image of a real finger and a prosthetic replica of the same finger. The imager's ability to distinguish between the two sample types is clear.
Another approach of the liveness detection using the wavelet analysis in images is presented in (Schuckers et al., 2004).

2.3 Ultrasonic technology

In this paragraph, a biometric system using an ultrasonic technology with inherent liveness testing capability will be described. This technique is being developed by the company Optel from Poland and is based on the phenomenon called contact scattering. Another ultrasonic biometric device is offered by the company Ultra-Scan from the USA, which is the second and last vendor of this technology principle in the market at the moment.
Standard ultrasonic methods (Kluz, 2005) use a transmitter, which emits acoustic signals toward the fingerprint, and a receiver, which detects the echo signals affected by the interaction with the fingerprint (Fig. 7). A receiver utilizes the fact that the skin (ridges) and the air (valleys) have difference in acoustic impedance; therefore the echo signals are reflected and diffracted differently in the contact area. This approach with inherent liveness testing capability among its foremost principles uses the fact that sound waves are not only reflected and diffracted, but are also subject to some additional scattering and

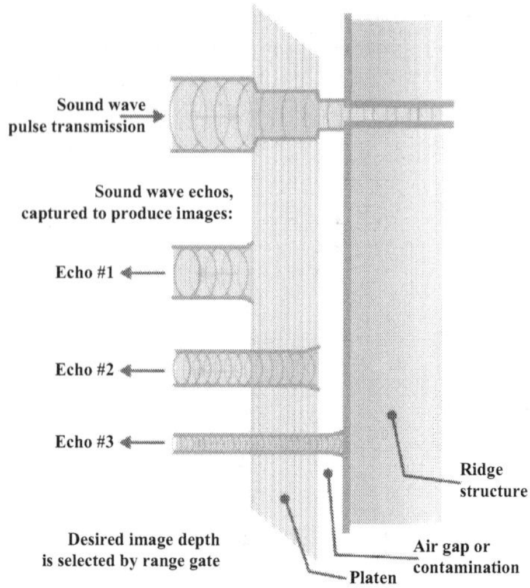

Fig. 7. Schematic of ultrasonic pulse/echo principle (UltraScan, 2004).

transformation. This phenomenon is called contact scattering (Kluz, 2005) and it was discovered that this scattering is, to a significant extent, affected by the subsurface structure of the acquired object. Hence, the class corresponding to the live tissue could be modeled and whenever the received acoustic waves are inconsistent with this class, they are rejected. The main problem here is not to obtain clear signals, but to analyze and to make a reconstruction of internal structures from signals which are very difficult to interpret.
The ultrasonic device reached the following conclusions (Kluz, 2005; Bicz, 2008):

- As the inner structure of the live skin compared with spoof samples differs, the character and the amplitude of acoustic signals also differ significantly. Hence, it is possible to distinguish between live and artificial fingers.
- There is no need to deal with the problem known as latent print reactivation because the signal level from the latent print is at least 30 dB lower than the signal given by the real finger. Even when the soot or metal powder is used in order to enhance the quality of signal, the previous is true.
- This method is much less sensitive to dirt, grease and water compared with other methods. In addition, fingers with damaged surface give a relatively clear image, because their inner structure seems to be visible.

Since this approach scans the inner structure of the object, it has the ability to check for pulse by measuring volumetric changes in the blood vessels (Bicz, 2008).

2.4 Physical characteristics: temperature

This simple method measures the temperature of the epidermis during a fingerprint acquisition. The temperature of the human epidermis of the finger moves in the range of approximately 25-37°C (see Fig. 8). However, this range usually has to be wider to make the system usable under different conditions. In addition, there are many people who have problems with blood circulation, a fact which leads to deviations in the body's temperature and hence to wrong liveness module decision. The only way how to improve such a situation is to make the working range broader again or simply warm the user's finger. The former will increase the likelihood that the system will be deceived while the latter can also be applied to fake samples. In the case where an attacker uses a wafer-thin artificial fingerprint glued on to his finger, this will result in a decrease by a maximum of 2°C (Drahanský, 2008) compared with an ordinary finger. Since the difference in temperature is small, the wafer-thin sample will comfortably fall within the normal working margin. In consequence, this method is not a serious security measure at all.

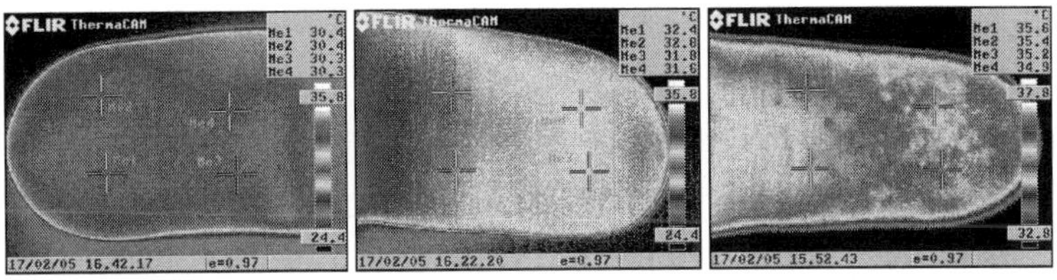

Fig. 8. Thermo-scans of the fingertips acquired using a thermo-camera FLIR.

2.5 Physical characteristics: hot and cold stimulus

This technique is based on the fact that the human finger reacts differently to thermal stimuli compared with other artificial, non-living material.

The designed liveness testing module (Kluz, 2005; U.S. Patent 6,314,195) is working as follows. A stimulus-giving section gives a stimulus (it may cover a cool and a hot stimulus) to the finger by a contact plate with which the finger makes contact. Next, typical information could be measured by an organism information-measuring section, which is produced by the live finger in response to the stimulus. Concretely, the amount of the fluctuation for the flow rate of the blood flowing in the peripheral vascular tracts varies according to the stimuli. Hence, as peripheral vascular tracts of the tip of the finger are extended or contracted, the amplitude value of the blood flow is measured and processed by an organism information-measuring section. Under hot stimulus the amplitude of the blood flow increases, while it decreases under cool stimulus. Moreover, according to the autonomic nervous system, the amplitude is delayed a little with respect to the application of the stimulus. Since these facts are typically observed when the live fingers are measured, they could be employed to distinguish live among artificial and dead samples. After the processing phase, such information is transferred to a determining section, where together with the other information related to stimulus (i.e. the time intervals, the strength of stimuli etc.) is evaluated. Finally, a determining section analyses how the amplitude of the blood flow fluctuates in response to the stimulus to make the right decision.

Since the human peripheral nervous system is very sensitive, it is able to react to weak cool and hot stimuli without being noticed by the person whose fingerprint is checked. This fact should also reduce success spoofing ratio. More information about the method discussed here can be found in (U.S. Patent 6,314,195).

2.6 Physical characteristics: pressure stimulus

The principle of this method lies in some changes in characteristics of the live skin, which are realized due to pressure applied to the finger (Kluz, 2005; U.S. Patent 5,088,817). Since the structure and the characteristics of artificial and dead samples are different, when compared with a live finger, this phenomenon could not be seen if such samples were used. The color of the live skin of the finger not under pressure is usually reddish but becomes whitish when pressure is applied to the skin of the finger. It has been shown that the spectral reflectance of the light in the red spectral range (i.e. the light wavelength of approximately 640–770 nm) (U.S. Patent 5,088,817) does not show a substantial difference between the pressed state and the non pressed state. On the other hand, the spectral reflectance of the light in the blue and green spectral range (i.e. the light wavelength of approximately 400–600 nm) (U.S. Patent 5,088,817) in the not pressed state is much smaller than in the pressed state. Hence, for the purposes of the device discussed in this section it is suitable to measure the spectral reflectance in the blue and green spectral range (see Fig. 9).

A liveness testing module is proposed in (U.S. Patent 5,088,817) and consists of a transparent plate, a light source, a light detection unit and a determining section. Since the light source and the light detection unit are placed under the plate, this plate has to be transparent to enable light to be sent towards the finger and receiving the reflected light. The light source projects a light beam towards the surface of the placed finger. Next, depending on the pressure or non-pressure state, the reflected light is measured by the light detection unit.

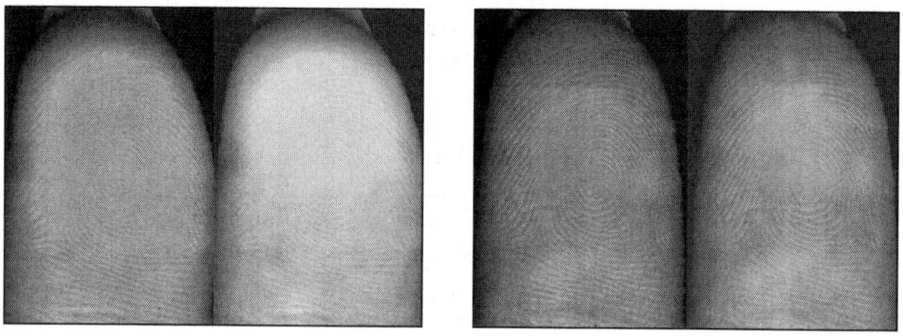

Fig. 9. Images of the fingertips pressed tightly (left subpart) and slightly (right subpart) to the sensor (Drahanský et al., 2008).

Based on such measurements the determining section returns the right decision, i.e. as the finger changes its state from non-pressure to pressure, the color of the skin changes from reddish to whitish, what leads to a change in the spectral reflectance. As a result, the light detection unit can detect that the spectral wavelength of the spectral ranges is increased.

Another method using pressure based characteristics is discussed in (U.S. Patent 6,292,576), but unlike the method described in the previous paragraph, this technique employs the change in fingerprint ridges width. When the fingerprint changes its state from non-pressure to pressure, the fingerprint ridges change, i.e. as the pressure becomes stronger, the fingerprint ridges flatten out, and therefore their change of width could be measured. Only objects which demonstrate the typical change in fingerprint ridge width due to pressure could be determined as live ones.

A new approach to the fake finger detection based on skin elasticity analysis has been introduced in (Jia et al., 2007). When a user puts a finger on the scanner surface, the scanner captures a sequence of fingerprint images at a certain frame rate. The acquired image sequence is used for the fake finger detection. One or more of them (see Fig. 10) can be used for fingerprint authentication.

Fig. 10. A sequence of fingerprint images describing the deformation of a real finger (Jia, 2007).

2.7 Physical characteristics: electrical properties

Some methods of liveness testing are based on the fact that the live human skin has different electrical properties compared with other materials (Kluz, 2005). The suitable fingerprint recognition system could be extended by an electrode system and an electrical evaluation unit. These sections are the main parts of the liveness testing module where the electrical evaluation unit can evaluate the change in the state in the electrode system. The sensing of the electrical change should take place simultaneously with the recognition of the fingerprint. Therefore, these parts of biometric systems should be designed in such a way that two simultaneous measurements cannot disturb each other. Furthermore, such a system may be able to measure more than one of the fingerprint liveness characteristics related to electrical properties (e.g. conductivity, dielectric constant).

The conductivity (Kluz, 2005) of the human skin is based on humidity, which is dependent on people's biological characteristics and environmental conditions: some people have dry fingers and others have sweaty ones; also during different seasons, climatic and environmental conditions, humidity differs significantly. As a result, the span of permissible resistance levels has to be big enough to make the system usable. In such a situation it is quite easy for an intruder to fool the system. Moreover, the intruder can use a salt solution of a suitable concentration or put some saliva on the fake finger to imitate the electric properties of the real finger.

The relative dielectric constant (RDC) (Kluz, 2005) of a specific material reflects the extent to which it concentrates the electrostatic lines of flux. Many advocates claim that the RDC has the ability to distinguish between real and artificial samples. However the RDC is highly dependent on the humidity of the sample, so the same situation as in the case of conductivity arises. To fool this method an attacker can simply use an artificial sample and dip it into a compound of 90% alcohol and 10% water. In (Pute et al., 2000) we can read that the RDC values of alcohol and water are 24 and 80, respectively, while the RDC of the normal finger is somewhere between these two values. Since the alcohol will evaporate faster than the water, the compound will slowly turn into the water. During evaporation, the RDC of spoof samples will soon be within the acceptance range of the sensor.

We have run a small test series with 10 people, each finger, horizontal and vertical measurement strips, and 5 measurements per finger – conductivity (resistance) measurements. The range of values we found was from 20 kΩ to 3 MΩ (Drahanský, 2008). A paper copy or an artificial finger made of non skin-like material have higher electrical resistance, but for example, soft silicon (moisturized) shows resistance values close to the range found in our experiments.

2.8 Physical characteristics: bio-impedance

Bio-impedance (Martinsen et al., 1999; Grimmes et al., 2006; BIA, 2007) describes the passive electrical properties of biological materials and serves as an indirect transducing mechanism for physiological events, often in cases where no specific transducer for that event exists. It is an elegantly simple technique that requires only the application of two or more electrodes. The impedance between the electrodes may reflect "seasonal variations in blood flow, cardiac activity, respired volume, bladder, blood and kidney volumes, uterine contractions, nervous activity, the galvanic skin reflex, the volume of blood cells, clotting, blood pressure and salivation."

Impedance Z (Grimmes et al., 2006) is a general term related to the ability to oppose AC (Alternating Current) flow, expressed as the ratio between an AC sinusoidal voltage and an AC sinusoidal current in an electric circuit. Impedance is a complex quantity because a biomaterial, in addition to opposing current flow, phase-shifts the voltage with respect to the current in the time-domain.

The conductivity of the body is ionic (electrolytic) (Grimmes et al., 2006), because of the presence of e.g. Na^+ and Cl^- in the body liquids. The ionic current flow is quite different from the electronic conduction found in metals: the ionic current is accompanied by a substance flow. This transport of substance leads to concentrational changes in the liquid: locally near the electrodes (electrode polarization), and in a closed-tissue volume during prolonged DC (Direct Current) current flow.

The body tissue is composed of cells with poorly conducting, thin-cell membranes. Therefore, the tissue has capacitive properties (Grimmes et al., 2006): the higher the frequency, the lower the impedance. The bio-impedance is frequency-dependent, and impedance spectroscopy, hence, gives important information about tissue and membrane structures as well as intra- and extracellular liquid distributions.

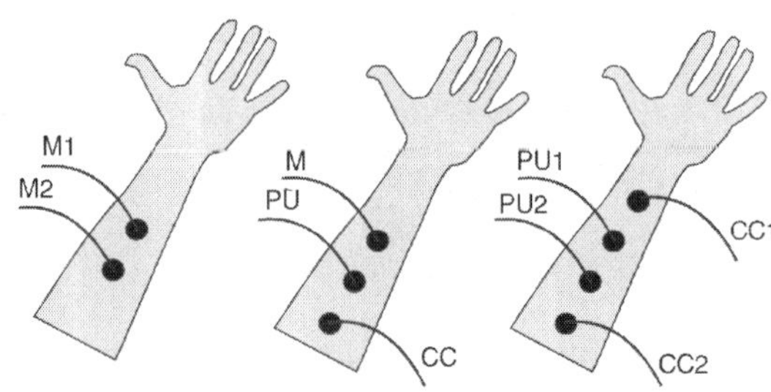

Fig. 11. Three skin surface electrode systems on an underarm (Grimmes et al., 2006). Functions: *M* – measuring and current carrying, *CC* – current carrying, *PU* – signal pick-up.

Fig. 11 shows three most common electrode systems. With two electrodes, the current carrying electrodes and signal pick-up electrodes are the same. If the electrodes are equal, it is called a bipolar lead, in contrast to a monopolar lead. With 3-(tripolar) or 4-(tetrapolar) electrode systems, separate current carrying and signal pick-up electrodes are used. The impedance is then transfer impedance (Grimmes et al., 2006): the signal is not picked up from the sites of current application.

Fig. 12 shows a typical transfer impedance spectrum obtained with the 4-electrode system from Fig. 11. It shows two dispersions (Grimmes et al., 2006). The transfer impedance is related to, but not solely determined by, the arm segment between the PU electrodes. The spectrum is determined by the sensitivity field of the 4-electrode system as a whole. The larger the spacing between the electrodes, the more the results are determined by deeper tissue volumes. Even if all the electrodes are skin surface electrodes, the spectrum is, in principle, not influenced by skin impedance or electrode polarization impedance.

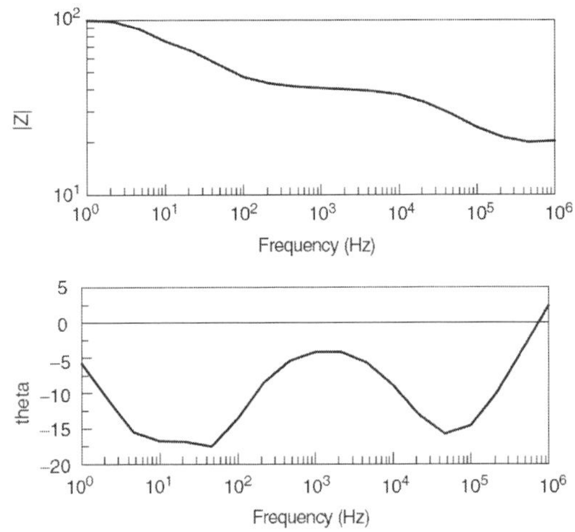

Fig. 12. Typical impedance spectrum obtained with four equal electrodes attached to the skin of the underarm (Grimmes et al., 2006).

2.9 Physical characteristics: pulse

Scanners based on this technique try to detect whether the scanned object exhibits characteristics of the pulse and blood flow consistent with a live human being (Kluz, 2005). It is not very difficult to determine whether the object indicates some kind of pulse and blood flow, but it is very difficult to decide if the acquired characteristics are coincident with a live sample. As a result, it is difficult to create an acceptance range of the sensor, which would lead to small error rates. The main problem is that the pulse of a human user varies from person to person – it depends on the emotional state of the person and also on the physical activities performed before the scanning procedure. In addition, the pulse and blood flow of the attacker's finger may be detected and accepted when a wafer-thin artificial sample is used.

One of the sensors usually detects variation in the levels of the reflected light energy from the scanned object as evidence of the pulse and blood flow (Kluz, 2005). First, the light source illuminates the object and then a photo-detector measures the light energy reflected from the object. Finally, there is the processing instrument (which also controls the light source) which processes the output from the photo-detector. Since there are some ways how to simulate pulse and blood flow characteristics (e.g. by flashing the light or by motion of the scanned object), scanners should have a deception detection unit (Kluz, 2005).

Our skin is semi-permeable for light, so that movements below the skin (e.g. blood flow) can be visualized. One example of an optical skin property is the skin reflection (Drahanský et

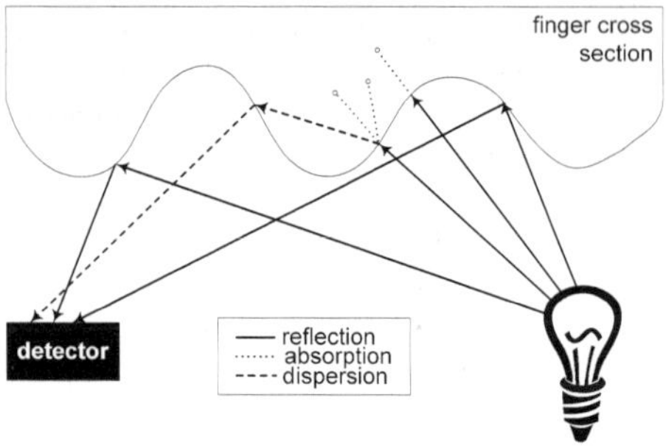

Fig. 13. Light absorption, dispersion and reflection by a fingerprint (Drahanský et al., 2006).

al., 2006; Drahanský et al., 2007). The light illuminating the finger surface is partly reflected and partly absorbed (Fig. 13). The light detector acquires the reflected light which has been changed in phase due to dispersion and reflection and thus has a slightly different wavelength compared to the original light source. One can try to link the change in wavelength to the specific characteristics of the skin with respect to light dispersion and reflection to detect whether the light has been scattered and reflected only from the fingerprint skin, or if there is some intermediate layer between the finger skin and the light source or detector.

Another example for optical skin feature is the saturation of hemoglobin (Drahanský et al., 2006; Drahanský et al., 2007), which binds oxygen molecules. When blood comes from the heart, oxygen molecules are bound to the hemoglobin, and vice versa, when blood is flowing back to the heart, it is less saturated by oxygen. The color of oxygenated blood is different from that of non-oxygenated blood. If we use a light source to illuminate the finger skin, we can follow the blood flow based on the detection of oxygenated and non-oxygenated blood, respectively (Drahanský et al., 2006; Drahanský et al., 2007). The blood flow exhibits a typical pattern for a live finger, i.e. the analysis of blood flow is well suited for finger liveness detection.

In both above mentioned examples, it is shown that the human skin has special characteristics which can be used for the liveness testing. It can be argued that it is possible to confuse such system, e.g. by using a substance with similar optical characteristics as a human skin, or, in the second example to simulate the blood flow. Even though the argument is correct, obviously the effort to be exerted for these attacks is much higher than for the other physical characteristics presented so far.

Another solution is proposed in (Drahanský et al., 2006; Drahanský et al., 2007) based on the analysis of movements of papillary lines of the fingertips and measurements of the distance of the fingertip surface to a laser sensor, respectively. The system is compact enough to be integrated with optical fingerprint sensors.

One advantage of this implementation is that the finger is not required to be in contact with a specific measuring device, and so it can be integrated with standard fingerprint sensors. Moreover, the implementation could be acceptably low. This is of particular importance, as in most cases the liveness detection will be an add-on that augments already existing robust and field-tested fingerprint scanners.

The method presented in (Drahanský et al., 2006; Drahanský et al., 2007) requires the analysis of at least one heart activity cycle, thus both the camera and the laser measurement method sketched in this section would add an extra time of at least one or two seconds to the overall authorization process interval.

In (Drahanský et al., 2006; Drahanský et al., 2007), two approaches for measuring of fine movements of papillary lines, based on optical principles, are suggested (Fig. 14). The first solution is based on a close-up view of the fingertip acquired with a CCD camera; the second one is distance measurement with a laser sensor. It should be noted that adding the proposed liveness detection solution (either camera or laser based) to a fingerprint recognition system, as proposed in Fig. 15 and Fig. 16, may significantly influence the hardware requirements imposed on the complete system.

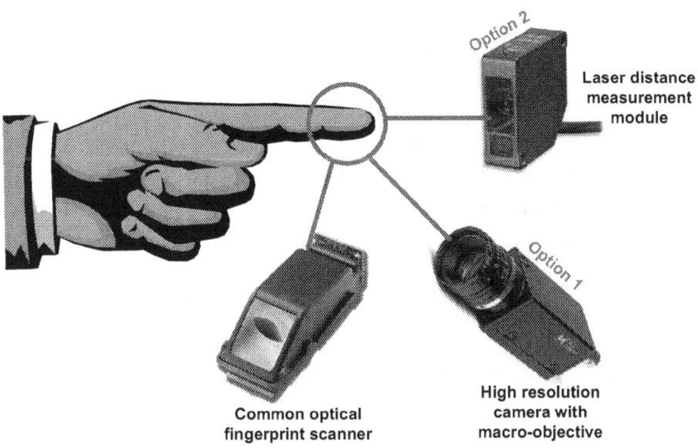

Fig. 14. Integrated liveness detection - scanner + optical and laser solution (Lodrová et al., 2008).

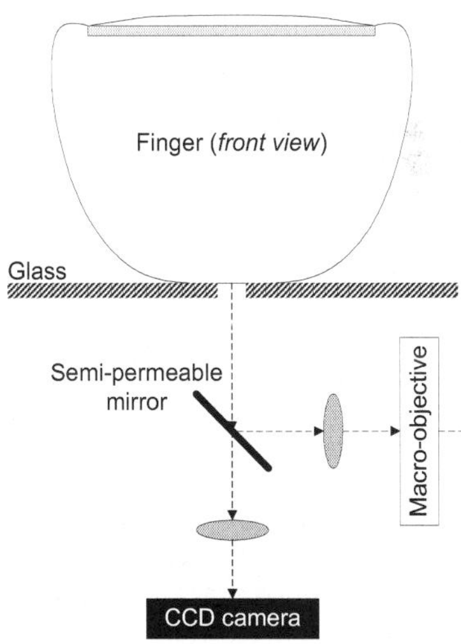

Fig. 15. Possible integration of a camera-based measurement system for liveness detection with optical fingerprint sensor (CCD/CMOS camera) (Drahanský et al., 2006).

2.9.1 Camera solution

The camera solution scheme is outlined in Fig. 15. The main idea is that a small aperture (approximately 6 mm) is created in the middle of a glass plate with an alternately functioning mirror below the plate.

First, during the fingerprint acquirement phase, the whole fingerprint is stored and the system operates as a classical fingerprint acquisition scanner (mirror permeable) by projecting the fingerprint on the CCD/CMOS camera. Next, in the liveness detection phase, the mirror is made impermeable for light and a part of the fingertip placed on the aperture is mirrored to the right and projected on the CCD/CMOS camera by a macro lens. The latter part of the system is used to acquire a video sequence for the liveness detection analysis.

2.9.2 Laser solution

The second optical method for the liveness testing is based on laser distance measurements (Drahanský et al., 2006; Drahanský et al., 2007). Fig. 16 outlines the laser distance measurement module, which could be integrated with a standard optical fingerprint sensor. The optical lens system and CCD camera for acquisition of the fingerprint are the same as in Fig. 15. However, unlike the solution shown in Fig. 15, the laser distance measurement module is placed to the right side of the glass plate, which is L-shaped here. The user places his finger in such a way that it is in contact with the horizontal and the vertical side of the glass plate.

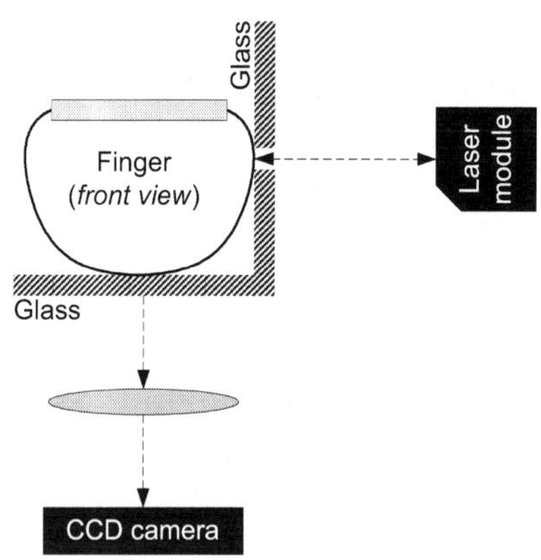

Fig. 16. Possible integration of laser distance measurement for liveness detection with optical fingerprint sensor (CCD/CMOS camera; aperture approx. 6 mm) (Drahanský et al., 2006).

The underlying physical measurement principle is the same as in the video camera solution. We assume volume changes (expansion and contraction) due to the heart activity, which causes fine movements of the skin. The laser sensor is able, based on the triangulation principle, to measure very small changes in distance down to several μm.

The comparison of the computed curve and a normalized standard curve (the template) will reveal whether the measurement corresponds to a standard live fingerprint or indicates a fake finger or another attempt of fraud. For example, the comparison between both curves can be realized by the normalization followed by the cross correlation.

There are other liveness detection methods based on optical principles - see (U.S. Patent 6,292,576) and (U.S. Patent 5,088,817). They coincide in principles (both are optical) but differ in monitored physical characteristics.

2.10 Physical characteristics: blood oxygenation

Sensors which measure blood oxygenation (Kluz, 2005) are mainly used in medicine and have also been proposed for use in liveness testing modules. The technology involves two physical principles. First, the absorption of light having two different wavelengths by hemoglobin differs depending on the degree of hemoglobin oxygenation. The sensor for the measurement of this physical characteristic contains two LEDs: one emits visible red light (660 nm) and the other infrared light (940 nm). When passing through the tissue, the emitted light is partially absorbed by blood depending on the concentration of oxygen bound on hemoglobin. Secondly, as the volume of arterial blood changes with each pulse, the light signal obtained by a photo-detector has a pulsatile component which can be exploited for the measurement of pulse rate.

The sensors mentioned above are able to distinguish between artificial (dead) and living samples but, on the other hand, many problems remain. The measured characteristics vary from person to person and the measurement is strongly influenced by dyes and pigments (e.g. nail varnish).

2.11 Other methods

There are some other methods based on the medical science characteristics which have been suggested for liveness testing purposes (Kluz, 2005). Nonetheless, they are mostly inconvenient and bulky. One example can be the measurement of blood pressure (Drahanský et al., 2006) but this technology requires to perform measurement at two different places on the body, e.g. on both hands.

We distinguish between the systolic and diastolic blood pressure (www.healthandage.com; Drahanský et al., 2006); these two levels characterize upper and lower blood pressure values, respectively, which depend on heart activity. For a healthy person the diastolic blood pressure should not be lower than 80 mm Hg (lower values mean hypotension) and the value of the systolic blood pressure should not be below 120 mm Hg (again, lower values mean hypotension). People with hypertension have higher blood pressure values, with critical thresholds 140 mm Hg for the diastolic blood pressure and 300 mm Hg for the systolic blood pressure. In fact, diastolic and systolic blood pressure values are bound up with the ranges from 80 mm Hg to 140 mm Hg and from 120 mm Hg to 300 mm Hg, respectively (www.healthandage.com). On one hand, blood pressure values outside these normal ranges can indicate a fake fingerprint (Drahanský et al., 2006). On the other hand we can think of configurations, where the blood pressure measurement of a fake fingerprint glued to the finger which significantly lowers the measured blood pressure value, can still give us a measurement value within the accepted range. An attacker with hypertension would be accepted as a registered person in such configuration (Drahanský et al., 2006).

3. Conclusion

The topic of this chapter is oriented towards the liveness detection in fingerprint recognition systems. At the beginning, certain basic threats, which can be used in an attack on the biometric system, are described in general. One of them is the use of fake finger(print)s. Of course, the security of the biometric system is discussed here too, however, this is rather out of scope of this thesis. This is followed by a detailed introduction to the liveness detection and to all known methods and related principles; these include perspiration, spectroscopic characteristics, ultrasonic principle and many physical characteristics.

4. Acknowledgment

This work is partially supported by the grant "*Information Technology in Biomedical Engineering*", GA102/09/H083 (CZ), by the grant "*Advanced secured, reliable and adaptive IT*", FIT-S-11-1 and the research plan "*Security-Oriented Research in Information Technology*", MSM0021630528 (CZ).

5. References

Ambalakat, P.: *Security of Biometric Authentication Systems*, In: 21st Computer Science Seminar, SA1-T1-1, 2005, p. 7.

Bicz, W.: *The Impossibility of Faking Optel's Ultrasonic Fingerprint Scanners*, Optel, Poland, http://www.optel.pl/article/english/livetest.htm, 2008.

Collins, C.G.: *Fingerprint Science*, Copperhouse/Atomic Dog Publishing, p. 192, 2001, ISBN 978-0-942-72818-7.

Das BIA-Kompendium - Data Input GmbH, *Body Composition*, 3rd Edition, 2007, p. 70, www.data-input.de.

Dessimoz, D., Richiardi, J., Champod, C., Drygajlo, A.: *Multimodal Biometrics for Identity Documents*, Research Report, PFS 341-08.05, Version 2.0, Université de Lausanne & École Polytechnique Fédérale de Lausanne, 2006, p. 161.

Drahanský M.: *Fingerprint Recognition Technology: Liveness Detection, Image Quality and Skin Diseases*, Habilitation thesis, Brno, CZ, 2010, p. 153.

Drahanský M., Lodrová D.: *Liveness Detection for Biometric Systems Based on Papillary Lines*, In: Proceedings of Information Security and Assurance, 2008, Busan, KR, IEEE CS, 2008, pp. 439-444, ISBN 978-0-7695-3126-7.

Drahanský M.: *Experiments with Skin Resistance and Temperature for Liveness Detection*, In: Proceedings of the Fourth International Conference on Intelligent Information Hiding and Multimedia Signal Processing, Los Alamitos, US, IEEE CS, 2008, pp. 1075-1079, ISBN 978-0-7695-3278-3.

Drahanský M., Funk W., Nötzel R.: *Method and Apparatus for Detecting Biometric Features*, International PCT Patent, Pub. No. WO/2007/036370, Pub. Date 05.04.2007, Int. Application No. PCT/EP2006/009533, Int. Filing Date 28.09.2006, http://www.wipo.int/pctdb/en/wo.jsp?wo=2007036370&IA=WO2007036370&DISPLAY=STATUS.

Drahanský M.: *Methods for Quality Determination of Papillary Lines in Fingerprints*, NIST, Gaithersburg, USA, 2007, p. 25.

Drahanský M., Funk W., Nötzel R.: *Liveness Detection based on Fine Movements of the Fingertip Surface*, In: IEEE - The West Point Workshop, West Point, New York, USA, 2006, pp. 42-47, ISBN 1-4244-0130-5.

Galbally, J., Fierrez, J., Ortega-Garcia, J.: *Vulnerabilities in Biometric Systems: Attacks and Recent Advances in Liveness Detection*, Biometrics Recognition Group, Madrid, Spain, 2007, p. 8.

Grimnes, S., Martinsen, O.G.: Bioimpedance, University of Oslo, Norway, *Wiley Encyclopedia of Biomedical Engineering*, John Wiley & Sons., Inc., 2006, p. 9.

Jain, A.K.: *Biometric System Security*, Presentation, Michigan State University, p. 57, 2005.

Jia, J., Cai, L., Zhang, K., Chen, D.: *A New Approach to Fake Finger Detection Based on Skin Elasticity Analysis*, In: S.-W. Lee and S.Z. Li (Eds.): ICB 2007, LNCS 4642, 2007, pp. 309-318, Springer-Verlag Berlin Heidelberg, 2007, ISSN 0302-9743.

Kluz, M.: *Liveness Testing in Biometric Systems*, Master Thesis, Faculty of Informatics, Masaryk University Brno, CZ, 2005, p. 57.

LN: *Němečtí hackeři šíří otisk prstu ministra (German Hackers Distribute the Minister's Fingerprint)*, Lidové noviny (newspaper), March 31, 2008.

Lodrová D., Drahanský M.: *Methods of Liveness Testing By Fingers*, In: Analysis of Biomedical Signals and Images, Brno, CZ, VUTIUM, 2008, p. 7, ISBN 978-80-214-3612-1, ISSN 1211-412X.

Martinsen, O.G., Grimnes, S., Haug, E.: *Measuring Depth Depends on Frequency in Electrical Skin Impedance Measurements*, In: Skin Research and Technology No. 5, 1999, pp. 179-181, ISSN 0909-752X.

Matsumoto, T., Matsumoto, H., Yamada, K., Hoshino, S.: *Impact of Artificial "Gummy" Fingers on Fingerprint Systems*, In: Proceedings of SPIE Vol. 4677, Optical Security and Counterfeit Deterrence Techniques IV, 2005, p. 11.

Putte, T., Keuning, J.: *Biometrical Fingerprint Recognition: Don't Get Your Fingers Burned*, In: IFIP TC8/WG8.8 4th Working Conference on Smart Card Research and Advanced Applications, Kluwer Academic Publishers, 2000, pp. 289-303.

Roberts, C.: *Biometric Attack – Vectors and Defences*, 2006, p. 25.

Rowe, R.K.: *Spoof Detection*, In: Summer School for Advanced Studies on Biometrics for Secure Authentication, Alghero, Italy, 2008, p. 43.

Rowe, R.K.: *A Multispectral Sensor for Fingerprint Spoof Detection*, www.sensormag.com, January 2005.

Schuckers, S., Abhyankar, A.: *Detecting Liveness in Fingerprint Scanners Using Wavelets: Results of the Test Dataset*, In: BioAW 2004, LNCS 3087, 2004, Springer-Verlag, pp. 100-110.

Schuckers, S., Hornak, L., Norman, T., Derakhshani, R., Parthasaradhi, S.: *Issues for Liveness Detection in Biometrics*, CITeR, West Virginia University, Presentation, 2003, p. 25.

Tan, B., Lewicke, A., Schuckers, S.: *Novel Methods for Fingerprint Image Analysis Detect Fake Fingers*, SPIE, 10.1117, 2.1200805.1171, p. 3, 2008.

Toth, B.: *Biometric Liveness Detection*, In: Information Security Bulletin, Vol. 10, 2005, pp. 291-297, www.chi-publishing.com.

UltraScan: *The Theory of Live-Scan Fingerprint Imaging (Breaking the Optical Barriers with Ultrasound)*, UltraScan, USA, 2004, p. 8.

U.S. Patent 6,314,195 – *Organism Identifying Method and Device*, November 2001.

U.S. Patent 6,292,576 - *Method and Apparatus for Distinguishing a Human Finger From a Reproduction of a Finger*, September 2001.

U.S. Patent 5,088,817 - *Biological Object Detection Apparatus*, February 1992.

Part 3

Advanced Methods and Algorithms

10

Fingerprint Recognition

Amira Saleh, Ayman Bahaa and A. Wahdan
Computer and systems engineering department
Faculty of Engineering /Ain Shams University
Egypt

1. Introduction

Recognition of persons by means of biometric characteristics is an emerging phenomenon in modern society. It has received more and more attention during the last period due to the need for security in a wide range of applications. Among the many biometric features, the fingerprint is considered one of the most practical ones. Fingerprint recognition requires a minimal effort from the user, does not capture other information than strictly necessary for the recognition process, and provides relatively good performance. Another reason for the popularity of fingerprints is the relatively low price of fingerprint sensors, which enables easy integration into PC keyboards, smart cards and wireless hardware (Maltoni et al., 2009).

Fig. 1 presents a general framework for a general fingerprint identification system (FIS) (Ji & Yi, 2008). Fingerprint matching is the last step in Automatic Fingerprint Identification System (AFIS). Fingerprint matching techniques can be classified into three types:

- Correlation-based matching,
- Minutiae-based matching, and
- Non-Minutiae feature-based matching.

Minutiae-based matching is the most popular and widely used technique, being the basis of the fingerprint comparison.

2. Previous work and motivation

In (Bazen & Gerez, 2003), a novel minutiae matching method is presented that describes elastic distortions in fingerprints by means of a thin-plate spline model, which is estimated using a local and a global matching stage. After registration of the fingerprints according to the estimated model, the number of matching minutiae can be counted using very tight matching thresholds. For deformed fingerprints, the algorithm gives considerably higher matching scores compared to rigid matching algorithms, while only taking 100 ms on a 1 GHz P-III machine. Furthermore, it is shown that the observed deformations are different from those described by theoretical models proposed in the literature.

In (Liang & Asano, 2006), minutia polygons are used to match distorted fingerprints. A minutia polygon describes not only the minutia type and orientation but also the minutia shape. This allows the minutia polygon to be bigger than the conventional tolerance box without losing matching accuracy. In other words, a minutia polygon has a higher ability to

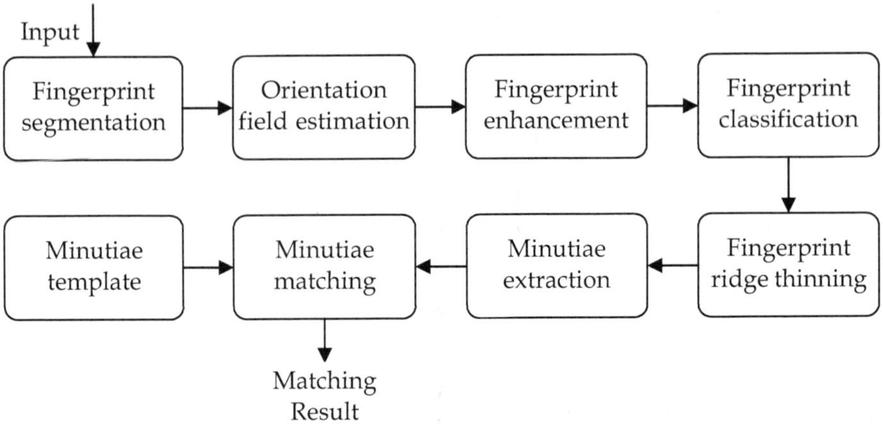

Fig. 1. General block diagram for a Fingerprint Identification System.

tolerate distortion. Furthermore, the proposed matching method employs an improved distortion model using a Multi-quadric basis function with parameters. Adjustable parameters make this model more suitable for fingerprint distortion. Experimental results show that the proposed method is two times faster and more accurate (especially, on fingerprints with heavy distortion) than the method in (Bazen & Gerez, 2003).

In (Jiang & Yau, 2000), a new fingerprint minutia matching technique is proposed, which matches the fingerprint minutiae by using both the local and global structures of minutiae. The local structure of a minutia describes a rotation and translation invariant feature of the minutia in its neighborhood. It is used to find the correspondence of two minutiae sets and increase the reliability of the global matching. The global structure of minutiae reliably determines the uniqueness of fingerprint. Therefore, the local and global structures of minutiae together provide a solid basis for reliable and robust minutiae matching. The proposed minutiae matching scheme is suitable for an on-line processing due to its high processing speed. Their experimental results show the performance of the proposed technique.

In (Jain et al., 2001), a hybrid matching algorithm that uses both minutiae (point) information and texture (region) information is presented for matching the fingerprints. Their results obtained show that a combination of the texture-based and minutiae-based matching scores leads to a substantial improvement in the overall matching performance. This work was motivated by the small contact area that sensors provide for the fingertip and, therefore, only a limited portion of the fingerprint is sensed. Thus multiple impressions of the same fingerprint may have only a small region of overlap. Minutiae-based matching algorithms, which consider ridge activity only in the vicinity of minutiae points, are not likely to perform well on these images due to the insufficient number of corresponding points in the input and template images.

In (Eckert et al., 2005), a new and efficient method for minutiae-based fingerprint matching is proposed, which is invariant to translation, rotation and distortion effects of fingerprint patterns. The algorithm is separated from a prior feature extraction and uses a compact description of minutiae features in fingerprints. The matching process consists of three major steps:

- Finding pairs of possibly corresponding minutiae in both fingerprint patterns,
- Combining these pairs to valid tuples of four minutiae each, containing two minutiae from each pattern.
- The third step is the matching itself.
  It is realized by a monotonous tree search that finds consistent combinations of tuples with a maximum number of different minutiae pairs. The approach has low and scalable memory requirements and is computationally inexpensive.

In (Yuliang et al., 2003), three ideas are introduced along the following three aspects:

- Introduction of ridge information into the minutiae matching process in a simple but effective way, which solves the problem of reference point pair selection with low computational cost;
- Use of a variable sized bounding box to make their algorithm more robust to non-linear deformation between fingerprint images;
- Use of a simpler alignment method in their algorithm.
  Their experiments using the Fingerprint Verification Competition 2000 (FVC2000) databases with the FVC2000 performance evaluation show that these ideas are effective.

In (Zhang et al., 2008), a novel minutiae indexing method is proposed to speed up fingerprint matching, which narrows down the searching space of minutiae to reduce the expense of computation. An orderly sequence of features is extracted to describe each minutia and the indexing score is defined to select minutiae candidates from the query fingerprint for each minutia in the input fingerprint. The proposed method can be applied in both minutiae structure-based verification and fingerprint identification. Experiments are performed on a large-distorted fingerprint database (FVC2004 DB1) to approve the validity of the proposed method.

In most existing minutiae-based matching methods, a reference minutia is chosen from the template fingerprint and the query fingerprint, respectively. When matching the two sets of minutiae, the template and the query, firstly, reference minutiae pair is aligned coordinately and directionally, and secondly, the matching score of the remaining minutiae is evaluated. This method guarantees satisfactory alignments of regions adjacent to the reference minutiae. However, the alignments of regions far away from the reference minutiae are usually not so satisfactory. In (Zhu et al., 2005), a minutia matching method based on global alignment of multiple pairs of reference minutiae is proposed. These reference minutiae are commonly distributed in various fingerprint regions. When matching, these pairs of reference minutiae are to be globally aligned, and those region pairs far away from the original reference minutiae will be aligned more satisfactorily. Their experiment shows that this method leads to improvement in system identification performance.

In (Jain et al., 1997a), the design and implementation of an on-line fingerprint verification system is described. This system operates in two stages: minutia extraction and minutia matching. An improved version of the minutia extraction algorithm proposed by (Ratha et al., 1995), which is much faster and more reliable, is implemented for extracting features from an input fingerprint image captured with an on-line inkless scanner. For minutia matching, an alignment-based elastic matching algorithm has been developed. This algorithm is capable of finding the correspondences between minutiae in the input image and the stored template without resorting to exhaustive search and has the ability of adaptively compensating for the nonlinear deformations and inexact pose transformations between fingerprints. The system has been tested on two sets of fingerprint images captured with inkless scanners. The verification accuracy is found to be acceptable. Typically, a complete fingerprint verification procedure takes, on an average, about eight seconds on a SPARC 20 workstation. These experimental results show that their system meets the response time requirements of on-line verification with high accuracy.

In (Luo et al., 2000), a minutia matching algorithm which modified (Jain et al., 1997a) algorithm is proposed The algorithm can better distinguish two images from different fingers and is more robust to nonlinear deformation. Experiments done on a set of fingerprint images captured with an inkless scanner shows that the algorithm is fast and of high accuracy.

In (Jie et al., 2006), a new fingerprint minutiae matching algorithm is proposed, which is fast, accurate and suitable for the real time fingerprint identification system. In this algorithm, the core point is used to determine the reference point and a round bounding box is used for matching. Experiments done on a set of fingerprint images captured with a scanner showed that the algorithm is faster and more accurate than that in (Luo et al., 2000) algorithm.

There are two major shortcomings of the traditional approaches to fingerprint representation (Jain et al., 2000):

1. For a considerable fraction of population, the representations based on explicit detection of complete ridge structures in the fingerprint are difficult to extract automatically. The widely used minutiae-based representation does not utilize a significant component of the rich discriminatory information available in the fingerprints. Local ridge structures cannot be completely characterized by minutiae.
2. Further, minutiae-based matching has difficulty in quickly matching two fingerprint images containing different number of unregistered minutiae points.

The filter-based algorithm in (Jain et al., 2000) uses a bank of Gabor filters to capture both local and global details in a fingerprint as a compact fixed length FingerCode. The fingerprint matching is based on the Euclidean distance between the two corresponding FingerCodes and hence is extremely fast. Verification accuracy achieved is only marginally inferior to the best results of minutiae-based algorithms published in the open literature (Jain et al., 1997b). Proposed system performs better than a state-of-the-art minutiae-based system when the performance requirement of the application system does not demand a very low false acceptance rate. Finally, it is shown that the matching performance can be improved by combining the decisions of the matchers based on complementary (minutiae-based and filter-based) fingerprint information.

Motivated by this analysis, a new algorithm is proposed in this chapter. This novel algorithm is minutiae-based matching algorithm. The proposed matching algorithm is described in section 3, the advantages are drawn in section 4, and finally, implementation, performance evaluation of the algorithm and conclusion are explained in section 5.

3. Proposed matching algorithm

Any Fingerprint Identification System (FIS) has two phases, fingerprint enrolment and fingerprint matching (identification or verification).

3.1 Enrolment phase

Fig. 2 shows the steps of the enrolment phase of the proposed matching algorithm, which is divided into the following steps:

1. Get the core point location of the fingerprint to be enrolled after applying enhancement process.

2. Extract all minutiae from the fingerprint image.
3. From output data of step2, get the minutiae locations (x, y coordinates) together with their type: type1 for termination minutiae and type2 for bifurcation minutiae.

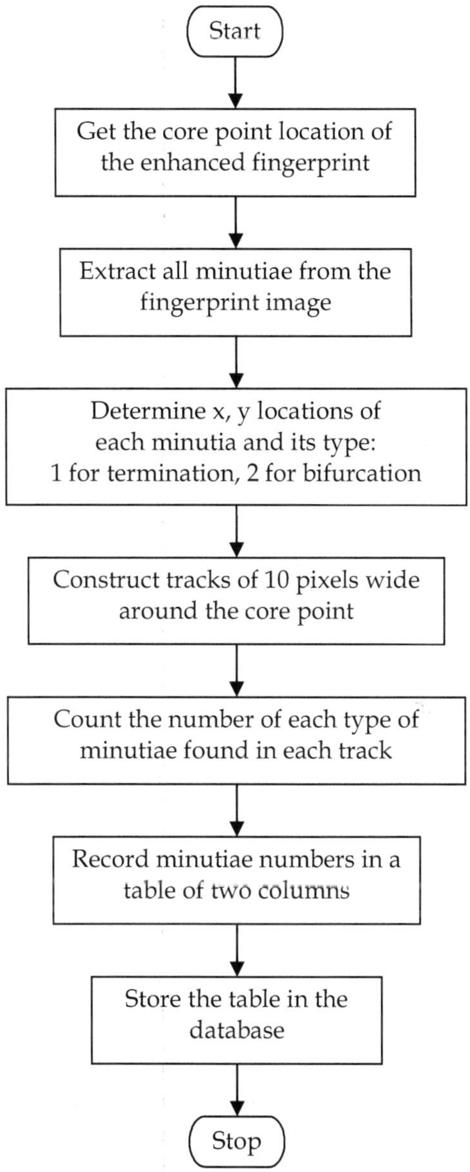

Fig. 2. Flowchart of the enrolment phase of the proposed matching algorithm.

4. Construct tracks of 10 pixels wide centred at the core point.
5. In each track, count the number of minutiae of type1 and the number of minutiae of type2.
6. Construct a table of two columns, column 1 for type1 minutiae and column 2 for type2 minutiae, having number of rows equal to number of found tracks.
7. In the first row, record the number of minutiae of type1 found in the first track in the first column, and the number of minutiae of type2 found in the first track in the second column.
8. Repeat step 7 for the remaining tracks of the fingerprint, and then store the table in the database.

This enrolment phase will be repeated for all prints of the same user's fingerprint. The number of prints depends on the application requirements at which the user registration takes place. For FVC2000 (Maio et al., 2002), there are 8 prints for each fingerprint. So, eight enrolments will be required for each user to be registered in the application. Finally, eight tables will be available in the database for each user.

3.2 Verification phase

For authenticating a user, verification phase should be applied on the user's fingerprint to be verified at the application. Fig. 3. shows the steps of the verification phase of the proposed matching algorithm, which is divided into the following steps:

1. Capture the fingerprint of the user to be verified.
2. Apply the steps of enrolment phase, described in section 3.1, on the captured fingerprint to obtain its minutiae table T.
3. Get all the minutiae tables corresponding to the different prints of the claimed fingerprint from the database.
4. Get the absolute differences, cell by cell, between minutiae table T and all minutiae tables of the claimed fingerprint taken from the database, now we have eight difference tables.
5. Get the summations of all cells in each of column1 (type1) and column2 (type2) for each difference table, now we have sixteen summations.
6. Get the geometric mean of the eight summations of type1 columns (gm1), and the geometric mean of the eight summations of type2 columns (gm2).
7. Check: if gm1<= threshold1 and gm2<=threshold2 then the user is genuine and accept him; else the user is imposter and reject him.

4. Advantages of the proposed matching algorithm

The proposed minutiae-based matching algorithm has the following advantages:

1. Since all cells in each minutiae table, representing the fingerprint in database, contain *just* the number of minutiae of type1 or type2 in each track around the core point of the fingerprint, neither position (x or y) nor orientation (θ) of the minutiae is considered; the algorithm is *rotation* and *translation invariant.*
2. The numbers of minutiae to be stored in the database *need less storage* than traditional minutiae-based matching algorithms which store position and orientation of each minutia. Experiments show that nearly 50% reduction in storage size is obtained.
3. Matching phase itself *takes less time* which, as will be shown in following sections, reaches 0.00134 sec.

5. Implementation of the proposed matching algorithm

Using MATLAB Version 7.9.0.529 (R2009b), both proposed enrolment and verification phases are implemented as described in next two subsections:

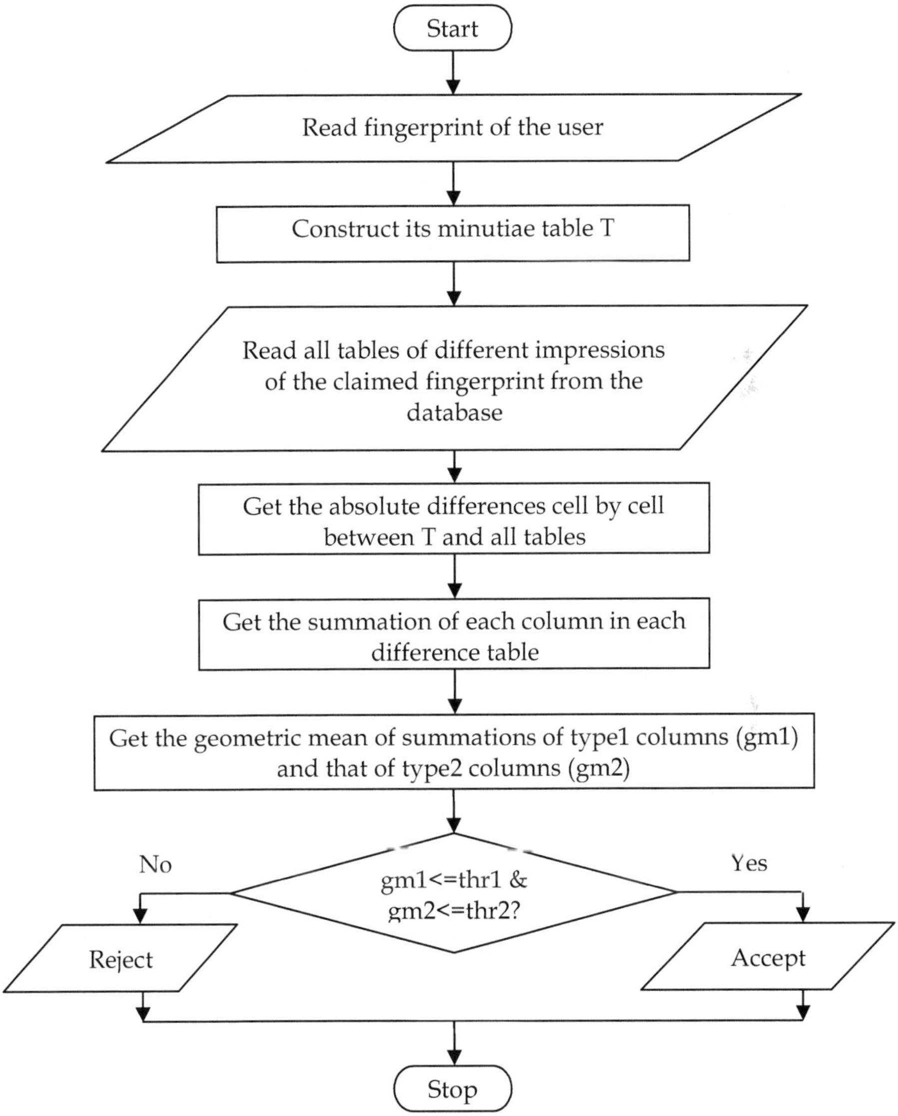

Fig. 3. Flowchart of the proposed verification phase.

5.1 Enrolment phase

5.1.1 Enhancement of the fingerprint image

The first step is to enhance the fingerprint image using Short Time Fourier Transform STFT analysis (O'Gorman, 1998). The performance of a fingerprint matching algorithm depends critically upon the quality of the input fingerprint image. While the quality of a fingerprint image may not be objectively measured, it roughly corresponds to the clarity of the ridge structure in the fingerprint image, and hence it is necessary to enhance the fingerprint image. Since the fingerprint image may be thought of as a system of oriented texture with non-stationary properties, traditional Fourier analysis is not adequate to analyze the image completely as the STFT analysis does (Yang & Park, 2008). Fingerprint enhancement MATLAB code is available at (http://www.hackchina.com/en/cont/18456).

The algorithm for image enhancement consists of two stages as summarized below:

Step 1. STFT analysis

1. For each overlapping block in an image, generate and reconstruct a ridge orientation image by computing gradients of pixels in a block, and a ridge frequency image through obtaining the FFT value of the block, and an energy image by summing the power of FFT value;
2. Smoothen the orientation image using vector average to yield a smoothed orientation image, and generate a coherence image using the smoothed orientation image;
3. Generate a region mask by thresholding the energy image;

Step 2. Apply Enhancement

For each overlapping block in the image, the next five sub-steps are applied:

1. Generate an angular filter Fa centered on the orientation in the smoothed orientation image with a bandwidth inversely proportional to coherence image;
2. Generate a radial filter Fr centered on frequency image;
3. Filter a block in the FFT domain, F=F×Fa×Fr;
4. Generate the enhanced block by inverse Fourier transform IFFT(F);
5. Reconstruct the enhanced image by composing enhanced blocks, and yield the final enhanced image with the region mask.

The result of the enhancement process is shown in Fig. 4, where Fig. 4.a is taken from FVC2000 DB1_B (108_5) and Fig. 4.b is the enhanced version of Fig. 4.a.

5.1.2 Get core point of the enhanced fingerprint

Core point MATLAB code is available at (http://www.hackchina.com/en/cont/18456) where the idea of determining the reference point is taken from (Yang & Park, 2008), which is described as follows:

The reference point is defined as "the point of the maximum curvature on the convex ridge (Liu et al., 2005)" which is usually located in the central area of fingerprint. The reliable detection of the position of a reference point can be accomplished by detecting the maximum curvature using complex filtering methods (Nilsson & Bigun, 2003).

They apply complex filters to ridge orientation field image generated from original fingerprint image. The reliable detection of reference point with the complex filtering methods is summarized below:

1. For each overlapping block in an image;
   a. Generate a ridge orientation image with the same method in STFT analysis;

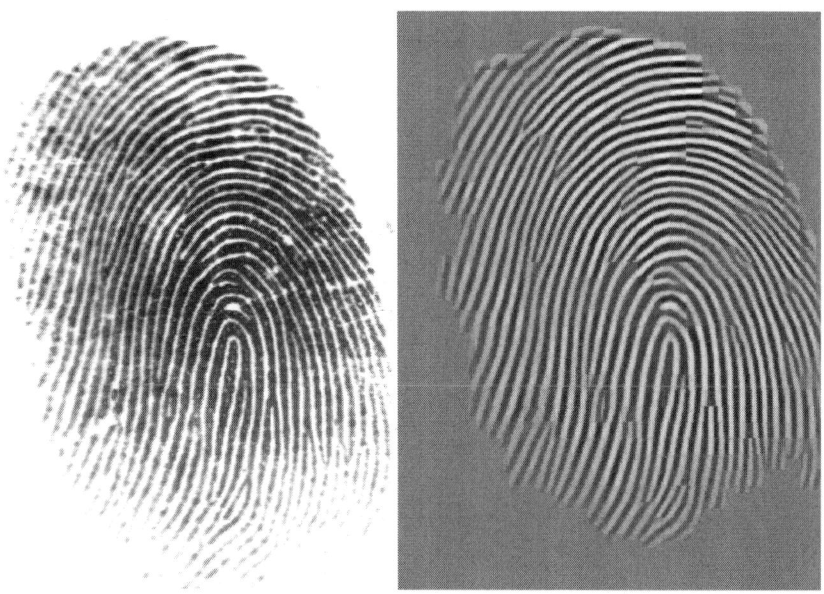

(a) (b)

Fig. 4. a) Fingerprint image 108_5 from DB1_B in FVC2000, b) Enhanced version of fingerprint image 108_5.

b. Apply the corresponding complex filter $h = (x + iy)^m g(x, y)$ centered at the pixel orientation in the orientation image, where m and $g(x, y) = \exp\{-((x^2 + y^2)/2\sigma^2))\}$ indicate the order of the complex filter and a Gaussian window, respectively;

c. For $m = 1$, the filter response of each block can be obtained by a convolution, $h * O(x, y) = g(y) * ((xg(x))^t * O(x,y))+ig(x)^t * ((yg(y) * O(x, y)))$
   where $O(x, y)$ represents the pixel orientation image;

2. Reconstruct the filtered image by composing filtered blocks.

The maximum response of the complex filter in the filtered image can be considered as the reference point. Since there is only one output, the unique output point is taken as the reference point (core point).

5.1.3 Minutiae extraction

To extract minutiae from the enhanced fingerprint image, the minutiae extraction method (Maltoni et al., 2003) is used. Hence we have three information for each minutia: x and y coordinates of its location, type of minutia (type1 if it is a termination, type2 if it is a bifurcation).

The result of this minutiae extraction stage is shown in Fig. 5, where Fig. 5.a is the same as Fig. 4.b, Fig. 5.b shows the termination minutiae in circles and the bifurcation minutiae in diamonds together with the core point of the fingerprint with an asterisk.

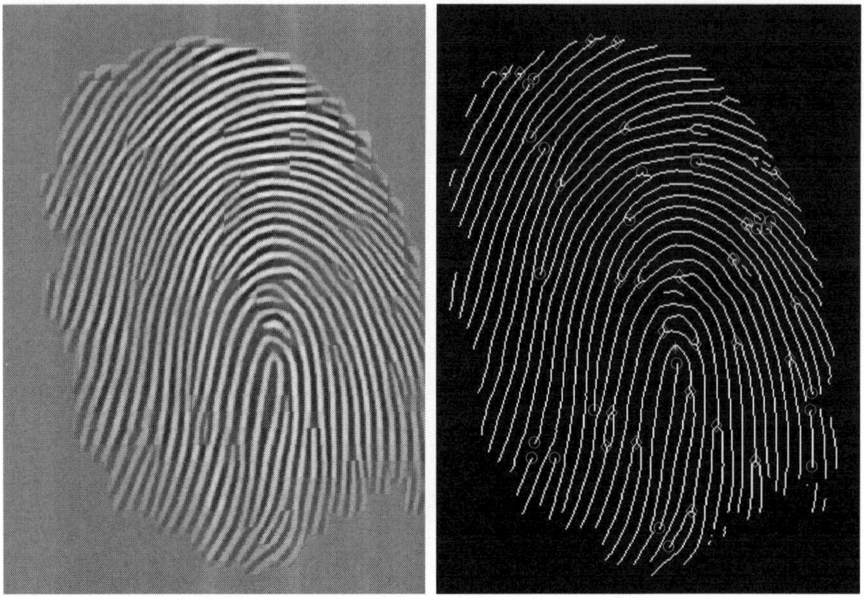

(a) (b)

Fig. 5. a) Enhanced Fingerprint image 108_5 from DB1_B in FVC2000, b) core point (asterisk), terminations (circles) and bifurcations (diamonds).

5.1.4 Construct the minutiae table

From the output of the minutiae extraction step, the proposed minutiae table is constructed as following:

a. Get all minutiae locations together with their types.
b. Using Euclidean distances, get the distances between all the minutiae and the core point of the fingerprint: if the core location is at (x_c, y_c) and a minutia location is at (x, y), the Euclidean distance between them will be:

$$\sqrt{(x-x_c)^2+(y-y_c)^2}$$

c. Construct tracks of 10×n pixels wide (where n=1…max_distance/10) centered at the core point until all minutiae are exhausted, the track width is chosen to be 10 as the average distance (in pixels) between two consecutive ridges is 10 pixels, this is achieved in the fingerprint 108_5 in Fig. 5.a. where it is of 96 dpi resolution.
d. In each track, count the number of existed minutiae of type1 and the number of existed minutiae of type2.
e. Construct a table of two columns, column 1 for type1 minutiae and column 2 for type2 minutiae, having a number of rows that is equal to the number of found tracks.
f. In the first row, record the number of minutiae of type1 found in the first track in the first column, and record the number of minutiae of type2 found in the first track in the second column.

g. Repeat last step for the remaining tracks of the fingerprint until all tracks are processed, and then store the minutiae table in the database.

The resulted minutiae table from Fig. 5.b is as shown in Table 1. The first column is for illustration only but in MATLAB, it does not exist and it is used as the index for each row of the minutiae table consisting of just two columns.

To validate the table's data, the total number of the minutiae of fingerprint 108_5 of both types: termination and bifurcation, in Fig. 5.b is found to be the total summation of both columns of minutiae table which is equal to 51 minutiae.

fingerprint	108_5		108_7	
Track number	# of type1 minutiae	# of type2 minutiae	# of type1 minutiae	# of type2 minutiae
1	0	0	1	0
2	1	2	1	2
3	0	0	1	0
4	0	1	0	0
5	0	1	1	2
6	0	2	1	1
7	0	3	6	1
8	2	2	5	1
9	0	2	2	1
10	1	2	4	0
11	6	1	3	1
12	3	1	2	0
13	3	0	5	1
14	4	1	1	1
15	1	2	3	1
16	0	1	2	1
17	0	0	0	2
18	1	0	2	0
19	1	0	4	1
20	0	0	4	1
21	0	0	3	2
22	1	0	1	0
23	1	4	1	1
24	0	1	--	--

Table 1. The minutiae table of fingerprint 108_5 and 108_7 from DB1_B in FVC2000

5.2 Verification phase

5.2.1 Capture the fingerprint to be verified

The user, who is claiming he is (e.g. M), will put his finger on the scanner to be captured at the application he wants to access. Now, a fingerprint will be available to be verified to check if he is actually M so he will be accepted or he is not so he will be rejected.

5.2.2 Construct the minutiae table of that fingerprint

The same steps of the enrolment explained in section 5.1 will be applied on the fingerprint obtained from the previous step. Now, a minutiae table T corresponding to the fingerprint under test will be built.

5.2.3 Get all corresponding minutiae tables from the database

In FVC2000 (Maio et al., 2002), each fingerprint has eight prints, so to verify a certain input fingerprint, all the corresponding minutiae tables of different prints of that claimed fingerprint stored in the database must be fetched.

Taking as an example the fingerprint 108_7 (see Fig. 6.a) taken from DB1_B in FVC2000 to be verified, and applying the same steps of enrolment, the results are shown in Fig. 6 where Fig. 6.b shows the enhanced version of Fig. 6.a, and Fig. 6.c shows its thinned version together with the terminations in circles, the bifurcations in diamonds and finally the core point in asterisk. As can be shown, because the image is of poor quality, the number of minutiae is so different.

Now the minutiae table is ready to be constructed as shown in Table 1. Summing all the minutiae of both types results in 73 minutiae which is different from the number before for fingerprint 108_5 which was 51 minutiae.

Following the same steps, fingerprints 108_1, 108_2, 108_3, 108_4, 108_6, and 108_8 are taken from DB1_B in FVC2000, and their corresponding minutiae tables are constructed in six tables.

5.2.4 Calculate the absolute differences between minutiae table of the input fingerprint and all minutiae tables of the claimed fingerprint

Now, the absolute differences between minutiae table corresponding to fingerprint 108_7 and all minutiae tables corresponding to fingerprints 108_1, 108_2, 108_3, 108_4, 108_5, 108_6, and 108_8 are calculated. Because the sizes (number of rows) of minutiae tables are not equal, the minimum size must be determined to be able to perform the absolute subtraction on the same size for different tables. The minimum number of tracks (rows) found in DB1_B is 14.

So, only the first 14 rows of each minutiae table will be considered during the absolute differences calculation. Table 2 shows the absolute differences between the minutiae table of fingerprint 108_7 and the minutiae tables of fingerprints 108_1 and 108_2.

5.2.5 Get the summation of each column in each difference table

At the bottom of Table 2 in the row titled "sum", the summation of each column in each difference table is drawn, applying the same steps for calculating the absolute differences between the minutiae table of fingerprint 108_7 and the minutiae tables of fingerprints 108_3, 108_4, 108_5, 108_6, and 108_8, will result in seven summations for type1 columns, and other seven summations for type2 columns.

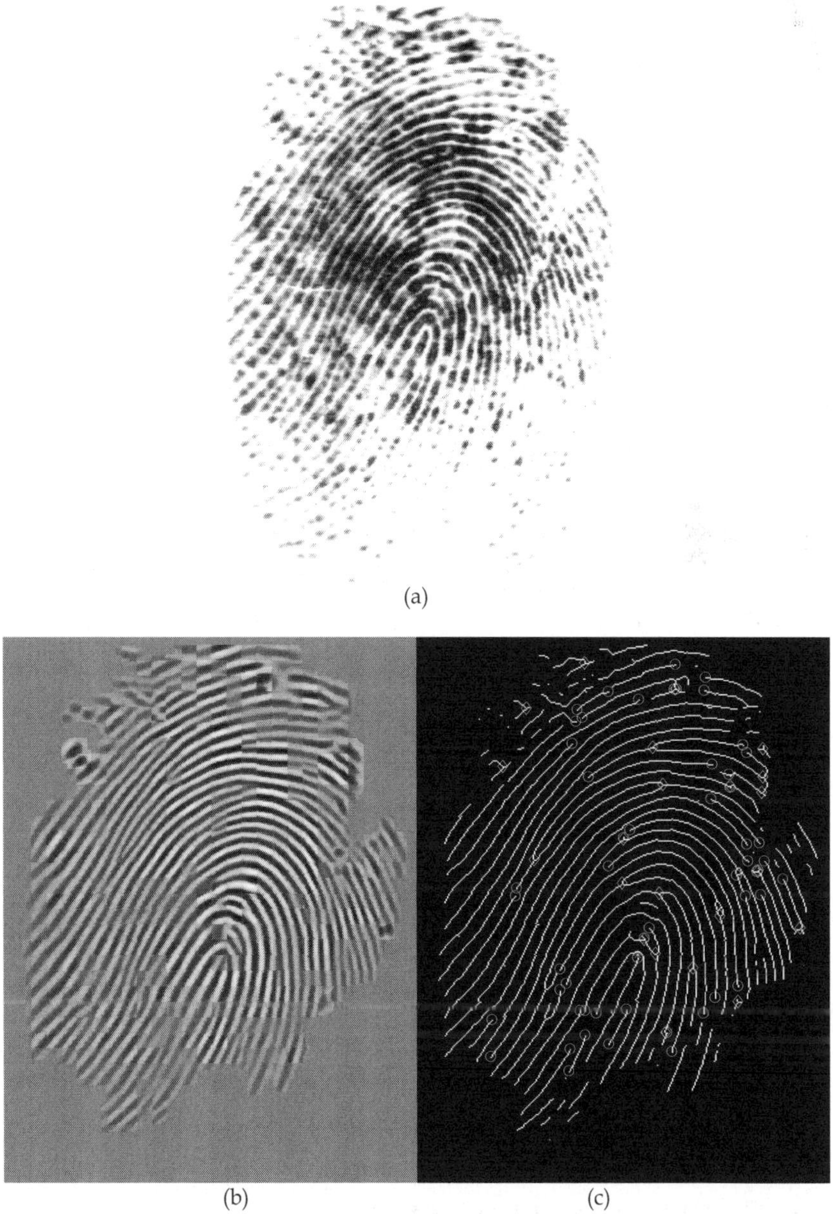

Fig. 6. a) Fingerprint image 108_7 from DB1_B in FVC2000, b) Enhanced version of fingerprint image 108_7, c) core point (asterisk), terminations (circles) and bifurcations (diamonds).

5.2.6 Get the geometric mean of the resulted summations for both of type1 and type2

Get the geometric mean of the summations of type1 columns in all difference tables. The geometric mean, in mathematics, is a type of mean or average, which indicates the central tendency or typical value of a set of numbers (http://en.wikipedia.org/wiki/Geometric_mean). It is similar to the arithmetic mean, which is what most people think of with the word "average", except that the numbers are multiplied and then the nth root (where n is the count of numbers in the set) of the resulting product is taken.

$$gm1 = \sqrt[7]{25 \times 24 \times 24 \times 26 \times 27 \times 17 \times 69} = 27.488$$

Get the geometric mean of the summations of type2 columns in all difference tables.

$$gm2 = \sqrt[7]{9 \times 9 \times 10 \times 15 \times 11 \times 7 \times 6} = 9.2082$$

Check the values of gm1 and gm2:

If gm1 <= threshold1 and gm2 <= threshold2 then

the user is genuine and accept him

else

the user is imposter and reject him

	abs(108_7-108_1)		abs(108_7-108_2)	
Track number	# of type1 minutiae	# of type2 minutiae	# of type1 minutiae	# of type2 minutiae
1	0	0	1	0
2	0	2	1	1
3	0	0	0	0
4	0	1	2	0
5	2	1	0	1
6	2	1	1	1
7	2	1	3	1
8	5	0	4	1
9	2	0	1	1
10	4	0	2	0
11	2	1	3	1
12	1	1	1	0
13	4	1	4	1
14	1	0	1	1
Sum	25	9	24	9

Table 2. Absolute differences between minutiae table of fingerprint 108_7 and both minutiae tables of fingerprints 108_1, 108_2.

5.3 Performance evaluation

To evaluate any matching algorithm performance, some important quantities have to be measured such as (Maio et al., 2002):

- False NonMatch Rate (FNMR) often referred to as False Rejection Rate (FRR)
- False Match Rate (FMR) often referred to as False Acceptance Rate (FAR)
- Equal Error Rate (EER)
- ZeroFNMR
- ZeroFMR
- Average enroll time
- Average match time

Because the presence of the fingerprint cores and deltas is not guaranteed in FVC2000 since no attention was paid on checking the correct finger position on the sensor (Maio et al., 2002), and the core point detection is the first step in the proposed matching algorithm, another group of fingerprints has been captured experimentally; this group contains the right forefinger of 20 different persons, each is captured three times, having 60 different fingerprint images. They are numbered as follows: 101_1, 101_2, 101_3, 102_1, …., 120_1, 120_2, 120_3. All these fingerprints have a core point. This group will be tested first and then the four databases DB1, DB2, DB3, and DB4 (from FVC2000) will be tested afterwards. All steps used to evaluate the performance of the proposed algorithm are implemented.

5.3.1 Calculate False NonMatch Rate (FNMR) or False Rejection Rate (FRR)

Each fingerprint template (minutiae table) T_{ij}, i=1…20, j=1…3, is matched against the fingerprint images (minutiae tables) of F_i, and the corresponding Genuine Matching Scores (GMS) are stored. The number of matches (denoted as NGRA - Number of Genuine Recognition Attempts (Maio et al., 2002)) is $20 \times 3 = 60$.

Now FRR(t) curve will be easily computed from GMS distribution for different threshold values. Given a threshold t, FRR(t) denotes the percentage of GMS $\geq t$. Here, because the input fingerprint is verified whether it gives less difference values between corresponding minutiae tables, lower scores are associated with more closely matching images. This is the opposite of most fingerprint matching algorithms in fingerprint verification, where higher scores are associated with more closely matching images. So, the FRR(t) (or FNMR(t)) curve will start from the left not from the right as usual. Also, it is worth to be noted that the curve of FRR(t) will be a 2D surface (FRR(t_1, t_2)) because there are two thresholds as mentioned in previous section.

For example, consider the fingerprint 101_1, or any slightly different version of it, is to be matched with its other prints 101_2, 101_3, this is considered as a genuine recognition attempt because they are all from the same fingerprint. Fig. 7 shows the fingerprint 101_1 together with its enhanced thinned version where core point is shown in a solid circle, terminations are shown with circles, and bifurcations are shown with diamonds.

As an example, some noise is applied on the minutiae table of fingerprint 101_1, to act as a new user's fingerprint. Noise is a sequence of Pseudorandom integers from a uniform discrete distribution used to randomly select tracks from the minutiae table that will be changed by adding '1' to values under termination (or bifurcation) column and subtracting '1' to values under bifurcation (or termination) column in each selected track. The sequence of numbers, produced using Matlab function called "randi", is determined by the internal state of the uniform pseudorandom number generator. The number of random selected tracks is a constant ratio (30%) from the overall number of tracks in each database.

Now, minutiae tables of fingerprints 101_1, 101_2, and 101_3 will be fetched from the database. The minimum number of rows (tracks) in all the minutiae tables in the database under study is found to be 13, so only the first 13 rows of any minutiae table are considered during calculations of absolute differences.

Fig. 7. Fingerprint 101_1 and its enhanced thinned version.

The second step is to calculate the geometric mean of the sum of each of type1 and type2 absolute differences:

$$gm1=\sqrt[3]{2\times 17\times 17}=8.33\,,$$

$$gm2=\sqrt[3]{3\times 8\times 7}=5.52$$

Then, since a user is accepted if the two geometric means satisfy that:

$$gm1<=threshold1(t_1)\wedge gm2<=threshold2(t_2)$$

Using Demorgan's law, the (FNMR) or false rejection rate will be computed as follows:

$$FNMR(t_1,t_2)=\frac{NGMS1s\ >\ t_1\vee NGMS2s\ >\ t_2}{NGRA}$$

Where NGMS1s is the number of genuine matching scores computed from gm1 values, NGMS2s is the number of genuine matching scores computed from gm2 values, and NGRA is the number of genuine recognition attempts which is 60. The threshold values, t_1 and t_2, vary from 1 to 100.

The same steps are performed for the remaining fingerprints, all 60 instances. The previous example is considered as a genuine recognition attempt as the comparison is held between a noisy version of the first of the three prints and the three true versions of them fetched from the database.

5.3.2 Calculate False Match Rate (FMR) or False Acceptance Rate (FAR)

Each fingerprint template (minutiae table) T_{ij}, $i = 1\ldots20$, $j = 1\ldots3$ in the database is matched against the other fingerprint images (minutiae tables) F_k, $k \neq i$ from different fingers and the corresponding *Imposter Matching Scores* ims are stored. Number of Impostor Recognition Attempts is $(20 \times 3) \times (20 - 1) = 60 \times 19 = 1140$.

Now FAR(t_1, t_2) surface will be easily computed from IMS distribution for different threshold values. Given thresholds t_1 and t_2, FAR(t_1, t_2) denotes the percentage of IMS1s <= t_1 and IMS2s<=t_2. Here, because the input fingerprint is rejected if it gives high difference values between corresponding minutiae tables; higher scores are associated with mismatching images. This is the opposite of most fingerprint matching algorithms in fingerprint verification, where lower scores are associated with mismatching images. So, the FAR(t_1, t_2) (or FMR(t_1, t_2)) surface will start from the right not from the left as usual.

For example, consider the noisy version of fingerprint 101_1 is to be matched with another fingerprint like 103, this is considered as an imposter recognition attempt because they are from different fingers. Now, all minutiae tables of fingerprints 103_1, 103_2, and 103_3 have to be fetched from the database. As before, because the minimum number of rows is 13, so only the first 13 rows of any minutiae table are considered during calculations of the absolute differences tables.

Geometric mean gm1 and gm2 are calculated as follows:

$$\text{gm1} = \sqrt[3]{26 \times 23 \times 16} = 21.23\,,$$

$$\text{gm2} = \sqrt[3]{14 \times 13 \times 13} = 13.33$$

The same steps are performed for the remaining fingerprints; all 60 instances (20 fingerprints, each having 3 impressions) will be matched against the other 19 fingerprints, so a total of 1140 IMSs.

FMR(t_1, t_2) will be calculated as follows:

$$\text{FMR}(t_1, t_2) = \frac{\text{NIMS1s} \le t_1 \wedge \text{NIMS2s} \le t_2}{\text{NIRA}}$$

Where NIMS1s is the number of imposter matching scores computed from gm1 values, NIMS2s is the number of imposter matching scores computed from gm2 values, and NIRA is the number of imposter recognition attempts which is 1140. The threshold values, t_1 and t_2, vary from 1 to 70.

Both surfaces FRR(t_1, t_2) and FAR(t_1, t_2) are drawn in Fig. 8 with blue and red colors respectively. The intersection between the two surfaces is drawn with a solid line used in the next section.

5.3.3 Equal Error Rate EER

The Equal Error Rate is computed as the point where FMR(t) = FNMR(t). From Fig. 8, to determine the equal error rate, the intersection line between the two surfaces is drawn and then the minimum value of error rates along this line is the EER from where the values of thresholds t_1 and t_2 can be determined.

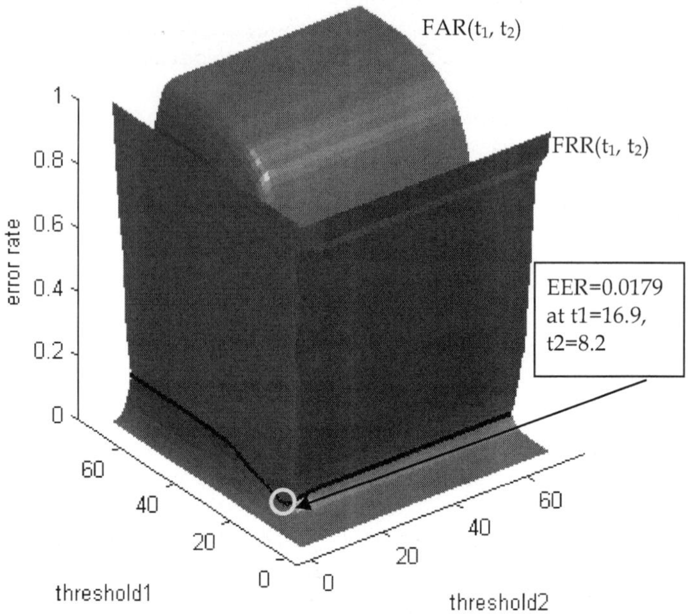

Fig. 8. FRR and FAR surfaces where the intersection between the two surfaces is drawn with a solid line.

In Fig. 8, it is shown that EER = 0.0179, where it corresponds to the threshold values of t_1 = 16.92 and t_2 = 8.21, values of thresholds are not always integers because it is not necessary for the two surfaces to intersect at integer values of thresholds.

Now to determine the integer values of thresholds that corresponds to error rates FRR and FAR, the four possible combinations of thresholds around the two thresholds given before are tested and the two values combination that gives the minimum difference between FRR and FAR (because EER is defined as the point where FRR and FAR are equal) are considered as the thresholds t_1 and t_2 that will be used for that database for any later fingerprint recognition operation.

So, the four possible combinations that threshold values t_1 and t_2 can take are: (16, 8), (16, 9), (17, 8), and (17, 9). It is found by experiment that the combination (17, 8) gives the minimum difference between FAR and FRR. So, when these thresholds are used in the proposed matching algorithm, the result is that FRR = 0.0167 and FAR = 0.0184.

True Acceptance Rate is

$$TAR = 1\text{-}FAR = 1\text{-}0.0184 = 0.9816$$

And the True Rejection Rate is

$$TRR = 1\text{-}FRR = 1\text{-}0.0167 = 0.9833$$

So, the recognition accuracy is ≈ 98% when thresholds values are t_1 = 17 and t_2 = 8.

5.3.4 ZeroFMR and ZeroFNMR

ZeroFMR is defined as the lowest FNMR at which no False Matches occur and ZeroFNMR as the lowest FMR at which no False Non-Matches occur(Maio et al., 2002):

$$\text{ZeroFMR}(t) = \min_t\{\text{FNMR}(t) \mid \text{FMR}(t) = 0\},$$

$$\text{ZeroFNMR}(t) = \min_t\{\text{FMR}(t) \mid \text{FNMR}(t) = 0\}.$$

Because now the FRR(FNMR) and FAR(FMR) are drawn as 2D surfaces, all locations of FAR points having zero values are determined and the minimum value of the corresponding FRR values at these locations is the ZeroFAR. Also, to calculate the ZeroFAR value, all locations of FRR points having zero values are determined and the minimum value of the corresponding FAR values at these locations is the ZeroFRR.
From Fig. 8, following values are drawn:

$$\text{ZeroFMR} = 0.3167 \text{ at } t_1 = 14 \text{ and } t_2 = 5,$$

$$\text{ZeroFRR} = 0.0316 \text{ at } t_1 = 16 \text{ and } t_2 = 10.$$

5.3.5 Drawing ROC curve

A ROC (Receiving Operating Curve) is given where FNMR is plotted as a function of FMR; the curve is drawn in log-log scales for better comprehension(Maio et al., 2002). To draw the curve in the positive portions of x- and y-axis, FMR and FNMR values are multiplied by 100 before applying the logarithm on them. Fig. 9 shows the ROC curve of the proposed matching algorithm. To get one curve, only one column of the FAR matrix is drawn against one column of the FRR matrix, after multiplying with 100 and applying the logarithm on both. As can be shown, the recognition performance is good by comparison with the curve of a good recognition performance system seen in (O'Gorman, 1998). It is noted that the curve in Fig. 9 is going to the top right portion of the plotting area whereas the good recognition performance curve in (O'Gorman, 1998) is going to the bottom left portion of the

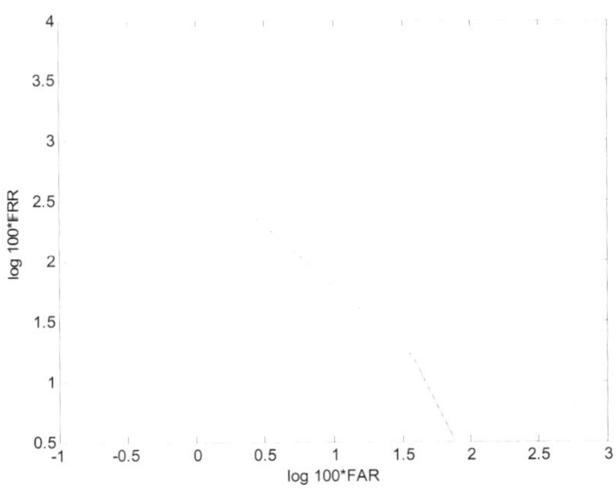

Fig. 9. ROC curve

plotting area, this is because in the proposed matching algorithm, lower scores are associated with matching fingerprints and higher scores are associated with mismatching fingerprints. This is the opposite of most fingerprint matching algorithms in fingerprint verification.

5.3.6 Applying the proposed matching algorithm on FVC2000

Applying the proposed matching algorithm and all above steps in previous sections on the database FVC2000, it is not expected to get good results compared with the results obtained in the previous section. This is due to the reasons mentioned in the beginning of section 5.3. Table 3 and Table 4 show the results of the proposed matching algorithm on FVC2000.

As shown, the recognition accuracy ranges from (1-0.2315 for DB2_B) 77% to (1 - 0.0882 for DB3_B) 91%.

Database	EER	t_1	t_2
DB1_A	0.2109	18	6.99
DB1_B	0.1988	31.68	10
DB2_A	0.1649	18.48	8
DB2_B	0.2315	24.096	14
DB3_A	0.1454	28	12.55
DB3_B	0.0882	28	12.85
DB4_A	0.1815	10	4.88998
DB4_B	0.1206	14.35	9

Table 3. Results of EER and its corresponding thresholds after applying the proposed matching algorithm on FVC2000

Database	FAR	FRR	t_1	t_2
DB1_A	0.2113	0.2105	18	7
DB1_B	0.2049	0.1944	32	10
DB2_A	0.1835	0.15	18	9
DB2_B	0.2403	0.2375	24	15
DB3_A	0.1325	0.1525	29	12
DB3_B	0.0944	0.075	28	13
DB4_A	0.1844	0.1788	10	5
DB4_B	0.1153	0.125	14	10

Table 4. Results of FAR and FRR and their corresponding thresholds after applying the proposed matching algorithm on FVC2000

Fig. 10 and Fig. 11 show the FRR and FAR surfaces at the left side and the ROC curves at the right side for databases DB1_A, DB2_A, DB2_B, DB3_A, and DB4_A respectively.

5.3.7 Calculating average enroll time

The average enroll time is calculated as the average CPU time taken by a single enrolment operation (Maio et al., 2002). The steps of enrolment are discussed in section 5.1. table 5 shows a detailed timing for each step in the enrolment phase. These results were implemented using MATLAB version 7.9.0529 (R2009b) as the programming platform. Programs were tested on a 2.00GHz personal computer with 1.99 GB of RAM.
Total enroll time is found to be 6.043 sec

Step	Average time taken (sec)
Enhancement of the fingerprint	3.7
Core point detection	0.54
Thinning and minutiae extraction	1.8
Minutiae table construction	0.003
Total enroll time	6.043

Table 5. Enroll timing details

5.3.8 Calculating average match time

The average match time is calculated as the average CPU time taken by a single match operation between a template and a fingerprint image (Maio et al., 2002). The steps of matching are discussed in section 5.2. Table 6 shows a detailed timing for each step in the matching phase after the construction of the minutiae table corresponding to the input fingerprint, which has been already estimated to be 6.043 sec from section 5.3.7.
Total match time is found to be 0.00134 sec

Step	Average time taken (sec)
Get all minutiae tables of the claimed fingerprint stored in the database	0.0011
Calculate absolute differences between the input fgp minutiae table and all minutiae tables obtained from the previous step and get the two geometric means	0.0002
Compare the resulting means with the two thresholds and decide if the user is accepted or rejected	0.00004
Total match time	0.00134

Table 6. Match timing details

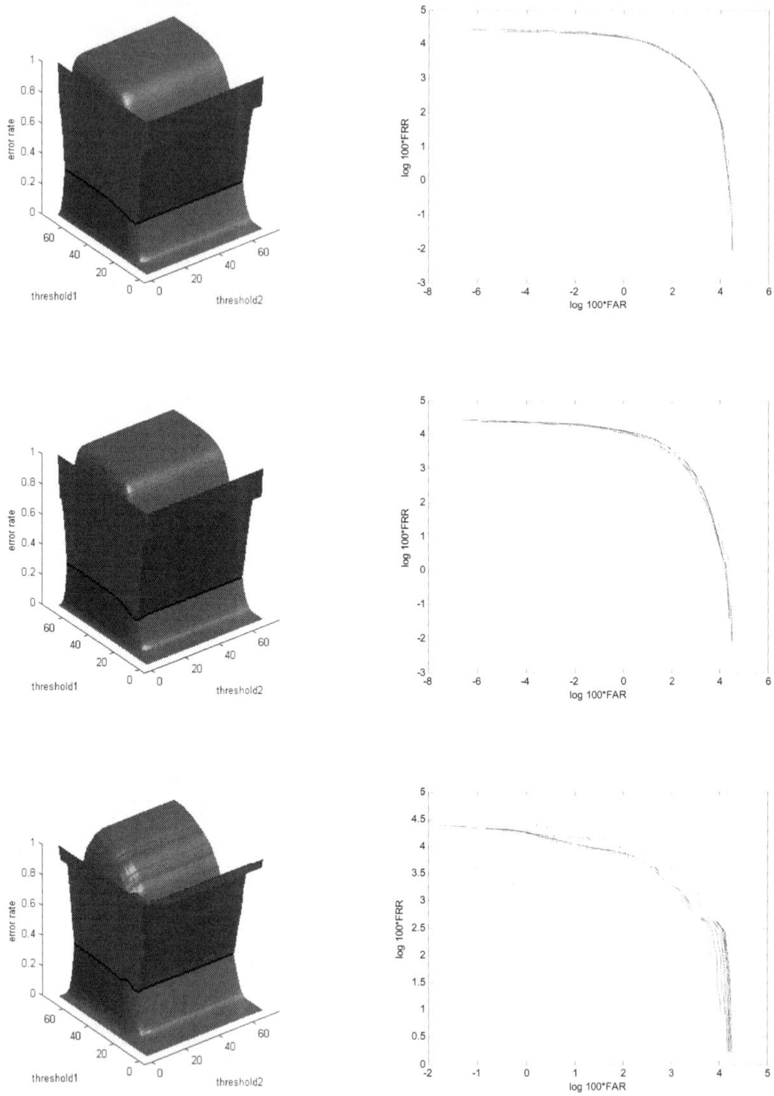

Fig. 10. FAR, FRR and ROC curves for DB1_A, DB2_A, and DB2_B respectively.

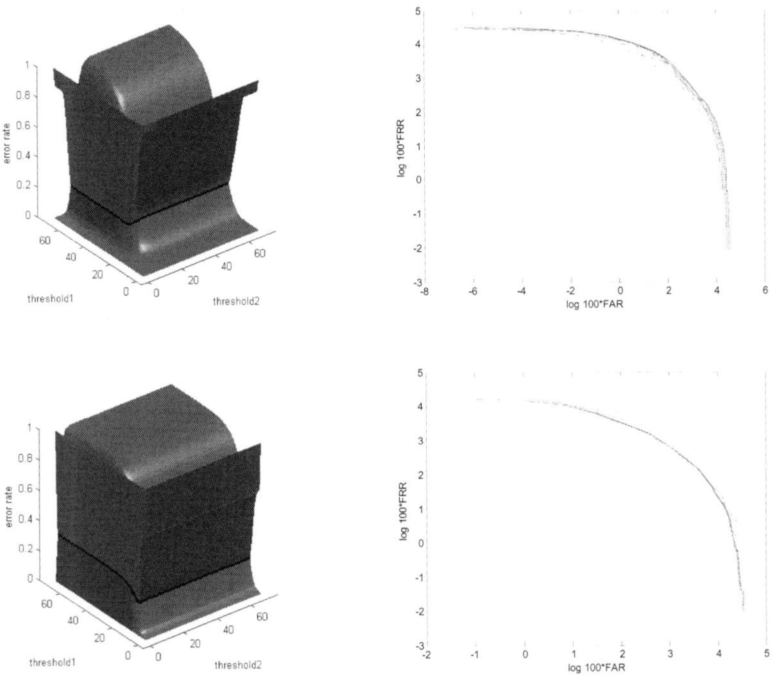

Fig. 11. FAR, FRR and ROC curves for DB3_A, DB4_A respectively.

5.4 Conclusion

As shown, the time for matching is extremely small using our algorithm as all the process is taking geometric mean of absolute differences. There is no need for any pre-alignment which is a very complicated and time consuming process. As a result, our algorithm is translation and rotation invariant.

Also, the space needed to store any minutiae table is in average 21 (as the average number of tracks in all database) $\times$ 2 $\times$ 4 = 168 bits = 168/8 bytes = 21 bytes which is small in comparison with the size of 85 bytes as in (Jain & Uludag, 2002) where the traditional method is storing locations and orientation for each minutia as a tuple <x, y, θ>.

6. References

Bazen, A. M., & Gerez, S. H. (2003). Fingerprint Matching by Thin-Plate Spline Modelling of Elastic Deformations. *Pattern Recognition,* Vol. 36, pp. 1859-1867

Eckert, G.; M¨uller, S., & Wiebesiek, T. (2005) Efficient Minutiae-Based Fingerprint Matching. *IAPR Conference on Machine VIsion Applications,* pp. 554-557, Tsukuba Science City, Japan, May 16-18

Jain, A.; Hong, L., & Bolle, R. (1997a). On-Line Fingerprint Verification. *IEEE Trans on Pattern Analysis and Machine Intelligence,* Vol. 19, No. 4, pp. 302-313

Jain, A.; Ross, A., & Prabhakar, S. (2001). Fingerprint Matching Using Minutiae and Texture Features. *Proc. Int. Conf. on Image Processing (ICIP)*, pp. 282-285, Thessaloniki, Greece, October 7-10

Jain, A. K.; Hong, L.; Pankanti, S., & Bolle, R. (1997b). An identity authentication system using fingerprints. *Proc. IEEE*, Vol. 85, pp. 1365-1388

Jain, A. K.; Prabhakar, S.; Hong, L., & Pankanti, S. (2000). Filterbank-Based Fingerprint Matching. *IEEE Transactions on Image Processing*, Vol. 9, No. 5, pp. 846-859

Jain, A. K., & Uludag, U. (2002). Hiding Fingerprint Minutiae in Images. *Proc. Workshop on Automatic Identification Advanced Technologies*, pp. 97-102

Ji, L., & Yi, Z. (2008). Fingerprint orientation field estimation using ridge projection. *Pattern Recognition*, Vol. 41, pp. 1491-1503

Jiang, X., & Yau, W. Y. (2000). Fingerprint Minutiae Matching Based on the Local and Global Structures. *in Proc. Int. Conf. on Pattern Recognition (15th)*, Vol. 2, pp. 1042-1045

Jie, Y.; fang, Y.; Renjie, Z., & Qifa, S. (2006). Fingerprint minutiae matching algorithm for real time system. *Pattern Recognition*, Vol. 39, pp. 143-146

Liang, X., & Asano, T. (2006). Fingerprint Matching Using Minutia Polygons. *Proc. Int. Conf. on Pattern Recognition (18th)*, Vol. 1, pp. 1046-1049

Liu, M.; Jiang, X. D., & Kot, A. (2005). Fingerprint reference-point detection. *EURASIP Journal on Applied Signal Processing*, Vol. 4, pp. 498-509

Luo, X.; Tian, J., & Wu, Y. (2000). A Minutia Matching Algorithm in Fingerprint Verification. *15th ICPR Int. Conf. on Pattern Recognition*, pp. 833-836, Barcelona, Spain, September 3-7

Maio, D.; Maltoni, D.; Cappelli, R.; Wayman, J. L., & Jain, A. K. (2002). FVC2000: Fingerprint Verification Competition. *IEEE Transactions on Pattern Analysis and Machine Intelligence*, Vol. 24, No. 3, pp. 402-412

Maltoni, D.; Maio, D.; Jain, A. K., & Prabhakar, S. (2003). *Handbook of Fingerprint Recognition*, Springer-Verlag New York

Maltoni, D.; Maio, D.; Jain, A. K., & Prabhakar, S. (2009). *Handbook of Fingerprint Recognition*, Springer-Verlag, London

Nilsson, K., & Bigun, J. (2003). Localization of corresponding points in fingerprints by complex filtering. *Pattern Recognition Letters*, Vol. 24, pp. 2135-2144

O'Gorman, L. (1998). An Overview of fingerprint verification technologies. *Elsevier Information Security Technical Report*, Vol. 3, No. 1, pp. 21-32

Ratha, N. K.; Chen, S. Y., & Jain, A. K. (1995). Adaptive Flow Orientation-Based Feature Extraction in Fingerprint Images. *Pattern Recognition*, Vol. 28, No. 11, pp. 1657-1672

Yang, J. C., & Park, D. S. (2008). Fingerprint Verification Based on Invariant Moment Features and Nonlinear BPNN. *International Journal of Control, Automation, and Systems*, Vol. 6, No. 6, pp. 800-808

Yuliang, H.; Tian, J.; Luo, X., & Zhang, T. (2003). Image enhancement and minutiae matching in fingerprint verification. *Pattern Recognition*, Vol. 24, pp. 1349-1360

Zhang, Y.; Tian, J.; Cao, K.; Li, P., & Yang, X. (2008). Improving Efficiency of Fingerprint Matching by Minutiae Indexing. *19th International Conference on Pattern Recognition*, Tampa, FL, December 8-11

Zhu, E.; Yin, J., & Zhang, G. (2005). Fingerprint matching based on global alignment of multiple reference minutiae. *Pattern Recognition*, Vol. 38, pp. 1685-1694

11

A Gender Detection Approach

Marcos del Pozo-Baños, Carlos M. Travieso,
Jaime R. Ticay-Rivas, and Jesús B. Alonso
*IDeTIC - Institute for Technological Development and Innovation in Communications,
ULPGC - Universidad de Las Palmas de Gran Canaria
Spain*

1. Introduction

Which are the brain processes that underlie facial identification? What information, among the available in the environment, is used to elaborate a response on a subject's identity? Certainly, our brain uses all the information in greater or less extent. Just focusing on that present on the human face, the system can obtain knowledge about gender, age and ethnicity. This demographic data may not be enough for subject identification, but it definitely gives us some valuable clues. The same can be applied for computer systems. For example, having gender information into account, the system can reduce the pool of the possible identities considerably, making the problem easier and enforcing the final response. Moreover, raw gender information can also be used in fields such as micromarketing and personalized services.

A practical example of this can be found on the work presented by Peng and Ding (Peng & Ding 2008). These authors proposed a tree structure system to increase the successful rate of a gender classification. In particular, the system first classify between Asian and Non-Asian ethnicities. Then, two specialized gender classification systems are trained, one for each ethnicity. This resulted in an increase of around 4% over an ordinary system (gender classifier without ethnicity specialization).

Therefore, demographic classification systems are as much important and valuables as face identification systems themselves. This is why they have received increasing attention in the last years. In particular, this chapter focuses its attention in facial-base gender-detection systems. A summary of the problem's characteristics is first given in section 2, along with an overview of the state of art. Section 3 introduces the structure of the system used for experiments of section 4, where we check the effect of preprocessing variations on the systems performance. Finally, conclusions derived from the obtained results are presented in section 5.

2. Biometric gender classification

In general, biometric problems can be classified in two groups: classification and verification. In the former, samples from a number of well defined classes are given to the system for training. When a testing sample is presented, the system must classify it in the corresponding class. In other words, the system answers the "who is this?" question.

On the other hand, during a verification problem training samples are divided in classes "A" (called positive) and "others" (called negative). Then, a testing sample claiming to be of class "A" is presented to the system and the system must verify that this sample corresponds to the claimed class. Obviously, a classification system can be built upon a combination of verification systems.

Tools show different behaviors in different situations. Some perform better in classification problems while others perform better during verification. This is because differences between classification and verification are not just a matter of the number of classes, but of how classes are built. In a classification problem, classes are well defined patterns coming from a common thing. However, this cannot be expected for the negative class of a verification problem. Usually this class is too wide in the sample space to be represented with a reasonable amount of samples, and other representation techniques must be used.

This does not mean that representation techniques that work on classification problems cannot perform on a verification scenario or vice versa. But usually you cannot expect them to work as well. Therefore it is important to define the problem before decide the approaching technique and the tools to be used. Gender classification systems find themselves in a rather special situation, as they only define two classes (male and female). Therefore classification and verification techniques can be used without penalties in this case.

2.1 Facial image databases

Due to the gaining importance of face identification systems on the security field, a great deal of facial databases has appeared in the last years. As any other component of a biometric system, databases' technology has experienced an important step forward too.

Some of the first widely used databases were the Olivetti Research Laboratory database (Samaria & Harter 1994), also known as AT&T, and the YALE database. These databases consist of images taken from a frontal or almost frontal facial poses. As it can be seen in figure 1, subjects on the ORL database, presents only smiling / not smiling facial expressions and images with smooth lighting variations. On the other hand, the YALE database presents subjects with a number of different configurations such as center / left / right lighting, with / without glasses, and happy / normal / sleepy / surprised / wink expressions. Some examples are given in figure 2.

In both cases, few subjects and few images per subject are provided. Thus, they represent a problem which can fit some practical situations such as access control systems with few authorized persons, as for these systems it may be easier to control lightning and to obtain good images in terms of pose and expression, as subjects are willing to get recognized. However, they are impossible to use in problems such as gender or ethnicity classification, where big pools of samples are necessary in order to obtain reliable results.

Fig. 1. Some examples of the ORL database

Fig. 2. Some examples of the YALE database

However, more powerful and complex situations involve huge security systems installed in airports or public buildings. These systems face uncontrolled lightning conditions, non collaborative subjects, and vast pools of identities. New databases incorporate some or all of these characteristics in order to test system against such situations. For example, the YALEb database (Georghiades et al. 2001) (Lee et al. 2005); examples in figure 3, contains images coming only from 10 persons, but each seen under 576 viewing conditions coming from combinations of 9 poses and 64 illumination conditions.

Fig. 3. Some examples of the YALEb database.

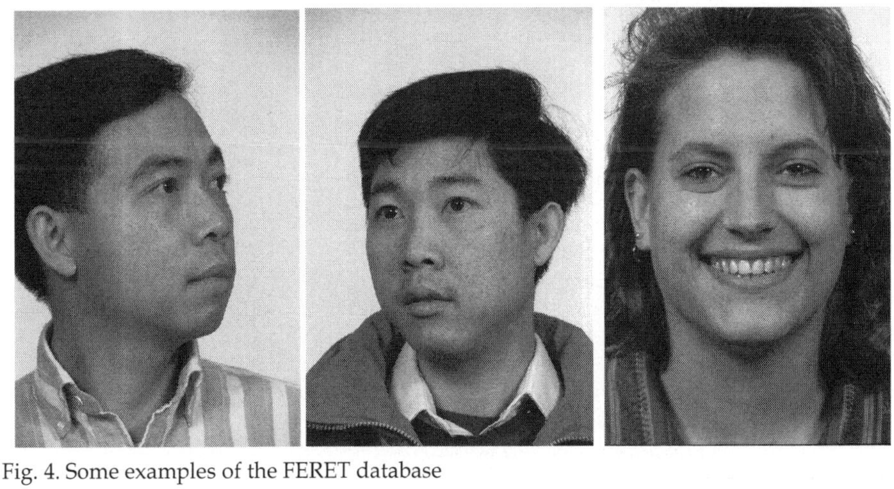

Fig. 4. Some examples of the FERET database

The FERET database (Phillips et al. 2000) consist of images collected in a semi-controlled environment, from 1199 subjects, and for different facial poses. Some examples can be seen in figure 4. An interesting property of this database is that it provides an extensive ground truth data specifying coordinates of facial organs, ethnicity, gender, and facial characteristics such as moustache, beard, and glasses information. Therefore, the FERET database may be easily used in almost any experimental situation. In fact, we will be using it for further experiments in this chapter.
The Face Recognition Grand Challenge (FRGC) database (Phillips et al. 2005) is another complete database. As FERET, the FRGC database consist of high resolution images from a pool of more than 1000 subjects and complete ground truth information files. However, this database provides images of full body, taken in different scenarios, which implies important changes in background and lightning conditions. Some examples can be seen in figure 5.

Fig. 5. Some examples of the FRGC database

In short, when testing a system it is important to keep in mind what type of database is been used. Using different databases provides different conditions, which allows us to test the system against different problems.

2.2 State of the art

A priori, techniques used for face identification or verification can also be used for gender identification. Finding inspiration in the biological system, S.L. Phung and A. Bouzerdoum proposed a system implementing a pyramidal neural network (Phung & Bouzerdoum 2007). This structure combines 1D and 2D neural network architectures with a resilient backpropagation learning algorithm, in such a way that some interesting properties arise. For example, neurons from the first layer are directly connected to image pixels, and the net's structure implements local receptive fields that are slightly overlapped. These two properties are somehow similar to the human eye. Using a set of 1152 male and 610 female images from the FERET database was used to test the system, with which a best classification rate of 89.8% was obtained.
On the other hand, B. Moghaddam and Ming-Hsuan Yang asserted that the Support Vector Machine (SVM) pattern classification outperforms traditional classifiers such as linear, quadratic, nearest neighbor, and Fisher linear discriminant, as well as more modern techniques such as Radial Basis Function (RBF) and large ensemble-RBF neural networks (Moghaddam & Ming-Hsuan 2002). The authors used a set of 1044 male and 711 female images from the FERET database for the experiments, and obtained a lowest error rate of 3.38% using a Gaussian RBF kernel on their SVM.
For the characterization of images, A. Jain et al. combined the Independent Component Analysis (ICA) feature extract technique with the SVM classifier (Jain & Huang 2004). They

tested the system using a set of 250 male and 250 female images from the FERET database, obtaining a classification rate of 95.67%. Then, Zhen-Hua Wang et al. applied a Genetic Algorithm (GA) search over the feature found by ICA, improving the system performance on a 7.5% (Zhen-Hua & Zhin-Chun 2009). Moreover, we have shown in (del Pozo Baños et al. 2011) that the ICA approach named Join Approximate Diagonalization of Eigenmatrices (JADE-ICA) outperforms the fast-ICA method in both error rate and stability on the gender classification problem.

Another interesting point is which face area provides the best information for gender classification. M. Castrillón Santana and Q.C. Vuong presented a psychological study on this aspect (Castrillon-Santana & Vuong 2007). They showed that when humans have no face information, the neck of males and the long hair of females provide the most diagnostic information. Moreover, in order to compare human and artificial systems they performed a series of experiments using different face masks. The system based on Incremental Principal Component Analysis (IPCA) and support vector machine (SVM) performed surprisingly similar using only face information (no neck and no hair) and face with hair line information. In a similar approach, Jing-Ming Guo et al. proposed the use of a mask to remove those pixels that are not discriminative as they are common for both classes or come from the background noise (Jing-Ming et al. 2010). This mask was based on the difference between the mean male image and the mean female image. Pixels selected by the mask were then used as inputs to a SVM classifier. Experiments were performed using a set of 1713 male and 1009 female images from the FERET database, and an accuracy of 88.89% was reported. J.R. Lyle et al. studied the validity of periocular images (area around eyes) for gender and ethnicity classification (Lyle et al. 2010). Images were rescaled to 251x251 pixels, converted to gray scale and their histograms equalized. The parameterization relied on the Local Binary Pattern (LBP) (Topi 2003) tool, and a SVM was applied for classification. Testing the system on the FERET database, the authors obtained a best accuracy of around 94%.

A more sophisticated system which performs score fusion of experts on different face areas is presented by F. Manesh et al. (Manesh et al. 2010). First, eyes and mouth coordinates are automatically extracted with the extended Active Shape Model (Milborrow & Nicolls 2008). The system aligns, crops, and rescaled face images to 80x85 pixels as a preprocessing stage. Faces are then divided in 16 regions based on a modification of the Golden ratio template proposed by K. Anderson et al. (Anderson & McOwan 2004). Each region has its own expert system. These experts use a family of Gabor filters (Gabor 1946) (Daugman 1980) with 5 scales and 8 orientations as a feature extractor method, and a SVM with a RBF kernel for classification. Score fusion is finally performed using the optimum data fusion rule, which weights the experts accordingly to their accuracy. For the experiments, a combination of 891 frontal images from the FERET database and 800 frontal images from the CAS-PEAL data base was used. This set was divided in 3 sub-sets labeled "training", "validation", and "test". Finally, the researchers reported an accuracy of 96% for the ethnicity problem (Asian vs. Non-Asian), and 94% for gender classification fusing the scores of eyes, nose, and mouth. S. Gutta et al. also highlighted how information such as gender, ethnicity or face pose can increase the performance of face identification systems (Butta et al. 2000). To automatically obtain this information from facial images, they proposed a mixture of experts' system, which uses the "divide-and-conquer" modularity principle. Therefore, the system is composed of several sub-systems or modules and it elaborates the final result based on the individual results. In particular, an architecture combining ensemble-RBF networks and decision trees techniques was used for gender classification. Using a set of 1906 male and 1100 female images from the FERET database the authors obtained a gender recognition rate of 96%.

As in any other face identification system, the preprocessing step is crucial. E. Makinen and R. Raisamo performed a set of experiments to evaluate the effect of face alignment (Makinen & Raisamo 2008). They reported no improvement when automatic face alignment techniques were used. However, manual alignment did increase the systems performance by a small factor. Giving these results, authors concluded that alignment methods must be improved in order to be of some use in the gender recognition problem. As they tested different classification techniques, they obtained the best performance with the SVM classifier, a classification rate of 84.39% using a set of 411 images from the FERET database. However, Adaboost with haar-like features offered very close results, while it was faster and more resistant to the in-plane rotation variations. Moreover, Jian-Gand Wang et al. also reported no significant improvement in terms of performance between manual, automatic, and none face alignment (Jian-Gan et al. 2010). Surprisingly, not only face alignment have none or little effect on gender classification, but many works has reported the same affect between low and high resolution images (Moghaddam & Ming-Hsuan 2002) (Lyle et al. 2010). In addition, many authors have reported no significant changes in performance when different image qualities were used (Moghaddam & Ming-Hsuan 2002) (Makinen & Raisamo 2008).

As we have experienced the same effect when quite different preprocessing methods were used (del Pozo-Baños et al. 2010), we decided to run here a further experiment using a common database to reinforce these results.

3. The proposed system model

The system used in this study has a block diagram composed of three main blocks: preprocessing, parameterization, and classification. Four quite different components were implemented for the preprocessing block, while two tools were used on the parameterization block. Figure 6 shows the aspect of this architecture, where only one preprocessing and one parameterization can be active at the same time.

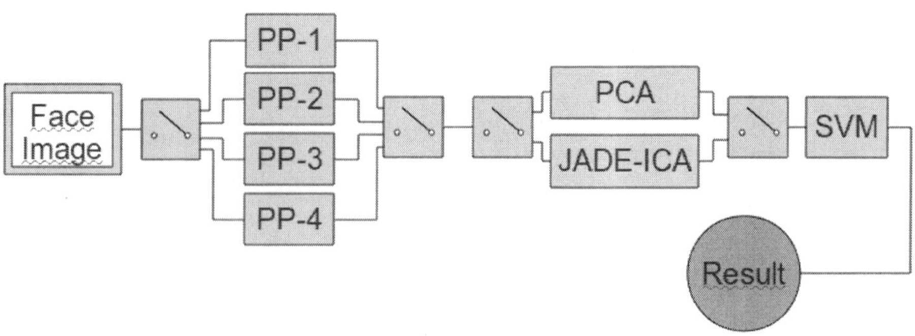

Fig. 6. Block diagram of the implemented system

3.1 Preprocessing methods

The first block at the system's entrance is the preprocessing block. This gets samples ready for the forthcoming blocks, reducing the noise and even transforming the original signal in a more readable one. Four different preprocessing components has been implemented.

- **PP-1.** This block normalizes the image histogram to a linear distribution before reducing its dimension to 15x20 pixels. Finally, an unsharpened filter is used to reduce the noise produced by the extreme reduction.
- **PP-2.** In this case, after histogram normalization a further local normalization (Xiong 2005) is performed. This normalization aims to reduce lighting effect through a double Gaussian filtering. Then, images are reduced to 15x20 pixels, and the unsharpened filter is applied.
- **PP-3.** The LBP (Topi 2003) is an invariant texture measure tool for gray scale images. When applied, it produces a matrix LBP were each point (x_c, y_c) corresponds to the differences between the centre pixel point (g_c) and its neighbours according to a given mask (g_m). The mathematical definition is:

$$LBP(x_c, y_c) = \sum_{m=0}^{M-1} S(g_m - g_c) \times 2^m, S(x) = \begin{cases} 1 & x \geq 0 \\ 0 & x < 0 \end{cases} \quad (1)$$

  Here, the factor power of two makes the result of every possible combination unique, so that the LBP transformation is reversible. After applying the LBP, the PP-2 component reduces the resulting matrix dimension to 15x20, and applies the unsharpened filter.
- **PP-4.** This component is similar to the previous one, although in this case the image is first reduced to 15x20 and the filtered before apply the LBP transformation.

At the end of every preprocessing component, an elliptical mask is applied to remove peripheral noise located on corners. Images are then vectorized considering only pixels falling within the elliptical mask, which provides further reduction of samples dimensions. The effects of applying each preprocessing component can be seen in figure 7.

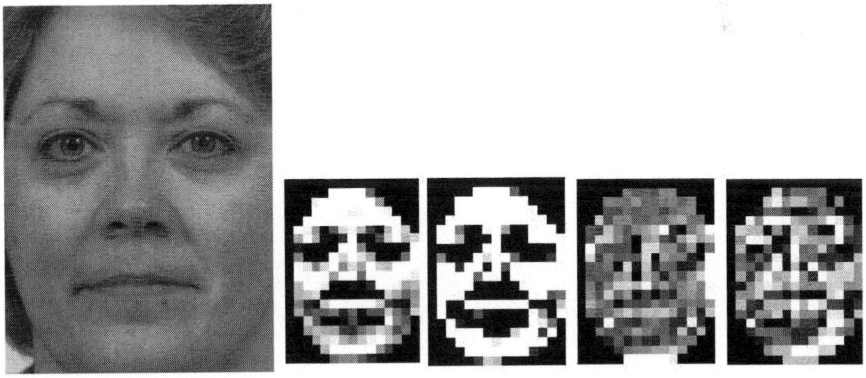

(a) (b) (c) (d) (e)

Fig. 7. Original image (A) and the resulting images for each preprocessing component: PP-1 (B), PP-2 (C), PP-3 (D), and PP-4 (E).

3.2 Parameterization techniques

The parameterization step analyzes samples and extracts relevant information. The proposed system uses both PCA and ICA appearance based methods.

3.2.1 Principal Component Analysis (PCA)

The PCA was introduced by Karl Pearson in 1901 (Jolliffe 2002), and then applied to face images by Kohonen (Kohonen 1989), Kirby y Sirovich (Kirby & Sirovich 1990). It was intended to extract information not viewable at first sight by projecting samples to a new space which maximizes variance. Moreover, by keeping only the first N coordinates of the new space, also called principal components (PCs) the system reduces sample dimensions keeping the most valuable information. Let X be a matrix of vectors x_i, each with p variables. PCA results in a set of projecting vectors α_i such that the transformation:

$$z_i = \alpha_i X = \sum_{j=1}^{p} \alpha_{ij} x_j \tag{2}$$

obtains a new set of vectors z_i representing the original x_i in a space maximizing its variance. Moreover, vectors α_i are uncorrelated to each other, so that new vectors appear in decreasing variance value order. By keeping the first N vectors z_i, the system remove redundant information and obtain an smaller representation of the data.

Projecting vectors α_i are computed by the eigenanalysis of the covariance matrix of X, referred to as S. Therefore, vector α_i corresponds to the i-th eigenvector of S, which when chosen to have unit length $(\alpha_i'\alpha_i = 1)$ proves to provide a vector z_i with variance equal to the corresponding i-th eigenvalue of S.

3.2.2 Joint Approximate Diagonalization of Eigen-matrices Independent Component Analysis (JADE-ICA)

ICA is a particularization of PCA to extract components that are, at the same time, non-gaussian and statistically independent (Hyvärinen 2000). When used on images, ICA obtains independent base images which are not necessarily orthogonal. Application of these base images extracts between pixels information related to high order statistics.

In this study, an approach named JADE-ICA has been used to implement this tool. JADE-ICA is based on joint diagonalization of cumulant matrices. For simplicity, the case of symmetric distributions is considered, where the odd-order cumulants vanish. Let $X_1,...,X_4$ be random variables, and defined $X_i^* = X_i - E(X_i)$. The second order cumulants can be written as:

$$C(X_1, X_2) = E(X_1^*, X_2^*) \tag{3}$$

And the fourth-order cumulants as:

$$\begin{aligned} C(X_1, X_2, X_3, X_4) = E(X_1^*, X_2^*, X_3^*, X_4^*) - E(X_1^*, X_2^*)E(X_3^*, X_4^*) - \\ E(X_1^*, X_3^*)E(X_2^*, X_4^*) - E(X_1^*, X_4^*)E(X_2^*, X_3^*) \end{aligned} \tag{4}$$

In addition, the definitions of variance and kurtosis of a random variable X are:

$$\sigma^2 = C(X, X) = E(X^{*2}) \tag{5}$$

$$kurt(X) = C(X, X, X, X) = E(X^{*4}) - 3E^2(X^{*4})$$

Now, under a linear transformation $Y = AX$, the cumulants of fourth-order transformation became:

$$C(Y_i, Y_j, Y_k, Y_l) = \sum_{p,q,r,s} a_{ip} a_{jq} a_{kr} a_{ls} C(X_p, X_q, X_r, X_s) \tag{6}$$

, with a_{ij} the i-th row and j-th column entry of matrix A. Since the ICA model ($X = AS$) is linear, using the assumption of independence by $C(S_p, S_q, S_r, S_s) = kurt(S_p)\delta_{pqrs}$ where:

$$\delta_{pqrs} = \begin{cases} 1 & if \quad p = q = r = s \\ 0 & otherwise \end{cases} \tag{7}$$

and S has independent entries:

$$C(Y_i, Y_j, Y_k, Y_l) = \sum_{m=1}^{n} kurt(S_m) a_{im} a_{jm} a_{km} a_{lm} \tag{8}$$

, the cumulants of the ICA model are obtained.
Given any $n\ x\ n$ matrix M and a random $n\ x\ 1$ vector X, we consider a cumulant matrix $Q_x(M)$ defined by:

$$Q_X(M) = \sum_{m=1}^{n} C(X_i, X_j, X_k, X_l) M_{ki} \tag{9}$$

If X is centered, the definition of (4) shows that:

$$\begin{aligned} Q_X(M) = E\{(X^T M X^T) X X^T\} - R^X tr(MR^X) - \\ R^X M R^X - R^X M^T R^X \end{aligned} \tag{10}$$

, where *tr(B)* denotes the trace of matrix B and $[R_X]_{ij} = C(X_i, X_j)$.
The structure of a cumulant $Q_x(M)$ in ICA model is easily deduced from (9) as:

$$Q_X(M) = A\Delta(M)A^T \tag{11}$$

With:

$$\Delta(M) = diag(kurt(S_1) a_1^T M a_1, ..., kurt(S_n) a_n^T M a_n \tag{12}$$

, where a_i is the i-th column of A.
Now, let W be a whitening matrix and $Z = WX$. Let us assume that the independent sources matrix S has unit variance, so that S is white. Thus $Z = WX = WAS$ is also white, and the matrix $U = WA$ is orthonormal. Similarly, the previous techniques can be applied into (13) for any $n\ x\ n$ matrix M.
First, the whitening matrix W and the cumulant matrix Z are estimated. Then, the estimation of an orthonormal matrix U, denoted by U, is calculated. Therefore, an estimated matrix A denoted by A is obtained from $W^{-1}U$, and the sources matrix S is calculated by $A^{-1}X$.
To measure non-diagonality of a matrix B, *off(B)* is defined as the sum of the squares of the non-diagonal elements:

$$off(B)=\sum_{i\neq j}(b_{ij})^2 \quad (13)$$

, where b_{ii} are elements of the matrix B. In particular $off(U^TQ_Z(M_i)U)=off\Delta_i=0$ since $Q_Z(M_i)=U\Delta_iU^T$ and U is orthonorgal. For any matrix set M and orthonormal matrix V, the joint diagonality criterion is defined as:

$$D_M(V)=\sum_{M_i\in M} off(V^TQ_Z(M_i)V) \quad (14)$$

, which measures diagonality far from the matrix V and bring the cumulants matrices from the set M.

3.3 Pattern classification

At this point, the system has retrieved and processed as much useful information from the input images as PCA or JADE-ICA can. Now, the classification component uses this information to take a decision on behalf the gender of the input face. To do so, this work uses the well known SVM (Schölkopf & Smola 2002).

The SVM is a structural risk minimization learning method of separating functions for patter classification, that was derived from the statistical learning theory elaborated by Vapnik and Chervonenkis (Vapnik 1995). In other words, SVM is a tool able to differ between classes characterized by parameters, after a training process.

What makes this tool powerful is the way it handles non-linearly separable problems. In these cases, the SVM transforms the problem into a linearly separable one by projecting samples into a higher dimensional space. This is done using an operator called kernel, which in this study is set to be a Radial Basis Function (RBF). Then, efficient and fast linear techniques can be applied in the transformed space. This technique is usually known as the kernel trick, and was first introduced by Boser, Guyon y Vapnik in 1992 (Yan et al. 2004).

For simplicity, we configure the SVM to work as a verification system. In this particular case, the negative class (-1) corresponds to males and the positive class (1) to females. As a result, the classifier answers the "is this female?" question. The output of the SVM is a numeric value between -1 and 1 named score. A threshold has to be set to define a border between male (-1) and female (1) responses.

However, if all samples are used for training, there are no new samples for setting the threshold, and using the training samples for this purpose will lead to bad adjustments. Therefore, a 20 iterations hold-4-out (2 from each class) cross-validation procedure is used over the training samples to obtain 80 scores. These scores are then used to set the system's threshold to the equal error rate (EER) point, which is the point where False Acceptance Rate (FAR) and False Rejection Rate (FRR) coincide. The system's margin, defined as the distance of the closest point to the threshold line, is also measured. All these measures are referred to as validation measures.

When the threshold is finally set, the SVM is trained using all available samples. Because no big differences exist in the number of training samples used for this final training and the validation, we can expect the system to have a very similar threshold than that computed before.

In particular, the Least Squares Support Vector Machines (LS-SVM) implementation is used (Suykens 2002). Given a training set of N data points $\{y_i, x_i\}_{k=1}^{N}$, where x_i is the k-th input sample and y_i its corresponding produced output, we can assume that:

$$\begin{cases} w^T\phi(x_i) + b \geq 1 & if \quad y_i = +1 \\ w^T\phi(x_i) + b \leq 1 & if \quad y_i = -1 \end{cases}, \tag{15}$$

where ϕ is the kernel function that maps samples into the higher dimensional space. The LS-SVM solves the classification problem:

$$\min L_2(w,b,e) = \frac{\mu}{2} w^T w + \frac{\zeta}{2} \sum_{i=1}^{N} e_{c,i}^2 , \tag{16}$$

where μ and ζ are hyper-parameters related to the amount of regularization versus the sum square error. Moreover, the solution of this problem is subject to the constraints:

$$y_i \left[w^T \phi(x_i) + b \right] = 1 - e_{c,i}, \quad i = 1, \ldots, N \tag{17}$$

3.4 System optimization

Because every preprocessing, parameterization, and classification technique may have its own optimal point in terms of configurable parameters, the system optimizes itself automatically using the validation results. Three parameters need to be optimized every time the system is trained: the number of kept principal/independent components for the parameterization component, and the regularization and kernel parameter for the LS-SVM. An exhaustive search is done along a configuration volume looking for the optimal point, defined as the point which provided a lower validation error rate and a larger validation system's margin.

4. Experiments and results

In order to obtain more reliable results, a 10-Folds cross-validation procedure was run on the experiments. Frontal facial images were taken from the FERET database, and cropped manually using ground information provided by the database. An example of this crop can be seen in figure 7. For each iteration the system was optimized and trained as it was explained in the previous section. Moreover, eight different systems, made out of every possible configuration between the four preprocessing components and the two parameterization tools were tested. All these facts made the experimental time impractical when the whole FERET database was used. To reduce this time a set of 1600 images (800 males and 800 females) were randomly selected.

Figures 8 and 9 show the results obtained when PCA and JADE-ICA were applied using all different preprocessing components. Numbers specified on the legend represent the areas under the curves. The EER points are given in table 1. In general terms, all preprocessing techniques provide similar vehabiors, although differences were magnified when JADE-ICA was used. In particular, PP-1 and PP-2 performed almost identical.

A second experiment was run combining the scores obtained with each preprocessing technique. In particular, the sum and the product score fusion techniques were applied. The former combines the scores by a sum before apply the decision threshold. The later performs a product after sifting the scores to range [1 3] instead of [-1 1], and then apply the threshold. The obtained results can be seen in table 1, and in figures 10 and 11.

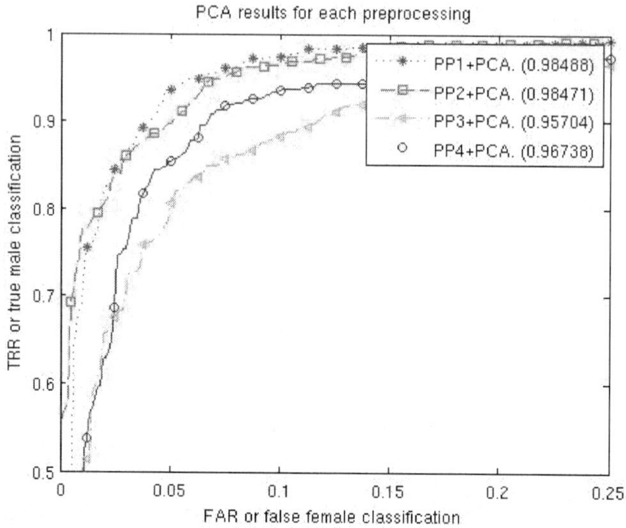

Fig. 8. Curves obtained for each preprocessing and PCA. In legend numbers represents the area under the curve.

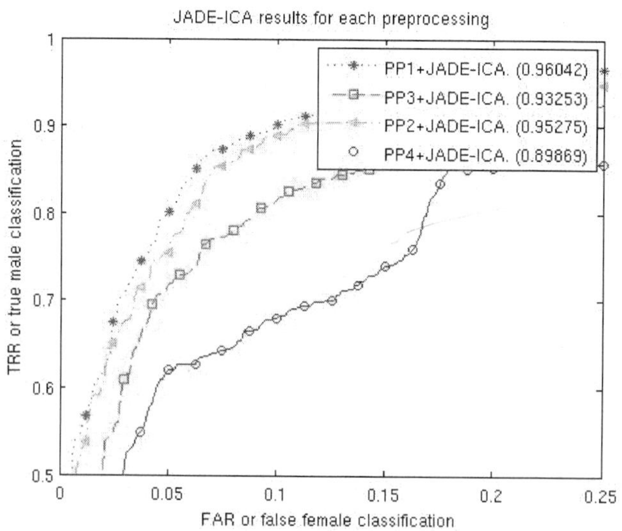

Fig. 9. Curves obtained for each preprocessing and JADE-ICA. In legend numbers represents the area under the curve.

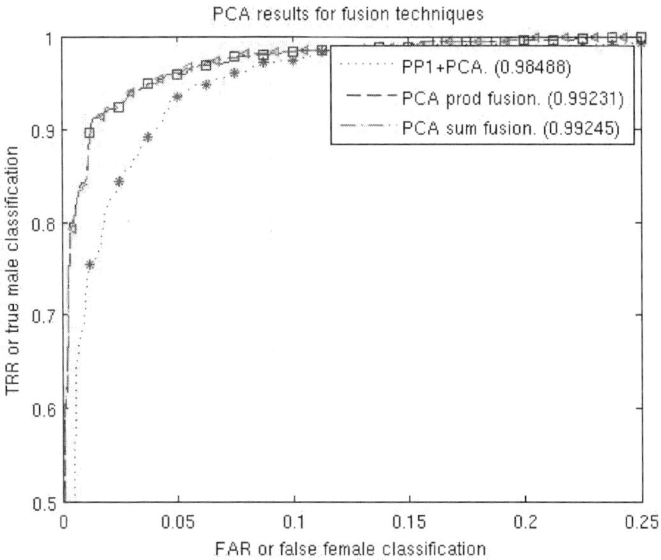

Fig. 10. Curves obtained for each score fusion of each preprocessing technique and PCA. In legend numbers represents the area under the curve.

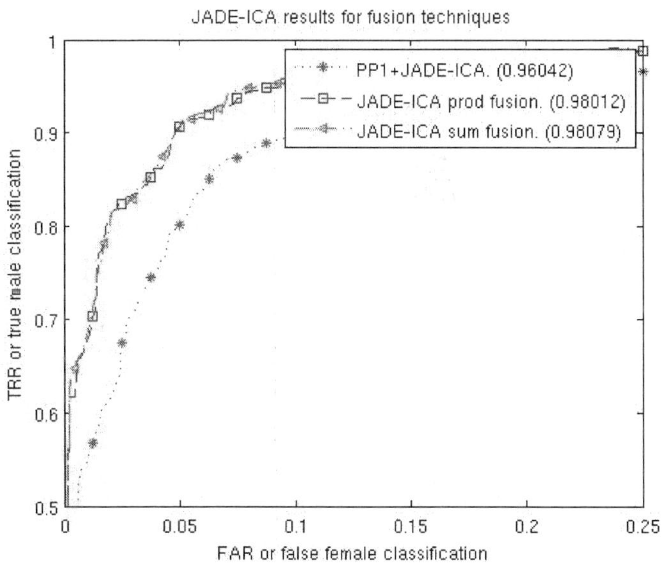

Fig. 11. Curves obtained for each score fusion of each preprocessing technique and PCA. In legend numbers represents the area under the curve.

Preprocessing	Parameterization	
	PCA	JADE-ICA
PP-1	5.54%	9.97%
PP-2	6.44%	10.54%
PP-3	10.83%	14.65%
PP-4	7.96%	16.50%
All preprocessing fusion		
Sum fusion	4.37%	6.99%
Prod fusion	4.44%	7.12%

Table 1. EER points for every combination of preprocessing and parameterization methods.

5. Conclusion

In this chapter, we have introduced the gender classification problem, from which a system automatically determines whether an input face corresponds to a female or a male. We have overview its characteristics as a bi-class problem and its relevance within the biometrics field. We have also introduced a biometric system with a simple architecture based on four preprocessing blocks, PCA and JADE-ICA parameterization, and an LS-SVM classifier. This system was used to test the variations of system's performance produced by wide changes on the preprocessing stage. The obtained results were consistent with other works, showing that in general there is little or no effect on the system's performance when these changes are applied.

Why do these big changes on the preprocessing stage provide similar results? Do they enhance different qualities of the facial images with similar level of discrimination? In a willing to through clarity on this intriguing characteristic of the gender recognition problem, we performed another experiment fusing scores obtained for each preprocessing technique. Both add- and product-fusion methods produced a small improvement in the system's behavior, of around 2% of reduction of EER comparing to the best preprocessing block.

This may suggests that all configurations are performing basing on the same or very alike information. Considering the massive differences between images resulting from each preprocessing block (figure 7), it is possible that this discriminant information is mostly related to very global and salient facial features, such as facial shape. This possibility is also consistent with the fact that image size does not affect gender classification performance. In fact, if facial shape is to be used, it does not matter whether the system has information coming from the inside of the face, or not.

6. Acknowledgment

This work has been partially supported by by "Cátedra Telefónica ULPGC 2009-10", and by the Spanish Government under funds from MCINN TEC2009-14123-C04-01.

7. References

Du, P.; & Xiaoqing, D. (2008). The Application of Decision Tree in Gender Classification. Congress on Image and Signal Processing, vol.4, no., pp.657-660, 27-30 May 2008.

Phung, S.L.; & Bouzerdoum, A. (2007). A Pyramidal Neural Network for Visual Pattern Recognition. IEEE Transactions on Neural Networks, vol.18, no.2, pp.329-343, March 2007.

Schölkopf, B.; & Smola, A.J. (2002). Learning with Kernels. Support Vector Machines, Regularization, Optimization, and Beyond. Published by The MIT Press, 2002.

Moghaddam, B.; & Ming-Hsuan, Y. (2002). Learning gender with support faces. IEEE Transactions on Pattern Analysis and Machine Intelligence, vol.24, no.5, pp.707-711, May 2002.

Hyvärinen, A. (2000). Independent Component Analysis: Algorithms and Applications. Neural Networks, 13(4-5): 411-430, 2000.

Jain, A.; & Huang, J. (2004). Integrating independent components and support vector machines for gender classification. Proceedings of the 17th International Conference on Pattern Recognition, vol.3, no., pp. 558- 561 Vol.3, 23-26 Aug. 2004.

Zhen-Hua, W.; & Zhi-Chun, M. (2009). "Gender classification using selected independent-features based on Genetic Algorithm," Machine Learning and Cybernetics, 2009 International Conference on , vol.1, no., pp.394-398, 12-15 July 2009.

del Pozo-Baños, M. ; Travieso, C.M. ; Alonso, J.B. ; & Ferrer, M.A. (2011) Gender Verification System based on JADE-ICA. International Conference on Bio-inspired Systems and Signal Processing, vol.7, no., pp. 570-576, ISSN: 978-989-8425-35-5, Rome, Italy, January 2011.

Castrillon-Santana, M ; & Vuong, Q.C. (2007). An Analysis of Automatic Gender Classification. *In proceedings of 12th Iberoamerican conference on Progress in Pattern Recognition, Image Analysis and Applications* (IbPRIA), pp.39-46, 2007.

Jing-Ming, G.; Chen-Chi, L.; & Hoang-Son, N. (2010). Face gender recognition using improved appearance-based Average Face Difference and support vector machine. *2010 International Conference on System Science and Engineering (ICSSE)*, vol., no., pp.637-640, 1-3 July 2010.

Lyle, J.R.; Miller, P.E.; Pundlik, S.J.; & Woodard, D.L. (2010). Soft biometric classification using periocular region features. *2010 Fourth IEEE International Conference on Biometrics: Theory Applications and Systems* (BTAS), vol., no., pp.1-7, 27-29 Sept. 2010.

Topi, M. (2003) The Local Binary Pattern Approach to Texture Analysis – Extensions and Applications. Infotech Oulu and Department of Electrical and Information Engineering, University of Oulu, P.O.Box 4500, FIN-90014 University of Oulu, Finland. 2003.

Manesh, F. Saei; Ghahramani, M.; & Tan, Y. P. (2010). Facial part displacement effect on template-based gender and ethnicity classification. *2010 11th International Conference on Control Automation Robotics & Vision* (ICARCV), vol., no., pp.1644-1649, 7-10 Dec. 2010.

Milborrow, S.; & Nicolls, F. (2008). Locating Facial Features with an Extended Active Shape Model. *Presented at the Proceedings of the 10th European Conference on Computer Vision: Part IV,* Marseille, France, 2008.

Anderson, K.; & McOwan, P. W. (2004). Robust real-time face tracker for cluttered environments. *Computer Vision Image Understanding*, vol. 95, pp. 184- 200, 2004.

Gabor, D. (1946). Theory of communication. *Journal of the Institute of Electrical Engineers,* 93, 429–457.

Daugman, J. (1980). Two-dimensional analysis of cortical receptive field profiles. Vision Research, 20, 846–856.

Gutta, S.; Huang, J.R.J.; Jonathon, P.; & Wechsler, H. (2000). Mixture of experts for classification of gender, ethnic origin, and pose of human faces. *IEEE Transactions on Neural Networks*, vol.11, no.4, pp.948-960, Jul 2000.

Makinen, E.; Raisamo, R. (2008). Evaluation of Gender Classification Methods with Automatically Detected and Aligned Faces. *IEEE Transactions on Pattern Analysis and Machine Intelligence*, vol.30, no.3, pp.541-547, March 2008.

Jian-Gang, W.; Hee, L.W.; Myint, Y.; Wei-Yun, Y. (2010). Real-time gender recognition with unaligned face images. *2010 the 5th IEEE Conference on Industrial Electronics and Applications* (ICIEA), vol., no., pp.376-380, 15-17 June 2010.

del Pozo-Baños, M. ; Travieso, C.M. ; Alonso, J.B. ; & Ferrer, M.A. (2010) A Gender classifier Problem. 1er Workshop de Tecnologías Multibiométricas para la Identificación de Personas, vol.1, no., pp. 25-29, ISSN: 978-84-693-3389-1, Las Palmas de Gran Canaria, Spain, July 2010.

Xiong, G. (2005). A Matlab implementation of the "localnormalize" function is available at http://www.mathworks.com/matlabcentral/fileexchange/8303-local-normalization . Last updated in 2005.

Jolliffe, I.T.; (2002). Principal Component Analysis. 2nd de. Springer Series in Statistics. 2002.

T. Kohonen, (1989). Self-organization and Associative Memory, Springer-Verlag, Berlin, 1989.

Kirby, M.; & Sirovich, L. (1990) Application of the karhunen-loeve procedure for the characterization of human faces. *IEEE Pattern Analysis and Machine Intelligence*, vol. 12, no. 1, pp. 103-108, 1990.

Vapnik, V. (1995) The Nature of Statistical learning Theory. Springer Verlag, New York, 1995.

Yan, F.; Qiang, Y.; Ruixiang, S.; Dequan L.; Rong, Z.; Ling, C.X.; & Wen, G. (2004) Exploiting the kernel trick to correlate fragment ions for peptide identification via tandem mass spectrometry. Bioinformatics, Vol. 20, Issue 12, pp. 1948-1954, March 25, 2004.

Georghiades, A.S.; Belhumeur, P.N.; & Kriegman, D.J.; 2001, From Few to Many: Illumination Cone Models for Face Recognition under Variable Lighting and Pose. *IEEE Trans. Pattern Analysis and Machine Intelligence*, vol. 23, n. 6, pp: 643-660, 2001.

Samaria, F.; & Harter, A. (1994). Parameterisation of a stochastic model for human face identification. *2nd IEEE Workshop on Applications of Computer Vision* . December 1994, Sarasota (Florida).

Phillips, P. J.; Moon, H.; Rizvi, S. A.; & Rauss, P. J. (2000). The FERET evaluation methodology for face-recognition algorithms. *IEEE Trans. Pattern Analysis and Machine Intelligence*, vol. 22, no. 10, pp. 1090–1104, Oct. 2000.

Phillips, P. J.; Flynn, P. J.; Scruggs, T.; Bowyer, K. W.; Chang, J.; Hoffman, K.; Marques, J.; Min, J.; & Worek. (2005) Overview of the Face Recognition Grand Challenge. *In Proceedings of IEEE Conference on Computer Vision and Pattern Recognition*, 2005.

Suykens, J.A.K.; Van Gestel, T.; De Brabanter, J.; De Moor, B.; & Vandewalle, J. (2002) Least Squares Support Vector Machines. World Scientific, ISBN 981-238-151-1, Singapore, 2002.

12

Improving Iris Recognition Performance Using Quality Measures

Nadia Feddaoui, Hela Mahersia and Kamel Hamrouni
LSTS Laboratory, National Engineering School of Tunis, University ElManar
Tunisia

1. Introduction

Biometric methods, which identify people based on physical or behavioural characteristics, are of interest because people cannot forget or lose their physical characteristics in the way that they can lose passwords or identity cards. Among these biometric methods, iris is currently considered as one of the most reliable biometrics because of its unique texture's random variation. Moreover, iris is proved to be well protected from the external environment behind the cornea, relatively easy to acquire and stable all over the person's life. For all of these reasons, iris patterns become interesting as an alternative approach to reliable visual recognition of persons. This recognition system involves four main modules: iris acquisition, iris segmentation and normalization, feature extraction and encoding and finally matching.

However, we noticed that almost all the iris recognition systems proceed without controlling the iris image's quality. Naturally, poor image's quality degrades significantly the performance of the recognition system. Thus, an extra module, measuring the quality of the input iris, must be added to ensure that only "good iris" will be treated by the system. The proposed module will be able to detect and discard the faulty images obtained in the segmentation process or which not have enough information to identify person. In literature, most of evaluation quality methods have developed indices to quantify occlusion, focus, contrast, illumination and angular deformation. These measurements are sensitive to segmentation errors. Only few methods have interested on the evaluation of iris segmentation.

This chapter aims to present, firstly a novel iris recognition method based on multi-channel Gabor filtering and Uniform Local Binary Patterns (ULBP), then to define a quality evaluation method which integrates additional module to the typical recognition system.

Proposed method is tested on Casia v3 iris database. Our experiments illustrate the effectiveness and robustness of ULBP to extract rich local and global information of iris texture when combined with simultaneously multi-blocks and multi-channel method. Also, obtained results show an improvement of iris recognition system by incorporating proposed quality measures in the typical system.

This chapter is organized as follows: Section 2 describe in details the proposed iris recognition system. The further represents the quality evaluation method. In section 4, we expose experiments, results and comparison. Finally, the conclusion is given in section 5.

2. Iris recognition system

In a typical iris recognition system, the eye image is preprocessed to obtain a segmented and normalized image, then its texture is analysed and encoded to form an iris features vector 'template'. Finally, we compare templates to estimate similarity between irises.

2.1 Iris preprocessing

Iris preprocessing step includes iris segmentation and iris normalization.

Iris segmentation aims to isolate iris texture from the acquired eye image, with exclusion of any obscuring elements such as eyelids, eyelashes (Fig. 1-a), and reflections from the cornea or possibly from eyeglasses. Various methods have been proposed to accomplish this task (Daugman, 1994, Wildes, 1997, Daugman, 2007, Liu, 2006, Krichen, 2007).

In the proposed method, we modelled the iris and pupil by two circles not necessarily concentric and eyelids by two segments of line. Different borders are located by the application of Hough transform. Then, we have applied the pseudo polar transformation of Daugman to transform the iris arc from raw coordinates (x,y) to a doubly dimensionless and non concentric coordinate system (r, θ) (Fig. 1-b). The details were described in (Feddaoui & Hamrouni, 2010).

Since, the result is not well contrasted, it is better to enhance the textured image before analyzing its texture (Fig. 1-c). According to result obtained in the segmentation step, we note that the area belonging to [π/6..11π/6] is generally disturbed by the presence of eyelids and eyelashes and consequently, the most discriminating information of texture is in the other portion of the iris. Moreover, in order to reduce the impact of reflection in this region, we don't consider texture present in the 1/6 internal portion (Fig. 1-d).

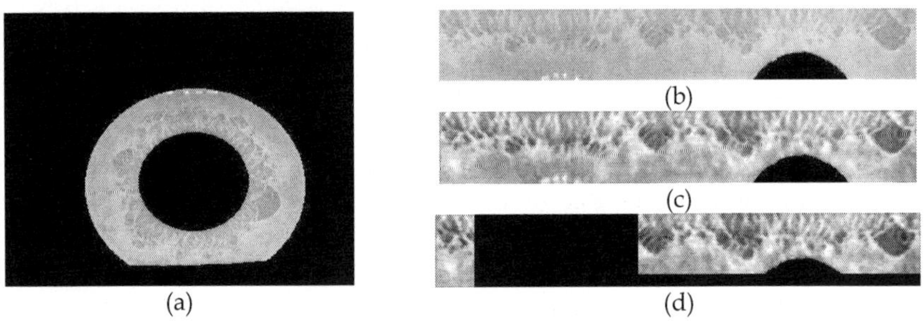

Fig. 1. Iris normalization (a) Segmented iris image (b) Rectangular iris image (c) Enhanced iris image (d) Region of interest in iris image

2.2 Iris texture analysis

In an iris recognition system, the feature extraction and encoding aims to represent the details of the iris texture by finding efficient and discriminative descriptors that are resistant to large variation in illumination, rotation, occlusions, deformation and other factors which disturb the iris texture. There are various methods proposed since the pioneer work of Daugman, in 1992 (Duagman, 1994). Gabor filters (Duagman, 2006) (Masek, 2003) (Ma et al., 2002) (Huang et al., 2007) (Feddaoui & Hamrouni, 2009, 2010) and wavelet (Lim et al., 2001) (Krichen et al., 2004) has shown very good performance because of their capability to multi-

scale representation. Gabor features encode edge information and texture shape of iris texture over a range of multiple narrow-band frequency and orientation components.
In this chapter, we proposed a novel method to extract iris features combining Gabor filters and ULBP operators. The LBP operator succeeds in many applications where it is combined with multi-resolution methods. It has proved its high discriminative power for texture analysis where employed with Cosine Transform (Ellaroussi et al., 2009) and Gabor wavelet in (Zhang et al., 2009) (Tan & Triggs, 2007) to face recognition, Wavelet Packet Transform (Qureshi, 2008) in medical image analysis, Haar wavelet transform (Wang et al., 2008) to ear recognition and Steerable Pyramid (Montoya et al., 2008) to texture analysis.

2.2.1 Proposed method

In the proposed method, we analyzed iris texture by Uniform Local Gabor Patterns ULGP which can be defined as an application of ULBP operators to the Gabor representation. The main components of the proposed method can be summarized as follows:

- Represent global spatial texture information by application of a set of Gabor filters on iris image

In spatial domain, the Gabor function in the spatial domain is a Gaussian modulated sinusoid. For a 2-D Gaussian curve with a spread of σ_x and σ_y in the x and y directions, the 2-D Gabor function is given by equation 1 and the real impulse response of the filter is presented in equation 2 (Bovic et al., 1990).

$$h(x,y) = \frac{1}{2\pi\sigma_x\sigma_y}\exp\left\{-\frac{1}{2}\left[\frac{x'^2}{\sigma_x^2}+\frac{y'^2}{\sigma_y^2}\right]\right\}\exp(2\pi i f x') \tag{1}$$

$$h_r(x,y) = \frac{1}{2\pi\sigma_x\sigma_y}\exp\left\{-\frac{1}{2}\left[\frac{x'^2}{\sigma_x^2}+\frac{y'^2}{\sigma_y^2}\right]\right\}\cos(2\pi f x') \tag{2}$$

Where

$$\begin{pmatrix} x' \\ y' \end{pmatrix} = \begin{pmatrix} \cos\theta & \sin\theta \\ -\sin\theta & \cos\theta \end{pmatrix}\begin{pmatrix} x \\ y \end{pmatrix}$$

In the frequency domain, the equations (3) and (4) represent respectively the frequency response of the complex and real Gabor filters.

$$G(u,v) = \exp\left\{-2\pi^2\left[(u-f)^2\sigma_x^2 + v^2\sigma_y^2\right]\right\} \tag{3}$$

$$\begin{aligned} G_r(u,v) = &\exp\left\{-2\pi^2\left[(u-f)^2\sigma_x^2 + v^2\sigma_y^2\right]\right\} \\ &+\exp\left\{-2\pi^2\left[(u+f)^2\sigma_x^2 + v^2\sigma_y^2\right]\right\} \end{aligned} \tag{4}$$

In our application, we have applied a bank of Gabor filters on iris image. We have chosen 4 frequencies (2, 4, 8, 16) and 4 orientations (0°, 45°, 90°, 135°) which generated a total of 16 filtered images. The space constant σ is chosen to be inversely proportional to the central frequency of the channels.

- Compute ULBP operators in each filtered image to encode local variation across different Gabor coefficients in defined radius.

Actually, LBP operator is one of the best texture feature descriptor and it has already proven its high discriminative power for texture analysis. The original version of the LBP operator was introduced by Ojiala *et al.*, it capture the micro-features in the image by encoding them in a 3*x*3 local window and thresholding eight neighbours pixels with the value of the central pixel, then a binomial factor of 2^i is assigned for each pixel. The LBP code is computed by summing the results of multiplying the thresholded value by a corresponding weight (Ojala et al., 1996). The process is summarized in Fig. 2.

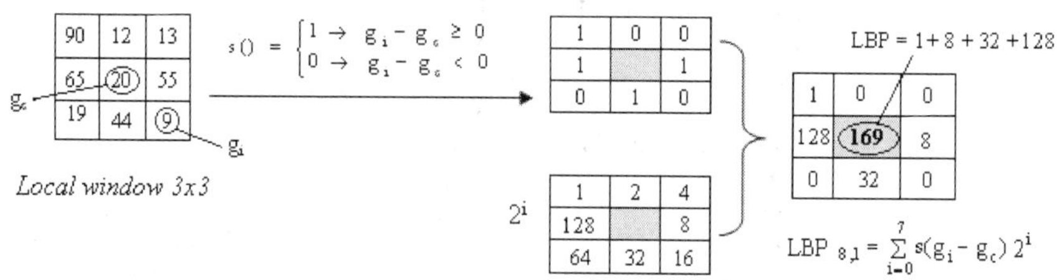

Fig. 2. The original version of LBP

Later, The conventional LBP operator has been extended to introduce a robust illumination and orientation invariant texture feature by computing the pixel values that lie on a circular pattern with a radius *r* around the central pixel. ULBP represent the most common LBP codes without significant loss in its discrimination capability. It reduces the 256 different local binary patterns defined in a 3 x 3 neighbourhoods to 10 by representing the number of bitwise spatial transitions (0/1) in a circular pattern. To quantify these ULBP (equation 5), an uniformity measure *U* was introduced (equation 6)

$$ULBP = \begin{cases} \sum_{i=0}^{7} s(g_i - g_c) & if \quad U(LBP) \leq 2 \\ 9 & otherwise \end{cases} \tag{5}$$

$$U(LBP) = \left| S(g_{i-1} - g_c) - S(g_0 - g_c) \right| + \sum_{i=0}^{7} \left| S(g_i - g_c) - S(g_{i-1} - g_c) \right| \tag{6}$$

- Extract the global and local signatures by first dividing each obtained image into non-overlapping blocks having a given size, then computing statistical features within each block to form a vector.

This step aims to extract appropriate texture features from ULGP representations. The idea is to divide the whole ULGP image into blocks having a given size. Then, to provide a good discriminating feature between irises. In this paper, we have proposed to compute statistical features on each ULGP representation. Finally, features are concatenated to form the complete feature vector which size varies depending on the block size (5x5, 10x10, 20x20, 30x30).

- Form iris template by encoding local relationship between measures of vector.

This step aims to generate iris template based on representing the features variations. First, we have linearized the matrix of statistical features then each coefficient is compared to the previous one and is encoded by 1 or 0: 1 if it is greater than the previous one and 0 otherwise. This process is similar to the computation of the derivative of the feature vector and the encoding of its sign. In practice, this binary iris code facilitates greatly the matching process.

Since persons are identified by their templates, the process of person verification needs a comparison between two templates in order to estimate their similarity. Considering that the iris is represented by a binary template, the Hamming distance is more suitable to estimate the difference between iris patterns with a bit-by-bit comparison.

The computation of the Hamming distance is given by the following expression:

$$HD = \frac{\|(A \otimes B) \cap MaskA \cap MaskB\|}{\|MaskA \cap MaskB\|} \tag{7}$$

Where {A,B} are the two templates and {MaskA, MaskB} their corresponding noise masks.

In our algorithm, we obtained rotation invariance by unwrapping the iris ring at different initial angles. Five initial angle values are used in experiments [-4°, -2°,0, 2°, 4°]. Thus, we defined five images for each iris class in the data base. And when matching the input feature code with a class, the minimum of the scores is taken as the final matching distance.

3. Quality evaluation method

Actually, the function of the iris recognition system is affected by the quality of used images. When the images used are not of a high quality, this affects significantly the performances, one has to discover the images of poor quality and to separate them.

An image of poor quality complicates the segmentation phase proved by the difficulty of detecting the different borders of iris, consequently the presence of the non detected noise in the iris texture. This noise affects the results of different stages of a recognition system particularly the stage of texture analysis, that can be also affected by the focus of the image and even the nature of the texture. In fact, the iris texture must include enough information to identify person.

To select images of good quality, we have developed a model quality that evaluates the results of segmentation and the richness of texture. The model is integrated in the recognition system after the phase of segmentation. Based on generated qualities measures, we process in the characterization stage, only image that surpass a certain threshold. The choice of its value depends of the security level of the intended application.

In the following section, we introduce the principal woks made to evaluate the quality of iris image.

3.1 State of art

During the last decade, many works have been interested in the evaluation of the quality of an iris image. Despite the diversity of the applied methods, they proved experimentally an improvement of the recognition system performance.

Most of these works defined the quality in terms of texture clarity, focus degree, occlusion rate, dilation degree, view angle, etc. The used techniques can be classified into 3 categories:

those operating in the Fourier Domain, those based on 2D wavelets transform and the statistical methods.

3.1.1 Quality measures in the frequential domain

The choice of indices quality measurement in the frequential domain is justified by the fact that an out-of-focus image can be considered as a result of the filtering of the ideal image by a low pass filter, and then the main part of information of texture is located in the low frequencies. On the contrary, this information is between the low and high frequencies for clear image.

Daugman estimated the clarity of iris image in term of the rate of the energy of high components. This energy increases proportionally to the degree of focus image (Daugman, 2004). Ma et al. analyzed the frequential distribution of the two areas of size 64x64 pixels around the pupil. Then, the quality indices are used by a SVM classifier for the training and the classification of images in 4 categories: clear images, out-of-focus images, blurred images due to the eye movement during the acquisition and images altered by the presence of eyelids and eyelashes (Ma et al., 2003).

Later, Kalka et al. studied many other factors on the system performance such as: the occlusion, the pupil dilation, the illumination, the percentage of significant pixels, the movement of eye, the reflections, the view angle and the distance from the camera. According to their study, the focus, the eye movement and the view angle degrade more the performances. They analyzed the high frequency components to measure the degree of blur due to camera distance and the directional properties of the Fourier spectrum for the blur due to the movement (Kalka et al., 2006). To evaluate the angle of view, they measured the circularity of iris by applying an integro-differential operator to different images obtained by projecting the original image at different angles. So, the angle of correction maximizes this operator.

Tissé has implemented 12 techniques proposed in the literature, to evaluate the clarity of iris image, based on: gradient , wavelets, filter, etc. He compared the obtained indices on an analysis region, he deduced that the method based on FSWM filters (Frequency Selective Weighted Median Filter) leads the better results (Tissé, 2007). In the same year, Ketchantang et al. proposed an index of image quality in a sequence of images acquired in real time. This measure combines the speed of the pupil moving between two successive frames, the density of dark pixels in the pupil area and the clarity of the collarette. The speed of displacement is estimated by using Kalman filter. This operator provides informations about the quantity of blur caused by the sudden movement of the eye during the acquisition.

The density of dark pixels in the pupil estimates the depth of focus. The clarity of the collarette is evaluated in the Fourier domain by measuring the energy of middle frequency components of a region around the pupil (Ketchantang et al., 2007).

3.1.2 Quality measures based on wavelet transform

Generally, the algorithms operating in the frequency domain are applied on the entire image (or an interest region), hence they are sensitive to the noise and give a global sight of the focus degree of the iris texture. To solve these problems, many solutions applied the 2D wavelets transform to produce a local descriptor of the iris quality. Chen et al. suggested a local measure of quality based on the « Mexican Hat » wavelet transform. The segmented iris image is divided into multiple concentric bands with a fixed width, around the pupil.

The degree of blur of each band is measured by the energy of the wavelet coefficients. Then, a global index of quality was defined as a weighted average of the local quality measurements. The weight reflects the distance of the candidate band relative to the pupil (Chen et al., 2006).

In order to enhance iris image, Vatsa et al. used a discret wavelet transform DWT and a SVM classifier on a set of 8 images that incorporates the original iris image and its transformed one by 7 known enhancement algorithms such as the histogram equalization, the entropy equalization, etc. the DWT is applied on each image, and the coefficients of the approximation and details bands are classified as coefficients of good quality by SVM classifier (Vatsa et al., 2008).

3.1.3 Quality measures based on statistical measures

In addition to the techniques described above, several researchers have considered statistical measures to assess the quality of iris images. Zhang et al. filed a patent concerning the process that determines whether the image is focused correctly. It is based on analyzing the shape and the continuity of the iris boundary. They considered a number of lines crossing the pupil boundary, for each line, statistical values are calculated for the pixels belonging either to the pupil and the iris (Zhang et al., 1999).

Proença et al. developed a method based on statistical measures and neural networks. The process consists in calculating 5 statistical measures in 7x7 windows derived from a segmented polar iris image. The measures commonly used are: ASM (Angular Second Moments), entropy, contrast, energy and inertia. Then, a simple thresholding of index computed in each analysis window permit the classification of central pixel into "noisy" or "significant" pixel (Proença & Alexandre, 2006).

Cambier et al. studied the impact of 7 quality measures on the performance of a recognition system for multi-cameras recognition system. These index are the iris and pupil radius, the pupil-iris ratio, the iris-sclera contrast, the iris intensity, the texture energy and the percentage of visible iris. Results showed that the texture energy and the rate of visible iris are the most important but specific to population (Cambier & Seelen, 2006).

Krichen introduced a statistical model GMM (Gaussian Markov Model) to define a global quality score estimating that iris image represents a good quality texture (Krichen., 2007).

To determine whether the image has enough information to identify person, Belcher et al. (Belcher & Du, 2008) and Zhou et al. (Zhou et al.,2009) developed a quality index by combining 3 index: the dilation score, the occlusion score and the feature information score. Iris image is segmented then the texture is analyzed by Log-Gabor filters. A global feature information score is estimated by averaging the entropy information distance between pairs of consecutive rows of the filtered image. This quality measure was used by Y. Du et al (Du et al, 2010) to evaluate the quality of a compressed iris image. In fact, during compression, iris patterns are replaced by new artificial patterns, and only the most distinctive iris patterns resist. These false patterns are too correlated compared to the original image and they become more important through compression rate. Consequently, the more compression rate is elevated, iris texture quality becomes weaker.

Based on the fact that the inner region of iris contain more discriminative patterns, Sung et al. improved the matching performance by merely weighting the inner and outer iris regions with respectively 1 and 0 (Sung et al., 2007).

In the cited works, the quality measures are often computed in the segmented images which makes them sensitive to the segmentation errors. In practice, no segmentation method has a rate of 100% of correct segmentation. Nevertheless, the study of Zhou et al. is among the rare that took into account the evaluation of the segmentation results of an iris image (Zhou et al., 2009). This is because the segmentation of the iris has specific particularities and existing methods for evaluating the segmentation (Chabier et al., 2005) (Foliguet & Guigues, 2006) are not applicable in this case. In (Zhou et al.,2009), Zhou et al evaluated the segmentation accuracy in terms of localizing correctly the center and boundary of pupil and the iris boundary including the limbic boundary and the eyelid boundaries. This measure is based on the analysis of the histogram of a rectangular horizontal area including the two centres, three sub-regions belonging to pupil, iris and sclera.

3.2 Proposed method for quality evaluation

A typical biometric system includes 4 stages: the capture of the eye, the segmentation and normalization of iris image, the texture analysis and encoding and finally the matching.
A too noisy image or poorly segmented is processed in all system steps which often leads to a false identity recognition. Thus, the proposed system consists of integrating a quality module after the segmentation step. The objective of this module is to select images to be processed in the system. For this, we developed two quality units: the first one assess the result of segmentation while the second estimates the richness of texture.
Fig. 3 illustrates different units of the given system and the following sections describe every module.

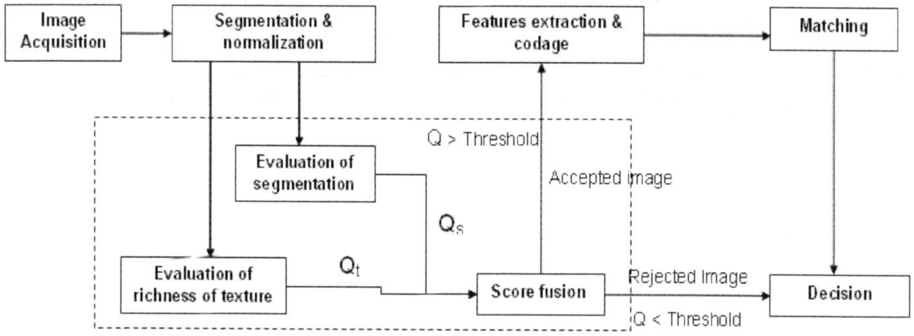

Fig. 3. Proposed iris recognition system: integration of quality units in typical system

3.3 Evaluation of the segmentation

This unit aims to verify if the image is correctly segmented. It takes as input the coordinates of the pupil and iris and the generated masks, then it analyzes each detected boundary in order to estimate the corresponding quality index.

3.3.1 Evaluation of the pupil boundary

The first stage in the segmentation process is the localisation of the pupil, which results influences the performance of the rest of system. In practice, an inadequate segmentation of the pupil alters the content of the iris arc by adding or eliminating a portion of the pupil located near the border. Fig. 4-a illustrates the relationship between the matching distance and the pupil radius. We can see that the correct segmented image (R) produces a low similarity distance whereas the bad localized pupil (R-8, R-4, R+4,R+8) degrades significantly the distance.

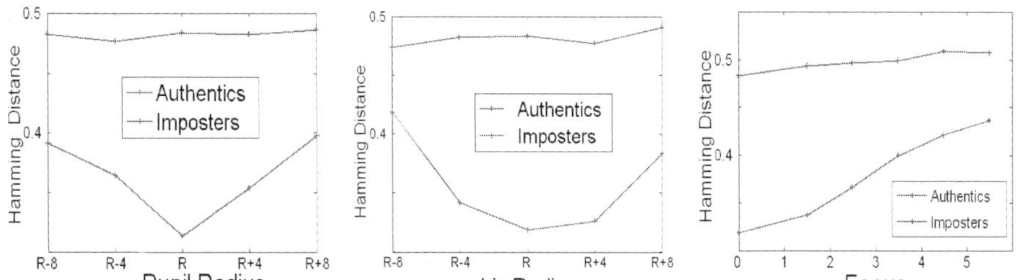

(a) (b) (c)

Fig. 4. Illustration of the relationship between the distance between two feature vectors and (a) the pupil radius (b) the iris radius (c) the richness of iris texture

These errors are mainly caused by the physiological nature of the iris and the lighting conditions during acquisition. In fact, the pupil is not perfectly circular or elliptical and always presents fluctuations and discontinuities along its border (Fig. 5). Also, the pupil is often partially hidden under the light spots and the eyelashes. the proximity of the noise to the pupil border increasingly complicates the segmentation process.

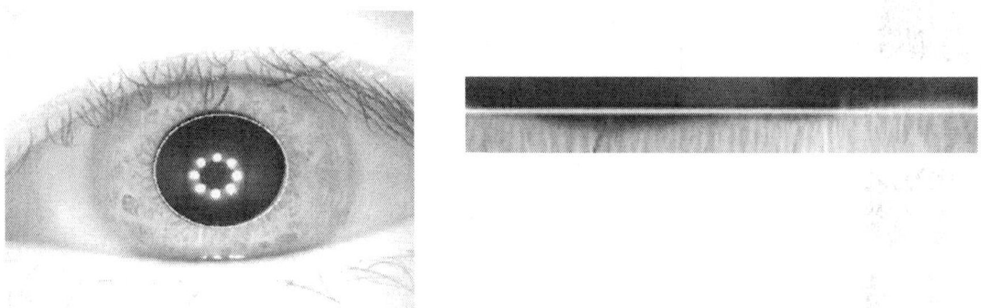

(a) (b)

Fig. 5. Illustration of the fluctuation of the pupil boundary relative to the detected border (white line): (a) segmented pupil image (b) border region in polar coordinates.

To estimate this defect, we generate an index evaluating the quality of segmentation through the following steps:

- Minimize the effect of light spots by filling the clear holes by the average of the image intensity
- Consider a region located on both sides of the detected border.
- Filter this region, then analyze the pupil fluctuations across the detected circle boundaries in order to determine the real mask of the pupil ($Mask_{real}$). In fact, we added to the detected mask ($Mask_{detect}$) during segmentation the non-detected pixels located outside the detected pupil, and we eliminate the invalid pixels belonging to the iris arc and located inside the detected pupil.
- Apply equation (8) to measure a quality index Q_p assessing the pupil segmentation:

$$Q_p = 1 - \frac{\|Mask_{real} \otimes Mask_{detect}\|}{\|Mask_{detect}\|} \quad (8)$$

3.3.2 Evaluation of iris/sclera boundary

The segmentation evaluation depends not only on the pupil localization but also on the iris boundary, which should separate the iris from eyelids and sclera. In this unit, we are interested in the iris/sclera boundary.

To show the impact of an adequate boundary detection on iris texture, we represent in Fig. 6, three polar images (b-c-d) that are relative to image (a) by considering 3 different iris radius. We can notice that the segmentation errors generally caused a significant destruction of texture patterns and consequently a wrong distance between iris codes. The relationship between the iris radius and this distance is illustrates in Fig. 6-b. We can see a clear degradation of distance with inadequate iris radius (R-4, R-8, R+4, R+8).

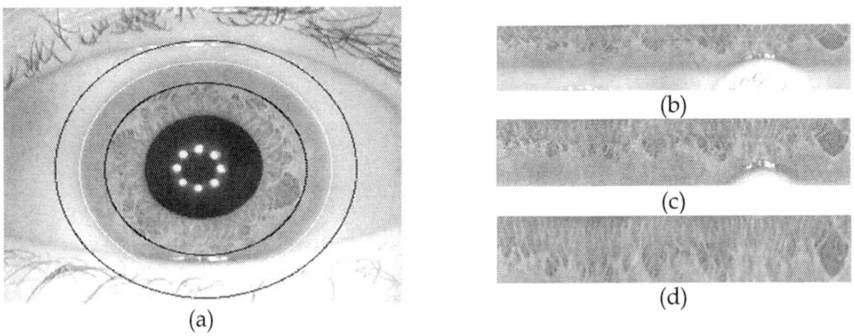

Fig. 6. The impact of the iris/sclera boundary on iris texture: original image (a) , polar image by considering adequate radius (c) and inadequate radius (c-d): the segmentation errors caused a significant destruction of texture patterns

In practice, the intensity variation between the region of iris and the sclera is important in perfect conditions of acquisition, which facilitates the border detection process. However, the reasons that might complicate this detection are diverse, we mainly cited the eyelashes occlusion, the lighting conditions, the percentage of visible iris and the view angle. In fact, the three first factors generates false contours that alter region boundaries. Despite the percentage of visible iris and the view angle should be reasonable to consider an important iris region in the segmentation process.

The evaluation of the iris boundary is performed in two eye regions respectively within the following ranges: [-30 ° .. 30 °] and [150 ° .. 210 °] relative to the iris center. The pixels are divided on both sides of the detected boundary (l_{detect}). Then, each region is subdivided into overlapping rectangular blocks of size hxw. A vertical projection is determined for each bloc. Then, we calculate the pixels average of every column. The obtained curve represents a minimum at the correct boundary of the iris (l_{thresh}). Its position can be localized by the maximum of the derivative of the curve. The distance separating the two positions l_{thresh} and l_{detect} is used to define a local index which assess the quality of iris boundary.

We notice that in the case of a correct segmentation, l_{thresh} and l_{detect} are confused. the process steps are described in the following algorithm:

- m=0: Number of blocks apt for the evaluation
- Subdivide the region of analysis R into n overlapping rectangular blocks B_i of size hxl. R={ B_i, i=1..n}.

For each block B_i **do**

- Count the number of noisy pixels N_{ps}

if $N_{ps} > \frac{h*l}{2}$ **then**

- Consider the B_i inapt for the evaluation

else

for each row $j \in [1..h]$ of the block B_i **do**

- Compute the average of the significant pixels m_j

end

- m=m+1
- Calculate the projection $hist_i$ of obtained averages{mj} (Fig. 7.a-c)
- Calculate the derivative of projection (Fig. 7. b-d).
- Detect the first peak l_{thresh} in the derivative
- Measure a local index of quality Q_i

$$Q_i = 1 - \frac{|l_{thresh} - l_{detect}|}{rayon_{iris}} \tag{9}$$

end

end

- Evaluate the global index of quality Q_{isc} based on the local index $\{Q_{ij}\}$

$$Q_{isc} = \frac{1}{m}\sum_{i=1}^{m} Q_i \tag{10}$$

Algorithm 1. Evaluation of iris/sclera boundary process

3.3.3 Evaluation eyelids boundaries

After the iris image segmentation, it is important to localize the eyelids occlusion. The quality of the result depends on the presence of eyelashes, glare and light spots on the eyelids borders and also the richness of texture.

In practice, these factors caused defects in contrast and false contours in the segmentation region, which don't lead to a perfect segmentation for all iris cases. To evaluate the error, we have developed a module to analyze the eyelid borders in the polar iris image. We concentrated on the low eyelid since we considered, in the next steps of the recognition process, a region of interest RI between [π/6..5π/6] (Fig. 8-b).

Algorithm 2. represents the proposed method, it is consist of analyzing the intensity variation in local windows of the regions of interest (RI) which are selected near and on the border of the detected eyelid boundary (Fig. 8-c). We computed statistical measures like standard deviation, variance, average absolute deviation, mean, entropy, etc. Tests showed that the variance gave the best results.

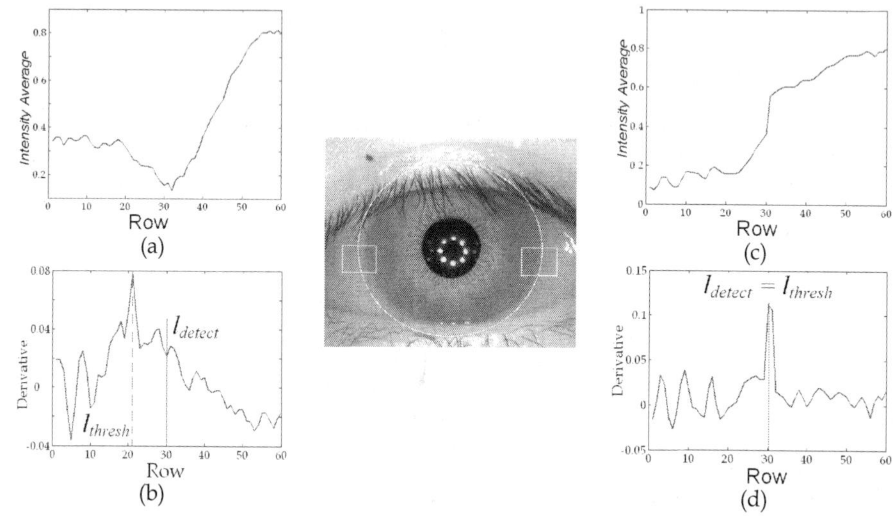

Fig. 7. Illustration of the iris boundary evaluation process: we analyze the vertical projection and its derivative of the local windows: (a-b) correspond to left region and (c-d) related to the right region. We notice that in the case of a correct segmentation, l_{thresh} and l_{detect} are confused.

---

if iris is not occluded by eyelids **then**

- Assign 100% to the eyelid quality score : $Q_e = 100\%$

else

- Subdivide the region of analysis of n overlapping blocks B_i^c (i=1..n) of size hxl. It is distributed on both sides ode detected boundary (Fig. 8-c).

For each block B_i^c **do**

- Consider a neighbour region B_i^v of the same size
- Compute V_i^c and V_i^v the variances respectively of B_i^c and B_i^v :

$$V(B_i) = \frac{1}{h*l} \sum_{j=1}^{h} \sum_{k=1}^{l} \left(B_i(j,k) - \overline{B_i}\right)^2 \tag{11}$$

$$\overline{B_i} = \frac{1}{h*l} \sum_{j=1}^{h} \sum_{k=1}^{l} B_i(j,k) \tag{12}$$

end

- Calculate the eyelid quality score:

$$Q_e = \frac{\sum_{i=1}^{n} V_i^c}{\sum_{i=1}^{n} V_i^c + \sum_{i=1}^{n} V_i^v} \tag{13}$$

End

Algorithm 2. Evaluation of eyelid boundary process

Based on local measures, we generated a global quality index Q_e evaluating the segmentation of the lower eyelid. The index penalizes under-segmentation of the eyelid. In fact, considering the noisy pixels in the region of analysis affects the system performance more than the elimination of pixels due to the over-segmentation.

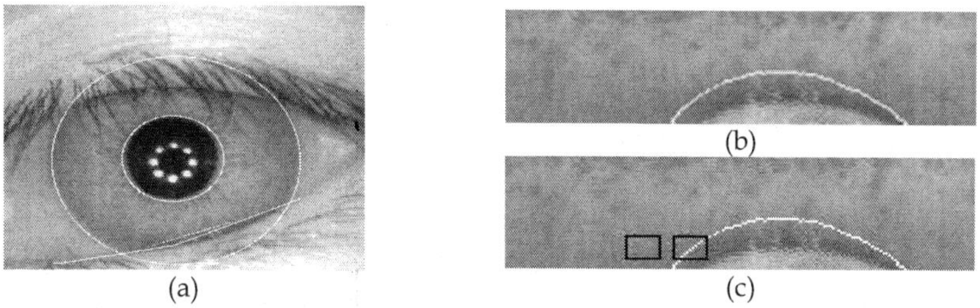

Fig. 8. Illustration of eyelids boundary evaluation (a) segmented iris image (b) segmented RI in polar coordinates (c) selection of two neighbours blocks near and on the detected boundary

3.3.4 Fusion of the segmentation scores

The global assessment of the segmentation depends on the 3 index Q_e, Q_p and Q_{isc} relating respectively to the boundary of eyelid, pupil and iris/sclera. Fig. 9 illustrates the correlation of different obtained scores. We notice that the measures are not grouped on the diagonal, which asserts that they are not correlated.

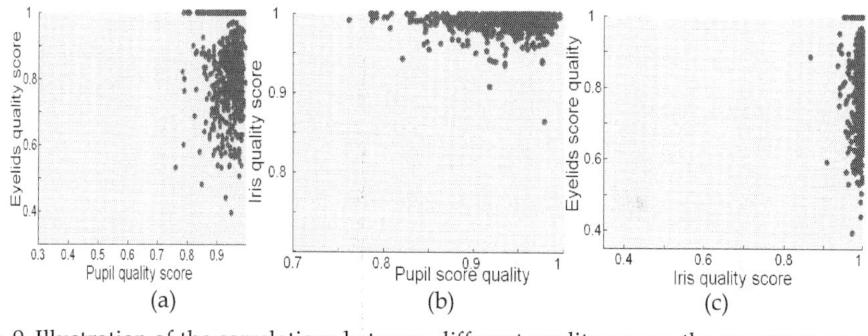

Fig. 9. Illustration of the correlations between different quality scores: the measures are not grouped on the diagonal, therefore they are uncorrelated

Since, they are all important to judge the segmentation, we average their values to generate a global index Q_s:

$$Q_s = (f_1(Q_p) + f_2(Q_{isc}) + f_3(Q_e)) / 3 \tag{14}$$

Inspired by the work (Zhou et al., 2009), f1, f2 and f3 are parameterized functions of normalization related to the function f defined as follows:

$$f(Q)\begin{cases} 1 & Q \geq \beta \\ e^{-\alpha(\beta - Q)} & 0 \leq Q < \beta \end{cases} \tag{15}$$

The choice of an exponential function is justified in (Zhou et al., 2009) (Du et al., 2010) (Belcher & Du, 2008) by proving that the relationship between the recognition rate and the degradation factors is not linear.
In this equation, parameter ß illustrates the minimum error above which the system performance is not degraded. In fact, tests have shown that we can neglect the weak error of segmentation because it doesn't affect significantly the recognition rate. We set ß to 0.9, 097 and 0.9 for respectively f1, f2 and f3. While the parameter α represents the weight of the error on the recognition results. We have assigned more weight to Qp because poor detection of the pupil affects mainly the area that contains the most discriminative iris texture.
Given that biometry by iris is often used in high security system, we chose $\alpha = 50$ for f1 in order to quickly extend the normalized score to 0 when Q_p is less than 80%. Also, we fixed α to 20 and 13 for f2 and f3 to obtain zero as normalized score when Q_{isc} and Q_e are respectively less than 70% and 50%.

3.4 Evaluation of the richness of texture

This module aims to generate a score expressing the richness of iris texture to verify if the RI of iris image has enough information to identify a person. This module is integrated into the recognition process after the image segmentation and normalization unit. It estimates a global quality index based on the following index: the occlusion score Q_o , the dilation score Q_d and the degree of texture information Q_f.

3.4.1 Occlusion score

The amount of available iris region can affect the recognition performance. In this unit, we developed an occlusion score Q_o estimating the percentage of significant pixels of the RI. This index is evaluated as follows:

$$Q_o = \frac{N_{ps}}{N} \tag{16}$$

Where N is the total number of pixels of RI and N_{ps} is the number of significant pixels of RI.

3.4.2 Dilation score

In a recognition system, the pupil dilation may affect performance. This is explained by the fact that the shape and the density of texture patterns are altered by the degree of iris deformation. In this paper, we developed an evaluation dilation unit which takes as input the pupil radius R_p and the iris radius R_i to compute the dilation score Q_d as follows:

$$Q_d = 1 - \frac{R_p}{R_i} \tag{17}$$

3.4.3 Estimation of texture information

When the iris image is slightly noisy and correctly segmented, the recognition problems related to the texture information. Whatever the adopted algorithm, its robustness depends on the richness of texture that varies from a subject to another. Fig. 10 gives examples of this diversity. The texture details are presented in the frequency domain by high frequency components.

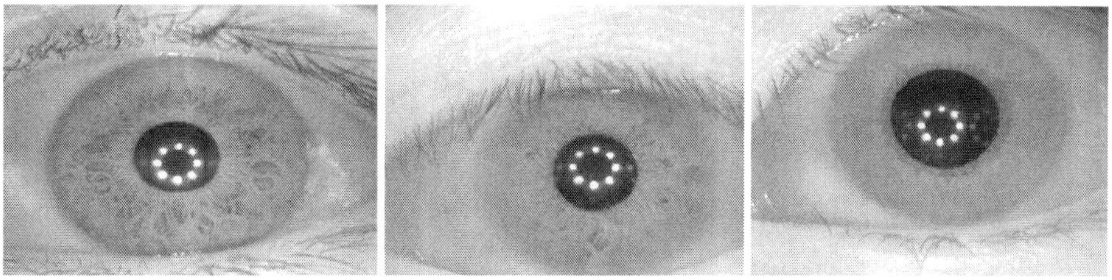

Fig. 10. Examples of the diversity of iris texture

Generally, the blur is appears as a loss of information as a smooth texture patterns, a bad separation between different iris patterns and a distortion of fine edges.

For that, we studied the evolution of the similarity distance between iris according to the different degradation of the texture information. We simulated this degradation by a low pass Gaussian filter with an increased mask size. Fig. 4-c shows a clear degradation of distances especially between genuine iris. In fact, an important amount of texture details are destroyed when the size of the filter mask is larger. This makes necessary to classify the iris in function of the amount of texture information and separate the clear images of blurred images. In addition, from a technical point of view, it is not possible to derive a clear iris image from a blurred one. Correcting the contrast can improve its quality.

All these problems require taking into account the texture information measure in the recognition process, to ensure a good selection of images.

In section (3.1), we presented some proposed texture information measures in literature. In order to ensure a good estimation of the texture richness and an acceptable computation time, we proposed a global quality index operating in the cosine transform (DCT) field.

This transform was used by Matej et al. to estimate the focus of a sequence of images taken in different conditions. They demonstrated the efficiency of the DCT compared to other existing methods and its robustness against the noise occlusion. In fact, a good evaluation of the focus has been achieved even when an artificial noise was added. (Matej et al., 2006)

This work helps us to develop a global index of texture information based on the shanon's entropy and the energy's distribution of DCT coefficients.

3.4.3.1 Cosine transform of an image

Cosine transform, (CT or DCT for digital cosine transform), is a linear mathematical transformation similar to the discrete Fourier transform, it was introduced in 1974 by Ahmed et al. to reduce redundancy information. A DCT expresses a sequence of cosine functions which generates real coefficients, and therefore it avoids the complex numbers as in the case of the Fourier transform. (Ahmed et al., 1974)

The DCT is often used in signal and image processing and especially in audio and video compression. The standards MPEG, JPEG and MJPEG apply the DCT in the compression process. However, DCT is commonly used in image processing to obtain the spectral representation of the digital image.

Given an image I of size MxN, the DCT transform provides an image R of the same dimension. This transformation represents an interest because it concentrates the major part

of energy in a minimum of low coefficients. Thus, an adequate use of this information leads to a good analysis of the candidate image. We can locate these low-frequency coefficients in the top left corner of the image (Fig. 14.d-e-f). Whereas the other coefficients in the corner right bottom of the image may be neglected and reduced to zero in the compression process. It is known that the Discrete Fourier Transform DFT is used by several algorithms for estimating the focus of an image. However, in the last decade, the DCT has been increasingly used by the visual systems that the DFT, and this is thanks to its high concentration of energy in some low-frequency coefficients. In addition, the basic function of DFT is exponential and generates complex coefficients therefore it requires a computation time greater than the DCT.
Moreover, Krotkov and Yeo et al. suggest in (Krottov, 1987) (Yeo et al., 1993) that the DFT contains information that is superfluous to the focus evaluation such as the phase information. For these reasons we preferred the DCT to measure the blur in image.
In practice, this optimal transformation is sensitive to contrast changes, so this defect may be limited by the image preprocessing or the normalization of the transformed image by DCT.

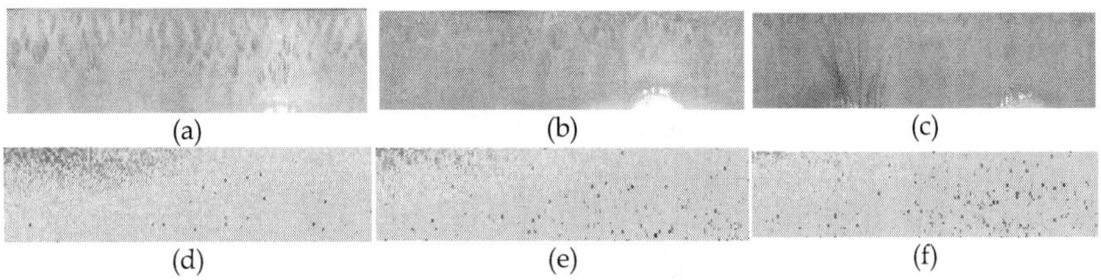

(a) (b) (c) (d) (e) (f)

Fig. 11. Illustration of the relationship between the richness of iris texture and the normalized DCT: for visualization purposes, we display in (d-e-f) the logarithm of the transformed images. The coefficients with higher values are shown in red. By comparing the images (d-e-f), we notice that the spectrum of sharp image is more uniform. Quality measures obtained are respectively: 0.76, 0.33 and 0.21.

3.4.3.2 Implementation

To evaluate the texture information, we developed a global quality index, based on the DCT representation of the iris image and the entropy of the energy distribution. For optimization purposes, the proposed method is based on an existing solution adopted by the compression standard JPEG, by applying the DCT in 8x8 blocks. Indeed, it was proven that the choice of this block size is a good compromise between quality and computation time (Pennebaker & Mitchell., 1992). Algorithm 3 summarizes the proposed procedure to evaluate a global texture information index Q_f. Fig. 11. illustrates the application of normalized DCT to three polar iris images (a-b-c). For the visualization purposes, we considered the logarithm of the transform in (d-e-f) images. When comparing these images, we can deduce that the image with highest texture information produces more uniform spectrum than others. In fact, the less sharp image concentrates more energy in a minimum coefficient of low coordinates. We obtained respectively the quality measures: 76%, 33% and 21%, which shows that the uniformity measure of the spectral representation of image is appropriate for estimating the texture information.

- Consider a region of analysis R of size 48x240
- Subdivide R into n non-overlapping blocks B_i of size 8x8. R={ B_i, i=1..n}.

for each block B_i **do**

- Compute R the DCT of B_i. $R(B_i)$={Rpq, p=1..8,q=1..8}
- Calculate R' the normalize DCT of B_i:. $R'(B_i)=\{R'_{pq}\}$ is defined as follows:

$$R'_{pq}(B_i) = \frac{R_{pq}(B_i)}{\sum_p \sum_q R_{pq}(B_i)} \tag{18}$$

- Estimate a local texture information score $Q_f(B_i)$ by computing the energy of B_i

$$Q_f(B_i) = \frac{1}{64} \sum_p \sum_q (R'pq(B_i))^2 \tag{19}$$

end

- Evaluate a global texture information score $Q_f(I)$ by computing the entropy of Shannon of local measures $\{Q_f(B_i), i = 1..n\}$. The entropy describe the energies distribution of the energies of the standardized TCD image.

$$Q_f(I) = -\sum_{i=1}^{nb} Q_f(B_i) \log(Q_f(B_i)) \tag{20}$$

Algorithm 3. Evaluation of texture information process

3.4.4 Scores fusion

To estimate the richness of texture, we generated a global quality index Q_t by averaging the three generated quality scores Q_o, Q_d and Q_f, respectively related to the occlusion, the dilation and the texture information. We started by normalizing these scores in [0 .. 1] by the function (6).

According to several experiments, we concluded that the image quality is not affected when the occlusion rate doesn't surpass 10%, the dilation score is less than 40% and the sharpness score overcomes the 50%. Despite that, the image is inadequate when the dilation score becomes 75% or the occlusion rate overcomes 60% or the sharpness score is less than 35%. So, the normalized scores tend to 0 when the corresponding threshold are reached.

3.5 Estimation of global quality score

In this part, we generated a global quality score Q which takes into consideration the amount of available region and the efficiency of information contained in the iris image. These criteria have already been evaluated by the quality scores Q_s and Q_t. The validation of these measures on the basis Casia v3 shows that both are important for quality assessment also they are uncorrelated, then they were combined by averaging their value.

4. Experimentations and results

4.1 Iris database

In order to evaluated the accuracy of the proposed method, extensive experiments on Casia v3 iris images are performed.

Currently, Casia v3 presents the largest iris database available in public domain. It has been released to more than 2 900 users from 70 countries since 2006. CASIAv3 includes three subsets which are labelled as CASIA-IrisV3-Interval, CASIA-IrisV3-Lamp, CASIA-IrisV3-Twins. All iris images are 8 bit gray-level JPEG files with 280x320 of resolution and collected under near infrared illumination in different times. Most of the images were captured in two sessions, with at least one month interval and the specula reflections from the NIR illuminators on the iris texture introduce more intra-class variations (Casia, 2006).
To validate our method, we have considered 2641 images of 249 people in CASIA-IrisV3-Interval database.

4.2 System evaluation

The experiments are completed in verification mode (one-to-one). In this system, the two compared templates are similar and represent the same iris if the distance is less than a given threshold value resulting from a training step. We measured the performance of the method in terms of four rates: False Acceptance Rate (FAR), False Rejection Rate (FRR), Equal Error Rate (ERR) which corresponds to the value where the FAR and FRR are equal and the Measure of Decidability (MD) which estimated the degree of separability between two distance distributions. An approximate measure of MD is given by equation (21):

$$MD = \frac{|\mu_1 - \mu_2|}{\sqrt{\frac{1}{2}({\sigma_1}^2 + {\sigma_2}^2)}} \tag{21}$$

Where (σ_1, μ_1) corresponds to the standard deviation and mean of intra-class distribution and (σ_2, μ_2) corresponds to the standard deviation and mean of inter-class distribution

4.3 Performance evaluation of the proposed recognition system

The proposed method is based on ULGP patterns and a quality measures which take into consideration the segmentation accuracy and the quality of texture. So, we have conducted two sessions of experiments. In the first session, we apply the typical system to demonstrate the discriminating properties of the ULGP method. In the second session, we used the proposed system to prove the importance of quality score to improve the accuracy. In the following sections, we will present the performed tests and the obtained rates.

4.3.1 Evaluation of ULGP method

To evaluate ULGP method, two series of experiments are performed. In the first series of experiments, we wanted to compare the performance of parameters which can be classified in three classes:

- Parameters of Gabor filters: frequencies and orientations
- Parameters of Local Binary Patterns features: radius and neighbours number
- Parameters of features extraction: statistical feature and the choice of blocks parameters (size and overlap).

For each fixed parameters, all possible comparisons between irises are made to obtain a total number of 1706849 on Casia v3 database including 8766 of intra-class comparisons and

1698083 of inter-class comparisons. We have illustrated results by plotting DET curves (Detection Error Tradeoff) which represents the evolution of FRR against FAR for each parameter (as shown in Fig. 12). Experimental results indicate that computing ULBP with R=2 and P=16 and considering standard deviation in non-overlapping blocks of 10x10 achieve high recognition performance. The intra-class and inter-class distance distribution of optimal parameters were illustrated in Fig. 13-a. This system achieved 0.68% of EER and 3.02 % of FRR where FAR=0.

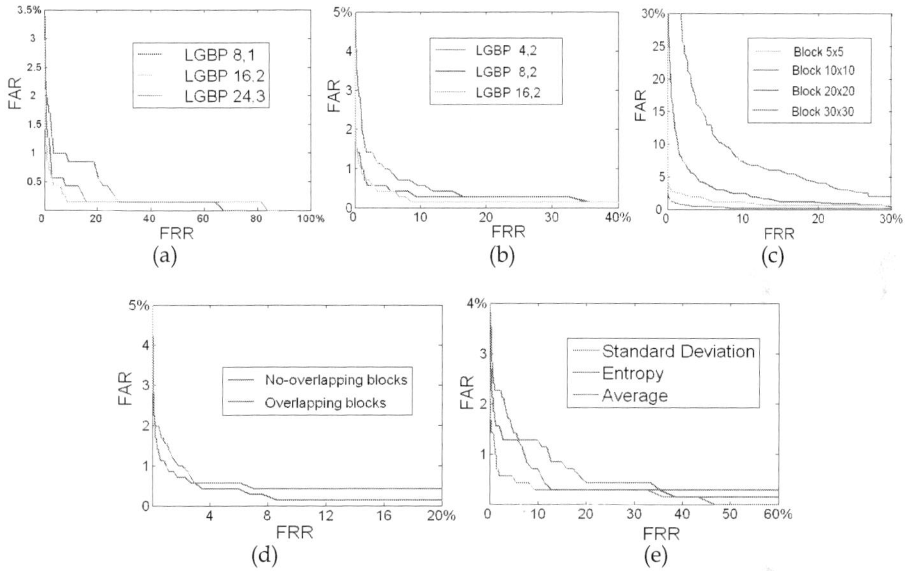

Fig. 12. DET curves for different parameters (a) ULBP rayon (b) Neighbours number (c) Block size (d) Choice of block (e) Statistical feature

In the next series of experiments, we have evaluated three other approaches:

- Application of ULBP operators on iris polar image and division of obtained image into non-overlapping blocks, then computation of statistical feature within each block.
- Generation of 16 Gabor filtered images, then for each output image, considering real part to compute statistical coefficient in extracted blocks.
- Description of iris texture by 16 Gabor filters and generation of iris signature according to "4 quadrants" coding phase.

Fig. 13 illustrate distance distribution of these experiments. As can be seen, Uniform Local Gabor Patterns ULGP perform better than others. Table 1. summarized results of evaluation of all approaches.

It has been confirmed experimentally that combining ULBP operators and Gabor filters provide a good discrimination between iris patterns and achieved higher accuracies than Gabor phase encoding.

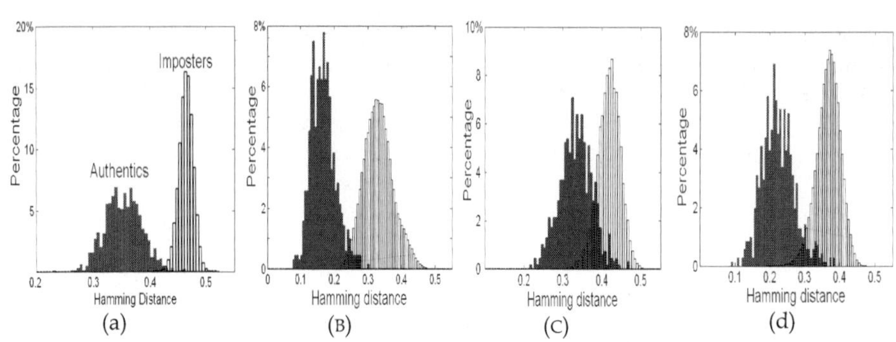

Fig. 13. Comparison between the results obtained by extraction of statistical features on (a) ULGP representation (b) Gabor outputs (c) ULBP on iris texture (d) Coding Gabor phase

	FRR/FAR=0 (%)	EER (%)	MD
Gabor phase	31.78	4.66	3.57
Gabor	8.47	1.97	4.17
ULBP	82.7	13.93	2.08
ULGP	3.02	0.68	4.67

Table 1. Reported results from different algorithms

4.3.2 Integration of quality scores to improve performance

4.3.2.1 Validation of quality scores on iris database

We calculated quality scores Q_s , Q_t and Q for all segmented images of Casia database. The Distributions of Fig. 14 show the percentage of iris images in function of the quality level. Fig. 14-a represents the distribution of irises according to Q_s. we can see that the quality score varies between 0.32 and 1 and most images are well segmented. Also, the distribution (Fig. 14-b) shows that most images contain enough information to be identified. The score Q_t varies between 0.37 and 1.According to Fig. 14-c, we can consider that the base Casia v3 has a medium quality and Q varies between 0.39 and 1.
Based on the distribution of the quality score Q, we fixed a number of decision thresholds specifying the minimum quality to exploit the image by the system.

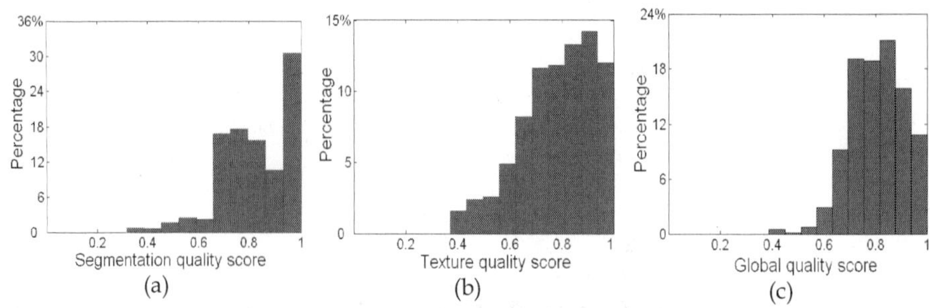

Fig. 14. Distribution of iris images based on quality index (a) Q_s (b) Q_t (c) Q

4.3.2.2 Evaluation of the proposed recognition system

After of the integration of the quality measures in the recognition system, we evaluated the system performance according to different quality thresholds. The tests are performed on iris image classes specified according to the quality thresholds defined experimentally. For each iris class, we calculated the similarity measures between all pairs of selected images. In Fig. 15, we have shown three distributions obtained by selection of images according to three quality thresholds Q: 70%, 75% and 80%.

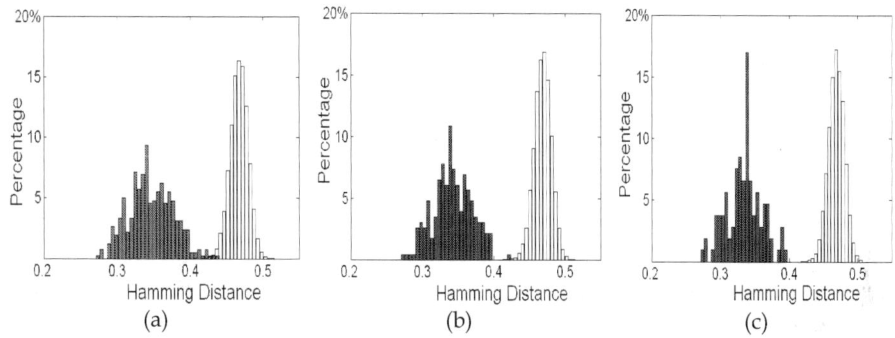

Fig. 15. Evolution of the distance distributions for different quality thresholds

By comparing these distributions with that obtained initially (Fig. 13-a) without considering the quality index, we can notice that the elimination of poor quality images improves the accuracy in the comparison. In fact, the overlap of two distributions (genuine and imposters) decreases by raising the quality level of selection. Since the threshold is set in this overlap area, we have to define a value for each class, then we verify the similarity of two irises by comparing the threshold to the distance between their feature vectors.
Based on the distributions obtained for different quality thresholds Q, we calculated the rate FAR, FRR, EER and MD for each distribution. We indicated in Table-2 the percentage of selected iris and the obtained rates for each quality threshold. We may notice an improvement of error rates by integrating the quality part in initial system and increasing the quality threshold value.

	Initial	S_1	S_2	S_3	S_4	S_5	S_6
% of database	100%	94%	85%	76%	70%	53%	36%
EER	0.68%	0.51%	0.4%	0.38%	0.16%	0.05%	0.02%
FAR/FRR=0	30.16%	28.95%	28.06%	27.22%	0.38%	0.04%	0.01%
FRR/FAR=0	3.02%	2.63%	1.48%	1.29%	0.98%	0.97%	0.85%
MD	4,67	4,83	5,01	5,04	5,17	5,34	5,52

Table 2. Reported results from proposed recognition system

5. Conclusion

In this chapter, we have presented a novel iris recognition system which combines Gabor filters and ULBP operators to characterize iris texture and a quality method to eliminate poor quality images. First, we have used Hough transform to segment iris. Then, we have evaluated different iris borders to generate a segmentation quality score. This score is combined with a quality texture score to define a global quality measure. So, we consider in the feature extraction step only iris images which are correctly segmented and has sufficient texture information for recognition. Then, we have applied a bank of Gabor filters to extract directional texture information of the accepted iris image, then, the real part of each Gabor transformed image is converted into ULGP index map and we have computed a set of statistical measures in local regions. These features are concatenated to form the complete feature vector. Finally, we have encoded relationship between values to generate an iris template of 1920 bits. The similarity between templates is estimated by computation of the Hamming distance.

The proposed method is tested on Casia v3 iris database. Our experiments illustrate the effectiveness of ULBP to extract rich local and global information of iris texture when combined with simultaneously multi-blocks and multi-channel method. Also, obtained results show an improvement of iris recognition system by incorporating proposed quality measures in the typical system.

6. References

A.C. Bovic, M.Clarck, W.S. Geiler (1990), *Multichannel texture analysis using localised spatial filters*, IEEE Transaction, Patt. Anal. Machine Intell, Vol 12, pp 55-73, 1990.

C. Belcher, Y. Du (2008), *A selective feature information approach for iris image quality measure*, IEEE Trans. Inf. Forensics Security, Vol. 3, N°. 3, pp. 572–577, 2008.

C. L. Tissé (2003), *Contribution à la Vérification Biométrique de Personnes par Reconnaissance de l'Iris*, Phd Thesis, Montpellier II University, Octobre 2003.

CASIA Iris database (2006), *www.sinobiometrics.com*, Chinese Academy of Sciences.

E. Krichen, A. Mellakh, S. Garcia-Salicetti, K. Hamrouni, N. Ellouze, B. Dorizzi (2004). *Iris Identification Using Wavelet Packet for Images in Visible Light illumination*, 1st International Conference on Biometric Authentication, China, July 15-17, 2004.

E. Krichen (2007), *Reconnaissance des personnes par l'iris en mode dégradé*, Phd Thesis, Ecole doctorale Sitevry / Evry-Val d'Essonne University, Octobre, 2007.

H. Proença, L.A. Alexandre (2006), *A Method for the Identification of Noisy Regions in Normalized Iris Images*, IEEE 18th International Conference on Pattern Recognition ICPR, Vol. 4, pp. 405-408, Hong Kong, August 2006.

H. Qureshi, O. Sertel, N. Rajpoot, R. Wilson, M. Gurcan (2008). Adaptive Discriminant Wavelet Packet Transform and Local Binary Patterns for Meningioma Subtype Classification, International Conference on Medical Image Computing and Computer-Assisted Intervention, pp.196-204, 2008.

H. Sung, J. Lim, J. Park, Y. Lee (2004), *Iris Recognition Using Collarette Boundary Localization*, 17th Int. Conf. on Pattern Recognition, Vol. 4, pp. 857-860, 2004

J. A. Montoya-Zegarra, J. Beeck, N. Leite,R. Torres, A. Falcao (2008). *Combining global with local texture information for image retrieval applications*, 10th IEEE International Symposium on Multimedia, 2008.

J. Cambier, U. C. Von Seelen (2006), *Multi-camera Iris Quality Study*, NIST Biometric Quality Workshop, 8-9 March, 2006.

J. Daugman (1994). Biometric personal identification system based on iris analysis, US PATENT, 5291560, Mars 1994.

J. Daugman (2007), *New Methods in Iris Recognition*, IEEE Trans. Systems, Man and Cybernics - Part B: Cybernics, Vol. 37, N°.5, pp. 1167–1175, October 2007.

J. Huang, L. Ma, T. Tan, Y. Wang (2007). *Learning Based Enhancement Model of Iris*, Scientific Literature Digital Library and SearchEngine, United States, 2007.

K. Matej, J. Pers, P. Matej, S. Kovaci (2006)c, *A Bayes-Spectral-Entropy-Based Measure of Camera Focus Using a Discrete Cosine Transform*, Int. Jour. Pattern Recognition Letters, Vol. 27, N°. 13, pp. 1431-1439, October 2006.

L. Ma, Y. Wang, T. Tan (2002). *Iris recognition using circular symmetric filters*, Proceeding of 16th International Conference on Pattern Recognition, Vol 2, , pp 414-417, August 11-15, 2002.

L. Ma, T. Tan, Y. Wang, D. Zhang (2003), *Personal Identification Based on Iris Texture Analysis*, IEEE Transactions on Pattern Analysis and Machine Intelligence, Vol. 25, N°. 12, December 2003.

L. Masek (2003). *Recognition of Human Iris Patterns for Biometric Identification*, Doctorate thesis, University of Western Australia, 2003.

M. El Aroussi (2009). *Information Fusion towards a Robust Face Recognition System*, Doctorate thesis, Sciences Faculty, Rabat, Maroc, 2009.

M. Vatsa, R. Singh, A. Noore (2008), *Improving Iris Recognition Performance Using Segmentation, Quality Enhancement, Match Score Fusion, and Indexing*, IEEE Transactions on Systems, Man and Cybernetics. Part B. Cybernetics Vol. 38, N° 4, pp. 1021-1035, 2008.

N. Ahmed, T. Natarajan, K. R. Rao (1974), *Discrete Cosine Transform*, IEEE Transactions on Computers, Vol. 23, pp. 90-93, January 1974.

N. D. Kalka, V. Dorairaj, Y. N. Shah, N. A. Schmid, B. Cukic (2006), *Image Quality Assessment for Iris Biometric*, Biometric Technology for Human Identification, Vol. 6202, pp. 1..11, Florida-USA, 17-18 April 2006.

N. Feddaoui, K. Hamrouni (2009). *An efficient and reliable algorithm for iris recognition based on Gabor filters*, Transactions on Systems, Signals & Devices, Vol.5, N°.3, pp.1-17, 2010.

N. Feddaoui, K. Hamrouni (2010). *Iris recognition based on multi-block Gabor statistical features encoding,* International Conference on *Soft Computing and Pattern Recognition* SoCPaR 2010, pp. 99-104, Cergy Pontoise/Paris France, 7-10 December, 2010.

R.P. Wildes (1997), *Iris recognition: an emerging biometric technology*, Proceedings of the IEEE , Vol. 85, N°. 9, pp. 1348 -1363, September 1997.

S. Chabrier, C. Rosenberger, B. Emile (2005), *Evaluation de la performance de la segmentation d'images par fusion de critères,* 9ème congrès jeunes chercheurs en Vision par ordinateur ORASIS, Clermont Ferrand, France, 2005.

S. Lim, K. Lee, O. Byeon, T. Kim (2001). *Efficient iris recognition through improvement of feature vector and classifier*, Electronics and Telecommunications Research Institute Journal, Vol. 23, N. 2, Korea, 2001.

S. P. Foliguet, L. Guigues (2006), *Evaluation of image segmentation:state of the art, new criteria and comparison,* Journal traitement de signal, Vol. 23, N°. 2, 2006.

T. Ojala, M. Pietikainen, and D. Harwood (1996), *A comparative study of texture measures with classification based on feature distributions*, Pattern Recognition, 1996, Vol. 29, N° 1, pp. 51-59. 1996

W. B. Pennebaker, J. L. Mitchell (1992), *JPEG Still Image Data Compression Standard,* 1st Kluwer Academic Publishers Norwell, USA 1992.

W. Ketchantang, S. Derrode, L. Martin, S. Bourennane (2007), *Nouveau Descripteur Local de Qualité des Images d'Iris dans les Séquences Vidéos*, GRETSI, Troyes, 11-14 septembre 2007.

X liu (2006), *Optimizations in Iris Recognition*, A Dissertation Submitted to the Graduate School of the University of Notre Dame in Partial Fulfillment of the Requirements for the Degree of Doctor of Philosophy in Computer Science, 2006.

X. Tan, B. Triggs (2007). *Fusing Gabor and LBP Feature Sets for Kernel-Based Face Recognition*, In: Analysis and Modeling of Faces and Gestures, pp. 235–249, 2007.

Y. Chen, S. C. Dass, A.K. Jain (2006), *Localized Iris Image Quality Using 2-D Wavelets*, International Conference on Biometrics, Vol.3832, pp.373-381, Chine, January 2006

Y. Du, C. Belcher, Z. Zhou, R. Ives, *Feature correlation evaluation approach for iris feature quality measure*, International Journal of Signal Processing, Vol. 90, N°. 4, pp. 1176–1187, April 2010.

Y. Wang, Z. Mu, H. Zeng (2008). *Block-based and Multi-resolution Methods for Ear Recognition Using Wavelet Transform and Uniform Local Binary Patterns*, 19th International Conference on Pattern Recognition, Dec. 2008.

Z. Zhang, Guang, Hua, Salganicoff, Marcos (1999), *Method of Measuring the Focus of Close-up Images of Eyes*, U.S. Patent 5 953 440, 14 September, 1999.

Z. Zhou, Y. Du, C. Belcher (2009), *Transforming Traditional Iris Recognition Systems to Work in Nonideal Situations*, IEEE Trans. on Industrial Electronics, Vol.56, N°.8, 2009.

13

Application of LCS Algorithm to Authenticate Users within Their Mobile Phone Through In-Air Signatures

Javier Guerra-Casanova, Carmen Sánchez-Ávila, Gonzalo Bailador-del Pozo and Alberto de Santos
Centro de Domótica Integral (CeDInt-UPM)
Universidad Politécnica de Madrid
Campus de Montegancedo, Madrid
Spain

1. Introduction

Authentication is one of the most important aspects regarding all the operations that may be performed from a mobile device. Starting up the device, making use of special functions, phoning reserved numbers, reading mail, accessing some Internet applications like e-commerce, electronic voting, e-learning, looking up the balance of a bank account or buying a product in an online shop are examples of operations that are nowadays performed from a mobile device and require authentication.

At present, most authentication procedures in mobile phones relies on handwritten passwords, with all their limitations. In this context, biometric techniques may offer a better solution to authenticate users according to their physical or behavioural characteristics.

Actually, there are already different approaches trying to join physical biometric techniques and mobile phones, as ho Cho et al. (2006), Jeong et al. (2005), Kurkovsky et al. (2010), Jeong et al. (2006) where users are authenticated through their iris or their faces Tao & Veldhuis (2006), yi Han et al. (2007), Ijiri et al. (2006). Besides, some work has been also developed with behavioral techniques in mobiles, authenticating users by means of their voice Shabeer & Suganthi (2007), Lapère & Johnson (1997), gait Mantyjarvi et al. (2005), Iso & Yamazaki (2006) or keystroke analysis Clarke & Furnell (2007), Saevanee & Bhatarakosol (2008).

In this chapter we propose to join handwritten signature, the most common biometric technique Nalwa (1997), in this mobile context. People are absolutely used to sign everyday when buying with their credit card, picking up a letter from the post office, authorizing operations on their name and lots of quotidian and legal scenarios else.

Consequently, we propose an adapted technique that allows people to authenticate themselves by a signature when they carry out, from their mobile phones, operations they used to perform in presence where they were used to authenticate themselves signing with their handwritten signature in a paper.

The biometric technique proposed consists of authenticating users when they execute an identifying gesture in the air (an in-air signature) while holding on their hand their mobile phone Guerra-Casanova et al. (2010). This identifying gesture should be easily remindful but complex enough to not being easily forgeable by other users. Creating a new personal gesture, easily to be repeated by the original user, but difficult to be reproduced by different people watching is not an easy task. Actually, most people who participated in this work chose their own handwritten signature as their identifying gesture, since it is a graph people is very used to repeat constantly.

To authenticate users within biometrics, it is necessary that users get enrolled in the system previously Jain et al. (2007). Actually, in the enrollment phase of the authentication technique based on gestures, users should repeat three times their identifying gesture to create their template of the gesture they will use as their signature to access the system.

When accessing the system, users only should repeat once their identifying gesture chose at enrollment phase. Then, the gesture performed is compared with the template stored, and if matches, the access is granted.

The main requirement of the technique is users should belong a mobile phone embedding an accelerometer since it is the sensor needed to extract the information of the performance of the gesture. This demand is not a problem since leader mobile phones manufacturers are marketing devices fulfilling this task with a very growing sales volume. It is expected that in several years, most mobile phones integrate an accelerometer resulting this proposed biometric technique accessible for most of the population. For example, Apple sold more than 4 million iPhone mobiles, embedding an accelerometer, just in the three first months of 2009 Steve Dowling (2009).

The in-air signature biometric technique provides several advantages:

- There are no additional widgets required to perform the signature, as any pen or any surface, so users only need their mobile phone to authenticate themselves by performing their signature in the air.
- The falsification of an in-air signature performed in this way is much harder to be forged as the gesture is performed in 3-D with no references as surfaces in handwritten signatures. Some experiments to verify this assumption are intended to be presented in this chapter.
- Users find this technique innovative but familiar so they are comfortable when performing their signature in this way whereas they are also grateful because they believe they are doing something novel.

According to these advantages, this authentication technique is expected to be widely accepted by people using their mobile phones to perform operations with a higher level of security. This assumption is based on the lack of invasiveness of the technique based on gestures, joined to the similarities to the high accepted and extended handwriting signature methods and the increase of security it provides Jain et al. (2002).

Although the in-air signature biometric technique seems quite similar to gesture recognition approaches, the point of view are radically different. In gesture recognition the crucial aim is to find a gesture from a database of known gestures that any person performs in a different manner, so it focus on finding similarities of samples of gestures. However, in our approach, we will find similar gestures performed by different people (for example one person trying to forge the authentic signature or another), but the main objective of this technique is to be able to differentiate from similar but different gestures, as they may come from impostor users Hsu et al. (2009), citeHe08.

This authentication technique provides an immediate application to the industry of mobile phones. It might be adopted to increase the security of different operations:

- Operations in the mobile phone: In this case, the template of the user should be stored in the mobile phone, and all the process to authenticate the user is executed in the device.
- Operations in a server: When users desire to execute an operation in a server through Internet, they should authenticate previously by performing their in-air signature holding their mobile. In this case, the in-air signature should be stored in the databases of the server, carrying out the authenticating process out of the mobile.
- E-commerce operations: Operations requiring authorization of the bank would be authenticated by performing an in-air signature. In this context, users should have been enrolled previously in the bank, in whose databases the in-air signatures are stored.

This chapter is divided into the following Sections:

- Section 2 describes how the in-air signature technique has been implemented on a mobile device. It starts with a motivation subsection where it is explained why a sequence alignment algorithm has been selected to analyze the in-air signatures. After that, Longest Common Subsequence algorithm is remembered to explain how has been adapted to this context in the different approaches evaluated in this chapter. Finally, it is described how the enrollment and authentication process is performed.
- Section 3 provides the description of the database of in-air signatures developed and considered in this chapter, as well as the results obtained for the different approaches assessed.
- Section 4 concludes the chapter with the conclusions obtained and the future work that may follow this study.

2. Implementation

This Section describes all the analysis method aspects included in the in-air signature biometric technique proposed in this work. It starts with the motivation of applying an algorithm based on obtaining the Longest Common Subsequence Bergroth et al. (2000). After that, a short review of LCS algorithm is included to be able to explain, afterwards, the generalization of LCS algorithm used in this work. Besides, another approach based on LCS is presented, consisting of using LCS to find the optimal global alignment and reconstruct two repetitions of an in-air signature to calculate the similarity value with a direct distance. Finally, the implementation of the algorithm in enrolling and verification phase is presented.

2.1 Motivation of applying Longest Common Subsequence algorithms to analyze acceleration signals of in-air signatures

Users authentication by means of gestures in the air involves a high intra-class deviation between the different performance of the in-air signatures, due to the fact that users are not able to repeat their identifying gestures with full precision. Consequently, applying a direct comparison method as Euclidean distance does not work properly.

Anytime users repeat their identifying gesture, they perform some parts of it faster or slower, more or less pronounced, or holding the mobile slightly different. In spite of all these variations, there exists an intrinsic part of the gestures remaining invariant that may be used to

recognize the person. Indeed, although users do not repeat exactly their identifying gesture, when they get used to repeat them, they are able to perform them naturally and in a quite similar manner but with some little differences.
Particularly, the following differences are found in acceleration signals when repeating gestures are found:

- Although the segmentation of the in-air signature is performed manually (there is a button to push at the start and stop of the performance of the in-air signature), the beginning of the gesture does not coincide. This happens because the time between users push the button and starts drawing their in-air signature is variable.
- In spite of performing the same in-air signature, there may exist peaks of acceleration more pronounced than others, relatives to more abrupt movements.
- Gestures do not long exactly the same, since any repetition is different from each other.
- Furthermore, it may happen that differences between repetitions occur only in certain parts of the performances of the gestures, where only some transitions are faster or slower than their respective.

Correcting most of those little deviations is the main aim of preprocessing acceleration signals of gestures through an alignment, keeping the intrinsic characteristics of the signals and rectifying slightly variations. As a consequence, in spite of the different, but quite similar, performance of the gesture, the system is able to assure the authenticity of the user.
According to this, the analysis algorithm required to compare the signals involved should correct slightly differences but this characteristic is as important as not compensating excessively. Otherwise, impostors would be able to imitate easily the gesture of another and forge the system, compromising the security of the authentication technique.
Indeed, the problem introduced to analyze acceleration signals of gestures has some analogies to global alignment algorithms, as Longest Common Subsequences, where the objective is to find the invariant information stored in two sequences independently of slightly differences. Consequently, both problems provide two signals to be compared which may have little differences but an important part of them remaining invariant.

2.2 LCS algorithm review

This algorithm provides directly a similarity value between two different sequences of values. LCS algorithm is based on finding the longest common subsequence of two sequences Wagner & Fischer (1974). This common subsequence is the one with lower edition distance, so both initial sequences can derived into the longest common subsequence within the lowest number of insertion and deletion operations Durbin et al. (2006).
Formally, a common subsequence of two sequences $\mathbf{v} = v_1 \ldots v_n$, $\mathbf{w} = w_1 \ldots w_m$ is defined as a sequence of positions in $\mathbf{v}$, so that $1 \leq i_1 \leq \ldots \leq i_k \leq n$ and a sequence of positions in $\mathbf{w}$, so that $1 \leq j_1 \leq \ldots \leq j_k \leq m$ fulfilling that the respective values of positions in $\mathbf{v}$ and $\mathbf{w}$ coincide, which means that, $v_{it} = w_{jt} for 1 \leq t \leq k$.
LCS algorithm is implemented defining a Matrix S that will be filled recursively with dynamic programming technique. The size of Matrix S is $n \times m$ and it is completed according to Equation 1:

$$s_{i,j} = \max \begin{cases} s_{i-i,j} + 0 \\ s_{i,j-1} + 0 \\ s_{i-1,j-1} + 1, \text{ if } v_i = w_j \\ s_{i-1,j-1} + 0, \text{ if } v_i \neq w_j \end{cases} \tag{1}$$

This equation includes the following aspects:

- The fist term of Equation 1 corresponds to the case when v_i is not present in the longest common subsequence of v and w (deletion in v_i or insertion in w_j).
- The second term represents the case when w_j is not present (deletion in w_j or insertion in v_i)
- The third term symbolizes the possible case when v_i and w_j are part of the longest common subsequence. Actually, when it happens that $v_i = w_j$ a 1 values is added in order to find the longest common subsequence.
- The forth term stands for the case when neither v_i nor w_j are part of the longest common subsequence (two deletions or insertions in v_i and w_j)

When matrix S is completed, the value of $\delta = s_{n,m}$ provides the length of the longest common subsequence. This value is itself a metric that compares the similarity of two sequences, the higher the more similar and vice versa.

2.3 Generalization of LCS algorithm

Classical LCS algorithm only assume that two points of the sequences in comparison belong to the longest common subsequence when their value is exactly equal. This premise is widely used when the alphabet of the values of the sequences is closed, which means that the possible values of the sequences are known and limited. For example, in genetic sequence alignment problems, the possible values of the genes are previously recognized and easily differentiabled. However, in the context of acceleration signals, quite similar values that belongs to the same point are classified as different, when it might not be.

According to this, a generalization of LCS algorithm is proposed by extending Equation 1 with Equation 2:

$$s_{i,j} = \max \begin{cases} s_{i-i,j} + 0 \\ s_{i,j-1} + 0 \\ s_{i-1,j-1} + 1, \text{ if } |v_i - w_j| \leq \psi \\ s_{i-1,j-1} + 0, \text{ if } |v_i - w_j| \nleq \psi \end{cases} \tag{2}$$

Consequently, in this approach it is not necessary that v_i and w_j are exactly the same values, but only with a lower difference than ψ.

This generalization may be represented by Equation 3:

$$s_{i,j} = \max \begin{cases} s_{i-i,j} + 0 \\ s_{i,j-1} + 0 \\ s_{i-1,j-1} + \zeta(v_i, w_j, \psi) \end{cases} \tag{3}$$

In Equation 3 ζ is a function of v_i and w_j according to Equation 4

$$\zeta(v_i, w_j, \psi) = \max \begin{cases} 1, if|v_i - w_j| \leq \psi \\ 0, if|v_i - w_j| \nleq \psi \end{cases} \quad (4)$$

According to this Equation, classical LCS algorithm is a particular method of this expression with $\psi = 0$. In this work, it has been proposed a step function to model ζ, however, some other functions may be useful in order to compare whether two points of two sequences belong to the longest common subsequence. Furthermore, an extension of this algorithm may be proposed including fuzzy logic so that two points have a percentage of probability to belong to the longest common subsequence, which may be modeled by a Gaussian, sigmoid or linear function.

2.4 An alignment and reconstructing approach

Previously, a metric based on the value of $s_{n,m}$ has been proposed to compare the similarity of two sequences. A different approach introduced in this work relies on utilizing the longest common subsequence as a method to find the optimal alignment between two sequences, and after that, trying to rebuild them in an optimal manner.
The longest common subsequence is obtained directly from matrix S. The procedure consists of finding the path joining element $s_{n,m}$ with $s_{1,1}$, considering that in the Equation of filling S:

- If the maximum value has been obtained through the first element, it correspond a $\leftarrow$ movement in S.
- If the maximum value has been obtained from the second element, the correspondent movement is vertical $\uparrow$.
- If the maximum value comes from the third element, it represents a diagonal movement in $S \nwarrow$.

The optimal alignment of two sequences when their longest common subsequence has been obtained consists of including a gap ("zero value") in v_i when a horizontal movement is required or a gap in w_j when the movement is vertical. Diagonal movements do not include gaps.
Then, every zero value is interpolated in order to reduce the differences between aligned signals. The interpolation method consists of substituting each zero value by the previous non-zero value of the sequence.
In this point, the initial sequences v and w have been aligned optimally and interpolated, deriving in v' and w'. The metric used to quantify the similarities between them implies calculating the absolute distance of v' and w' following Equation 5.

$$\delta = \sum_{i=0}^{L'} (v'_i - w'_i)^2 \quad (5)$$

Because of including gaps in the sequences in the optimal alignment procedure, the length of the signals may have increased to L', whose maximum values is $L' \leq m + n$.
In conclusion, after the analysis of two sequences, a score value δ has been obtained quantifying the similarity of both sequences. In the previous approach, the higher δ is the more similar sequences are, but in this one, it happens the opposite; the more similar sequences are those with a lower δ.

2.5 Application to in-air signatures

According to the previous subsections, when two sequences are compared a value δ is obtained in order to quantify the similarity of both sequences. This is also extensive to in-air signature acceleration signals, with only one consideration: Each in-air signature repetition consists of three signals, representing the accelerations on axes X, Y and Z.

In this work we propose to compare two repetitions of in-air signatures by evaluating the acceleration signal of each axis separately. Consequently, for each in-air signature comparison, three signal comparisons are required, deriving in three values δ_x, δ_y and δ_z representing the similarity of the signal of each axis separately.

Finally, the value of similarity of the complete signatures, Δ, is obtained as the average of the values obtained on each axis separately.

2.6 Implementation of enrolling process

Users enrolling with their mobile phones should repeat three times their identifying gesture, as precisely as they are able.

These three repetitions are analyzed by pairs, obtaining three values of the similarities between the different performances of their in-air signatures ($\Delta_{1,2}$, $\Delta_{1,3}$ and $\Delta_{2,3}$).

From these values, the parameter μ_T is calculated following Equation 6:

$$\mu_T = \frac{\Delta_{1,2} + \Delta_{1,3} + \Delta_{2,3}}{3} \tag{6}$$

Parameter μ_T provides information about the ability of the user to repeat his/her in-air signature, which would be essential in the access process to infer whether the accessing attempt belongs to the real user or not.

The in-air signature template of each user is composed by:

- The signals of accelerations on each axis of each repetition of the in-air signature the user performed at enrollment phase.
- Parameter μ_T representing the similarity between both three repetitions.

2.7 Implementation of accessing process

Users should authenticate themselves in their mobile phone by repeating once the in-air signatures they chose at enrolment phase.

This accessing in-air signature includes three acceleration signals, corresponding to the three axes X, Y and Z. This access attempt is compared with each of the three samples composing the template, deriving in three similarity values ($\Delta_{A,1}$, $\Delta_{A,2}$ and $\Delta_{A,3}$) obtained as described previously.

Finally, a global score is calculated including the similarities with each sample of the template as well as parameter μ_T obtained at enrollment phase. This global score follows Equation 7

$$\Psi = \frac{\Delta_{A,1} + \Delta_{A,2} + \Delta_{A,3}}{3\mu_T} \tag{7}$$

Finally, depending on the behavior of the similarity values, the accessing attempt is rejected or accepted according to a threshold value Θ following these considerations:

- If a high δ stands for high similarity, the accessing attempt is accepted when $\Psi > \Theta$, otherwise is rejected.

- If a high δ represents low similarity, the accessing attempt is accepted when $\Psi < \Theta$, otherwise is rejected.

It is remarkable that the selection of an optimal Θ is crucial in order to reduce false positives and negatives errors. In this approach, Θ value will be set up to the value of Equal Error Rate, when the False Rejection Rate is equal to the False Acceptance Rate. In spite of this, Θ value might be modified in order to reduce one of the rates at the expense of increase the other. This might be interesting if users do not care about repeating sometimes twice their in-air signature.

3. Experiments

3.1 Database

According to the knowledge of the authors, there are no public databases of in-air signatures performed a mobile embedding an accelerometer. Therefore, a private database has been created in order to obtain a significant number of samples in order to evaluate the algorithms previously proposed.
This database contains in-air signatures of 50 different users, who have created their identifying gesture trying to select in-air signatures easily remindful and complex enough to not be forged automatically. Most of the users chose to perform in the air their own handwritten signature as their identifying gesture in this biometric technique.
Users repeated 8 times their in-air signature in front of a video camera. Afterwards, 6 people tried to falsify each original in-air signature from studying those records. Each falsifier tried 7 times to forge each in-air signature.
All of the original and falsifying samples of the database have been obtained sampling the in-air signatures at a rate of 50 Hz.

3.2 Results

From the analysis of the samples in the database created, an "active impostor attack" scenario has been represented, to evaluate the performance of the in-air signature biometric technique with real attempts of falsification.
Three random samples of each user would be considered as the repetitions performed at enrollment phase. With these samples, the enrollment procedure is execute, obtaining parameter μ_T for each user. The rest of samples of each user would be considered as original access attempts whereas the falsifying samples of each user will symbolize fraudulent access attempts to the system.
Each experiment presented in this work has been repeated five times; each repetition implies a different selection of the samples composing the template of each user. Results present the average and deviation of all of them.
The metric used to evaluate the performance of the algorithm will be Equal Error Rate Wayman et al. (2004). This rate is obtained as follows:

- Analysis of original samples: The original access attempt samples of each gesture are used to calculate False Rejection Rate, since they are authentic attempts of accessing the system. For each original trial Ψ is obtained when comparing the accessing gesture with the three gestures of the original user template, considering the correspondent μ_T of the user.

- Analysis of falsified samples: All the impostor attempts trying to access the system are used to evaluate False Acceptance Rate. For each falsification trial, Ψ is also obtained as the value of comparing the sample with the template considering, as well, parameter μ_T of the user.
- Obtention of False Acceptance Rate (FAR) and False Rejection Rate (FRR): FAR and FRR are obtained in terms of Θ as follows:
  - If a high δ value stands for high similarity, the % of original samples that are under Θ in case of False Rejection Rate and the % of falsified samples that are over Θ in case of False Acceptance Rate. In this case, it is accomplished that when Θ is very high, most falsifications are rejected but so do some original access. However, the lower Θ, the more original access are authentic allowed but also the more falsifications are granted.
  - If a low δ value stands for high similarity, the % of original samples that are over Θ in case of False Rejection Rate and the % of falsified samples that are under Θ in case of False Acceptance Rate. In this case, the behaviour of FAR and FRR in respect with Θ is the opposite.
- Obtention of Equal Error Rate (EER): EER is defined as the value of the error when False Acceptance Rate is equal to False Rejection Rate, and it is the metric commonly used to measure the performance of the biometric technique.

Usually, the performance of the biometric systems is represented by a Receiver Operating Characteristic figure (ROC) Fawcett (2006), where axis X represents False Matching Rate (FMR) and axis Y True Matching Rate (TMR). When Failure-to-acquire (FTA) rate is 0 (as in the experiments presented in this article, since the samples evaluated come from a closed database), it is verified that FMR=FAR and FNMR=FRR. Besides, FNMR is defined as FNMR=1-TMR. Consequently, when FTA=0, it is equivalent to represent a ROC figure within FMR vs. TMR and FAR vs. (1-FRR). Moreover, Equal Error Rate is defined as the intersection of the line where FAR=FRR so it is also equivalent to calculate EER as 1-EER′ considering EER′ the intersection between ROC curve and FAR=1-FRR line.
Assuming all these considerations, the following approaches to calculate the similarity score between two acceleration signals are evaluated:

- Calculating similarity score applying LCS algorithm.
- Calculating similarity score through absolute distance, aligning previously the sequences with LCS.
- Calculating similarity score applying Generalized LCS algorithm.
- Calculating similarity score through absolute distance, aligning previously the sequences with Generalized LCS.

3.2.1 Calculating similarity score applying LCS algorithm

When applying Longest Common Subsequence algorithm to analyze the in-air signatures of the database created as explained in Section 2.2, the following results are obtained.
In Table 1 the Equal Error Results obtained on each of the repetition of the experiment is presented, as well as the average value:
These values are represented in Figure 1:
An average value of EER of 5.28±0.63% is finally obtained with this approach.

EER_1	EER_2	EER_3	EER_4	EER_5	$EER_{average}$	EER_{std}
5.37%	6.09%	5.22%	5.40%	4.33%	5.28%	0.63%

Table 1. EER results when obtaining the similarity score of in-air signatures applying LCS algorithm

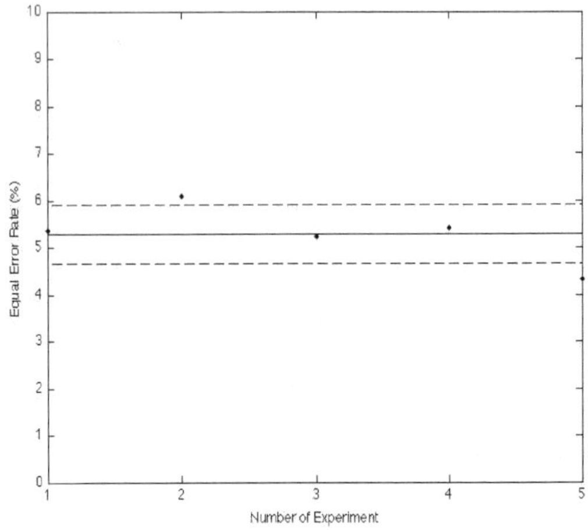

Fig. 1. Representation of EER results when obtaining the similarity score of in-air signatures applying LCS algorithm

EER_1	EER_2	EER_3	EER_4	EER_5	$EER_{average}$	EER_{std}
14.51%	10.63%	12.48%	14.09%	12.96%	12.93%	1.52%

Table 2. EER results when finding the optimal alignment within LCS algorithm, interpolating and calculating absolute distance as the similarity score of in-air signatures

3.2.2 Calculating similarity score through absolute distance, aligning previously the sequences with LCS

When utilizing LCS algorithm to find the optimal alignment of two in-air signatures, and then interpolating as explained in 2.4, the results of each repetition of the experiment are presented in Table 2:

These values are represented in Figure 4:

This approach obtains an EER value of 12.93±1.52 %, much worse than obtaining directly the score of LCS.

3.2.3 Calculating similarity score applying Generalized LCS algorithm

One more approach previously explained is based on generalize LCS algorithm with a step function $\zeta(v_i, w_j, \psi)$. In Table 3 the average EER results for different values of ψ are presented,

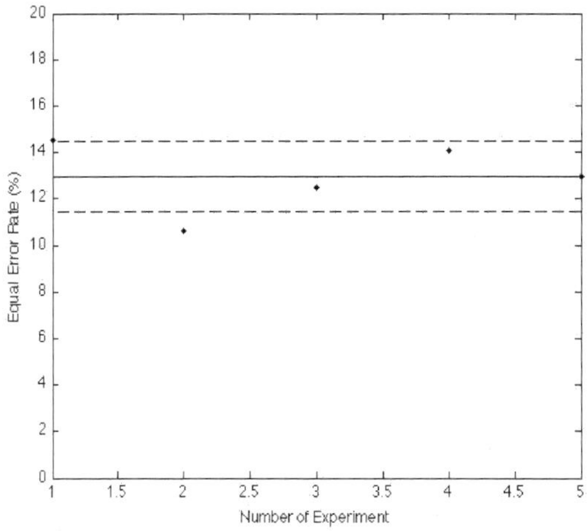

Fig. 2. Representation of EER results when obtaining the similarity score of in-air signatures applying LCS algorithm to align the signals, interpolating and calculating the absolute distance of the signals

ψ	$EER(\%)$
0.05	5.65±0.47
0.1	3.94±0.21
0.15	3.58±0.78
0.2	4.27±0.52
0.25	5.05±1.12
0.3	4.92±0.45
0.35	5.50 ±0.84
0.4	6.41±0.45
0.45	6.40±0.31

Table 3. EER results when obtaining the similarity score of in-air signatures applying a generalized LCS algorithm

utilizing LCS algorithm as a direct manner to obtain a similarity score between two in-air signature repetitions.

These values including average and deviation results are represented in Figure 3:

Some configurations of generalized LCS improve the performance results of classic LCS. In particular, an optimal EER value of 3.58±0.78 has been obtained when utilizing $\psi = 0.15$ as the maximum value that two points are considered that belongs to the longest common subsequence.

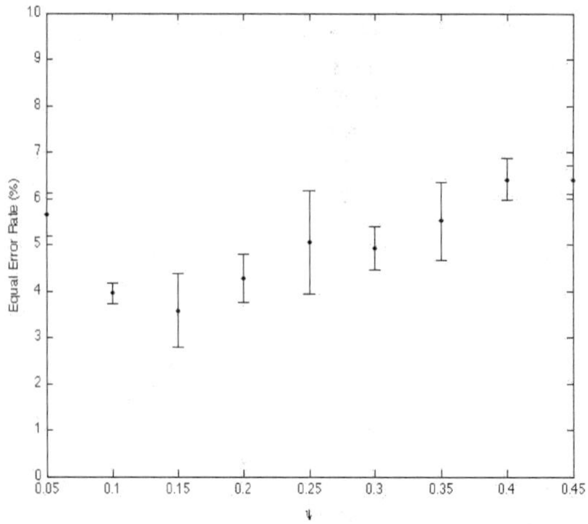

Fig. 3. Representation of EER results when obtaining the similarity score of in-air signatures applying a generalized LCS algorithm

ψ	$EER(\%)$
0.05	9.12±1.05
0.1	6.63±1.07
0.15	6.11±0.82
0.2	6.24±0.89
0.25	5.67±0.63
0.3	6.25±0.81
0.35	7.08±0.87
0.4	6.84±1.25
0.45	7.06±0.98

Table 4. EER results when finding the optimal alignment within a generalized LCS algorithm, interpolating and calculating absolute distance as the similarity score of in-air signatures

3.2.4 Calculating similarity score through absolute distance, aligning previously the sequences with generalized LCS

The generalization of LCS depending on ψ has been also applied to the optimal alignment and interpolation approach, obtaining the average EER results presented in Table 4:
These values with their respective deviations are represented in Figure 4:

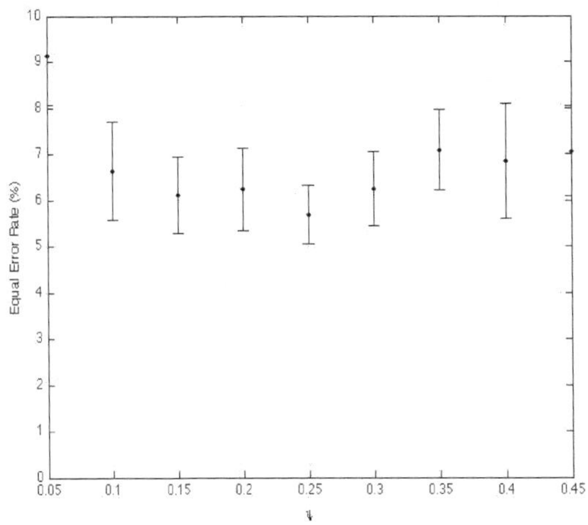

Fig. 4. Representation of EER results when obtaining the similarity score of in-air signatures applying a generalized LCS algorithm to align the signals, interpolating and calculating the absolute distance of the signals

In this case, utilizing a $\psi > 0$ value in the generalization of LCS implies a much better performance in respect to utilizing LCS to align and calculating afterwards a direct distance. In spite of this, the results when obtaining the score directly from the LCS algorithm are much better than this interpolating and calculating distance approach.

4. Conclusion and future work

Security in mobile devices may take advantage of the use of biometrics in order to authenticate the identity of the real person behind a mobile device. Nowadays, most of the applications requiring authentication rely on the use of passwords, with all their limitations.

At present, there are already other works trying to utilize biometric characteristics in mobile devices to authenticate users. In this article, a handwritten signature technique adapted to mobiles is proposed. This biometric technique is based on recognizing an identifying gesture carried out in the air. To accomplish this aim, users are authenticated by a gesture they perform moving their hand holding an accelerometer-embedded mobile device.

Authentication procedure requires uses to be enrolled in the system by repeating three times their in-air signature, invented by them. Afterwards, they are able to entry the system by performing it again.

As users are not able to repeat their in-air signatures accurately, different algorithm based on sequence alignment have been proposed to correct slightly differences between different repetitions of a gesture and provide a metric to quantify the similarities between them.

In particular, we have utilized four approaches based on the Longest Common Subsequence:

- The LCS algorithm to obtain a score of similarity of two sequences.

- The LCS algorithm to perform an optimal global alignment between two sequences, an interpolation the gaps previously introduced and a calculation of the absolute distance.
- A generalized LCS algorithm to obtain the score.
- A generalized LCS to align, interpolate and calculate absolute distance.

All these approaches have been evaluated within a database of 50 users who repeated 8 times their in-air signature holding a mobile device, and 6 people trying to forge each original gesture from video records.
From the results presented it can be concluded that:

- The performance of the LCS algorithm to obtain the score is better than to align the signals, interpolate and calculate absolute distance.
- Including the generalization proposed improves the results of classical LCS algorithm.
- An optimal EER value of 3.58 ± 0.78 (%) has been obtained with utilizing the generalized ($\psi = 0.15$) LCS algorithm to obtain the similarity score to compare two performances of in-air signatures.

As future work, some other time series distances, as Dynamic Time Warping Berndt & Clifford (1994), may be used or other different approaches based on statistical method should be evaluated Suk et al. (2010), Lee & Kim (1999).

5. References

Bergroth, L., Hakonen, H. & Raita, T. (2000). A survey of longest common subsequence algorithms, *String Processing and Information Retrieval, 2000. SPIRE 2000. Proceedings. Seventh International Symposium on*, pp. 39–48.

Berndt, D. J. & Clifford, J. (1994). Using dynamic time warping to find patterns in time series, *KDD Workshop*, pp. 359–370.

Clarke, N. & Furnell, S. (2007). Authenticating mobile phone users using keystroke analysis, *International Journal of Information Security* 6: 1–14.

Durbin, R., Eddy, S., Krogh, A. & Mitchison, G. (2006). *Biological sequence analysis: Probabilistic Models of Proteins and Nucleic Acids*, eleventh edn, Cambridge University Press.

Fawcett, T. (2006). An introduction to roc analysis, *Pattern Recogn. Lett.* 27: 861–874.

Guerra-Casanova, J., Sánchez-Ávila, C., de Santos-Sierra, A., del Pozo, G. B. & Jara-Vera, V. (2010). A real-time in-air signature biometric technique using a mobile device embedding an accelerometer., *in* F. Zavoral, J. Yaghob, P. Pichappan & E. El-Qawasmeh (eds), *NDT (1)*, Vol. 87 of *Communications in Computer and Information Science*, Springer, pp. 497–503.

ho Cho, D., Park, K. R., Rhee, D. W., Kim, Y. & Yang, J. (2006). Pupil and iris localization for iris recognition in mobile phones, *Software Engineering, Artificial Intelligence, Networking and Parallel/Distributed Computing, International Conference on & Self-Assembling Wireless Networks, International Workshop on* 0: 197–201.

Hsu, W.-H., Chiang, Y.-Y., Lin, W.-Y., Tai, W.-C. & Wu, J.-S. (2009). Integrating lcs and svm for 3d handwriting recognition on handheld devices using accelerometers, *Proceedings of the 3rd International Conference on Communications and information technology*, CIT'09, World Scientific and Engineering Academy and Society (WSEAS), Stevens Point, Wisconsin, USA, pp. 195–197.

Ijiri, Y., Sakuragi, M. & Lao, S. (2006). Security management for mobile devices by face recognition, *Mobile Data Management, 2006. MDM 2006. 7th International Conference on*, pp. 49 – 49.

Iso, T. & Yamazaki, K. (2006). Gait analyzer based on a cell phone with a single three-axis accelerometer, *MobileHCI '06: Proceedings of the 8th conference on Human-computer interaction with mobile devices and services*, ACM, New York, NY, USA, pp. 141–144.

Jain, A. K., Flynn, P. & Ross, A. A. (2007). *Handbook of Biometrics*, Springer-Verlag New York, Inc., Secaucus, NJ, USA.

Jain, A. K., Griess, F. D. & Connell, S. D. (2002). On-line signature verification, *Pattern Recognition* 35(12): 2963 – 2972.

Jeong, D., Park, H.-A., Park, K. & Kim, J. (2005). Iris recognition in mobile phone based on adaptive gabor filter, *in* D. Zhang & A. Jain (eds), *Advances in Biometrics*, Vol. 3832 of *Lecture Notes in Computer Science*, Springer Berlin Heidelberg, pp. 457–463.

Jeong, D. S., Park, H.-A., Park, K. R. & Kim, J. (2006). Iris recognition in mobile phone based on adaptive gabor filter, *Advances in Biometrics, International Conference, ICB 2006, Hong Kong, China, January 5-7, 2006, Proceedings*, pp. 457–463.

Kurkovsky, S., Carpenter, T. & MacDonald, C. (2010). Experiments with simple iris recognition for mobile phones, *Information Technology: New Generations, Third International Conference on* 0: 1293–1294.

Lapère, M. & Johnson, E. (1997). User authentication in mobile telecommunication environments using voice biometrics and smartcards, *IS&N '97: Proceedings of the Fourth International Conference on Intelligence and Services in Networks*, Springer-Verlag, London, UK, pp. 437–443.

Lee, H.-K. & Kim, J. H. (1999). An hmm-based threshold model approach for gesture recognition, *IEEE Transactions on Pattern Analysis and Machine Intelligence* 21: 961–973.

Mantyjarvi, J., Lindholm, M., Vildjiounaite, E., Makela, S. M. & Ailisto, H. A. (2005). Identifying users of portable devices from gait pattern with accelerometers, *Acoustics, Speech, and Signal Processing, 2005. Proceedings. (ICASSP '05). IEEE International Conference on*, Vol. 2, pp. ii/973–ii/976 Vol. 2.

Nalwa, V. S. (1997). Automatic on-line signature verification, *Proceedings of the IEEE*, pp. 215–239.

Saevanee, H. & Bhatarakosol, P. (2008). User authentication using combination of behavioral biometrics over the touchpad acting like touch screen of mobile device, *Computer and Electrical Engineering, International Conference on* 0: 82–86.

Shabeer, H. A. & Suganthi, P. (2007). Mobile phones security using biometrics, *Computational Intelligence and Multimedia Applications, International Conference on* 4: 270–274.

Steve Dowling, Nancy Paxton, J. H. (2009). Apple reports first quarter results. URL: *http://www.apple.com/pr/library/2009/01/21results.html*

Suk, H. I., Sin, B. K. & Lee, S. W. (2010). Hand gesture recognition based on dynamic bayesian network framework, *Pattern Recogn.* 43: 3059–3072.

Tao, Q. & Veldhuis, R. (2006). Biometric authentication for a mobile personal device, *Mobile and Ubiquitous Systems, Annual International Conference on* 0: 1–3.

Wagner, R. A. & Fischer, M. J. (1974). The string-to-string correction problem, *J. ACM* 21: 168–173.

Wayman, J. L., Jain, A. K., Maltoni, D. & Maio, D. (2004). *Biometric Systems: Technology, Design and Performance Evaluation*, Springer-Verlag New York, Inc., Secaucus, NJ, USA.

yi Han, S., Park, H.-A., Cho, D. H., Park, K. R. & Lee, S. (2007). Face recognition based on near-infrared light using mobile phone, *Adaptive and Natural Computing Algorithms, 8th International Conference, ICANNGA 2007, Warsaw, Poland, April 11-14, 2007, Proceedings, Part II*, pp. 440–448.

14

Performance Comparison of Principal Component Analysis-Based Face Recognition in Color Space

Seunghwan Yoo[1], Dong-Gyu Sim[2], Young-Gon Kim[1] and Rae-Hong Park[1]
[1]Sogang University,
[2]Kwangwoon University,
South Korea

1. Introduction

Light reflected from an object is multi-spectral, and human beings recognize the object by perceiving color spectrum of the visible light (Wyszecki & Stiles, 2000). However, most of face recognition algorithms have used only luminance information (Bartlett et al., 2002; Belhumeur et al., 1997; Etemad & Chellappa, 1997; Liu & Wechsler, 2000; Turk & Pentland, 1991a, 1991b; Wiskott et al., 1997; Yang, 2002). Many face recognition algorithms convert color input images to grayscale images by discarding their color information.

Only a limited number of face recognition methods made use of color information. Torres et al. proposed a global eigen scheme to make use of color components as additional channels (Torres et al., 1999). They reported color information could potentially improve performance of face recognition. Rajapakse et al. proposed a non-negative matrix factorization method to recognize color face images and showed that the color image recognition method is better than grayscale image recognition approaches (Rajapakse et al., 2004). Yang et al. presented the complex eigenface method that combines saturation and intensity components in the form of a complex number (Yang et al., 2006). This work shows that the multi-variable principal component analysis (PCA) method outperforms traditional grayscale eigenface methods. Jones III and Abbott showed that the optimal transformation of color space into monochrome form can improve the performance of face recognition (Jones III & Abbott, 2004), and Neagoe extended the optimal transformation to two-dimensional color space (Neagoe, 2006).

Color images include more visual clues than grayscale images, and the above-mentioned work showed effectiveness of color information for face recognition. However, there is lack of analysis and evaluation regarding the recognition performance in various color spaces. A large number of face recognition algorithms (Bartlett et al., 2002; Belhumeur et al., 1997; Etemad & Chellappa, 1997; Liu & Wechsler, 2000; Turk & Pentland, 1991a, 1991b; Wiskott et al., 1997; Yang, 2002) have been presented.

This paper is an extended version of the paper (Yoo et al., 2007), in which analysis of the recognition rate in various color spaces with two different approaches in CMU PIE database (Sim et al., 2003; Zheng et al., 2005) and color FERET database (Phillips et al., 1998, Phillips et al., 2000) is supplemented. Note that PCA-based algorithms are employed since they are the most fundamental and prevalent approaches. Recognition performance is evaluated in various

color spaces with two different approaches (independent and concatenated processing). SV, RGB, YCg'Cr', YUV, YCbCr, and YCgCb color spaces are used for investigation of performance analysis. Experimental results show that use of color information can give significant improvement in terms of the recognition rate in CMU and FERET database which contain a large number of face images with wide variation of illumination, facial expressions, and aging for test sets. To use color information for PCA-based face recognition, we adopt two kinds of approaches: independent and concatenated PCA-based face recognition.
The rest of the paper is organized as follows. In Section 2, a fundamental eigenface method is introduced. In Section 3, two schemes for color PCA-based face recognition are introduced and in Section 4, six color spaces for face recognition are described. Performance comparison of the face recognition for six color spaces is presented in Section 5. Finally, Section 6 gives conclusions and future work.

2. Eigenface face recognition

Turk and Pentland proposed the eigenface-based face analysis that is based on the PCA for efficient face recognition (Turk & Pentland, 1991a, 1991b). The algorithm consists of two phases: training and recognition phases. In the training phase of the eigenface method, eigenvectors are calculated with a large number of training faces. The computed eigenvectors are called as eigenfaces. Then, faces are enrolled in a face recognition system by their projection onto the eigenface space. In the recognition phase, an unknown input face can be identified by measuring the distances of the projected coefficients between the input face and the enrolled faces in database.

2.1 Eigenface space decomposition

Dimension of an image space is so high that it is often not only impractical but also inefficient to deal with all the data of images in their own dimensions. PCA enables to optimally reduce the dimensionality of images by constructing the eigenface space which is composed of eigenvectors (Turk & Pentland, 1991a, 1991b). An algorithmic procedure of eigenface decomposition is briefly described in the following.
Let $\{x_1, x_2, \ldots, x_{M_t}\}$ be a training set of face images, and $\mathbf{x}_i$ represent the ith training face image which is expressed as an N×1 vector. Note that M_t signifies the number of training images and N denotes the total number of pixels in an image. The mean vector $\boldsymbol{\mu}$ of the dataset is defined by

$$\boldsymbol{\mu} = \frac{1}{M}\sum_{i=1}^{M_t} x_i . \quad (1)$$

Then, the N×N covariance matrix C of the dataset is computed by

$$C = \frac{1}{M}\sum_{i=1}^{M_t}(x_i - \boldsymbol{\mu})(x_i - \boldsymbol{\mu})^T , \quad (2)$$

where the superscript T denotes a transpose operation. The eigenvalues and the corresponding eigenvectors of C can be computed with the singular value decomposition (SVD). Let $\lambda_1, \lambda_2, \ldots, \lambda_N$ be eigenvalues of C, where the eigenvalues are ordered in decreasing order, and $\mathbf{u}_1, \mathbf{u}_2, \ldots, \mathbf{u}_N$ represent N eigenvectors of C. Note that the ith eigenvalue, λ_i is associated with the ith eigenvector, $\mathbf{u}_i$. The eigenvectors having larger λ_i are considered as be more dominant axis to represent the training face images. We can choose N' eigenvectors as the eigenface space for face recognition ($N' << N$).

2.2 Projection onto the eigenface space

A face image is transformed by projecting it onto the eigenface space. Let $\{\mathbf{y}_1, \mathbf{y}_2, ..., \mathbf{y}_{M_g}\}$ be a gallery set of face images, where M_g is the size of the gallery set. Then, the weight ω_{ik} of $\mathbf{y}_i$ with respect to the kth eigenface can be obtained by

$$\omega_{ik} = \mathbf{u}_k^t(\mathbf{y}_i - \boldsymbol{\mu}) \tag{3}$$

and all the weights are represented by a weight vector, $\boldsymbol{\Omega}_i = [\omega_{i1}\ \omega_{i2}\ \cdots \omega_{iN'}]^T$.

2.3 Classification

Given an unknown face image, we obtain the weight vector, $\boldsymbol{\Omega} = [\omega_1\ \omega_2\ \cdots \omega_{N'}]^T$, by projecting it onto the eigenface space. Then, the input face image can be classified using the nearest neighborhood classifier. The distances between the input face and the other faces in the gallery are computed in the eigenface space. The Euclidean distance between the input face and the ith face image in the gallery set is defined by

$$d_e(\boldsymbol{\Omega}, \boldsymbol{\Omega}_i) = \sum_{k=1}^{N'} |\omega_k - \omega_{ik}| , \tag{4}$$

whereas the Mahalanobis distance is defined by

$$d_m(\boldsymbol{\Omega}, \boldsymbol{\Omega}_i) = \sum_{k=1}^{N'} \frac{1}{\sqrt{\lambda_k}} |\omega_k - \omega_{ik}| . \tag{5}$$

The identity of the input face image can be determined by finding the minimum distance with a distance measure such as the above-mentioned distance function. The decision rule for face recognition can be expressed as

$$i_{matching} = \arg \min_{1 \le i \le M_g} d(\boldsymbol{\Omega}, \boldsymbol{\Omega}_i) , \tag{6}$$

where $i_{matching}$ is the index indicating the identified person.

3. Face recognition in different color spaces

In general, color images have three components or channels: red (R), green (G), and blue (B). To apply the eigenface method to color facial images, two methods are employed. One way is to combine outcomes of independent PCA for each color component (independent processing), whereas the other is to serially concatenate three color components into a single component (concatenated processing). In this section, we will describe these two approaches for face recognition in different color spaces.

3.1 Independent color face recognition

Each color component of a signal can be independently fed into an eigen-face method, as shown in Fig. 1(a). The final decision is made with the distances from three independent eigenface modules (Torres et al., 1999). Fig. 1(a) shows the block diagram of the face recognition system (independent processing) for multi-channel face images. First, color space conversion is performed, i.e., three components of RGB color space, $\mathbf{x}_R$, $\mathbf{x}_G$, and $\mathbf{x}_B$, are converted into three other color components $\mathbf{x}_{C1}$, $\mathbf{x}_{C2}$, and $\mathbf{x}_{C3}$. At the second stage, the eigenface analysis is performed for each component independently. Then, the three distance vectors, $\mathbf{d}_{C1}$, $\mathbf{d}_{C2}$, and $\mathbf{d}_{C3}$ are consolidated with weighting factors and a person in the database is finally identified.

3.2 Concatenated color face recognition

The simple way to process a multi-channel signal is to concatenate independent multiple components into a single component (concatenated processing) and process it as if it is obtained from a single channel, as shown in Fig. 1(b). $\mathbf{x}_R$, $\mathbf{x}_G$, and $\mathbf{x}_B$ are N×1 vectors, denoting red, green, and blue components of an input face image, respectively, while $\mathbf{x}_C$ is a $3N$×1 vector, representing a serially combined input for a color eigenface system. $\mathbf{d}_C$ is an M_g×1 vector that represents the distance between the input and M_g persons in a gallery. In this way, the multiple-component signal is converted into a single channel signal. The number of components becomes one, whereas the length of the component increases as many times as the original number of components. Then, the eigenface method is applied to the combined signal.

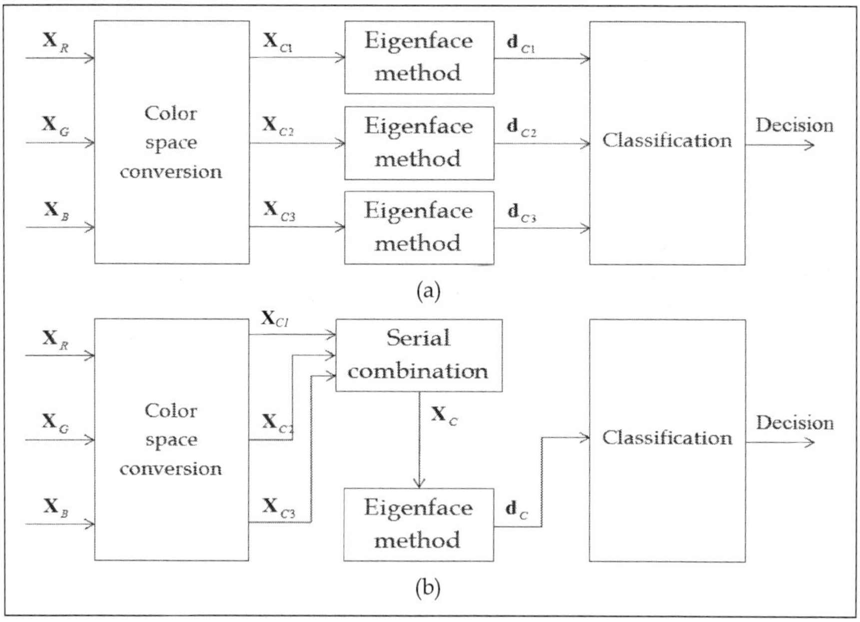

Fig. 1. Block diagram of the face recognition system using color information. (a) Independent processing, (b) Concatenated processing.

In the case of color images which consist of three channels, (RR…), (GG…), and (BB…), the concatenated signal will be expressed as (RGBRGB…).

4. Color spaces for face recognition

Even though most of digital image acquisition devices produce R, G, and B components, the RGB color space is converted into different color spaces for each application. For face recognition, the eigenface analysis in the RGB color space domain is known not to be effective, because R, G, and B components are largely correlated with each other. Some literatures also pointed that the RGB domain is inadequate for face recognition (Torres et al., 1999). Instead of RGB color space, other color spaces that are less correlated between their components should be investigated for face recognition. In this work, performance evaluation is conducted on SV, RGB, YCg'Cr', YUV, YCbCr, and YCgCb color spaces.
The HSV and HSI color spaces are the well-known color spaces reflecting the human visual perception and they are composed of hue (H), saturation (S), and value (V)/ intensity (I) (Jack, 2001). The conversion equations are given by

$$H = \begin{cases} \theta, & \text{if } B \le G \\ 360 - \theta, & \text{if } B \le G \end{cases} \tag{7}$$

$$S = 1 - \frac{3}{(R+G+B)}\left[\min(R,G,B)\right] \tag{8}$$

$$V = \max(R,G,B), \quad I = \frac{R+G+B}{3} \tag{9}$$

where θ is computed by

$$\theta = \cos^{-1}\left\{ \frac{0.5\left[(R-G)+(R-B)\right]}{\left[(R-G)^2+(R-B)(G-B)\right]^{1/2}} \right\}. \tag{10}$$

The YUV color space consisting of luminance (Y) and chrominance (U, V) components has been widely used for video transmission systems. The black-and-white video systems use only Y information and U and V components are added for color systems. RGB to YUV conversion can be performed by

$$\begin{bmatrix} Y \\ U \\ V \end{bmatrix} = \begin{bmatrix} 0.299 & 0.587 & 0.114 \\ -0.148 & -0.291 & 0.436 \\ 0.615 & -0.515 & -0.100 \end{bmatrix} \begin{bmatrix} R \\ G \\ B \end{bmatrix}. \tag{11}$$

The YCbCr color space is an alternative to the YUV color space by employing an offset value for each component. It is used for multiple coding standards. This color space is also known as an effective space for skin color segmentation (Chai & Ngan, 1999) and the conversion matrix is defined by

$$\begin{bmatrix} Y \\ Cb \\ Cr \end{bmatrix} = \begin{bmatrix} 0.257 & 0.504 & 0.098 \\ -0.148 & -0.291 & 0.439 \\ 0.439 & -0.368 & -0.071 \end{bmatrix} \begin{bmatrix} R \\ G \\ B \end{bmatrix} + \begin{bmatrix} 16 \\ 128 \\ 128 \end{bmatrix}. \tag{12}$$

The YIQ color space is related to the YUV color space. The 'I' represents 'inphase' and the 'Q' does 'quadrature', which is based on quadrature amplitude modulation. I and Q from U and V are computed by

$$\begin{bmatrix} I \\ Q \end{bmatrix} = \begin{bmatrix} 0 & 1 \\ 1 & 0 \end{bmatrix} \begin{bmatrix} \cos 33^\circ & \sin 33^\circ \\ -\sin 33^\circ & \cos 33^\circ \end{bmatrix} \begin{bmatrix} U \\ V \end{bmatrix}. \tag{13}$$

The YCgCr color space was proposed for fast face segmentation (De Dios & Garcia, 2003). This color space produces another chrominance component Cg instead of Cb in YCbCr. Moreover, the YCg'Cr' color space was derived by rotating the CgCr plane for face segmentation (Dios & Garcia, 2004). YCgCr and YCg'Cr' are defined by

$$\begin{bmatrix} Y \\ Cg \\ Cr \end{bmatrix} = \begin{bmatrix} 0.257 & 0.504 & 0.098 \\ -0.3178 & 0.438 & -0.121 \\ 0.439 & -0.368 & -0.071 \end{bmatrix} \begin{bmatrix} R \\ G \\ B \end{bmatrix} + \begin{bmatrix} 16 \\ 128 \\ 128 \end{bmatrix}. \tag{14}$$

$$\begin{aligned} Cg' &= Cg \cos 30^\circ + Cr \sin 30^\circ - 48 \\ Cr' &= -Cg \sin 30^\circ + Cr \cos 30^\circ + 80 \end{aligned}. \tag{15}$$

The YCgCb color space was also proposed for face segmentation (Zhang & Shi, 2009). This color space produces another chrominance component Cb instead of Cr in YCbCr, expressed as,

$$\begin{bmatrix} Y \\ Cg \\ Cb \end{bmatrix} = \begin{bmatrix} 0.257 & 0.504 & 0.098 \\ -0.3178 & 0.438 & -0.121 \\ -37.797 & -74.203 & 112 \end{bmatrix} \begin{bmatrix} R \\ G \\ B \end{bmatrix}. \tag{16}$$

Among various color spaces described in this section, only six color spaces that give high face recognition rates are presented in next section.

5. Experimental results and discussions

5.1 Database and preprocessing

For experiments, we used CMU PIE and FERET databases. CMU database was used in order to test face recognition performance in illumination variation because it has significant change of lighting conditions. FERET database has smaller variation of illuminations than CMU database. Instead, it includes expression changes and aging.

To remove the effect of background and hair style variations, face regions were cropped to exclude the background and hair regions. All the face images in CMU database were rescaled to 150×150 pixels while those in FERET database were done to 50×50 pixels, and rotated so that the line connecting two eyes is aligned horizontally. Then the color component of each transformed image was normalized to set mean and variance to have zero mean and unit variance.

CMU database used in our experiments consists of three gallery sets (Subset-1, Subset-2, and Subset-3) and three probe sets (Subset-4, Subset-5, and Subset-6), as shown in Fig. 2. Each gallery set consists of 24 face images with various poses while each probe set consists of 1632 face images with various illuminations. Other 412 face images were used as a training set to construct an eigenface space. Fig. 2 shows example face images for each data set from CMU database used in our experiments. Fig. 2(a) shows example face images in three gallery sets with no illumination change: from left to right, frontal face image (Subset-1), half right profile face image (Subset-2), and full right profile face image (Subset-3). Figs. 2(b)-2(d) show three probe sets with illumination variation: frontal face images (Subset-4), half right profile face images (Subset-5), and full right profile face images (Subset-6), with five face images in each probe set.

FERET database used in our experiments consists of one gallery set (Fa) and three probe sets (Fb, Dup1, and Dup2). We used 194 images of set Fa as gallery set of our system, while three sets Fb, Dup1, and Dup2, which consist of 194, 269, and 150 face images, respectively, were used as probe sets. Other 386 face images were used as the training set to construct an eigenface space. Fig. 3 shows example faces of each data set in FERET database used in our experiments. Fig. 3(a) shows an example face image in the gallery set with no facial expression. Figs. 3(b)-3(d) show three example sets: face images with different facial expression (Fb), additional short-term aging (Dup1), and additional long-term aging (Dup2).

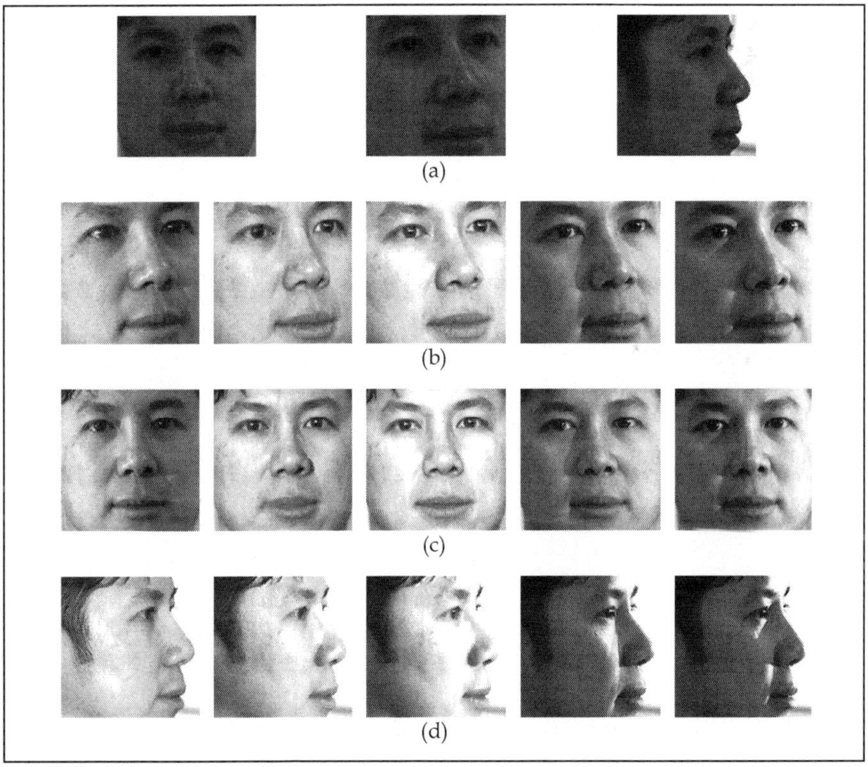

Fig. 2. Color CMU database: (a) Gallery sets (Subset-1, Subset-2, and Subset-3), (b) Probe set 1 (Subset-4), (c) Probe set 2 (Subset-5), (d) Probe set 3 (Subset-6).

In this section, the PCA-based color face recognition system with various color spaces including SV, RGB, YCg'Cr', YUV, YCbCr, and YCgCb is investigated using CMU database and FERET database. We compare recognition performance of independent and concatenated processing with that of the conventional eigenface method employing only luminance information. Note that luminance component images are generated with two different conversions, i.e., $Y = 0.3R + 0.59G + 0.11B$ and $I = (R + G + B) / 3$.

Figs. 4 and 5 illustrate the recognition rate of probe sets in CMU database and FERET database, respectively, in different color spaces with independent and concatenated processing when the number of features is set from 10 to 200. From all the graphs shown in Figs. 4 and 5, it is noted that the more features we use, the higher the recognition rate is. The recognition rate becomes saturated when the number of features is large enough, i.e., 180.

The recognition rates on the saturation range are influenced by color space and data set used for the probe set. Tables 1 and 2 show the maximum recognition rates in each color space for probe sets in CMU database and FERET database, respectively.

5.2 Different color spaces (CMU database)

The performance of face recognition in various lighting conditions is presented, in this subsection. The performance of the PCA-based face recognition algorithm in six different color spaces is evaluated, with independent and concatenated processing for CMU database images. The performance is compared in terms of the recognition rate as a function of the number of features (Fig. 4) and in terms of the maximum recognition rate (Table 1).

For probe set 1 consisting of frontal face images with illumination variations, the best performance is observed in the SV color space, with independent and concatenated processing, as shown in Fig. 4 (probe set 1). For probe set 2 consisting of half profile face images with illumination variations, the recognition rate in the SV color space, with independent and concatenated processing, also gives the best performance, as shown in Fig. 4 (probe set 2). For probe set 3 consisting of full profile face images with illumination variations, the recognition rate in the SV color space with independent processing also gives the best performance, as shown in Fig. 4(a) (probe set 3), whereas the recognition rate in the RGB color space with concatenated processing gives the best performance, as shown in Fig. 4(b) (probe set 3).

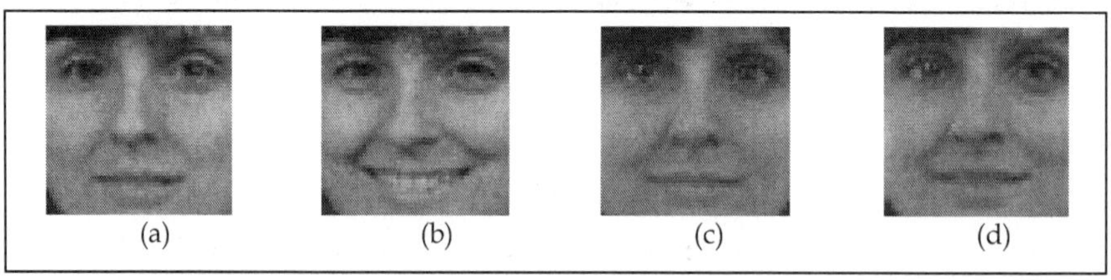

Fig. 3. Color FERET database: (a) Gallery set (Fa), (b) Probe set 1 (Fb), (c) Probe set 2 (Dup1), (d) Probe set 3 (Dup2).

As shown in Table 1(a) with independent processing, for probe set 1, the maximum recognition rate in the SV color space is 18.3% and 22.3% higher than that in the RGB and YCg'Cr' color spaces, respectively, whereas for probe set 2, 17.1% and 22.8% higher, respectively, and for probe set 3, 5.5% and 11.2% higher, respectively.

As shown in Table 1(b) with concatenated processing, for probe set 1, the maximum recognition rate in the SV color space is 16.8% and 26.9% higher than that in the RGB and YCbCr color spaces, respectively, while for probe set 2, 13.3% and 19% higher, respectively, and for probe set 3, 2.8% lower and 0.3% higher, respectively.

Not using H component in the HSV color space improves the recognition rate, as shown in Fig. 4 and Table 1. Because S component is not sensitive to illumination change, robustness to illumination variation can be observed. Various experiments show that the recognition rate in all the color spaces with independent processing is higher than that with concatenated processing, as shown in Fig. 4 and Table 1.

Probe \ Color	SV	RGB	YCg′Cr′	YUV	YCbCr	YCgCb
Probe set 1	90.5	72.2	68.2	66.1	65.7	56.9
Probe set 2	81.7	64.6	58.9	57.3	56.9	55.2
Probe set 3	71.6	66.1	60.4	60.7	60.2	51.5

(a)

Probe \ Color	SV	RGB	YCg′Cr′	YUV	YCbCr	YCgCb
Probe set 1	85.2	68.4	53.5	56.0	58.3	56.8
Probe set 2	73.0	59.7	43.7	48.7	54.0	48.7
Probe set 3	59.8	62.6	52.7	57.1	59.5	31.3

(b)

Table 1. Maximum recognition rate in different color spaces (CMU database, unit: %). (a) Independent processing, (b) Concatenated processing.

5.3 Different color spaces (FERET database)

The performance of face recognition in various expressions and aging is shown in this subsection. The performance of the PCA-based face recognition algorithm in six different color spaces is evaluated, with independent and concatenated processing for FERET database images. The performance is compared in terms of the recognition rate as a function of the number of features (Fig. 5) and in terms of the maximum recognition rate (Table 2).

For probe set 1 with facial expression variations, the best performance is observed in the YUV/YCbCr color spaces with independent processing, as shown in Fig. 5(a) (probe set 1). The recognition rate in the YUV space gives the best performance with concatenated processing, as shown in Fig. 5(b) (probe set 1). Fig. 5 (probe set 2) shows the recognition rate of face images with short-term aging as well as facial expression variations. As shown in Fig. 5 (probe set 2), the recognition rate in the YCg′Cr′ color space, with independent and

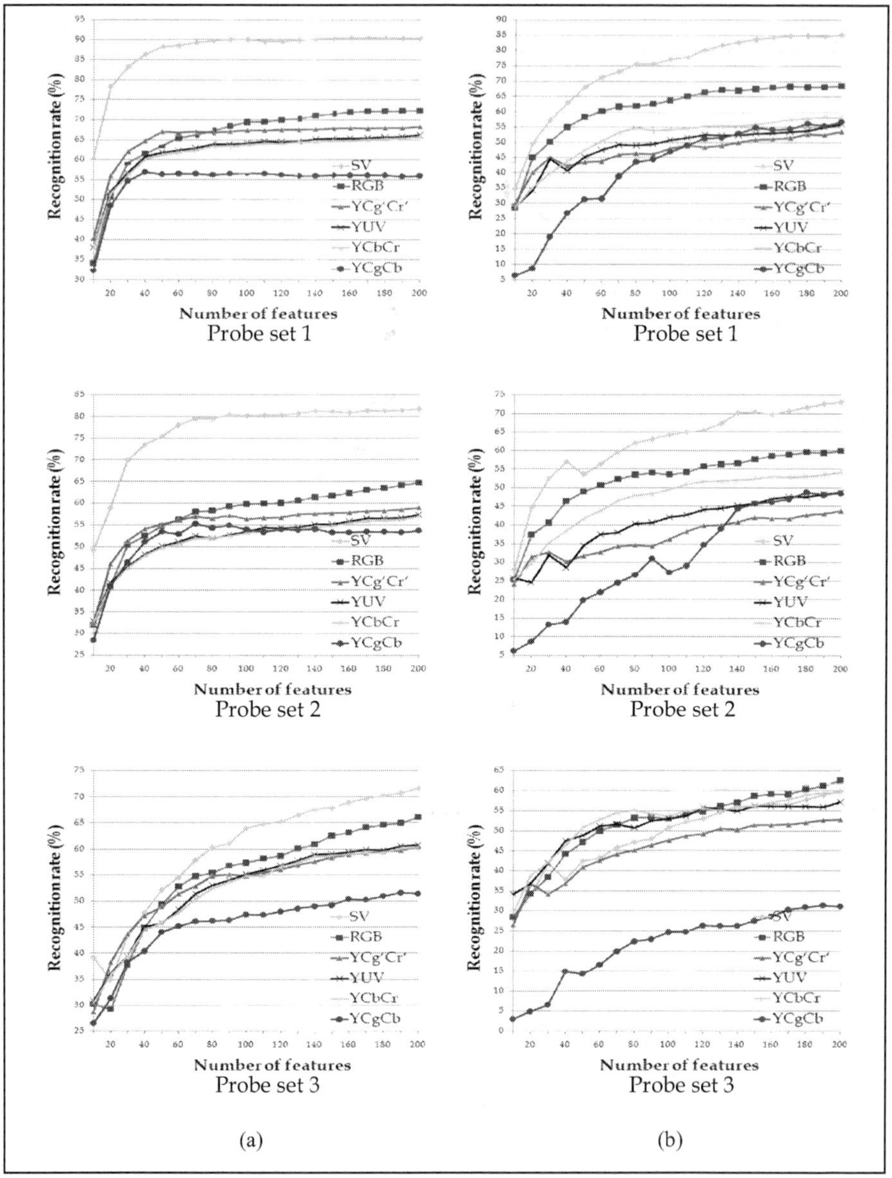

Fig. 4. Performance comparison in terms of the recognition rate as a function of the number of features in different color spaces (CMU database). (a) Independent processing, (b) Concatenated processing.

Probe \ Color	SV	RGB	YCg′Cr′	YUV	YCbCr	YCgCb
Probe set 1	91.2	85.6	88.1	92.3	92.3	80.9
Probe set 2	64.3	59.5	69.5	65.4	65.1	64.7
Probe set 3	56.7	51.3	62.0	58.0	58.0	59.3

(a)

Probe \ Color	SV	RGB	YCg′Cr′	YUV	YCbCr	YCgCb
Probe set 1	88.7	86.1	85.1	89.7	84.0	80.9
Probe set 2	60.6	52.0	66.9	62.8	56.1	63.2
Probe set 3	51.3	58.7	59.3	50.0	48.7	35.3

(b)

Table 2. Maximum recognition rate in different color spaces (FERET database, unit: %). (a) Independent processing, (b) Concatenated processing.

concatenated processing, gives the best performance. For probe set 3 consisting of full profile face images with long-term aging as well as facial expression variation, the recognition rate in the YCg′Cr′ color space, with independent and concatenated processing, also gives the best performance, as shown in Fig. 5 (probe set 3).

As shown in Table 2(a), for probe set 1 with independent processing, the maximum recognition rate in the YUV/YCbCr color spaces is 1.1% and 4.2% higher than that in the SV and YCg′Cr′ color spaces, respectively. For probe set 2, the maximum recognition rate in the YCg′Cr′ color space is 4.1% and 4.4% higher than that in the YUV and YCbCr color spaces, respectively. For probe set 3, the maximum recognition rate in the YCg′Cr′ color space is 2.7% and 4% higher than that in the YCgCb and YUV/YCbCr color spaces, respectively.

As shown in Table 2(b) with concatenated processing, for probe set 1, the maximum recognition rate in the YUV color space is 1% and 3.6% higher than that in the SV and RGB color spaces, respectively. For probe set 2, the maximum recognition rate in the YCg′Cr′ space is 3.7% and 4.1% higher than that in the YCgCb and YUV color spaces, respectively. For probe set 3, the maximum recognition rate in the YCg′Cr′ color space is 0.6% and 8% higher than that in the RGB and SV color spaces, respectively.

Noted that the Cg′Cr′ components are more robust to illumination variations and short- and long-term aging than the CbCr components, in the sense that the YCg′Cr′ color space is more efficient than the YCbCr and color spaces for probe sets 2 and 3 that consist of face images with short and long-term aging, respectively, as well as illumination changes.

We found that the recognition rate in all the color spaces with independent processing is higher than that with concatenated processing, as shown in Fig. 5 and Table 2.

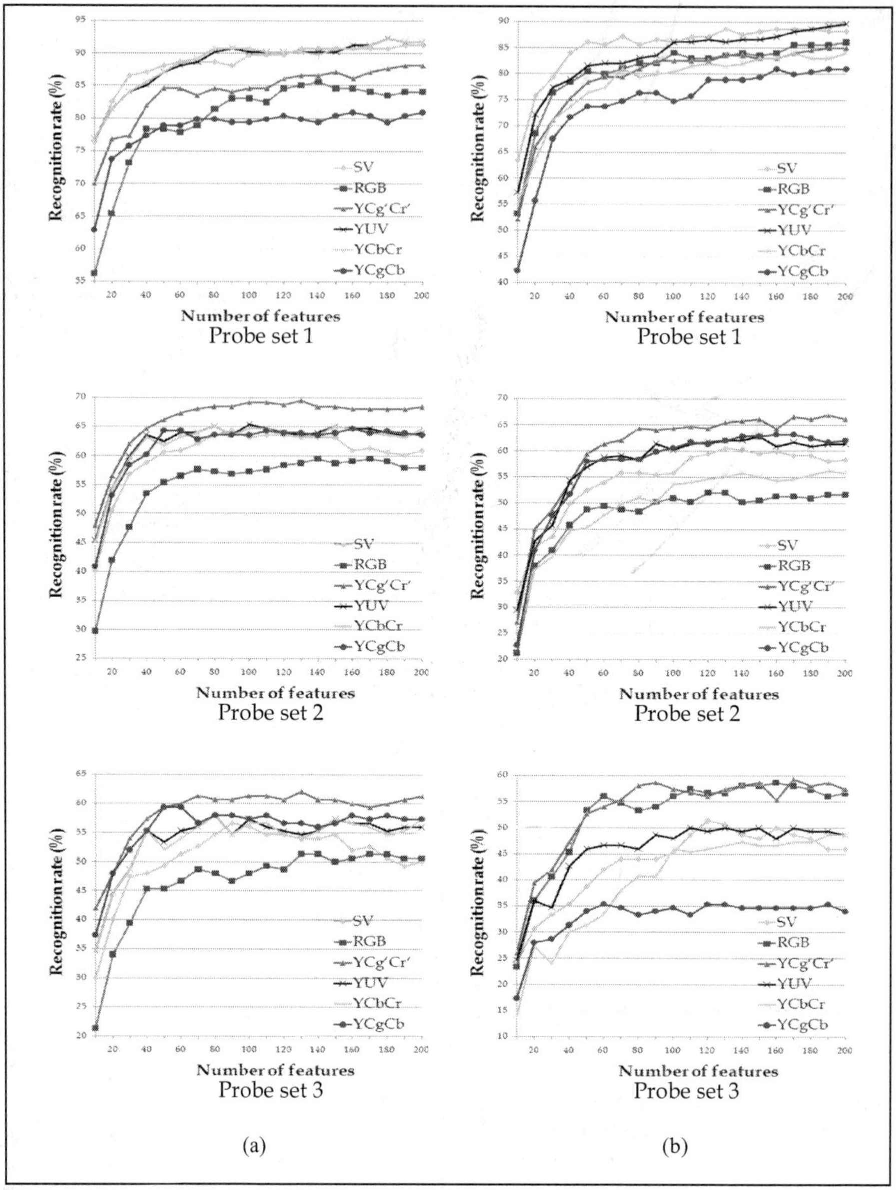

Fig. 5. Performance comparison in terms of the recognition rate as a function of the number of features in different color spaces (FERET database). (a) Independent processing, (b) Concatenated processing.

5.4 Color space vs. gray space

Fig. 6 shows the importance of color information for face recognition. The performance of face recognition with color information is significantly improved compared with that using only grayscale information. We used Subset-4 in CMU database and Fb in FERET database as a probe set (independent processing) and compared face recognition performances in color spaces and gray spaces. The recognition rate in the SV color space is approximately 20 % and 5% higher than that in the gray space (luminance space, i.e., Y and I), in CMU and FERET database images, respectively. Note that the performance of the RGB color space is similar to that of the luminance space. The use of RGB components gives little benefit in generating distinguishable features for effective face recognition, since all the three components of the RGB color space are strongly correlated with each other. On the other hand, the SV color space is effective because its components are less correlated with each other through separation of luminance and chrominance components.

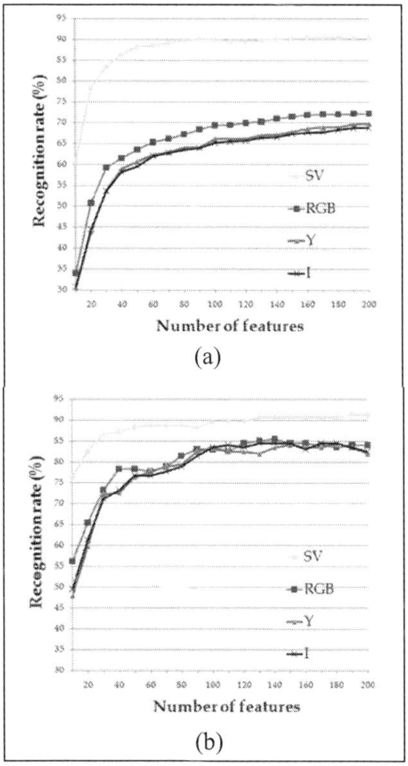

Fig. 6. Performance comparison in terms of the recognition rate as a function of the number of features in different color spaces (independent processing) and gray spaces. (a) CMU database (Subset-4), (b) FERET database (Fb).

6. Conclusions

In this paper, we evaluate the PCA-based face recognition algorithms in various color spaces and analyze their performance in terms of the recognition rate. Experimental results with a large number of face images (CMU and FERET databases) show that color information is beneficial for face recognition and that the SV, YCbCr, and YCg′Cr′ color spaces are the most appropriate spaces for face recognition. The SV color space is shown to be effective to illumination variation, the YCbCr color to facial expression variation, and the YCg′Cr′ color space to aged faces. From experiments, we found that the recognition rate in all the color spaces with independent processing is higher than that with concatenated processing. Further work will focus on the analysis of inter-color correlation and investigation of illumination-invariant color features for effective face recognition.

7. Acknowledgment

This work was supported in part by Brain Korea 21 Project. Portions of the research in this paper use CMU database of facial images collected by Carnegie Mellon University and FERET database of facial images collected under FERET program.

8. References

Bartlett, M. S.; Movellan, J. R. & Sejnowski, T. J. (2002). Face recognition by independent component analysis. *IEEE Transactions on Neural Networks*, Vol. 13, No. 6, pp. 1450–1464, ISSN 1045-9227

Belhumeur, P. N.; Hespanha, J. P. & Kriegman, D. J. (1997). Eigenface vs. Fisherfaces: Recognition using class specific linear projection. *IEEE Transactions on Pattern Analysis and Machine Intelligence,* Vol. 19, No. 7, (August 2002), pp. 711–720, ISSN 0162-8828

Chai, D. & Ngan, K. N. (1999). Face segmentation using skin-color map in videophone applications. *IEEE Transactions on Circuits and Systems for Video Technology,* Vol. 9, No. 4, (August 2002), pp. 551–564, ISSN 1051-8215

De Dios, J. J. & Garcia, N. (2003). Face detection based on a new color space YCgCr. *Proceedings of 2003 International Conference on Image Processing,* Vol. 3, pp. 909–912, ISBN 0-7803-7750-8, Barcelona, Spain, September 14-17, 2003

De Dios, J. J. & Garcia, N. (2004). Fast face segmentation in component color space. *Proceedings of 2004 International Conference on Image Processing,* Vol. 1, pp. 191–194, ISBN 0-7803-8554-3, Singapore, October 24-27, 2004

Etemad, K. & Chellappa, R. (1997). Discriminant analysis for recognition of human face images. *Proceedings of the First International Conference on Audio- and Video-Based Biometric Person Authentication*, pp. 1724–1733, ISBN 3-540-62660-3, Crans Montana, Switzerland, March 12-14, 1997

Jack, K. (2001). *Video Demystified–A Handbook for the Digital Engineer* (3rd). LLH Technology Publishing, 0750683953, LLH Technology Publishing, Eagle Rock, VA, USA

Jones III, C. F. & Abbott, A. L. (2004). Optimization of color conversion for face recognition. *EURASIP Journal on Applied Signal Processing,* Vol. 1, No. 4, (October 2003), pp. 522–529, January 2004

Liu, C. & Wechsler, H. (2000). Evolutionary pursuit and its application to face recognition. *IEEE Transactions on Pattern Analysis and Machine Intelligence,* Vol. 22, No. 6, (August 2002), pp. 570–582, ISSN 0162-8828

Neagoe, V.-E. (2006). An optimum 2D color space for pattern recognition. *Proceedings of the 2006 International Conference on Image Processing, Computer Vision, & Pattern Recognition,* Vol. 2, pp. 526–532, ISBN 0162-8828, Las Vegas, NV, USA, August 26-29, 2006

Phillips, P. J.; Moon, H.; Rizvi, S. A. & Rauss, P. J. (1998). The FERET database and evaluation procedure for face recognition algorithms. *Image and Vision Computing,* Vol. 16, No. 5, pp. 295–306, ISSN 0262-8856

Phillips, P. J.; Moon, H.; Rizvi, S. A. & Rauss, P. J. (2000). FERET evaluation methodology for face recognition algorithms. *IEEE Transactions on Pattern Analysis and Machine Intelligence,* Vol. 22, No. 10, (August 2002), pp. 1090–1104, ISSN 0162-8828

Rajapakse M.; Tan, J. & Rajapakse, J. (2004). Color channel encoding with NMF for face recognition. *Proceedings of 2004 International Conference on Image Processing,* Vol. 3, pp. 2007–2010, ISBN 0-7803-8554-3, Singapore, October 24-27, 2004

Sim, T.; Baker, S. & Bast, M. (2003). The CMU pose, illumination and expression database. *IEEE Transactions on Patten Analysis and Machine Intelligence,* Vol. 25, No. 12, (December 2003), pp. 1615–1618, ISSN 0162-8828

Torres, L.; Reutter, J. Y. & Lorente, L. (1999). The importance of the color information in face recognition. *Proceedings of 1999 International Conference on Image Processing,* Vol. 3, pp. 627–631, ISBN 0-7803-5467-2, Kobe, Japan, October 24-28, 1999

Turk, M. & Pentland, A. (1991). Face recognition using eigenfaces. *Proceedings of CVPR 1991 IEEE Computer Society Conference on Computer Vision and Pattern Recognition,* pp. 586–591, ISBN 0-8186-2148-6, Maui, HI, USA, June 3-6, 1991

Turk, M. & Pentland, A. (1991). Eigenfaces for recognition. *International Journal of Cognitive Neuroscience,* Vol. 3, No. 1, pp. 71–86, ISSN 0898-929X

Wiskott, L.; Fellous, J.-M.; Krueuger, N. & von der Malsburg, C. (1997). Face recognition by elastic bunch graph matching. *IEEE Transactions on Pattern Analysis and Machine Intelligence,* Vol. 19, No. 7, (August 2002), pp. 775–779, ISSN 0162-8828

Wyszecki, G. & Stiles, W. S. (2000). *Color Science: Concepts and Methods, Quantitative Data and Formulae* (2nd). John Wiley & Sons, New York, USA

Yang, J.; Zhang, D.; Xu, Y. & Yang, J.-Y. (2006). Recognize color face images using complex eigenfaces, In Zhang, D. & Jain, A. K. (eds.): *Advances in Biometrics,* Lecture Notes in Computer Science, Vol. 3832, Springer-Verlag Berlin Heidelberg, pp. 64–68, ISSN 03202-9743

Yang, M.-H. (2002). Kernel eigenfaces vs. kernel Fisherfaces: Face recognition using kernel methods. *Proceeding of the 5th IEEE International Conference on Automatic Face and Gesture Recognition,* pp. 215–220, ISBN 0-7695-1602-5, Washington, D.C., USA, May 20-21, 2002

Yoo, S.; Park, R.-H. & Sim, D.-G. (2007). Investigation of color spaces for face recognition. *Proceedings of IAPR Conference on Machine Vision Application*, pp. 106–109, ISBN 978-4-901122-07-8, Tokyo, Japan, May 16-18, 2007

Zhang, Z & Shi, Y. (2009). Skin color detecting unite YCgCb color space with YCgCr color space. *Proceedings of 2009 International Conference on Image Analysis and Signal Processing*, pp. 221–225, ISBN 978-1-4244-3987-4, Taizhou, China, April 11-12, 2009

Zheng, W.-S.; Lai, J.-H. & Yuen, P. C. (2005). GA-fisher: A new LDA-based face recognition algorithm with selection of principal components. *IEEE Transactions on Systems, Man, and Cybernetics, Part B: Cybernetics*, Vol. 35, No. 5, pp. 1065–1078, ISSN 1083-4419

Part 4

Other Biometric Technologies

15

Block Coding Schemes Designed for Biometric Authentication

Vladimir B. Balakirsky[1] and A. J. Han Vinck[2]
[1]*Data Security Association "Confident",*
American University of Armenia
[2]*Institute for Experimental Mathematics*
[1]*Russia, Armenia*
[2]*Germany*

1. Introduction

We address the biometric authentication setup where the outcomes of biometric observations received at the verification stage are compared with the sample data formed at the enrollment stage. The result of comparison is either the acceptance or the rejection of the identity claim. The acceptance decision corresponds to the case when the analyzed values belong to the same person.

A possible solution to the problem, called the direct authentication, is implemented when the outcomes of biometric observations at the enrollment stage are stored in the database, and they are available to the verifier. The possible incorrect verifier's decisions are caused by the fact that these observations are noisy. The probabilities of errors are called the false rejection and the false acceptance rates. The features of the direct authentication are as follows: 1) data compression is not included at the enrollment stage; 2) the scheme does not require an additional external randomness; 3) if the stored data become available to an attacker, then he knows the outcomes of biometric observations of the person and can pass through the verification stage with the acceptance decision by presenting these data to the verifier. The considered below coding approaches to the problem require an external randomness and relax the constraint that the database has to be protected against reading. These approaches include the additive and the permutation coding schemes.

Both the direct authentication and an additive coding scheme are illustrated using a proposed mathematical model for the DNA measurements. We present the model and describe a data compression method that can be used to approach a uniform probability distribution over the obtained data for their further use in the additive scheme and other purposes. The processing of the DNA data also serves as an example of possible processing data generated by an arbitrary memoryless source.

The additive block coding scheme can be viewed as a variant of stream ciphering scheme where the data, to be hidden, are added to a key. The subtraction of the noisy version of the data creates a corrupted version of the key. If the key is a codeword of a code having certain error–correcting property, then the fact, whether the key can be reconstructed or not, characterizes the level of the noise. In the permutation scheme, the enciphering of the input data is organized by choosing a permutation, which maps the biometric vector to a key vector. There are many permutations that can be used for this purpose, and it gives additional possibilities to the designer of the verification scheme.

The efficiency of cryptographic schemes, like the additive and the permutation schemes, is measured by the difference between the probabilities of the successful attack by an attacker, who either knows the content of the database or ignorant about these data. The additive scheme is efficient when the probability distribution over the input vectors is close to a uniform distribution. This requirement is less critical for the permutation scheme, but input vectors have to be represented by binary vectors having a fixed number of ones. We will present a simple numerical example of the implementation of the permutation scheme and describe an algorithm for the transformation of an arbitrary binary vector to a balanced vector having the same number of zeroes and ones.

There is a number of open problems in the implementation of coding schemes. One of the main problems is the representation of real biometric data in digital format, which allows one to use the memoryless assumption about the data and the Hamming distance as the measure of closeness of two observations. Another class of problems is constructing the specific codes and the decoding algorithms having a low computational complexity. We also believe that there is a request for a general theory of processing noisy data, since the known solutions in biometrics are mostly oriented to specific measurements (fingerprints, iris, palmprints, etc.) and a particular application.

The authentication problem belongs to the list of basic problems that have to be solved in the biometric direction, and it is included in the most of the books on biometrics (see Bolle et. al (2004), for example). The additive block coding scheme was suggested in Juels & Wattenberg (1999). The close relationships between the additive scheme and the wiretap channel, introduced in Wyner (1975), where the verifier receives the signals from the outputs of two parallel channels in the legitimate case and the signals from only one of channels in the case of the presence of an attacker. It implies the relevance of information and coding theory results (see Cohen & Zemor (2006), for example) to the investigation of the scheme. The permutation scheme was proposed in Dodis, et. al (2004) under the uniform probability distribution over the permutations. The algorithm for the mapping of an arbitrary binary vector to a balanced vector, which can be used in the permutation scheme, was described in Knuth (1986). The available DNA measurement data were received in the BioKey–STR project (Korte et. al (2008)).

The text of the chapter is a compressed version of the results in Balakirsky, Ghazaryan & Han Vinck (2006–2011). The general principles of constructing biometric authentication, which also include the points of rate–distortion coding, were presented in (2006a), (2006b). The described mathematical model for the DNA data was introduced in (2008a), and the data processing scheme was studied in (2009b) as an extension of the transformations for continuous random variables described in (2007). The similar analysis is relevant to the constructing passwords from biometric data, as it is indicated in (2010). The general expressions for the additive and the permutation block coding schemes for an arbitrary probability distribution over the biometric vectors are given in (2008a), (2009a). The standard technique of probability and coding theory, which is used in the chapter, can be found in Gallager (1968).

2. Notation and basic assumptions

Let $\mathcal{B} = \mathcal{B}_1 \times \cdots \times \mathcal{B}_n$, where $\mathcal{B}_t = \{0, \ldots, K_t - 1\}$ is a finite set containing K_t elements. We say that $\mathbf{b} = (b_1, \ldots, b_n) \in \mathcal{B}$ is a biometric vector and assume that the probability distribution

$$\omega = \Big(\omega(\mathbf{b}) = \Pr_{\text{bio}}\{ B = \mathbf{b} \},\, \mathbf{b} \in \mathcal{B} \Big)$$

is known. Moreover, let ω be a memoryless probability distribution, i.e.,

$$\omega(\mathbf{b}) = \prod_{t=1}^{n} \omega_t(b_t) \tag{1}$$

for all $\mathbf{b} \in \mathcal{B}$. We also write

$$\omega_t(b) = \Pr_{\text{bio}}\{ B_t = b \}$$

for all $b \in \mathcal{B}_t$. Denote the most likely biometric vector by $\mathbf{b}^* = (b_1^*, \ldots, b_n^*)$,

$$\mathbf{b}^* = \arg\max_{\mathbf{b} \in \mathcal{B}} \omega(\mathbf{b}).$$

Then, by (1),

$$b_t^* = \arg\max_{b \in \mathcal{B}_t} \omega_t(b), \quad t = 1, \ldots, n,$$

and

$$\omega(\mathbf{b}^*) = \prod_{t=1}^{n} \omega_t^*$$

where

$$\omega_t^* = \max_{b \in \mathcal{B}_t} \omega_t(b). \tag{2}$$

Furthermore, let

$$\overline{\omega}_t = \sum_{b \in \mathcal{B}_t} \omega_t^2(b) \tag{3}$$

and

$$H(\omega_t) = -\sum_{b=0}^{q_t - 1} \omega_t(b) \log \omega_t(b). \tag{4}$$

Then $\overline{\omega}_t$ is the probability that two independent runs of the t-th biometric source result in two equal symbols, and $H(\omega_t)$ is the entropy of the probability distribution ω_t, which can be understood as the number of random bits at the output of the t-th biometric source.

We will use the component–wise transformation of the vector $\mathbf{b}$ to another vector $\mathbf{z}$ and organize it in such a way that the probability distribution over the vectors $\mathbf{z}$ is close to a uniform distribution. Introduce the following notation. Let us fix $q_t \le K_t$ as an integer power of 2 and let $\mathcal{Z}_t = \{0, \ldots, q_t - 1\}$. Let us map $b \in \mathcal{B}_t$ to $z \in \mathcal{Z}_t$ if and only if $b \in \mathcal{B}_{t,z}$, where $\mathcal{B}_{t,0}, \ldots, \mathcal{B}_{t,q_t-1}$ are pairwise disjoint sets whose union coincides with $\mathcal{B}_t$. One can see that such a specification uniquely determines z and we denote it by $z(b|q_t)$. Let

$$\mathbf{z}_{\mathbf{b}} = (z(b_1|q_1), \ldots, z(b_n|q_n)) \tag{5}$$

denote the result of the mapping $\mathcal{B} \to \mathcal{Z} = \mathcal{Z}_1 \times \cdots \times \mathcal{Z}_n$, which is parameterized by the vector $\mathbf{q} = (q_1, \ldots, q_n)$ and the partitionings of the sets $\mathcal{B}_1, \ldots, \mathcal{B}_n$. We also denote

$$\Omega_t(z) = \sum_{b \in \mathcal{B}_{t,z}} \omega_t(b)$$

for all $z \in \mathcal{Z}_t$ and

$$\Omega(\mathbf{z}) = \prod_{t=1}^{n} \Omega_t(z_t)$$

for all $\mathbf{z} \in \mathcal{Z}$. Furthermore, let

$$\rho_t = \frac{\max_{z \in \mathcal{Z}_t} \Omega(z)}{\min_{z \in \mathcal{Z}_t} \Omega(z)}. \tag{6}$$

Let the noisy observations of the biometric vector $\mathbf{b}$ be specified by the conditional probability distributions

$$\Big(V(\mathbf{b}'|\mathbf{b}) = \Pr_{\text{err}}\{ B' = \mathbf{b}' \mid B = \mathbf{b} \},\ \mathbf{b}' \in \mathcal{B} \Big),\ \mathbf{b} \in \mathcal{B},$$

and let

$$V(\mathbf{b}'|\mathbf{b}) = \prod_{t=1}^{n} V_t(b_t'|b_t) \tag{7}$$

for all $\mathbf{b}, \mathbf{b}' \in \mathcal{B}$. We also write

$$V_t(b'|b) = \Pr_{\text{err}}\{ B_t' = b' \mid B_t = b \}$$

for all $b, b' \in \mathcal{B}_t$ and pay special attention to the conditional probability distributions such that

$$V_t(b|b) = 1 - \varepsilon, \text{ for all } b \in \mathcal{B}_t, \tag{8}$$

where $\varepsilon > 0$ is a given constant.
The transformation $\mathcal{B} \to \mathcal{Z}$ preserves the V channel in a sense that (8) implies

$$V_t(z_b|b) = \sum_{b' \in \mathcal{B}_{t,z_b}} V_t(b'|b) \geq V_t(b|b) = 1 - \varepsilon$$

for all $b \in \mathcal{B}_t$. Therefore, the V_t channel $\mathcal{B}_t \to \mathcal{B}_t$ is transformed to another V_{t,q_t} channel $\mathcal{Z}_t \to \mathcal{Z}_t$ such that

$$V_{t,q_t}(z|z) \geq 1 - \varepsilon, \text{ for all } z \in \mathcal{Z}_t. \tag{9}$$

Let

$$\text{Ham}(\mathbf{b}, \mathbf{b}') = \Big| \Big\{ t \in \{1, \ldots, n\} : b_t \neq b_t' \Big\} \Big|$$

denote the Hamming distance between the vectors $\mathbf{b}, \mathbf{b}' \in \mathcal{B}$ and let

$$\mathcal{D}_T(\mathbf{b}) = \Big\{ \mathbf{b}' \in \mathcal{B} : \text{Ham}(\mathbf{b}, \mathbf{b}') \leq T \Big\} \tag{10}$$

denote the set of biometric vectors located at distance T or less from the vector $\mathbf{b}$. The conditional probability of generating a vector belonging to the set $\mathcal{D}_T(\mathbf{b})$, given the vector

$\mathbf{b}$, is defined as

$$V(\mathcal{D}_T(\mathbf{b})|\mathbf{b}) = \sum_{\mathbf{b}' \in \mathcal{D}_T(\mathbf{b})} V(\mathbf{b}'|\mathbf{b}). \tag{11}$$

Notice that if conditions (8) are satisfied, then

$$V(\mathcal{D}_T(\mathbf{b})|\mathbf{b}) = \sum_{d=0}^{T} \binom{n}{d} (1-\varepsilon)^{n-d} \varepsilon^d \tag{12}$$

for all $\mathbf{b} \in \mathcal{B}$.

3. Mathematical model for the DNA measurements

The most common DNA variations are Short Tandem Repeats (STR): arrays of 5 to 50 copies (repeats) of the same pattern (the motif) of 2 to 6 pairs. As the number of repeats of the motif highly varies among individuals, it can be effectively used for identification of individuals. The human genome contains several 100,000 STR loci, i.e., physical positions in the DNA sequence where an STR is present. An individual variant of an STR is called allele. Alleles are denoted by the number of repeats of the motif. The genotype of a locus comprises both the maternal and the paternal allele. However, without additional information, one cannot determine which allele resides on the paternal or the maternal chromosome. If the measured numbers are equal to each other, then the genotype is called homozygous. Otherwise, it is called heterozygous. The STR measurement errors are usually classified into three groups: *(1) allelic drop–in*, when in a homozygous genotype, an additional allele is erroneously included, e.g. genotype (10,10) is measured as (10,12); *(2) allelic drop–out*, when an allele of a heterozygous genotype is missing, e.g. genotype (7,9) is measured as (7,7); *(3) allelic shift*, when an allele is measured with a wrong repeat number, e.g. genotype (10,12) is measured as (10,13).

The points above can be formalized as follows. Suppose that there are n sources. For all $t = 1, \ldots, n$, there is a probability distribution

$$\pi_t = \Big(\pi_t(i),\, i \in \{c_t, \ldots, c_t + k_t - 1\} \Big),$$

where c_t, k_t are given positive integers. Let the probability that the t-th source generates the pair (i,j), where $i,j \in \{c_t, \ldots, c_t + k_t - 1\}$, be defined as

$$\Pr_{\mathrm{DNA}} \Big\{ (A_{t,1}, A_{t,2}) = (i,j) \Big\} = \pi_t(i)\pi_t(j).$$

Thus, we assume that $A_{t,1}$ and $A_{t,2}$ are independent random variables that contain information about the number of repeats of the t-th motif in the maternal and the paternal allele. We also assume that $(A_{1,1}, A_{1,2}), \ldots, (A_{n,1}, A_{n,2})$ are independent pairs of random variables, i.e.,

$$\Pr_{\mathrm{DNA}} \Big\{ (A_1, A_2) = (\mathbf{i}, \mathbf{j}) \Big\} = \prod_{t=1}^{n} \Pr_{\mathrm{DNA}} \Big\{ (A_{t,1}, A_{t,2}) = (i_t, j_t) \Big\},$$

where $A_1 = (A_{1,1}, \ldots, A_{n,1})$, $A_2 = (A_{1,2}, \ldots, A_{n,2})$ and $\mathbf{i} = (i_1, \ldots, i_n)$, $\mathbf{j} = (j_1, \ldots, j_n)$.

Let

$$S_t = \Big(\min\{A_{t,1}, A_{t,2}\}, \max\{A_{t,1}, A_{t,2}\} \Big).$$

Then

$$\Pr_{\mathrm{DNA}} \Big\{ S_t = (i,j) \Big\} = \tilde{\pi}_t(i,j),$$

where

$$\tilde{\pi}_t(i,j) = \begin{cases} \pi_t^2(i), & \text{if } j = i, \\ 2\pi_t(i)\pi_t(j), & \text{if } j > i, \\ 0, & \text{if } j < i. \end{cases}$$

Denote $\mathcal{B}_t = \{0, \ldots, K_t - 1\}$, where $K_t = k_t(k_t+1)/2$, order K_t probabilities belonging to the distribution

$$\tilde{\pi}_t = \Big(\tilde{\pi}_t(i,j), i,j \in \{c_t, \ldots, c_t + k_t - 1\}, j \geq i \Big)$$

in the decreasing order, assign them indices $b = 0, \ldots, K_t - 1$, and replace $\tilde{\pi}_t$ with the probability distribution

$$\omega_t = \Big(\omega_t(b), b \in \{0, \ldots, K_t - 1\} \Big),$$

i.e., the probability distributions $\tilde{\pi}_t$ and ω_t contain the same entries in different order.
The transformations below are illustrated for the **TH01** allele (see Tables 2, 3), where $t = 12$, $c_t = 6$, $k_t = 4$, and

$$(\pi_t(6), \ldots, \pi_t(9)) = (.234, .192, .085, .487).$$

Then

$$\Big[\pi_t(i)\pi_t(j) \Big]_{i,j=6,\ldots,9} =$$

	$j=6$	$j=7$	$j=8$	$j=9$
$i=6$	.0550	.0452	.0200	.1143
$i=7$	.0452	.0371	.0165	.0939
$i=8$	.0200	.0165	.0073	.0416
$i=9$	.1143	.0939	.0416	.2376

To compute the entries of the probability distribution $\tilde{\pi}_t$, we transform this matrix to the right triangular matrix below. The entries above the diagonal are doubled, and the entries below the diagonal are replaced with the zeroes.

$$\Big[\tilde{\pi}_t(i,j) \Big]_{\substack{i,j=6,\ldots,9 \\ j \geq i}} =$$

	$j=6$	$j=7$	$j=8$	$j=9$
$i=6$	.0550	.0903	.0401	.2286
$i=7$		.0371	.0329	.1878
$i=8$			.0073	.0833
$i=9$				.2376

The ordering of the non-zero entries of this matrix brings the probability distribution ω_t. Its entries and parameters $\omega_t^*, \overline{\omega}_t$, defined in (2), (3), are given below.

i,j	9,9	6,9	7,9	6,7	8,9	6,6	6,8	7,7	7,8	8,8
$\tilde{\pi}_t(i,j)$	.2376	.2286	.1878	.0903	.0833	.0550	.0401	.0371	.0329	.0073
b	0	1	2	3	4	5	6	7	8	9
$\omega_t(b)$	.2376	.2286	.1878	.0903	.0833	.0550	.0401	.0371	.0329	.0073
ω_t^*	.2376									
$\overline{\omega}_t$	.2376 .2376 + ... + .0073 .0073 = .0609									

b	0	9	1	8	2	5	3	4	6	7
$\omega_t(b)$	.2376	.0073	.2286	.0329	.1878	.0550	.0903	.0833	.0401	.0371
z	0		1		2		3			
$\Omega_t(z)$	.2449		.2615		.2428		.2508			
$\rho_t(z)$	.2615/.2428 = 1.08									

Table 1. Example of the mapping $\{0,\ldots,9\} \to \{0,\ldots,3\}$.

Let q_t be the maximum integer power of 2 such that

$$1/q_t \geq \omega_t^*,$$

where ω_t^* is defined in (2). Then one can partition the set $\mathcal{B}_t$ in q_t subsets in such a way the resulting probability distribution over these subsets is close to a uniform distribution. An example of the partitioning is given in Table 1. Notice that the entropy of the distribution ω_t is equal to 2.851 (see Table 3), while the entropy of the distribution Ω_t is less and it is close to $\log q_t$.

The available experimental data consist of probability distributions $\pi_1,\ldots,\pi_{28}$, and they are given in Table 2. The computed parameters are shown in Table 3. We conclude that results of the DNA measurements can be represented by a binary vector of length 140 bits. However the probability distribution over these vectors is non–uniform and, roughly speaking, only 109 bits carry information about the measurements. The most likely vector of pairs has the probability $0.124\ldots0.243 = 10^{-23}$, and the probability that the sources independently generate two equal vectors is equal to $0.013\ldots0.046 = 10^{-50}$. The greedy algorithm for partitioning the sets $\mathcal{B}_1,\ldots,\mathcal{B}_n$ in $q_1,\ldots,q_n$ brings the vectors that can be expressed by $\log q_1 + \cdots + \log q_n = 68$ bits with the property that $\rho_1\ldots\rho_n \approx 16$, where $\rho_1,\ldots,\rho_n$ are defined in (6). Therefore, the most likely vector of length 68 bits has the probability 2^{-64}. Notice that the spectrum of components of the vector $\mathbf{q}$ can be presented the as the sequence $(q \times N_q)$, $q = 2^1,\ldots,2^6$, where N_q is the number of indices t with $q_t = q$. Namely, the constructed vector $\mathbf{q}$ has the spectrum

$$(2\times 7), (4\times 8), (8\times 9), (16\times 3), (32\times 0), (64\times 1) \tag{13}$$

and

$$28 = 7+8+9+3+0+1,$$
$$68 = 7\cdot\log 2 + 8\cdot\log 4 + 9\cdot\log 8 + 3\cdot\log 16 + 0\cdot\log 32 + 1\cdot\log 64.$$

4. Direct authentication schemes

Let us consider the following setup. Suppose that $\mathbf{b},\mathbf{b}' \in \mathcal{B}$ are given vectors of length n. If the Hamming distance between these vectors is not greater than a fixed threshold T, then the verifier has to make the acceptance decision. Otherwise, the verifier has to make the rejection decision. Hence, the rules are as follows:

$\mathrm{R}_{\mathrm{Acc}}$: *if* $\mathbf{b}' \in \mathcal{D}_T(\mathbf{b})$, *then accept the identity claim* (Acc);

$\mathrm{R}_{\mathrm{Rej}}$: *if* $\mathbf{b}' \notin \mathcal{D}_T(\mathbf{b})$, *then reject the identity claim* (Rej).

t	Name	π_t
1	**D8S1179**	.319 .194 .173 .119 .105 .086
2	**D3S1358**	.265 .257 .218 .154 .104
3	**VWA**	.283 .202 .202 .111 .105 .095
4	**D7S820**	.248 .211 .180 .168 .155 .035
5	**ACTBP2**	.089 .080 .073 .072 .070 .064 .062 .053 .051 .049 .047 .046 .043 .039 .037 .034 .033 .028 .012 .009
6	**D7S820**	.243 .207 .177 .165 .152 .034 .018
7	**FGA**	.223 .192 .139 .139 .129 .072 .053 .026 .023
8	**D21S11**	.308 .200 .183 .160 .091 .028 .026
9	**D18S51**	.162 .142 .142 .135 .130 .129 .078 .039 .022 .016
10	**D19S433**	.382 .259 .173 .086 .082 .015
11	**D13S317**	.339 .248 .124 .112 .074 .051 .048
12	**TH01**	.487 .234 .192 .085
13	**D2S138**	.182 .146 .122 .117 .114 .093 .079 .041 .038 .033 .029
14	**D16S539**	.326 .321 .145 .112 .056 .019 .018
15	**D5S818**	.389 .365 .142 .052 .050
16	**TPOX**	.537 .244 .119 .056 .041
17	**CF1PO**	.365 .305 .219 .097 .011
18	**D8S1179**	.304 .185 .165 .114 .100 .082 .031 .011 .003
19	**VWA1**	.283 .202 .202 .111 .105 .095
20	**PentaD**	.265 .214 .189 .156 .089 .060 .014 .010
21	**PentaE**	.180 .170 .110 .105 .102 .080 .056 .051 .051 .034 .029 .010 .010 .007
22	**DYS390**	.422 .282 .164 .103 .014 .011
23	**DYS429**	.445 .325 .118 .096 .013
24	**DYS437**	.528 .317 .154
25	**DYS391**	.513 .451 .018 .016
26	**DYS385**	.551 .124 .097 .087 .059 .037 .030 .012
27	**DYS389I**	.663 .186 .150
28	**DYS389II**	.446 .272 .167 .081 .032

Table 2. The entries of the probability distributions $\pi_1, \ldots, \pi_{28}$, which are greater than 0.001, given in the decreasing order.

"The identity claim" in the description above appears because we assume that the vectors **b** and **b**$'$ contain outcomes of measurements of some biometric parameters of two people. The verification is understood as a procedure, which checks whether the difference between the results is caused by the observation noise or by the fact that people are different.
The direct implementation of the authentication procedure includes the enrollment and the verification stages (see Figure 1).

The enrollment stage.

– *Store the biometric vector* **b** *in the database.*

t	Name	$\log K_t$	$\lceil \log K_t \rceil$	ω_t^*	$\log q_t$	$H(\omega_t)$	$\overline{\omega}_t$
1	**D8S1179**	4.392	5	0.124	3	4.083	0.013
2	**D3S1358**	3.907	4	0.137	2	3.714	0.012
3	**VWA**	4.392	5	0.115	3	4.127	0.010
4	**D7S820**	4.392	5	0.105	3	4.074	0.008
5	**ACTBP2**	7.714	8	0.014	6	7.426	0.000
6	**D7S820**	4.807	5	0.101	3	4.241	0.008
7	**FGA**	5.492	6	0.086	3	4.916	0.005
8	**D21S11**	4.807	5	0.124	3	4.130	0.013
9	**D18S51**	5.781	6	0.046	4	5.279	0.002
10	**D19S433**	4.392	5	0.199	2	3.593	0.027
11	**D13S317**	4.807	5	0.169	2	4.151	0.018
12	**TH01**	3.322	4	0.238	2	2.851	0.061
13	**D2S138**	6.044	7	0.053	4	5.601	0.002
14	**D16S539**	4.807	5	0.210	2	3.776	0.023
15	**D5S818**	3.907	4	0.285	1	3.111	0.041
16	**TPOX**	3.907	4	0.289	1	2.909	0.087
17	**CF1PO**	3.907	4	0.223	2	3.157	0.029
18	**D8S1179**	5.492	6	0.113	3	4.487	0.011
19	**VWA1**	4.392	5	0.115	3	4.127	0.010
20	**PentaD**	5.170	6	0.114	3	4.325	0.009
21	**PentaE**	6.907	7	0.062	4	5.870	0.002
22	**DYS390**	4.392	5	0.239	2	3.238	0.039
23	**DYS429**	3.907	4	0.290	1	2.972	0.051
24	**DYS437**	2.585	3	0.335	1	2.259	0.089
25	**DYS391**	3.322	4	0.464	1	1.902	0.111
26	**DYS385**	5.170	6	0.304	1	3.607	0.093
27	**DYS389I**	2.585	3	0.440	1	2.008	0.195
28	**DYS389II**	3.907	4	0.243	2	3.145	0.046
		128.6	140	10^{-23}	68	109.1	10^{-50}

Table 3. Some characteristics of the probability distributions $\omega_1, \ldots, \omega_{28}$ that describe the DNA measurements.

The verification stage.

– *Read the biometric vector* $\mathbf{b}$ *associated with the claimed person from the database. If* $\mathbf{b}' \in \mathcal{D}_T(\mathbf{b})$, *then make the acceptance decision* (Acc). *If* $\mathbf{b}' \notin \mathcal{D}_T(\mathbf{b})$, *then make the rejection decision* (Rej).

The basic parameters of the scheme are the false rejection rate FRR, the false acceptance rate FAR, and the average false acceptance rate $\overline{\text{FAR}}$, introduced as

$$\text{FRR} = \sum_{\mathbf{b},\mathbf{b}' \in \mathcal{B}} \omega(\mathbf{b}) V(\mathbf{b}'|\mathbf{b}) \chi\{\mathbf{b}' \notin \mathcal{D}_T(\mathbf{b})\}, \tag{14}$$

$$\text{FAR} = \max_{\mathbf{b}' \in \mathcal{B}} \sum_{\mathbf{b} \in \mathcal{B}} \omega(\mathbf{b}) \chi\{\mathbf{b}' \in \mathcal{D}_T(\mathbf{b})\}, \tag{15}$$

$$\overline{\text{FAR}} = \sum_{\mathbf{b},\mathbf{b}' \in \mathcal{B}} \omega(\mathbf{b}) \omega(\mathbf{b}') \chi\{\mathbf{b}' \in \mathcal{D}_T(\mathbf{b})\}, \tag{16}$$

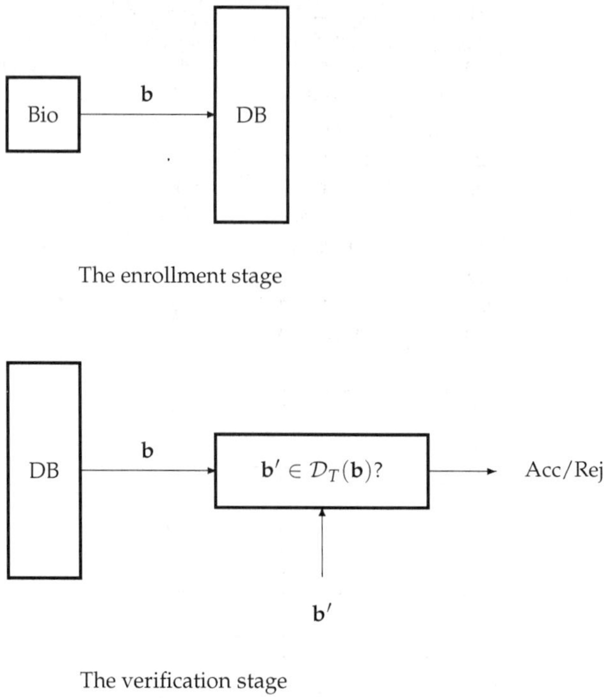

Fig. 1. The data processing in a direct authentication scheme.

where χ denotes the indicator function: $\chi\{\mathcal{S}\} = 1$ is the statement $\mathcal{S}$ is true and $\chi\{\mathcal{S}\} = 0$ otherwise. The false rejection rate is the probability of the event that the verifier makes the rejection decision when the observations belong to the same person. The false acceptance rate is the probability of the event that the verifier makes the acceptance decision when the vector $\mathbf{b}'$ is generated by an attacker. The average false acceptance rate is the probability of the event that the verifier makes the acceptance decision when the vector $\mathbf{b}'$ contains outcomes of biometric observations of a randomly chosen person.
If the V channel satisfies (8), then the false rejection rate is expressed using (12),

$$\mathrm{FRR} = \sum_{d=T+1}^{n} \binom{n}{d} (1-\varepsilon)^{n-d} \varepsilon^{d}. \tag{17}$$

To compute the false acceptance rates, we use the generating functions technique.
Let us consider the problem of computing $\overline{\mathrm{FAR}}$ and introduce the generating function

$$\overline{G}_t(z) = \overline{\omega}_t + (1 - \overline{\omega}_t)z,$$

where z is a formal variable and $\overline{\omega}_t$ is defined in (3) as the probability that two independent runs of the t-th source result in two equal symbols. Furthermore, denote

$$\overline{G}(z) = \prod_{t=1}^{n} \overline{G}_t(z)$$

and represent the polynomial $\overline{G}(z)$ as

$$\overline{G}(z) = \sum_{d=0}^{n} \mathrm{Coef}_d\Big[\,\overline{G}(z)\,\Big] z^d.$$

Then the d-th term of the sum at the right-hand side is equal to the probability that two independent runs of n sources result in vectors that differ in d components. Hence,

$$\overline{\mathrm{FAR}} = \sum_{d=0}^{T} \mathrm{Coef}_d\Big[\,\overline{G}(z)\,\Big].$$

Similar manipulations bring the formula

$$\sum_{\mathbf{b}\in\mathcal{B}} \omega(\mathbf{b})\chi\{\mathbf{b}' \in \mathcal{D}_T(\mathbf{b})\} = \sum_{d=0}^{T} \mathrm{Coef}_d\Big[\,G(z|\mathbf{b}')\,\Big], \tag{18}$$

where

$$G(z|\mathbf{b}') = \prod_{t=1}^{n}\Big(\omega_t(b_t') + (1-\omega_t(b_t'))z\Big).$$

One can easily see that the sum at the right-hand side of (18) is maximized when $\mathbf{b}' = \mathbf{b}^*$ and

$$\mathrm{FAR} = \sum_{d=0}^{T} \mathrm{Coef}_d\Big[\,G(z|\mathbf{b}^*)\,\Big],$$

where

$$G(z|\mathbf{b}^*) = \prod_{t=1}^{n}\Big(\omega_t^* + (1-\omega_t^*)z\Big)$$

and $\omega_1^*, \ldots, \omega_n^*$ are defined in (2).

Some numerical results for the DNA data are given in Table 4. We conclude that the probability of successful attack in the case when the attacker does not know the content of the database can be very small. However, the main problem with the direct authentication scheme is caused by the point that the biometric vector itself is stored in the database. If an attacker would have an access to the database, then he does not have any difficulties with the passing through the verification stage with the acceptance decision. Moreover, the biometrics, being compromized, is compromized forever and it can be also used for any other purposes. A possible solution to the hiding problem is the use of the cryptographic "one–way" hash function Hash : it is assumed that the value of the function can be easily computed for a given argument, but the value of the argument is hard to get for a given value of the function. If only Hash($\mathbf{b}$) is known to the verifier, then he can compute the values of Hash($\tilde{\mathbf{b}}$) for all vectors $\tilde{\mathbf{b}}$ located at the Hamming distance at most T from the vector $\mathbf{b}'$ and make the acceptance

T	FRR $\varepsilon = 0.05$	FRR $\varepsilon = 0.01$	FAR	$\overline{\text{FAR}}$	$\widehat{\text{FAR}}$
0	$7.6 \cdot 10^{-1}$	$2.5 \cdot 10^{-1}$	$7.7 \cdot 10^{-24}$	$2.5 \cdot 10^{-50}$	$3.4 \cdot 10^{-21}$
1	$4.1 \cdot 10^{-1}$	$3.2 \cdot 10^{-2}$	$1.9 \cdot 10^{-21}$	$1.7 \cdot 10^{-46}$	$6.9 \cdot 10^{-19}$
2	$1.6 \cdot 10^{-1}$	$2.7 \cdot 10^{-3}$	$2.0 \cdot 10^{-19}$	$3.4 \cdot 10^{-43}$	$6.1 \cdot 10^{-17}$
3	$4.9 \cdot 10^{-2}$	$1.7 \cdot 10^{-4}$	$1.3 \cdot 10^{-17}$	$3.7 \cdot 10^{-40}$	$3.2 \cdot 10^{-15}$
4	$1.2 \cdot 10^{-2}$	$8.1 \cdot 10^{-6}$	$5.8 \cdot 10^{-16}$	$2.5 \cdot 10^{-37}$	$1.2 \cdot 10^{-13}$
5	$2.3 \cdot 10^{-3}$	$3.1 \cdot 10^{-7}$	$1.9 \cdot 10^{-14}$	$1.2 \cdot 10^{-34}$	$3.1 \cdot 10^{-12}$
6	$3.6 \cdot 10^{-4}$	$9.8 \cdot 10^{-9}$	$4.8 \cdot 10^{-13}$	$4.2 \cdot 10^{-32}$	$6.4 \cdot 10^{-11}$
7	$4.9 \cdot 10^{-5}$	$2.6 \cdot 10^{-10}$	$9.7 \cdot 10^{-12}$	$1.1 \cdot 10^{-29}$	$1.0 \cdot 10^{-9}$
8	$5.6 \cdot 10^{-6}$	$5.8 \cdot 10^{-12}$	$1.6 \cdot 10^{-10}$	$2.3 \cdot 10^{-27}$	$1.3 \cdot 10^{-8}$
9	$5.6 \cdot 10^{-7}$	$1.1 \cdot 10^{-13}$	$2.1 \cdot 10^{-9}$	$3.7 \cdot 10^{-25}$	$1.4 \cdot 10^{-7}$

Table 4. The false rejection and the false acceptance rates for the DNA measurements.

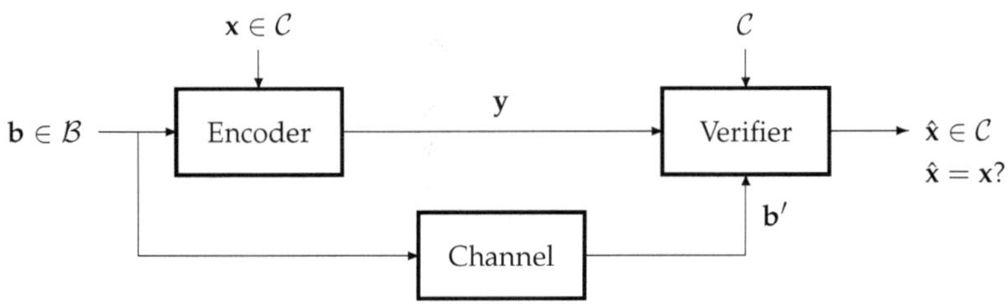

Fig. 2. General authentication scheme.

decision if one of them is equal to Hash$(\mathbf{b})$. Such a scheme is secure up to the security of hashing, but requires the hash function to be defined over the set of $|\mathcal{B}|$ vectors and very large computational complexity. The block coding schemes can be viewed as solutions introduced to relax these requirements.

5. Block coding approach to the authentication problem

The coding problem for biometric verification can be presented as designing codes for the scheme in Figure 2. Let $\mathcal{C} \subset \mathcal{B}$ be a subset whose entries are codewords assigned by the designer. The encoding is the transformation of a pair $(\mathbf{x}, \mathbf{b}) \in \mathcal{C} \times \mathcal{B}$, where the vector $\mathbf{b}$ is generated by the source and $\mathbf{x}$ is chosen according to a uniform probability distribution over the code $\mathcal{C}$, to another vector $\mathbf{y} = (y_1, \ldots, y_n)$ belonging to some finite set $\mathcal{Y}$. The mappings

$$(\mathbf{x}, \mathbf{b}) \to \mathbf{y}, \quad (\mathbf{y}, \mathbf{b}') \to \mathbf{x}$$

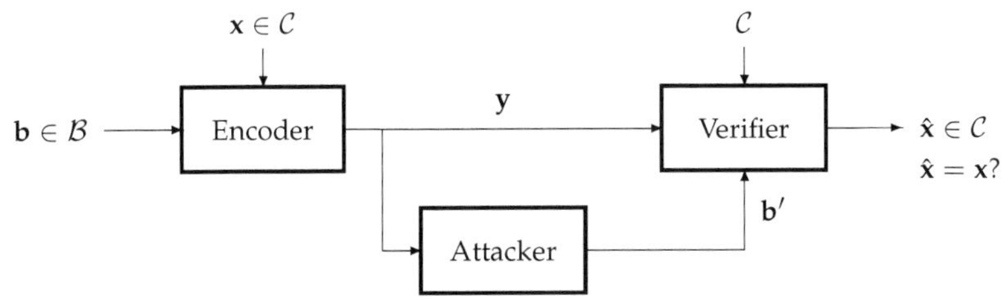

Fig. 3. General authentication scheme from the attacker's prospective.

are called the encoding and the decoding, respectively. The general requirement to the these mappings can be presented as

$$(\mathbf{x}, \mathbf{b}) \to \mathbf{y} \;\Rightarrow\; \begin{cases} \mathbf{b}' \in \mathcal{D}_T(\mathbf{b}) \Rightarrow (\mathbf{y}, \mathbf{b}') \to \mathbf{x}, \\ \mathbf{b}' \notin \mathcal{D}_T(\mathbf{b}) \Rightarrow (\mathbf{y}, \mathbf{b}') \not\to \mathbf{x}. \end{cases} \tag{19}$$

In other words, the results of the decoding for the vectors $\mathbf{b}$ and $\mathbf{b}'$ have to coincide if and only if $\mathbf{b}' \in \mathcal{D}_T(\mathbf{b})$.

Both the vector $\mathbf{y}$ and the value of $\text{Hash}(\mathbf{x})$ are stored in the database under the name of the person whose biometric characteristics are expressed by the vector $\mathbf{b}$. Having received the vector $\mathbf{b}'$ and the name of the person, the decoder reads $(\mathbf{y}, \text{Hash}(\mathbf{x}))$ from the database and uses the error–correcting capabilities of the code to decode "the transmitted codeword" $\mathbf{x}$ as $\hat{\mathbf{x}}$. If $\text{Hash}(\hat{\mathbf{x}}) = \text{Hash}(\mathbf{x})$, then the identity claim is accepted. Otherwise, the claim is rejected.

From the attacker's prospective, the authentication scheme can be viewed as the scheme in Figure 3. The attacker reads the content of the database associated with a person, presents the name of the person, and generates the vector $\mathbf{b}'$. The goal of the attacker is generating of a vector leading to the verifier's acceptance decision. The coding problem can be formulated as constructing codes that simultaneously satisfy the constraint (19) and guarantee a low probability of the attacker's success.

6. Additive block coding schemes

Given a positive integer q, let $\oplus_q$ and $\ominus_q$ denote the addition and the subtraction modulo q, respectively,

$$z \oplus_q z' = \begin{cases} z + z', & \text{if } z + z' \le q, \\ z + z' - q, & \text{if } z + z' > q \end{cases}$$

$$z \ominus_q z' = \begin{cases} z - z', & \text{if } z + z' \ge 0, \\ z - z' + q, & \text{if } z + z' < 0. \end{cases}$$

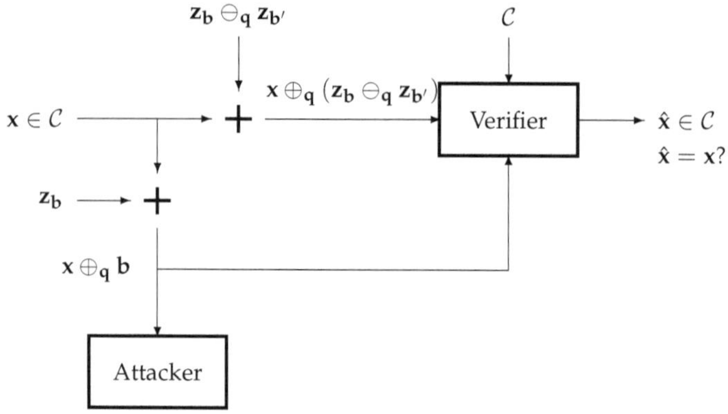

Fig. 4. Wiretap-type additive block coding scheme.

The operations $\oplus_\mathbf{q}$ and $\ominus_\mathbf{q}$, where $\mathbf{q} = (q_1, \ldots, q_n)$, being applied to the vectors of length n, are understood as component–wise addition and subtraction modulo $q_1, \ldots, q_n$, i.e.,

$$\mathbf{z} \oplus_\mathbf{q} \mathbf{z}' = (z_1 \oplus_{q_1} z_1', \ldots, z_n \oplus_{q_n} z_n'),$$
$$\mathbf{z} \ominus_\mathbf{q} \mathbf{z}' = (z_1 \ominus_{q_1} z_1', \ldots, z_n \ominus_{q_n} z_n').$$

Let us consider the biometric vector $\mathbf{b}$ as an additive noise that corrupts the transmitted codeword $\mathbf{x}$ and the received vector is defined as

$$\mathbf{y} = \mathbf{x} \oplus_\mathbf{q} \mathbf{z_b},$$

where $\mathbf{z_b}$ is the result of the transformation of the biometric vector $\mathbf{b}$ defined in (5). The decoding is based on the observation:

$$\left.\begin{array}{l} \mathbf{y} = \mathbf{x} \oplus_\mathbf{q} \mathbf{z_b} \\ \mathrm{Ham}(\mathbf{z_b}, \mathbf{z_{b'}}) \leq T \end{array}\right\} \Rightarrow \mathrm{Ham}(\mathbf{y}, \mathbf{x} \oplus_\mathbf{q} \mathbf{z_{b'}}) \leq T.$$

Notice also that

$$\mathbf{y} = \mathbf{x} \oplus_\mathbf{q} \mathbf{z_b} \Rightarrow \mathrm{Ham}(\mathbf{y}, \mathbf{x} \oplus_\mathbf{q} \mathbf{z_{b'}}) = \mathrm{Ham}(\mathbf{y} \ominus_\mathbf{q} \mathbf{z_{b'}}, \mathbf{x}) = \mathrm{Ham}(\mathbf{x} \oplus_\mathbf{q} (\mathbf{z_b} \ominus_\mathbf{q} \mathbf{z_{b'}}), \mathbf{x}).$$

Thus, the verifier analyzes the outcomes of transmission of the codeword $\mathbf{x}$ over two parallel channels,

$$\mathbf{x} \to \mathbf{x} \oplus_\mathbf{q} (\mathbf{z_b} \ominus_\mathbf{q} \mathbf{z_{b'}}) \quad \text{(the observation channel)},$$
$$\mathbf{x} \to \mathbf{x} \oplus_\mathbf{q} \mathbf{z_b} \quad \text{(the biometric channel)},$$

while the attacker analyzes only the output of the biometric channel (see Figure 4).

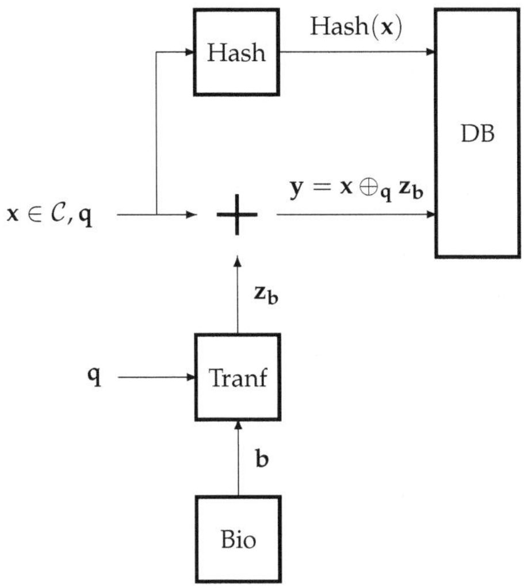

The enrollment stage

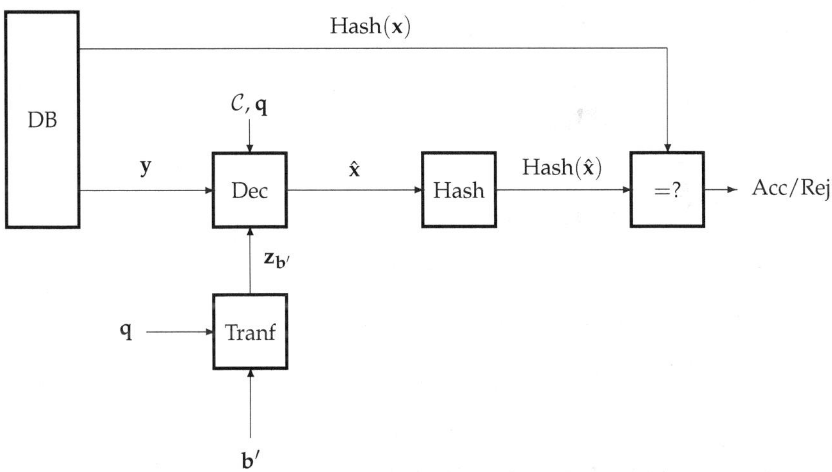

The verification stage

Fig. 5. The data processing in an additive block coding scheme.

Processing of a given biometric vector $\mathbf{b}$ at the enrollment stage and processing of data at the verification stage when the verifier considers only the output of the observation channel is illustrated in Figure 5.

The enrollment stage.

- *Choose a key codeword* $\mathbf{x}$ *according to a uniform probability distribution over the code* $\mathcal{C}$ *and compute the value of* $\text{Hash}(\mathbf{x})$.
- *Store* $(\text{Hash}(\mathbf{x}), \mathbf{x} \oplus_{\mathbf{q}} \mathbf{z_b})$ *in the database.*

The verification stage.

- *Read the data* $(\text{Hash}(\mathbf{x}), \mathbf{y})$ *associated with the claimed person from the database.*
- *Decode the key codeword, given a received vector* $\mathbf{z} = \mathbf{y} \ominus_{\mathbf{q}} \mathbf{z_{b'}}$, *as* $\hat{\mathbf{x}}$. *If* $\text{Hash}(\hat{\mathbf{x}}) = \text{Hash}(\mathbf{x})$, *then make the acceptance decision* (Acc). *If* $\text{Hash}(\hat{\mathbf{x}}) \neq \text{Hash}(\mathbf{x})$, *then make the rejection decision* (Rej).

Let us illustrate the additive block coding and the decoding algorithms that will be described in a general form by the numerical example. Let $q_1 = \cdots = q_6 = 2$, $n = 6$, and let $\mathcal{C}$ be a binary code consisting of 8 codewords,

$\mathbf{x}_1$	$\mathbf{x}_2$	$\mathbf{x}_3$	$\mathbf{x}_4$	$\mathbf{x}_5$	$\mathbf{x}_6$	$\mathbf{x}_7$	$\mathbf{x}_8$
000000	001011	010101	011110	100110	101101	110011	111000

For example,

$$\left.\begin{array}{r} \mathbf{z_b} = 011011 \\ \mathbf{x} = 011110 \end{array}\right\} \rightarrow \mathbf{y} = 000101,$$

and the vector $\mathbf{y}$ is stored in the database. Having received another vector $\mathbf{z_{b'}}$, the verifier tries to find a codeword $\hat{\mathbf{x}}$ located at distance at most 1 from the vector $\mathbf{y} \ominus_{\mathbf{q}} \mathbf{z_{b'}}$. For example,

$$\left.\begin{array}{r} \mathbf{z_{b'}} = 111011 \\ \mathbf{y} = 000101 \end{array}\right\} \rightarrow \mathbf{y} \ominus_{\mathbf{q}} \mathbf{z_{b'}} = 111110 \rightarrow \hat{\mathbf{x}} = 011110,$$

and the verifier makes the acceptance decision, since $\hat{\mathbf{x}} = \mathbf{x}$ implies $\text{Hash}(\hat{\mathbf{x}}) = \text{Hash}(\mathbf{x})$. An attacker wants to submit some vector $\mathbf{b}'$, which also leads to the acceptance. He constructs the list of candidate vectors as $\mathbf{y} \ominus_{\mathbf{q}} \mathbf{x}$, $\mathbf{x} \in \mathcal{C}$, and finds the vector $\hat{\mathbf{x}}$ such that $\Omega(\mathbf{y} \ominus_{\mathbf{q}} \mathbf{x})$ is the maximum. For example,

$\mathbf{y} \ominus_{\mathbf{q}} \mathbf{x}_1$	$\mathbf{y} \ominus_{\mathbf{q}} \mathbf{x}_2$	$\mathbf{y} \ominus_{\mathbf{q}} \mathbf{x}_3$	$\mathbf{y} \ominus_{\mathbf{q}} \mathbf{x}_4$	$\mathbf{y} \ominus_{\mathbf{q}} \mathbf{x}_5$	$\mathbf{y} \ominus_{\mathbf{q}} \mathbf{x}_6$	$\mathbf{y} \ominus_{\mathbf{q}} \mathbf{x}_7$	$\mathbf{y} \ominus_{\mathbf{q}} \mathbf{x}_8$
000101	001110	**010000**	011011	100011	101000	110110	111101

In particular, if the probabilities $\Omega(\mathbf{z})$ decrease when the weight of the vector $\mathbf{z}$ increases, then this algorithm brings the vector $\hat{\mathbf{x}} = \mathbf{x}_3$, and the attacker's vector $\mathbf{b}'$ is such that $\mathbf{z_{b'}} = \mathbf{z}_{\mathbf{y} \ominus_{\mathbf{q}} \mathbf{x}_3}$. Suppose that $\mathcal{C}$ is a block code consisting of M codewords $\mathbf{x}_1, \ldots, \mathbf{x}_M \in \mathcal{Z}_1 \times \cdots \times \mathcal{Z}_n$ and having the minimum distance greater than $2T$, i.e.,

$$\left.\begin{array}{r} \mathbf{x}, \mathbf{x}' \in \mathcal{C} \\ \mathbf{x} \neq \mathbf{x}' \end{array}\right\} \Rightarrow \text{Ham}(\mathbf{x}, \mathbf{x}') \geq 2T + 1. \tag{20}$$

Then the Hamming balls of radius T centered at codewords, $\mathcal{D}_T(\mathbf{x})$, $\mathbf{x} \in \mathcal{C}$, are pairwise disjoint sets. As a result, for any $\mathbf{y}, \mathbf{z_{b'}} \in \mathcal{Z}$, there is at most one codeword $\mathbf{x} \in \mathcal{C}$ such that

$$\mathrm{Ham}(\mathbf{y}, \mathbf{x} \oplus_{\mathbf{q}} \mathbf{z}_{\mathbf{b}'}) \leq T. \tag{21}$$

Let us denote this codeword by $\hat{\mathbf{x}}(\mathbf{y}, \mathbf{z}_{\mathbf{b}'})$. If the inequality (21) does not hold for all codewords, we assume that $\hat{\mathbf{x}}(\mathbf{y}, \mathbf{z}_{\mathbf{b}'})$ is a fixed vector (for example, the all–zero vector). Thus,

$$\mathrm{Ham}(\mathbf{z}_{\mathbf{b}}, \mathbf{z}_{\mathbf{b}'}) \leq T \Rightarrow \mathrm{Ham}(\mathbf{x} \oplus_{\mathbf{q}} \mathbf{z}_{\mathbf{b}}, \mathbf{x} \oplus_{\mathbf{q}} \mathbf{z}_{\mathbf{b}'}) \leq T \Rightarrow \hat{\mathbf{x}}(\mathbf{x} \oplus_{\mathbf{q}} \mathbf{z}_{\mathbf{b}}, \mathbf{z}_{\mathbf{b}'}) = \mathbf{x}.$$

Hence, if $\mathbf{x}$ is the codeword, which was used to encode the vector $\mathbf{z}_{\mathbf{b}}$, and the vector $\mathbf{z}_{\mathbf{b}'}$ differs from the vector $\mathbf{z}_{\mathbf{b}}$ in at most T components, then the codeword is decoded. Therefore the false rejection rate is expressed by (14),

$$\mathrm{FRR} = \sum_{\mathbf{b}, \mathbf{b}' \in \mathcal{B}} \omega(\mathbf{b}) V(\mathbf{b}' | \mathbf{b}) \chi\{\mathbf{z}_{\mathbf{b}'} \notin \mathcal{D}_T(\mathbf{z}_{\mathbf{b}})\}.$$

The similar conclusion is valid for the false acceptance rate of a randomly chosen person,

$$\overline{\mathrm{FAR}} = \sum_{\mathbf{b}, \mathbf{b}'} \omega(\mathbf{b}) \omega(\mathbf{b}') \chi\{\mathrm{Ham}(\mathbf{z}_{\mathbf{b}}, \mathbf{z}_{\mathbf{b}'}) \leq T\}.$$

Let us analyze the situation when an attacker is present. He receives only the result of transmission of the codeword over the biometric channel and his action can be presented as the mapping

$$(\mathbf{z}_{\mathbf{b}_1} = \mathbf{y} \ominus_{\mathbf{q}} \mathbf{x}_1, \ldots, \mathbf{z}_{\mathbf{b}_M} = \mathbf{y} \ominus_{\mathbf{q}} \mathbf{x}_M) \rightarrow \mathbf{b}' = \mathbf{b}_{\hat{m}},$$

where $\hat{m} \in \{1, \ldots, M\}$ is chosen in such a way that

$$\Omega(\mathbf{z}_{\mathbf{b}_{\hat{m}}}) = \max_{1 \leq m \leq M} \Omega(\mathbf{z}_{\mathbf{b}_m}). \tag{22}$$

The submission of the vector $\mathbf{b}_{\hat{m}}$ to the verifier implies $\hat{\mathbf{x}} = \mathbf{x}_{\hat{m}}$, and the acceptance decision is made if and only if $\mathbf{x}_{\hat{m}}$ is the codeword that was used to encode the biometric vector at the enrollment stage. The probability of the attacker's success, given the vectors $\mathbf{z}_{\mathbf{b}_1}, \ldots, \mathbf{z}_{\mathbf{b}_M}$, is equal to

$$\frac{\Omega(\mathbf{z}_{\mathbf{b}_{\hat{m}}})}{\sum_{m=1}^{M} \Omega(\mathbf{z}_{\mathbf{b}_m})} \leq \frac{\max_{1 \leq m \leq M} \Omega(\mathbf{z}_{\mathbf{b}_m})}{M \min_{1 \leq m \leq M} \Omega(\mathbf{z}_{\mathbf{b}_m})} \leq \frac{\max_{\mathbf{z} \in \mathcal{Z}} \Omega(\mathbf{z})}{M \min_{\mathbf{z} \in \mathcal{Z}} \Omega(\mathbf{z})} = \frac{1}{M} \prod_{t=1}^{n} \rho_t, \tag{23}$$

where $\rho_1, \ldots, \rho_n$ are defined in (6). Since the upper bound (23) holds for any received vector $\mathbf{y}$, which determines the vectors $\mathbf{z}_{\mathbf{b}_1}, \ldots, \mathbf{z}_{\mathbf{b}_M}$,

$$\mathrm{FAR} \leq \frac{1}{M} \prod_{t=1}^{n} \rho_t. \tag{24}$$

Let us evaluate the bound (24) using the standard covering arguments of coding theory. Given the vector $\mathbf{q}$, introduce the generating function

$$G(z) = \prod_{t=1}^{n} G_t(z),$$

where

$$G_t(z) = \frac{1}{q_t} + \frac{q_t - 1}{q_t} z.$$

For example, for the DNA data (see (13)),

$$G_{\text{DNA}}(z) = \Big(\frac{1}{2} + \frac{1}{2}z\Big)^7 \Big(\frac{1}{4} + \frac{3}{4}z\Big)^8 \Big(\frac{1}{8} + \frac{7}{8}z\Big)^9 \Big(\frac{1}{16} + \frac{1}{15}z\Big)^3 \Big(\frac{1}{64} + \frac{63}{64}z\Big)^1.$$

One can easily see that the d-th coefficient of the polynomial $G(z)$ is equal to the ratio of the number of vectors $\mathbf{x}' \in \mathcal{Z}$ located at the Hamming distance d from any fixed vector $\mathbf{x} \in \mathcal{Z}$ and $q_1 \dots q_n$, i.e.,

$$\frac{1}{\prod_{t=1}^n q_t} \Big| \Big\{ \mathbf{x}' \in \mathcal{Z} : \text{Ham}(\mathbf{x}, \mathbf{x}') = d \Big\} \Big| = \text{Coef}_d[\, G(z) \,].$$

Therefore,

$$\frac{1}{\prod_{t=1}^n q_t} |\mathcal{D}_T(\mathbf{x})| = \sum_{d=0}^{T} \text{Coef}_d[\, G(z) \,]. \tag{25}$$

Since $\mathcal{D}_T(\mathbf{x}_1), \dots, \mathcal{D}_T(\mathbf{x}_M)$ are pairwise disjoint sets,

$$\sum_{m=1}^{M} |\mathcal{D}_T(\mathbf{x}_m)| \le \prod_{t=1}^{n} q_t,$$

and (25) implies

$$\frac{1}{M} \ge \sum_{d=0}^{T} \text{Coef}_d[\, G(z) \,]. \tag{26}$$

By assuming that there is a code such that (26) holds with the equality and by replacing the parameters $\rho_1, \dots, \rho_M$ with 1's, we evaluate the false acceptance rate, estimated in (24), as

$$\text{FAR} \approx \widehat{\text{FAR}} = \sum_{d=0}^{T} \text{Coef}_d[\, G(z) \,].$$

The values of $\widehat{\text{FAR}}$ are given in Table 4 for the DNA data. As a result, one can conclude that the additive coding scheme can give a very efficient solution to the authentication problem provided that there is a class of specific codes having the certain minimum distance and corresponding decoding algorithms that require a low computational complexity.

7. Permutation block coding schemes

The permutation block coding scheme can be viewed as a modification of the scheme in Figure 4 where the sum modulo $\mathbf{q}$ in the link to the attacker is replaced by a stochastic mapping $f(\mathbf{x}, \mathbf{b})$, as it is shown in Figure 6. In this section, we will assume that $q = 2$. In particular, the modification of a wiretap-type block coding scheme is possible when both the vector $\mathbf{x}$ and $\mathbf{b}$ have equal weights and $f(\mathbf{x}, \mathbf{b})$ stands for the binary representation of a permutation π that transforms the vector $\mathbf{x}$ to the vector $\mathbf{b}$. Formally, let $\mathcal{B} = \{0,1\}^n_w$, where $\{0,1\}^n_w$ is the set consisting of binary vectors of the Hamming weight w. Thus, the biometric

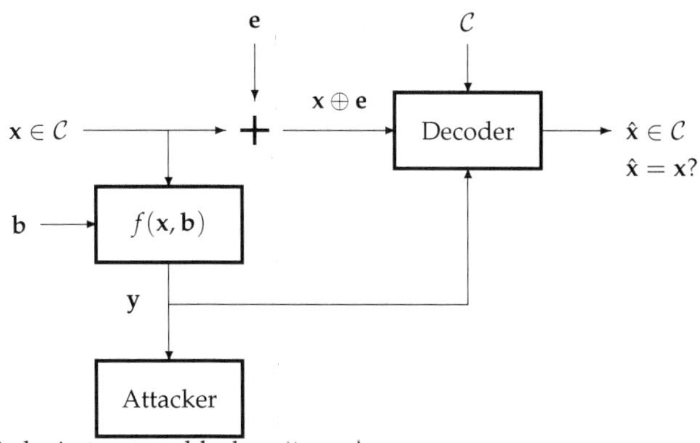

Fig. 6. Modified wiretap-type block coding scheme.

vector is a binary vector $\mathbf{b}$ of length n chosen by a combinatorial (n, w)-source, i.e.,

$$\text{wt}(\mathbf{b}) \neq w \ \Rightarrow \ \Pr_{\text{bio}}\{B = \mathbf{b}\} = 0. \tag{27}$$

Let $\mathcal{C}$ denote a binary code consisting of M different codewords of length n and weight w, i.e., $\mathcal{C} \subseteq \{0,1\}^n_w$ and $|\,\mathcal{C}\,| = M$.
The permutation of components of some vector $\mathbf{x} = (x_1, \ldots, x_n) \in \{0,1\}^n_w$ is determined by a vector $\pi \in \mathcal{P}$ in such a way that $\pi(\mathbf{x}) = (x_{\pi_1}, \ldots, x_{\pi_n})$, where $\mathcal{P}$ is the set of all possible permutations of components of the vector $(1, \ldots, n)$. Given a vector $\mathbf{b} \in \{0,1\}^n_w$ and a permutation $\pi \in \mathcal{P}$, let $\pi^{-1} \in \mathcal{P}$ denote the inverse permutation, i.e., $\pi^{-1}(\mathbf{b}) = (b_{i_1(\pi)}, \ldots, b_{i_n(\pi)})$, where $i_j(\pi) \in \{1, \ldots, n\}$ is the index determined by the equation $\pi_{i_j(\pi)} = j$.
For all vectors $\mathbf{x}, \mathbf{b} \in \{0,1\}^n_w$, let

$$\mathcal{P}(\mathbf{x} \to \mathbf{b}) = \{\, \pi \in \mathcal{P} : \ \pi(\mathbf{x}) = \mathbf{b} \,\} \tag{28}$$

denote the set of permutations that transform the vector $\mathbf{x}$ to the vector $\mathbf{b}$. Let us introduce the probability distribution

$$\gamma_{\mathbf{x},\mathbf{b}} = (\,\gamma(\pi|\mathbf{x}, \mathbf{b}), \ \pi \in \mathcal{P}\,)$$

in such a way that $\gamma(\pi|\mathbf{x}, \mathbf{b})$ can be positive only if $\pi \in \mathcal{P}(\mathbf{x} \to \mathbf{b})$. Let us also denote a uniform probability distribution over the set $\mathcal{P}(\mathbf{x} \to \mathbf{b})$ by

$$\overline{\gamma}_{\mathbf{x},\mathbf{b}} = (\,\overline{\gamma}(\pi|\mathbf{x}, \mathbf{b}), \ \pi \in \mathcal{P}\,),$$

where

$$\overline{\gamma}(\pi|\mathbf{x}, \mathbf{b}) = \begin{cases} |\,\mathcal{P}(\mathbf{x}, \mathbf{b})|^{-1}, & \text{if } \pi \in \mathcal{P}(\mathbf{x} \to \mathbf{b}), \\ 0, & \text{if } \pi \notin \mathcal{P}(\mathbf{x} \to \mathbf{b}). \end{cases}$$

For example, let $n = 4, k = 2$. The set $\{0,1\}^4_2$ consists of $\binom{4}{2} = 6$ binary vectors of length 4 having the weight 2 and $\mathcal{P}$ is the set consisting of $4! = 24$ permutations of components of the vector $(1,2,3,4)$. For all $\mathbf{x}, \mathbf{b} \in \{0,1\}^4_2$, the set $\mathcal{P}(\mathbf{x} \to \mathbf{b})$ consists of $2!2! = 4$ permutations. In particular,

$$\mathcal{P}(1100 \to 1010) = \{\, 1324, 1423, 2314, 2413 \,\}.$$

Notice that

$$\left.\begin{array}{l} \mathbf{b} \;=\; \pi(\mathbf{x}) \\ \mathbf{b}' \;=\; \mathbf{b} \oplus \mathbf{e} \end{array}\right\} \Rightarrow \pi^{-1}(\mathbf{b}') = \pi^{-1}(\mathbf{b}) \oplus \pi^{-1}(\mathbf{e}) = \mathbf{x} \oplus \pi^{-1}(\mathbf{e}) \tag{29}$$

and

$$\mathrm{wt}(\pi^{-1}(\mathbf{e})) = \mathrm{wt}(\mathbf{e}), \tag{30}$$

i.e., the decoder observes "the transmitted codeword" $\mathbf{x}$ as $\mathbf{x} \oplus \pi^{-1}(\mathbf{e})$. If the source generating the noise vectors is assumed to be a memoryless source, then (30) implies that the presence of the permutation π^{-1} does not affect the decoding strategy, and the scheme is equivalent to the one in Figure 6.

Processing of a given biometric vector $\mathbf{b}$ at the enrollment stage and processing data at the verification stage when the verifier considers only the output of the observation channel is illustrated in Figure 7.

The enrollment stage.

- *Choose a key codeword* $\mathbf{x}$ *according to a uniform probability distribution over the code* $\mathcal{C}$ *and compute the value of* $\mathrm{Hash}(\mathbf{x})$.
- *Given a pair of vectors* $(\mathbf{x}, \mathbf{b}) \in \{0,1\}^n_w \times \{0,1\}^n_w$, *choose a permutation* $\pi \in \mathcal{P}$ *according to the probability distribution* $\gamma_{\mathbf{x},\mathbf{b}}$.
- *Store* $(\mathrm{Hash}(\mathbf{x}), \pi)$ *in the database.*

The verification stage.

- *Read the data* $(\mathrm{Hash}(\mathbf{x}), \pi)$ *associated with the claimed person from the database.*
- *Apply the inverse permutation* π^{-1} *to the vector* $\mathbf{b}'$ *and decode the key codeword given a received vector* $\pi^{-1}(\mathbf{b}')$ *as* $\hat{\mathbf{x}}$. *If* $\mathrm{Hash}(\hat{\mathbf{x}}) = \mathrm{Hash}(\mathbf{x})$, *then accept the identity claim* (Acc). *If* $\mathrm{Hash}(\hat{\mathbf{x}}) \neq \mathrm{Hash}(\mathbf{x})$, *then reject the identity claim* (Rej).

One can easily see that if the code $\mathcal{C}$ satisfies (20), then (29), (30) guarantee that the false rejection rate FRR and the false acceptance rate for a randomly chosen person $\overline{\mathrm{FAR}}$ are the same as for the additive block coding scheme. Therefore, the reasons for introducing the more advanced permutation scheme are caused by possible decrease of the false acceptance rate for an attacker. We will derive a general formula for the FAR and demonstrate the effects for a specific assignment of input data.

Let

$$\gamma = (\, \gamma_{\mathbf{x},\mathbf{b}}, \; \mathbf{x}, \mathbf{b} \in \{0,1\}^n_w \,)$$

denote the list of conditional probability distributions over the set $\mathcal{P}$. In general, the attacker applies a fixed function $\psi : \mathcal{P} \to \{0,1\}^n$ to the permutation π stored in the DB and submits the vector $\mathbf{b}' = \psi(\pi)$ to the verifier. Let us assume that the verifier decodes the key codeword as the vector $\hat{\mathbf{x}}[\pi^{-1}(\mathbf{b}')]$. The probability of successful attack can be expressed as

$$\mathrm{FAR} = \frac{1}{M} \sum_{\mathbf{x} \in \mathcal{C}} \sum_{\mathbf{b}} \omega(\mathbf{b}) \sum_{\pi \in \mathcal{P}} \gamma(\pi | \mathbf{x}, \mathbf{b}) \chi\{\, \hat{\mathbf{x}}[\pi^{-1}(\psi(\pi))] = \mathbf{x} \,\}, \tag{31}$$

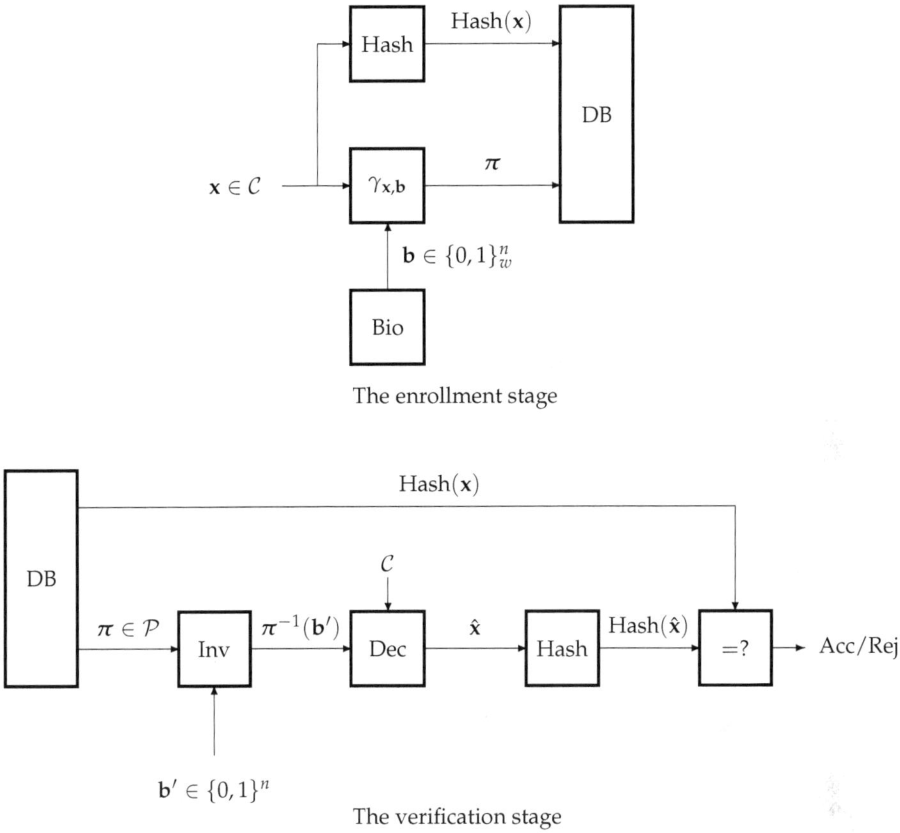

Fig. 7. The data processing in a permutation block coding scheme.

and one can easily see that FAR is maximized when the attacker applies the maximum *a posteriori* probability decoding, which results in

$$\psi(\pi) = \pi\Big(\arg\max_{\mathbf{x}\in\mathcal{C}} \gamma_{\text{bio}}(\pi|\mathbf{x}) \Big),$$

where

$$\gamma_{\text{bio}}(\pi|\mathbf{x}) = \sum_{\mathbf{b}} \omega(\mathbf{b})\gamma(\pi|\mathbf{x},\mathbf{b}). \tag{32}$$

Then

$$\text{FAR} = \frac{1}{M} \sum_{\pi\in\mathcal{P}} \max_{\mathbf{x}\in\mathcal{C}} \gamma_{\text{bio}}(\pi|\mathbf{x}).$$

Notice that $(\gamma_{\rm bio}(\pi|\mathbf{x}),\ \pi \in \mathcal{P})$ is the conditional probability distribution over the set $\mathcal{P}$ and

$$\sum_{\pi \in \mathcal{P}} \gamma_{\rm bio}(\pi|\mathbf{x}) = 1.$$

Notice also that the vector $\mathbf{x} \in \{0,1\}^n_w$ and the permutation $\pi \in \mathcal{P}$ uniquely determine the vector $\mathbf{b}^0 \in \{0,1\}^n_w$ such that $\pi \in \mathcal{P}(\mathbf{x} \to \mathbf{b}^0)$. Namely, $\mathbf{b}^0 = \pi(\mathbf{x})$, and the sum at the right-hand side of (32) contains at most one non–zero term.

The attacker has two simple possibilities: 1) fix a codeword $\mathbf{x}' \in \mathcal{C}$ and submit the vector $\mathbf{b}' = \pi(\mathbf{x}')$; 2) submit the most likely biometric vector. In the first case, the attacker has to know the code $\mathcal{C}$ and the stored permutation π. In the second case, he does not know these data and equivalent to an attacker, who does not have access to the database and ignorant about the code. One can easily see that the probabilities of successful attacks are equal to $1/M$ and ω^*, respectively. Therefore the probability of successful attack under the maximum *a posteriori* probability decoding of the key codeword is bounded from below as follows:

$$\text{FAR} \geq \max\Big\{\frac{1}{M}, \omega^*\Big\}.$$

Let $n = 8$, $w = 4$, $M = 4$. Let the codewords $\mathbf{x}_1, \ldots, \mathbf{x}_4$ and the biometric vectors that can be processed at the enrollment stage be specified as

$$\begin{bmatrix}\mathbf{x}_1\\ \mathbf{x}_2\\ \mathbf{x}_3\\ \mathbf{x}_4\end{bmatrix} = \begin{bmatrix}00110011\\ 01010101\\ 10101010\\ 11001100\end{bmatrix}, \quad \begin{bmatrix}\mathbf{b}_1\\ \cdot\\ \cdot\\ \mathbf{b}_6\end{bmatrix} = \begin{bmatrix}00001111\\ 00110011\\ 01010101\\ 10101010\\ 11001100\\ 11110000\end{bmatrix},$$

i.e., $\mathcal{C} = \{\mathbf{x}_1, \mathbf{x}_2, \mathbf{x}_3, \mathbf{x}_4\}$ and $\mathcal{B} = \{\mathbf{b}_1, \ldots, \mathbf{b}_6\}$. Then, for all pairs of vectors $(\mathbf{x}, \mathbf{b}) \in \mathcal{C} \times \mathcal{B}$,

$$|\,\mathcal{P}(\mathbf{x} \to \mathbf{b})\,| = (4!)^2 = 576 \tag{33}$$

and

$$|\,\mathcal{P}_{\mathcal{C}\to\mathcal{B}}(\mathbf{x} \to \mathbf{b})\,| = 4(2!)^4 = 64, \tag{34}$$

where $\mathcal{P}_{\mathcal{C}\to\mathcal{B}}(\mathbf{x} \to \mathbf{b})$ denotes the set of permutations $\pi \in \mathcal{P}(\mathbf{x} \to \mathbf{b})$ such that $\pi(\mathbf{x}') \in \mathcal{B}$ for all $\mathbf{x}' \in \mathcal{C}$.

Let us illustrate our considerations by the following examples:

$$\begin{bmatrix}\pi'\\ \pi'(\mathbf{x}_1)\\ \pi'(\mathbf{x}_2)\end{bmatrix} = \begin{bmatrix}1\,2\,5\,6\,3\,4\,7\,8\\ 0\,0\,0\,0\,1\,1\,1\,1\\ 0\,1\,0\,1\,0\,1\,0\,1\end{bmatrix}, \quad \begin{bmatrix}\pi''\\ \pi''(\mathbf{x}_1)\\ \pi''(\mathbf{x}_2)\end{bmatrix} = \begin{bmatrix}1\,2\,6\,5\,3\,4\,7\,8\\ 0\,0\,0\,0\,1\,1\,1\,1\\ 0\,1\,1\,0\,0\,1\,0\,1\end{bmatrix}.$$

The permutations π' and π'' belong to the set $\mathcal{P}$. Furthermore, $\pi'(\mathbf{x}_1) = \pi''(\mathbf{x}_1) = \mathbf{b}_1$. However $\pi'(\mathbf{x}_2) \in \mathcal{B}$, while $\pi''(\mathbf{x}_2) \notin \mathcal{B}$. Suppose that π' is the permutation stored in the database. The attacker applies this permutation to all codewords of the code $\mathcal{C}$ and constructs the list $\pi'(\mathbf{x}_1), \ldots, \pi'(\mathbf{x}_4)$. All entries of this list are possible biometric vectors. If the permutation π'' is stored in the database, then the list $\pi'(\mathbf{x}_1), \ldots, \pi'(\mathbf{x}_4)$ contains only

2 biometric vectors. The probability of successful attack is greater in the second case, and the permutation π' can be considered as "a bad" permutation.

The most of the permutations are bad permutations (see (33), (34)). This observation leads to the statement that the uniform probability distribution over the set $\mathcal{P}(\mathbf{x} \to \mathbf{b})$, where $\mathbf{x}$ is the selected codeword and $\mathbf{b}$ is the biometric vector, can bring a rather poor performance. Namely, suppose that the probability distribution over the set $\mathcal{B}$ is uniform, i.e., $\omega(\mathbf{b}) = 1/6$ for all $\mathbf{b} \in \mathcal{B}$. Let $\mathbf{x}$ be the codeword of the code $\mathcal{C}$ used at the enrollment stage. If $\gamma_{\mathbf{x},\mathbf{b}} = \overline{\gamma}_{\mathbf{x},\mathbf{b}}$, then the permutation is uniformly chosen from the set containing 576 entries. Only 64 of these permutations have the property that the set $\pi(\mathbf{x})$, $\mathbf{x} \in \mathcal{C}$ contains 4 biometric vectors, and the probability of successful attack is equal to $1/4$. For the other 512 permutations, the set $\pi(\mathbf{x})$, $\mathbf{x} \in \mathcal{C}$, contains 2 biometric vectors, and the probability of successful attack is equal to $1/2$. Thus

$$\mathrm{FAR} = \frac{64}{576}(1/4) + \frac{512}{576}(1/2) = 17/36.$$

Let us assign $\gamma_{\mathbf{x},\mathbf{b}}$ as a uniform probability distribution over the set $\mathcal{P}_{\mathcal{C}\to\mathcal{B}}(\mathbf{x} \to \mathbf{b})$ consisting of 64 entries. In all cases, the list $\pi(\mathbf{x})$, $\mathbf{x} \in \mathcal{C}$, contains 4 biometric vectors, and the probability of successful attack is equal to $1/4$. As a result, the probability of successful attack is expressed as

$$\mathrm{FAR} = \frac{64}{64}(1/4) = 1/4,$$

which is approximately twice less the value obtained with the uniform probability distribution. Moreover, we obtain that the lower bound $1/M$ on the probability FAR is attained with the equality.

Let us consider a non–uniform probability distribution over the set $\mathcal{B}$. Namely, let $a \in [1/4, 1/2]$ be a fixed parameter and let

$$\omega(\mathbf{b}) = \begin{cases} a, & \text{if } \mathbf{b} \in \{00001111, 11110000\}, \\ 1/4 - a/2, & \text{if } \mathbf{b} \in \mathcal{B} \backslash \{00001111, 11110000\}. \end{cases}$$

Notice that the set $\mathcal{P}_{\mathcal{C}\to\mathcal{B}}(\mathbf{x}_1 \to \mathbf{b}_1)$ contains 32 permutations π such that

$$\{\pi(\mathbf{x}_1), \pi(\mathbf{x}_2), \pi(\mathbf{x}_3), \pi(\mathbf{x}_4)\} = \{\mathbf{b}_1, \mathbf{b}_2, \mathbf{b}_5, \mathbf{b}_6\}$$

and 32 permutations π such that

$$\{\pi(\mathbf{x}_1), \pi(\mathbf{x}_2), \pi(\mathbf{x}_3), \pi(\mathbf{x}_4)\} = \{\mathbf{b}_1, \mathbf{b}_3, \mathbf{b}_4, \mathbf{b}_6\}.$$

Let us denote the subsets of these permutations by $\mathcal{P}'_{\mathcal{C}\to\mathcal{B}}(\mathbf{x}_1 \to \mathbf{b}_1)$ and $\mathcal{P}''_{\mathcal{C}\to\mathcal{B}}(\mathbf{x}_1 \to \mathbf{b}_1)$, respectively. Let

(a) $\gamma_{\mathbf{x}_1,\mathbf{b}_1}, \gamma_{\mathbf{x}_1,\mathbf{b}_6}$ be uniform probability distributions over the set $\mathcal{P}_{\mathcal{C}\to\mathcal{B}}(\mathbf{x}_1 \to \mathbf{b}_1)$;

(b) $\gamma_{\mathbf{x}_1,\mathbf{b}_2}, \gamma_{\mathbf{x}_1,\mathbf{b}_5}$ be uniform probability distributions over the set $\mathcal{P}'_{\mathcal{C}\to\mathcal{B}}(\mathbf{x}_1 \to \mathbf{b}_1)$;

(c) $\gamma_{\mathbf{x}_1,\mathbf{b}_3}, \gamma_{\mathbf{x}_1,\mathbf{b}_4}$ be uniform probability distributions over the set $\mathcal{P}''_{\mathcal{C}\to\mathcal{B}}(\mathbf{x}_1 \to \mathbf{b}_1)$.

If $\pi \in \mathcal{P}'_{\mathcal{C}\to\mathcal{B}}(\mathbf{x}_1 \to \mathbf{b}_1)$, then the *a posteriori* probabilities associated with the biometric vectors $\mathbf{b}_1, \mathbf{b}_2, \mathbf{b}_5, \mathbf{b}_6$ are equal to

$$\frac{1}{32}(a/2, 1/2 - a/2, 1/2 - a/2, a/2).$$

$\tilde{\mathbf{b}}$	$\tilde{w}$	i	$\mathbf{b}$	w	$\tilde{\mathbf{b}}$	$\tilde{w}$	i	$\mathbf{b}$	w	$\tilde{\mathbf{b}}$	$\tilde{w}$	i	$\mathbf{b}$	w	$\tilde{\mathbf{b}}$	$\tilde{w}$	i	$\mathbf{b}$	w	$\tilde{\mathbf{b}}$	$\tilde{w}$	i	$\mathbf{b}$	w
0000	0	2	1100	2	0001	1	1	1001	2	0010	1	1	1010	2	0100	1	1	1100	2	1000	1	3	0110	2
1111	4	2	0011	2	1110	3	1	0110	2	1101	3	1	0101	2	1011	3	1	0011	2	0111	3	3	1001	2

Table 5. Transformation of vectors of length $n = 4$ and weights 0,1,3,4 to balanced vectors, where $\tilde{w}, w$ are the Hamming weights of the vectors $\tilde{\mathbf{b}}, \mathbf{b}$ and i is the length of the prefix of the vector $\tilde{\mathbf{b}}$, which has to be inverted to obtain the vector $\mathbf{b}$.

However $a/2 \geq 1/2 - a/2$, and the attacker outputs either the key codeword, which is mapped to the vector $\mathbf{b}_1$, or the key codeword, which is mapped to the vector $\mathbf{b}_6$. Similar considerations can be presented for the permutations belonging to the set $\mathcal{P}''_{\mathcal{C}\to\mathcal{B}}(\mathbf{x}_1 \to \mathbf{b}_1)$. As a result, we conclude that

$$\text{FAR} = 64(a/64) = a,$$

i.e., the lower bound ω^* on the false acceptance rate is attained with the equality.
Let us consider the error–correcting capabilities of the verifier, who processes data of a legitimate user. Let P_w denote the probability that the vector $\mathbf{b}'$ differs from the vector $\mathbf{b}$ in w positions, $w = 0, \ldots, 8$. Then, assuming that the vectors $\mathbf{b}'$ are uniformly distributed over the set of vectors located at a fixed distance from the vector $\mathbf{b}$, we obtain that the probability of correct decoding for the code C and the threshold $T = 2$ is equal to

$$1 - \text{FRR} = P_0 + P_1 + (16/28)P_2,$$

since the decoder makes the correct decision for all error patterns of weight at most 1 and for 16 error patterns of weight 2 (the total number of error patterns of weight 2 is equal to 28). Suppose that the processed biometric vectors are constructed as a concatenation of L vectors $\mathbf{b}^{(1)}, \ldots, \mathbf{b}^{(L)} \in \mathcal{B}$, i.e., the total length of the vector is equal to $8L$. Suppose also that the vectors $\mathbf{b}^{(1)}, \ldots, \mathbf{b}^{(L)}$ are independently generated according to a uniform probability distribution over the set $\mathcal{B}$. Let the verifier make the acceptance decision if and only if such a decision is made for all L entries. Then the probability of correct decision is equal to $(1 - \text{FRR})^L$. On the other hand, the probability of successful attack, when the probability distributions $\gamma_{\mathbf{x},\mathbf{b}}$ are used is equal to $(1/4)^L$. This example illustrates the possibility of constructing the desired probability distribution over the permutations only for the subblocks of input data, and the search for good distributions is computationally feasible.
Notice that the fixed Hamming weight of the possible biometric vectors is the constraint that has to be satisfied to implement the permutation block coding scheme. It can be done if the observer takes into account only a fixed number of the most reliable biometric parameters. For example, in the case of processing fingerprints, one can put an $n_1 \times n_2$ grid on the 2-dimensional plane (in this case, $n = n_1 n_2$) and register the w most reliable minutiae points in the cells of that grid. In general case, the biometric binary vector of length n can be viewed as a vector of n features where positions of 1's index the features that are present in the outcomes of the measurements. The total number of the most reliable features taken into account by the authentication scheme can be fixed in advance.
Another useful possibility is known as balancing arbitrary binary vector by the inversion of its prefix in such a way that the obtained vector has weight $\lfloor n/2 \rfloor$. The corresponding statement is presented below, and the examples of the transformation are given in Table 5. One can see that, for any binary vector $\tilde{\mathbf{b}} \in \{0,1\}^n$, one can find an index $i \in \{0, \ldots, n\}$ in such a way that the vector $\tilde{\mathbf{b}}$ is transformed to a balanced vector by the inversion of the first i components,

i.e., $(i - \tilde{w}_i) + \tilde{w} - \tilde{w}_i = \lfloor n/2 \rfloor$, where $\tilde{w}$ and $\tilde{w}_i$ denote the Hamming weight of the vector $\tilde{\mathbf{b}}$ and the Hamming weight of the prefix of length i of the vector $\tilde{\mathbf{b}}$, respectively. The proof directly follows from the observation that the path on the plane whose coordinates are defined as $(j, \tilde{w}_j)$, $j = 0, \ldots, n$, starts at the point $(0, \mathrm{wt}(\mathbf{b}))$, ends at the point $(n, n - \mathrm{wt}(\mathbf{b}))$, and has increments ± 1. Therefore, there is at least one index i such that $\tilde{w}_i = \lfloor n/2 \rfloor$. Notice that the case $w = \lfloor n/2 \rfloor$ can be viewed as the most interesting one meaning the characteristics of the permutation block coding scheme. The claim above shows that an additional storage of the value of the parameter i used to transform an arbitrary binary vector to a vector belonging to the set $\{0,1\}^n_{\lfloor n/2 \rfloor}$ makes the implementation of such a scheme possible in general.

The mapping of the pair $(\mathbf{x}, \mathbf{b})$ to a binary string stored in the database can be viewed as the encryption of the message $\mathbf{b}$, which is parameterized by a key codeword $\mathbf{x} \in \mathcal{C}$ chosen at random. An interesting point is the possibility of decreasing the probability of successful attack, when an attacker tries to pass through the authentication stage with the acceptance decision, by using a randomized mapping, although *the values of additional random parameters are public.* In the permutation block coding scheme, a randomly chosen permutation that transforms the vector $\mathbf{x}$ to the vector $\mathbf{b}$ is used for these purposes. As the set of possible permutations has the cardinality, which is exponential in the length of the vectors, the designer has good chances to hide many of biometric vectors that differ from the most likely vector $\mathbf{b}^*$ into the information that can correspond to the vector $\mathbf{b}^*$. Thus, one can even reach exactly the same secrecy of the coded system as the secrecy of the blind guessing of the biometric vector, when the attacker does not have access to the database and ignorant about the code. In other words, one can talk about the possibility of constructing permutation block coding schemes that have *a perfect algorithmic secrecy*. This notion is different from the usual definition of perfectness, which is understood as the point that the conditional entropy of the probability distribution over the key codewords, given the content of the database, is equal to $\log M$. In our example presented in the previous subsection, the *a posteriori* probability distribution over the key codewords certainly depends on a particular permutation, and the conditional entropies of these distributions can be much less than the entropy of a uniform probability distribution. Nevertheless, an optimum attacker cannot use this fact, and his observations do not introduce changing in the decoding algorithm.

8. References

Bolle, R. M., Connell, J. H., Pankanti S., Ratha, N. K. & Senior A. W. (2004). *Guide to Biometrics,* Springer.

Cohen, G. & Zemor G. (2006). Syndrome–coding for the wiretap channel revisited, *Proceedings of IEEE Information Theory Workshop,* IEEE Press, China, pp. 33–36.

Dodis Y., Reyzin L. & Smith, A. (2004). Fuzzy extractors: How to generate strong keys from biometrics and other noisy data, *Advances in Cryptography: Lecture Notes in Computer Science,* no. 3027, Springer, pp. 523–540.

Gallager, R. (1968). *Information Theory and Reliable Communication,* Willey.

Juels, A. & Wattenberg, M. (1999). A fuzzy commitment scheme, *Proceedings of ACM Conference on Computer and Communication Security,* ACM Press, Singapore, pp. 28–36.

Knuth, D. E. (1986). Efficient balanced codes, *IEEE Transactions on Information Theory,* vol. 32, no. 1, pp. 51–53.

Korte, U., Krawczak, M., Merkle, J., Plaga, R., Niesing, M., Tiemann, C., Han Vinck, A. J., Martini, U. (2008). A cryptographic biometric authentication system based on genetic fingerprints, *Proceedings of Sicherheit*, Springer, Germany, pp. 263–276.

Wyner, A. (1975). The wiretap channel, *Bell System Technical Journal*, vol. 54, no. 8, pp. 1355–1387.

Balakirsky, V. B., Ghazaryan, A. R. & Han Vinck, A. J. (2006a). Processing fingerprints via binary codes: The BMW algorithm, *Proceedings of the 27th Symposium on Information Theory in the Benelux*, Lagendijk, R. L. & Weber, J. H. (Eds.), The Netherlands, pp. 267–274.

Balakirsky, V. B., Ghazaryan, A. R. & Han Vinck, A. J. (2006b). General principles of constructing biometric authentication schemes using block codes, *Proceedings of the International Workshop "Algorithms and Mathematical Methods in Networking"*, Han Vinck, A. J. (Ed.), Institute fur Experimentelle Mathematik Press, Germany, pp. 8–18.

Balakirsky, V. B., Ghazaryan, A. R. & Han Vinck, A. J. (2007). Testing the independence of two non–stationary random processes with applications to biometric authentication, *Proceedings of the International Symposium on Information Theory*, IEEE Press, France, pp. 2671–2675, 2007.

Balakirsky, V. B., Ghazaryan, A. R. & Han Vinck, A. J. (2008a). Additive block coding schemes for biometric authentication with the DNA data, *Lecture Notes in Computer Science*, vol. 5372, Schouten, B., et al. (Eds.), Springer, pp. 160–169.

Balakirsky, V. B., Ghazaryan, A. R. & Han Vinck, A. J. (2008b). Performance of additive block coding schemes oriented to biometric authentication, *Proceedings of the 29th Symposium on Information Theory in the Benelux*, Van de Perre, L. et. al (Eds.), Belgium, pp. 19–26.

Balakirsky, V. B., Ghazaryan, A. R. & Han Vinck, A. J. (2009a). Secrecy of permutation block coding schemes designed for biometric authentication, *Proceedings of the 30th Symposium on Information Theory in the Benelux*, Willems, F. M. J., & Tjalkens, T. J. (Eds.), The Netherlands, pp. 11–19.

Balakirsky, V. B., Ghazaryan, A. R. & Han Vinck, A. J. (2009b). Mathematical model for constructing passwords from biometrical data, *Security and Communication Networks*, vol. 2, no. 1, Wiley, pp. 1–9.

Balakirsky, V. B. & Han Vinck, A. J. (2010). A simple scheme for constructing fault–tolerant passwords from biometric data, *EURASIP Journal on Information Security*, vol. 2010, Article ID 819376, doi:10.1155/2010/819376.

16

Perceived Age Estimation from Face Images

Kazuya Ueki[1], Yasuyuki Ihara[1] and Masashi Sugiyama[2]
[1]*NEC Soft, Ltd.*
[2]*Tokyo Institute of Technology*
Japan

1. Introduction

In recent years, demographic analysis in public places such as shopping malls and stations is attracting a great deal of attention. Such demographic information is useful for various purposes, e.g., designing effective marketing strategies and targeted advertisement based on customers' gender and age. For this reason, a number of approaches have been explored for age estimation from face images (Fu et al., 2007; Geng et al., 2006; Guo et al., 2009), and several databases became publicly available recently (FG-Net Aging Database, n.d.; Phillips et al., 2005; Ricanek & Tesafaye, 2006). It has been reported that age can be accurately estimated under controlled environment such as frontal faces, no expression, and static lighting conditions. However, it is not straightforward to achieve the same accuracy level in a real-world environment due to considerable variations in camera settings, facial poses, and illumination conditions. The recognition performance of age prediction systems is significantly influenced by such factors as the type of camera, camera calibration, and lighting variations. On the other hand, the publicly available databases were mainly collected in semi-controlled environments. For this reason, existing age prediction systems built upon such databases tend to perform poorly in a real-world environment.

In this chapter, we address the problem of perceived age estimation from face images, and describe our new approaches proposed in Ueki et al. (2010) and Ueki et al. (2011), which involve three novel aspects.

The first novelty of our proposed approaches is to take the heterogeneous characteristics of human age perception into account. It is rare to misjudge the age of a 5-year-old child as 15 years old, but the age of a 35-year-old person is often misjudged as 45 years old. Thus, magnitude of the error is different depending on subjects' age. We carried out a large-scale questionnaire survey for quantifying human age perception characteristics, and propose to utilize the quantified characteristics in the framework of weighted regression.

The second is an efficient active learning strategy for reducing the cost of labeling face samples. Given a large number of unlabeled face samples, we reveal the cluster structure of the data and propose to label cluster-representative samples for covering as many clusters as possible. This simple sampling strategy allows us to boost the performance of a manifold-based semi-supervised learning method only with a relatively small number of labeled samples.

The third contribution is to apply a recently proposed machine learning technique called *covariate shift adaptation* (Shimodaira, 2000; Sugiyama & Kawanabe, 2011; Sugiyama et al.,

2007; 2008) to alleviating lighting condition change between laboratory and practical environment.

Through real-world age estimation experiments, we demonstrate the usefulness of the proposed approaches.

2. Age estimation based on age perception characteristics

In this section, we mathematically formulate the problem of age estimation, and show how human age perception characteristics can be incorporated systematically.

2.1 Formulation

Throughout this chapter, we perform age estimation based not on subjects' real age, but on their *perceived* age. Thus, the 'true' age of the subject y is defined as the average perceived age evaluated by those who observed the subject's face images (the value is rounded-off to the nearest integer).

Let us consider a regression problem of estimating the age y^* of subject $\boldsymbol{x}$ (face features). Suppose we are given labeled training data

$$\{(\boldsymbol{x}_i^{\mathrm{tr}}, y_i^{\mathrm{tr}})\}_{i=1}^{l}.$$

We use the following kernel model for age regression.

$$f(\boldsymbol{x};\boldsymbol{\alpha}) = \sum_{i=1}^{l} \alpha_i K(\boldsymbol{x}, \boldsymbol{x}_i^{\mathrm{tr}}), \tag{1}$$

where $\boldsymbol{\alpha} = (\alpha_1, \ldots, \alpha_l)^\top$ is a model parameter, $^\top$ denotes the transpose, and $K(\boldsymbol{x}, \boldsymbol{x}')$ is a *positive definite kernel* (Schölkopf & Smola, 2002). We use the Gaussian kernel:

$$k(x, x') = \exp\left(-\frac{\|x - x'\|^2}{2\sigma^2}\right),$$

where σ^2 is the Gaussian variance.

A standard approach to learning the model parameter $\boldsymbol{\alpha}$ would be *regularized least-squares* (Hoerl & Kennard, 1970).

$$\min_{\boldsymbol{\alpha}} \left[\frac{1}{l}\sum_{i=1}^{l}(y_i^{\mathrm{tr}} - f(\boldsymbol{x}_i^{\mathrm{tr}};\boldsymbol{\alpha}))^2 + \lambda\|\boldsymbol{\alpha}\|^2\right], \tag{2}$$

where $\|\cdot\|$ denotes the Euclidean norm, and $\lambda(> 0)$ is the regularization parameter to avoid overfitting.

Below, we explain that merely using regularized least-squares is not appropriate in real-world perceived age prediction, and show how to cope with this problem.

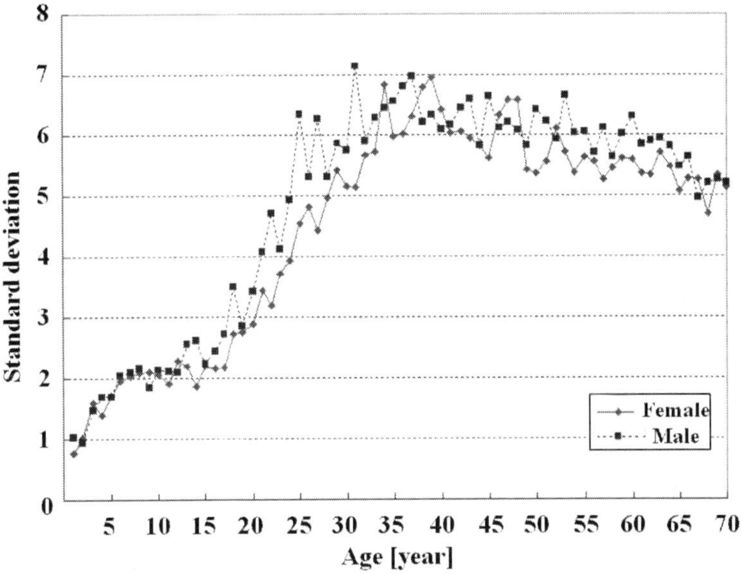

Fig. 1. The relation between subjects' true age y^* (horizontal axis) and the standard deviation of perceived age (vertical axis).

2.2 Incorporating age perception characteristics

Human age perception is known to have heterogeneous characteristics, e.g., it is rare to misjudge the age of a 5-year-old child as 15 years old, but the age of a 35-year-old person is often misjudged as 45 years old.

In order to quantify this phenomenon, we investigated human age perception characteristics through a large-scale questionnaire survey. We used an in-house face image database consisting of approximately 500 subjects whose age almost uniformly covers the range of our interest (i.e., age 1 to 70). For each subject, 5 to 10 face images with different face poses and lighting conditions were taken. We asked each of 72 volunteers to give age labels y to the subjects. The *'true'* age of a subject is defined as the average of estimated age labels y (rounded-off to the nearest integer) for that subject, and denoted by y^*. Then the standard deviation of age labels y is calculated as a function of y^*, which is summarized in Figure 1.

The standard deviation is approximately 2 (years) when the true age y^* is less than 15. The standard deviation increases and goes beyond 6 as the true age y^* increases from 15 to 35. Then the standard deviation decreases to around 5 as the true age y^* increases from 35 to 70. This graph shows that the perceived age deviation tends to be small in younger age brackets and large in older age groups. This would well agree with our intuition considering the human growth process.

Now let us incorporate the above survey result into the perceived age estimation framework described in Section 2.1. When the standard deviation is small (large), making an error is regarded as more (less) critical. This idea follows a similar line to the *Mahalanobis distance* (Duda et al., 2001), so it would be reasonable to incorporate the above survey result into the framework of *weighted regression analysis*. More precisely, weighting the goodness-of-fit term in Eq.(2) according to the inverse of the error variance optimally adjusts to the characteristics of human perception:

$$\min_{\alpha} \left[\frac{1}{l} \sum_{i=1}^{l} \frac{(y_i^{\mathrm{tr}} - f(x_i^{\mathrm{tr}}; \alpha))^2}{w_{\mathrm{age}}(y_i^{\mathrm{tr}})^2} + \lambda \|\alpha\|^2 \right], \tag{3}$$

where $w_{\mathrm{age}}(y)$ is the value given in Figure 1.

2.3 Evaluation criterion

Conventionally, the performance of an age prediction function $f(x)$ for test samples $\{(x_j^{\mathrm{te}}, y_j^{\mathrm{te}})\}_{j=1}^{t}$ was evaluated by the mean absolute error (MAE) (Geng et al., 2006; Lanitis et al., 2004; 2002; Ueki et al., 2008):

$$\mathrm{MAE} = \frac{1}{t} \sum_{j=1}^{t} \left| y_j^{\mathrm{te}} - f(x_j^{\mathrm{te}}) \right|.$$

However, as explained above, this does not properly reflect human age perception characteristics.

Here we propose to use the weighted criterion also for performance evaluation in experiments. More specifically, we evaluate the prediction performance by the *weighted mean squared error* (WMSE):

$$\mathrm{WMSE} = \frac{1}{t} \sum_{j=1}^{t} \frac{(y_j^{\mathrm{te}} - f(x_j^{\mathrm{te}}))^2}{w_{\mathrm{age}}(y_j^{\mathrm{te}})^2}. \tag{4}$$

The smaller the value of WMSE is, the better the age prediction function would be.

3. Semi-supervised approach

In this section, we give an active learning strategy and a semi-supervised age regression method within the age-weighting framework described in the previous section.

3.1 Clustering-based active learning strategy

First, we explain our active learning strategy for reducing the cost of labeling face samples.

Face samples contain various diversity such as individual characteristics, angles, lighting conditions, etc. They often possess cluster structure, and face samples in each cluster tend to have similar ages (Fu et al., 2007; Guo et al., 2008; Ueki et al., 2008). Based on these empirical observations, we propose to label the face images which are closest to cluster centroids.

For revealing the cluster structure, we apply the k-means clustering method (MacQueen, 1967) to a large number of unlabeled samples. Since clustering of high-dimensional data is often unreliable, we first apply *principal component analysis* (PCA) (Jolliffe, 1986) to the face images for dimension reduction, which is a well-justified preprocessing for k-means clustering (Ding & He, 2004). The proposed active learning strategy is summarized as follows.

1. For a set of d-dimensional unlabeled face image samples $\{X_i\}_{i=1}^{n}$, we compute $\{x_i\}_{i=1}^{n}$ of r ($\ll d$) dimensions by the PCA projection.
2. Using the k-means clustering algorithm, we compute the l ($\ll n$) cluster centroids $\{m_i\}_{i=1}^{l}$.

3. We choose $\{x_i^{\text{tr}} \mid x_i^{\text{tr}} = x_{\tau(i)}\}_{i=1}^{l}$ from $\{x_i\}_{i=1}^{n}$ as samples to be labeled, where

$$\tau(i) = \underset{i'}{\text{argmin}} \|x_{i'} - m_i\|,$$

and $\|\cdot\|$ denotes the Euclidean norm.

Let $\{y_i^{\text{tr}}\}_{i=1}^{l}$ be the labels for $\{x_i^{\text{tr}}\}_{i=1}^{l}$, and let the remaining samples of size u $(= n - l)$ that were not chosen to be labeled be denoted as

$$\{x_i^{\text{tr}}\}_{i=l+1}^{n} = \{x_i\}_{i=1}^{n} \backslash \{x_i^{\text{tr}}\}_{i=1}^{l}.$$

Thus, the first l training samples $\{x_i^{\text{tr}}\}_{i=1}^{l}$ are labeled, and the remaining u training samples $\{x_i^{\text{tr}}\}_{i=l+1}^{l+u}$ are unlabeled.

3.2 Semi-supervised age regression with manifold regularization

Face images often possess cluster structure, and face samples in each cluster tend to have similar ages. Here we utilize this cluster structure by employing a method of semi-supervised regression with manifold regularization (Sindhwani et al., 2006).
For age regression, we use the following kernel model:

$$f(x) = \sum_{i=1}^{l+u} \alpha_i k(x, x_i^{\text{tr}}), \tag{5}$$

where $\alpha = (\alpha_1, \ldots, \alpha_{l+u})^\top$ are parameters to be learned, $\top$ denotes the transpose, and $k(x, x')$ is a *reproducing kernel function*. We included $(l + u)$ kernels in the kernel regression model (5), but u can be very large in age prediction. In practice, we may only use c $(\leq u)$ elements randomly chosen from the set $\{k(x, x_i^{\text{tr}})\}_{i=l+1}^{l+u}$ for reducing the computational cost; then the total number of basis functions is reduced to $b = l + c$. However, we stick to Eq.(5) below for keeping the explanation simple.
We employ a manifold regularizer (Sindhwani et al., 2006) in our training criterion, i.e., the parameter α is learned so that the following criterion is minimized.

$$\frac{1}{l}\sum_{i=1}^{l} \frac{(y_i^{\text{tr}} - f(x_i^{\text{tr}}))^2}{w_{\text{age}}(y_i^{\text{tr}})^2} + \lambda\|\alpha\|^2 + \frac{\mu}{4(l+u)^2}\sum_{i,i'=1}^{l+u} A_{i,i'}(f(x_i^{\text{tr}}) - f(x_{i'}^{\text{tr}}))^2, \tag{6}$$

where λ and μ are non-negative regularization parameters. $A_{i,i'}$ represents the affinity between x_i^{tr} and $x_{i'}^{\text{tr}}$, which is defined by

$$A_{i,i'} = \exp\left(-\frac{\|x_i - x_{i'}\|^2}{2\nu^2}\right) \tag{7}$$

if x_i^{tr} is a h-nearest neighbor of $x_{i'}^{\text{tr}}$ or vice versa; otherwise $A_{i,i'} = 0$.
The first term in Eq.(6) is the goodness-of-fit term and the second term is the ordinary regularizer for avoiding overfitting. The third term is the manifold regularizer. The weight $A_{i,i'}$ tends to take large values if x_i^{tr} and $x_{i'}^{\text{tr}}$ belong to the same cluster. Thus, the manifold regularizer works for keeping the outputs of the function $f(x)$ within the same cluster close to each other.

An important advantage of the above training method is that the solution can be obtained *analytically* by

$$\widehat{\alpha} = \left(K^\top DK + l\lambda I_{l+u} + \frac{l\mu}{(l+u)^2} K^\top LK \right)^{-1} K^\top Dy, \tag{8}$$

where K is the $(l+u) \times (l+u)$ kernel Gram matrix whose (i,i')-th element is defined by

$$K_{i,i'} = k(\boldsymbol{x}_i^{\mathrm{tr}}, \boldsymbol{x}_{i'}^{\mathrm{tr}}).$$

D is the $(l+u) \times (l+u)$ diagonal weight matrix with diagonal elements defined by

$$w_{\mathrm{age}}(y_1^{\mathrm{tr}})^{-2}, \ldots, w_{\mathrm{age}}(y_l^{\mathrm{tr}})^{-2}, 0, \ldots, 0.$$

L is the $(l+u) \times (l+u)$ Laplacian matrix whose (i,i')-th entry is defined by

$$L_{i,i'} = \delta_{i,i'} \left(\sum_{i''=1}^{l+u} A_{i,i''} \right) - A_{i,i'},$$

where $\delta_{i,i'}$ is the Kronecker delta. I_{l+u} denotes the $(l+u) \times (l+u)$ identity matrix. y is the $(l+u)$-dimensional label vector defined as

$$y = (y_1^{\mathrm{tr}}, \ldots, y_l^{\mathrm{tr}}, 0, \ldots, 0)^\top.$$

If u is very large (which would be the case in age prediction), computing the inverse of the $(l+u) \times (l+u)$ matrix in Eq.(8) is not tractable. To cope with this problem, reducing the number of kernels from $(l+u)$ to a smaller number b would be a realistic option, as explained above. Then the matrix K becomes an $(l+u) \times b$ rectangular matrix and the identity matrix in Eq.(8) becomes I_b. Thus the size of the matrix we need to invert becomes $b \times b$, which would be tractable when b is kept moderate. We may further reduce the computational cost by numerically computing the solution by a *stochastic gradient-decent method* (Amari, 1967).

3.3 Empirical evaluation

Here, we apply the above age prediction method to in-house face-age datasets, and experimentally evaluate its performance.

3.3.1 Data acquisition and experimental setup

Age prediction systems are often used in public places such as shopping malls or train stations. In order to make our experiments realistic, we collected face image samples from video sequences taken by ceiling-mounted surveillance cameras with depression angle 5–10 degrees. The recording method, image resolution, and the image size are diverse depending on the recording conditions—for example, some subjects were illuminated by dominant light sources, walking naturally, seated on a stool, and keeping their heads still. The subjects' facial expressions were typically subtle, switching between neutral and smiling. We used a face detector for localizing the two eye-centers, and then rescaled the image to 64×64 pixels. Examples of face images are shown in Figure 2. Faces whose age ranges from 1 to 70 were used in our experiments.

As pre-processing, we extracted 100-dimensional features from the 64×64 face images using a neural network feature extractor proposed in Tivive & Bouzerdoumi (2006a) and

Fig. 2. Examples of face images.

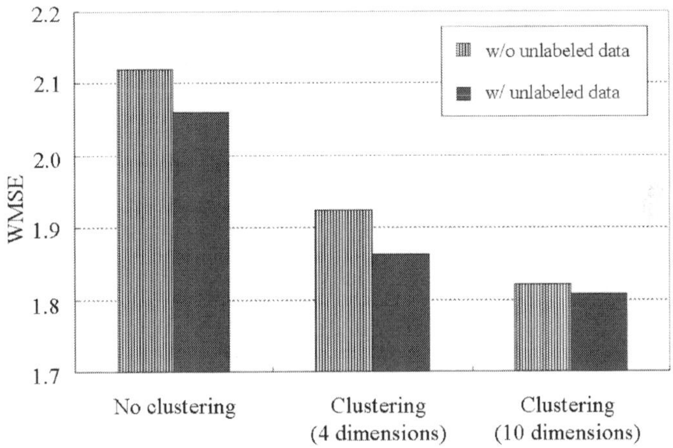

Fig. 3. Comparison of WMSE Eq.(4).

Tivive & Bouzerdoumi (2006b). In total, we have 28500 face samples in our database. Among them, $n = 27000$ are treated as unlabeled samples and the remaining $t = 1500$ are used as test samples. From the 27000 unlabeled samples, we choose $l = 200$ samples to be labeled by active learning. The Gaussian-kernel variance σ^2 and the regularization parameters λ and μ were determined so that WMSE for the test data is minimized (i.e., they are optimally tuned). For manifold regularization, we fixed the number of nearest neighbors and the decay rate of the similarity to $h = 5$ and $\nu = 1$, respectively (see Eq.(7)).

3.3.2 Results

We applied the k-means clustering algorithm to 27000 unlabeled samples in the 4-dimensional or 10-dimensional PCA subspace and extracted 200 clusters. We chose 200 samples that are closest to the 200 cluster centroids and labeled them; then we trained a regressor using the weighted manifold-regularization method described in Section 2.2 with the 200 labeled samples and 5000 unlabeled samples randomly chosen from the pool of 26800 (= 27000 − 200) unlabeled samples. We compared the above method with random sampling strategy. Figure 3 summarizes WMSE obtained by each method; in the comparison, we also included supervised regression where unlabeled samples were not used (i.e., $\mu = 0$).

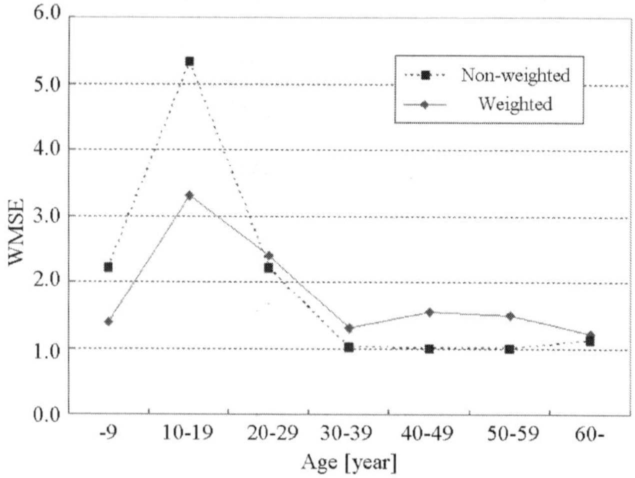

Fig. 4. WMSE for each age-group.

Figure 3 shows that the proposed active learning method gave smaller WMSE than the random sampling strategy; the use of unlabeled samples also improved the performance. Thus the proposed active learning method combined with manifold-based semi-supervised learning is shown to be effective for improving the age prediction performance.

In order to more closely understand the effect of age weighting, we investigated the prediction error for each age bracket. Figure 4 shows age-bracket-wise WMSE when the age-weighted learning method or the non-weighted learning method is used. The figure shows that the error in young age groups (less than 20 years old) is significantly reduced by the use of the age weights, which was shown to be highly important in practical human evaluation (see Section 2.2). On the other hand, the prediction error for middle/older age groups is slightly increased, but a small increase of the error in these age brackets was shown to be less significant in our questionnaire survey. Therefore, the experimental result indicates that our approach qualitatively improves the age prediction accuracy.

4. Coping with lighting condition change

In this section, we consider another semi-supervised learning setup where training and test samples follow different distributions. Such a situation often happens in real-world age prediction tasks, and we describe a systematic method to cope with such distribution change.

4.1 Lighting condition change as covariate shift

When designing age estimation systems, the environment of recording training face images is often different from the test environment in terms of lighting conditions. Typically, training data are recorded indoors such as a studio with appropriate illumination. On the other hand, in a real-world environment, lighting conditions have considerable varieties, e.g., strong sunlight might be cast from a side of the face or there is no enough light. In such situations, age estimation accuracy is significantly degraded.

Let $p_{\mathrm{tr}}(\boldsymbol{x})$ be the probability density function of training face features and $p_{\mathrm{te}}(\boldsymbol{x})$ be the probability density function of test face features. When these two densities are different, it would be natural to emphasize the influence of training samples $(\boldsymbol{x}_i^{\mathrm{tr}}, y_i^{\mathrm{tr}})$ which have high similarity to data in the test environment. Such adjustment can be systematically carried out as follows (Shimodaira, 2000; Sugiyama & Kawanabe, 2011; Sugiyama et al., 2007; 2008):

$$\min_{\boldsymbol{\alpha}} \left[\frac{1}{l} \sum_{i=1}^{l} w_{\mathrm{imp}}(\boldsymbol{x}_i^{\mathrm{tr}}) \frac{(y_i^{\mathrm{tr}} - f(\boldsymbol{x}_i^{\mathrm{tr}}; \boldsymbol{\alpha}))^2}{w_{\mathrm{age}}(y_i^{\mathrm{tr}})^2} + \lambda \|\boldsymbol{\alpha}\|^2 \right], \tag{9}$$

i.e., the goodness-of-fit term in Eq.(3) is weighted according to the *importance function* (Fishman, 1996):

$$w_{\mathrm{imp}}(\boldsymbol{x}) = \frac{p_{\mathrm{te}}(\boldsymbol{x})}{p_{\mathrm{tr}}(\boldsymbol{x})}.$$

The solution of Eq.(9) can be obtained analytically by

$$\widehat{\boldsymbol{\alpha}} = (K^\top W K + l\lambda I_l)^{-1} K^\top W \boldsymbol{y}, \tag{10}$$

where K is the kernel matrix whose (i, i')-th element is defined by

$$K_{i,i'} = K(\boldsymbol{x}_i^{\mathrm{tr}}, \boldsymbol{x}_{i'}^{\mathrm{tr}}),$$

W is the l-dimensional diagonal matrix with (i, i)-th diagonal element defined by

$$W_{i,i} = \frac{w_{\mathrm{imp}}(\boldsymbol{x}_i^{\mathrm{tr}})}{w_{\mathrm{age}}(y_i^{\mathrm{tr}})^2},$$

I_l is the l-dimensional identity matrix, and

$$\boldsymbol{y} = (y_1^{\mathrm{tr}}, \ldots, y_l^{\mathrm{tr}})^\top.$$

When the number of training data l is large, we may reduce the number of kernels in Eq.(1) so that the inverse matrix in Eq.(10) can be computed with limited memory; or we may compute the solution numerically by a stochastic gradient-decent method (Amari, 1967).

4.2 Importance-Weighted Cross-Validation (IWCV)

In supervised learning, the choice of models (for example, the basis functions and the regularization parameter) is crucial for obtaining better prediction performance. *Cross-validation* (CV) would be one of the most popular techniques for model selection (Stone, 1974). CV has been shown to give an *almost* unbiased estimate of the generalization error with finite samples (Schölkopf & Smola, 2002), but such almost unbiasedness is no longer fulfilled under covariate shift.

To cope with this problem, a variant of CV called *importance-weighted CV* (IWCV) has been proposed (Sugiyama et al., 2007). Let us randomly divide the training set

$$\mathcal{Z} = \{(\boldsymbol{x}_i^{\mathrm{tr}}, y_i^{\mathrm{tr}})\}_{i=1}^{l}$$

into M disjoint non-empty subsets $\{\mathcal{Z}_m\}_{m=1}^{M}$ of (approximately) the same size. Let $f_{\mathcal{Z}_m}(\boldsymbol{x})$ be a function learned from $\mathcal{Z}\backslash\mathcal{Z}_m$ (i.e., without $\mathcal{Z}_m$). Then the *M-fold IWCV* (IWCV) estimate of the generalization error is given by

$$\frac{1}{M}\sum_{m=1}^{M}\frac{1}{|\mathcal{Z}_m|}\sum_{(\boldsymbol{x},y)\in\mathcal{Z}_m}\frac{w_{\mathrm{imp}}(\boldsymbol{x})}{w_{\mathrm{age}}(y)^2}(f_{\mathcal{Z}_m}(\boldsymbol{x})-y)^2,$$

where $|\mathcal{Z}_m|$ denotes the number of samples in the subset $\mathcal{Z}_m$.

It was proved that IWCV gives an *almost* unbiased estimate of the generalization error even under covariate shift (Sugiyama et al., 2007).

4.3 Kullback-Leibler Importance Estimation Procedure (KLIEP)

In order to compute the solution (10) or performing IWCV, we need to know the values of the importance weights

$$w_{\mathrm{imp}}(\boldsymbol{x}_i^{\mathrm{tr}}) = \frac{p_{\mathrm{te}}(\boldsymbol{x}_i^{\mathrm{tr}})}{p_{\mathrm{tr}}(\boldsymbol{x}_i^{\mathrm{tr}})},$$

which include two probability densities $p_{\mathrm{tr}}(\boldsymbol{x})$ and $p_{\mathrm{te}}(\boldsymbol{x})$.

In addition to the training samples $\{(\boldsymbol{x}_i^{\mathrm{tr}}, y_i^{\mathrm{tr}})\}_{i=1}^{l}$, suppose we are given unlabeled test samples $\{\boldsymbol{x}_j^{\mathrm{te}}\}_{j=1}^{t}$ which are drawn independently from the density $p_{\mathrm{te}}(\boldsymbol{x})$. Then, performing density estimation of $p_{\mathrm{tr}}(\boldsymbol{x})$ and $p_{\mathrm{te}}(\boldsymbol{x})$ gives an approximation of $w_{\mathrm{imp}}(\boldsymbol{x})$. However, since density estimation is a hard problem, the two-stage approach of first estimating $p_{\mathrm{tr}}(\boldsymbol{x})$ and $p_{\mathrm{te}}(\boldsymbol{x})$ and then taking their ratio may not be reliable.

Here we describe a method called *Kullback-Leibler Importance Estimation Procedure* (KLIEP) (Sugiyama et al., 2008), which allows us to directly estimate the importance function $w_{\mathrm{imp}}(\boldsymbol{x})$ without going through density estimation of $p_{\mathrm{tr}}(\boldsymbol{x})$ and $p_{\mathrm{te}}(\boldsymbol{x})$.

Let us model $w_{\mathrm{imp}}(\boldsymbol{x})$ using the following model:

$$\widehat{w}_{\mathrm{imp}}(\boldsymbol{x}) = \sum_{k=1}^{b}\beta_k \exp\left(-\frac{\|\boldsymbol{x}-\boldsymbol{c}_k\|^2}{2\gamma^2}\right), \tag{11}$$

where $\boldsymbol{\beta} = (\beta_1,\ldots,\beta_b)^{\top}$ is a parameter, and $\{\boldsymbol{c}_k\}_{k=1}^{b}$ is a subset of test input samples $\{\boldsymbol{x}_j^{\mathrm{te}}\}_{j=1}^{t}$AD Using the model $\widehat{w}_{\mathrm{imp}}(\boldsymbol{x})$, we can estimate the test input density $p_{\mathrm{te}}(\boldsymbol{x})$ by

$$\widehat{p}_{\mathrm{te}}(\boldsymbol{x}) = \widehat{w}_{\mathrm{imp}}(\boldsymbol{x})p_{\mathrm{tr}}(\boldsymbol{x}). \tag{12}$$

We determine the parameter $\boldsymbol{\beta}$ in the model (12) so that the Kullback-Leibler divergence from p_{te} to $\widehat{p}_{\mathrm{te}}$ is minimized:

$$KL(p_{\mathrm{te}}\|\widehat{p}_{\mathrm{te}}) = \int p_{\mathrm{te}}(\boldsymbol{x}) \log \frac{p_{\mathrm{te}}(\boldsymbol{x})}{\widehat{p}_{\mathrm{te}}(\boldsymbol{x})} \mathrm{d}\boldsymbol{x}$$

$$= \int p_{\mathrm{te}}(\boldsymbol{x}) \log \frac{p_{\mathrm{te}}(\boldsymbol{x})}{p_{\mathrm{tr}}(\boldsymbol{x})} \mathrm{d}\boldsymbol{x} - \int p_{\mathrm{te}}(\boldsymbol{x}) \log \widehat{w}_{\mathrm{imp}}(\boldsymbol{x}) \mathrm{d}\boldsymbol{x}.$$

Since the first term is a constant with respect to the parameter $\boldsymbol{\beta}$, we ignore it and define the second term as

$$\mathrm{KL}' = \int p_{\mathrm{te}}(\boldsymbol{x}) \log \widehat{w}_{\mathrm{imp}}(\boldsymbol{x}) \mathrm{d}\boldsymbol{x}.$$

We would like to determine the parameter $\boldsymbol{\beta}$ so that KL' is maximized. Let us impose $\widehat{w}_{\mathrm{imp}}(\boldsymbol{x})$ to be non-negative and normalized. Then we obtain the following convex optimization problem:

$$\max_{\boldsymbol{\beta}} \left[\sum_{j=1}^{t} \log \left(\sum_{k=1}^{b} \beta_k \exp\left(-\frac{\|\boldsymbol{x}_j^{\mathrm{te}} - \boldsymbol{c}_k\|^2}{2\gamma^2} \right) \right) \right]$$

$$\text{s.t.} \begin{cases} \beta_k \geq 0 \ \text{ for } k = 1, \ldots, b, \\ \dfrac{1}{l} \displaystyle\sum_{i=1}^{l} \left(\sum_{k=1}^{b} \beta_k \exp\left(-\frac{\|\boldsymbol{x}_i^{\mathrm{tr}} - \boldsymbol{c}_k\|^2}{2\gamma^2} \right) \right) = 1. \end{cases}$$

This is a convex optimization problem and the global solution—which tends to be sparse (Boyd & Vandenberghe, 2004)—can be obtained, e.g., by simply performing gradient ascent and feasibility satisfaction iteratively. A pseudo code of KLIEP is described in Table 1.

Input: Kernel width γ, training inputs $\{\boldsymbol{x}_i^{\mathrm{tr}}\}_{i=1}^{l}$, and test inputs $\{\boldsymbol{x}_j^{\mathrm{te}}\}_{j=1}^{t}$
Output: $\widehat{w}(\boldsymbol{x})$

Randomly choose $\{\boldsymbol{c}_k\}_{k=1}^{b}$ from $\{\boldsymbol{x}_j^{\mathrm{te}}\}_{j=1}^{t}$;

$$B_{j,k} \leftarrow \exp\left(-\frac{\|\boldsymbol{x}_j^{\mathrm{te}} - \boldsymbol{c}_k\|^2}{(2\gamma^2)} \right)$$

$$b_k \leftarrow \frac{1}{l} \sum_{i=1}^{l} \exp\left(-\frac{\|\boldsymbol{x}_i^{\mathrm{tr}} - \boldsymbol{c}_k\|^2}{(2\gamma^2)} \right);$$

Initialize $\boldsymbol{\beta}$ $(> \boldsymbol{0})$ and ε $(0 < \varepsilon \ll 1)$;
Repeat until convergence

$\boldsymbol{\beta} \leftarrow \varepsilon B^\top (\mathbf{1}./B\boldsymbol{\beta})$;
$\boldsymbol{\beta} \leftarrow \boldsymbol{\beta} + (1 - \boldsymbol{b}^\top \boldsymbol{\beta})\boldsymbol{b}/(\boldsymbol{b}^\top \boldsymbol{b})$;
$\boldsymbol{\beta} \leftarrow \max(\mathbf{0}, \boldsymbol{\beta})$;
$\boldsymbol{\beta} \leftarrow \boldsymbol{\beta}/(\boldsymbol{b}^\top \boldsymbol{\beta})$;

end

Table 1. Pseudo code of KLIEP. './' indicates the element-wise division. Inequalities and the 'max' operation for vectors are applied in an element-wise manner.

The tuning parameter γ in KLIEP can be optimized based on *cross-validation* (CV) as follows (Sugiyama et al., 2008). First, divide the test samples $\mathcal{X}^{\mathrm{te}} = \{\boldsymbol{x}_j^{\mathrm{te}}\}_{j=1}^{t}$ into M disjoint subsets

Input: Kernel width candidates $\{\gamma\}$, training inputs $\{\boldsymbol{x}_i^{\mathrm{tr}}\}_{i=1}^{l}$, and test inputs $\{\boldsymbol{x}_j^{\mathrm{te}}\}_{j=1}^{t}$
Output: $\widehat{w}(\boldsymbol{x})$

Split $\mathcal{X}^{\mathrm{te}} = \{\boldsymbol{x}_j^{\mathrm{te}}\}_{j=1}^{t}$ into M disjoint subsets $\{\mathcal{X}_m^{\mathrm{te}}\}_{m=1}^{M}$;
for each model γ
  for each split $m = 1, \ldots, M$
    $\widehat{w}_{\mathcal{X}_m^{\mathrm{te}}}(\boldsymbol{x}) \longleftarrow \mathrm{KLIEP}(\gamma, \{\boldsymbol{x}_i^{\mathrm{tr}}\}_{i=1}^{l}, \mathcal{X}^{\mathrm{te}} \backslash \mathcal{X}_m^{\mathrm{te}})$;
    $\widehat{\mathrm{KL}}'_m(\gamma) \longleftarrow \frac{1}{|\mathcal{X}_m^{\mathrm{te}}|} \sum_{\boldsymbol{x} \in \mathcal{X}_m^{\mathrm{te}}} \log \widehat{w}_{\mathcal{X}_m^{\mathrm{te}}}(\boldsymbol{x})$;
  end
  $\widehat{\mathrm{KL}}'(\gamma) \longleftarrow \frac{1}{M} \sum_{m=1}^{M} \widehat{\mathrm{KL}}'_m(\gamma)$;
end
$\widehat{\gamma} \longleftarrow \operatorname*{argmax}_{\gamma} \widehat{\mathrm{KL}}'(\gamma)$;
$\widehat{w}(\boldsymbol{x}) \longleftarrow \mathrm{KLIEP}(\widehat{\gamma}, \{\boldsymbol{x}_i^{\mathrm{tr}}\}_{i=1}^{l}, \mathcal{X}^{\mathrm{te}})$;

Table 2. Pseudo code of CV-based model selection for KLIEP.

$\{\mathcal{X}_m^{\mathrm{te}}\}_{m=1}^{M}$ of (approximately) the same size. Then obtain an importance estimate $\widehat{w}_{\mathcal{X}_m^{\mathrm{te}}}(\boldsymbol{x})$ from $\mathcal{X}^{\mathrm{te}} \backslash \mathcal{X}_m^{\mathrm{te}}$ (i.e., without $\mathcal{X}_m^{\mathrm{te}}$), and approximate KL$'$ using $\mathcal{X}_m^{\mathrm{te}}$ as

$$\widehat{\mathrm{KL}}'_r := \frac{1}{|\mathcal{X}_m^{\mathrm{te}}|} \sum_{\boldsymbol{x} \in \mathcal{X}_m^{\mathrm{te}}} \log \widehat{w}_{\mathcal{X}_m^{\mathrm{te}}}(\boldsymbol{x}).$$

This procedure is repeated for $m = 1, \ldots, M$, and the average $\widehat{\mathrm{KL}}'$ is used as an estimate of KL$'$:

$$\widehat{\mathrm{KL}}' := \frac{1}{m} \sum_{m=1}^{M} \widehat{\mathrm{KL}}'_m. \tag{13}$$

For model selection, we compute $\widehat{\mathrm{KL}'}$ for all model candidates (the Gaussian kernel width γ in the current setting), and choose the one that minimizes $\widehat{\mathrm{KL}'}$. A pseudo code of the CV procedure is summarized in Figure 2.

One of the potential limitations of CV in general is that it is not reliable in small sample cases since data splitting by CV further reduces the sample size. On the other hand, in our CV procedure, the data splitting is performed only over the *test input samples* $\mathcal{X}^{\mathrm{te}} = \{\boldsymbol{x}_j^{\mathrm{te}}\}_{j=1}^{t}$, not over the training samples. Therefore, even when the number of training samples is small, our CV procedure does not suffer from the small sample problem as long as a large number of test input samples are available.

4.4 Empirical evaluation

Here, we experimentally evaluate the performance of the proposed method using in-house face-age datasets.

We use the face images recorded under 17 different lighting conditions: for instance, average illuminance from above is approximately 1000 lux and 500 lux from the front in the standard lighting condition, 250 lux from above and 125 lux from the front in the dark setting, and 190 lux from left and 750 lux from right in another setting (see Figure 5). Note that these

Fig. 5. Examples of face images under different lighting conditions (left: standard lighting, middle: dark, right: strong light from a side)

17 lighting conditions are diverse enough to cover real-world lighting conditions. Images were recorded as movies with camera at depression angle 15 degrees. The number of subjects is approximately 500 (250 for each gender). We used a face detector for localizing the two eye-centers, and then rescaled the image to 64 $\times$ 64 pixels. The number of face images in each environment is about 2500 (5 face images $\times$ 500 subjects).

As pre-processing, a neural network feature extractor (Tivive & Bouzerdoumi, 2006a;b) was used to extract 100-dimensional features from 64 $\times$ 64 face images. We constructed the male/female age prediction models only using male/female data, assuming that gender classification had been correctly carried out.

We split the 250 subjects into the *training set* (200 subjects) and the *test set* (50 subjects). The training set was used for training the kernel regression model (1), and the test set was used for evaluating its generalization performance. For the test samples $\{(\boldsymbol{x}_j^{\mathrm{te}}, y_j^{\mathrm{te}})\}_{j=1}^{t}$ taken from the test set in the environment with strong light from a side, age-weighted mean square error (WMSE)

$$\mathrm{WMSE} = \frac{1}{t}\sum_{j=1}^{t}\frac{(y_j^{\mathrm{te}} - f(\boldsymbol{x}_j^{\mathrm{te}}; \widehat{\boldsymbol{\alpha}}))^2}{w_{\mathrm{age}}(y_j^{\mathrm{te}})^2}$$

was calculated as a performance measure. The training and test sets were shuffled 5 times in such a way that each subject was selected as a test sample once. The final performance was evaluated based on the average WMSE over the 5 trials.

We compared the performance of the proposed method with the two baseline methods:

Baseline method 1: Training samples were taken only from the standard lighting condition and age-weighted regularized least-squares (3) was used for training.

Baseline method 2: Training samples were taken from all 17 different lighting conditions and age-weighted regularized least-squares (3) was used for training.

The importance weights were not used in these baseline methods. The Gaussian width σ and the regularization parameter λ were determined based on 4-fold CV over WMSE, i.e., the training set was further divided into a training part (150 subjects) and a validation part (50 subjects).

In the proposed method, training samples were taken from all 17 different lighting conditions (which is the same as the baseline method 2). The importance weights were estimated by KLIEP using the training samples and additional *unlabeled* test samples; the hyper-parameter γ in KLIEP was determined based on 2-fold CV (Sugiyama et al., 2008). We then computed the average importance score over different samples for each lighting condition and used the average importance score for training the regression model. The Gaussian width σ and the

	Male	Female
Baseline method 1	2.83	6.51
Baseline method 2	2.64	4.40
Proposed method	**2.54**	**3.90**

Table 3. The test performance measured by WMSE.

regularization parameter λ in the regression model were determined based on 4-fold IWCV (Sugiyama et al., 2007).

Table 3 summarizes the experimental results, showing that, for both male and female data, the baseline method 2 is better than the baseline method 1 and the proposed method is better than the baseline method 2. This illustrates the effectiveness of the proposed method. Note that WMSE for female subjects is substantially larger than that for male subjects. The reason for this would be that female subjects tend to have more diversity such as short/long hair and with/without makeup, which makes prediction harder (Ueki et al., 2008).

5. Conclusion

We introduced three novel ideas for perceived age estimation from face images: taking into account the human age perception for improving the prediction accuracy (Section 2), clustering-based active learning for reducing the sampling cost (Section 3), and alleviating the influence of lighting condition change (Section 4).

We have incorporated the characteristics of human age perception as weights—error in younger age brackets is treated as more serious than that in older age groups. On the other hand, our framework can accommodate *arbitrary* weights, which opens up new interesting research possibilities. Higher weights lead to better prediction in the corresponding age brackets, so we can improve the prediction accuracy of arbitrary age groups (but the price we have to pay for this is a performance decrease in other age brackets). This property could be useful, for example, in cigarettes and alcohol retail, where accuracy around 20 years old needs to be enhanced but accuracy in other age brackets is not so important. Another possible usage of our weighted regression framework is to combine learned functions obtained from several different age weights, which we would like to pursue in our future work.

Lighting condition change is one of the critical causes of performance degradation in age prediction from face images. In this chapter, we proposed to employ a machine learning technique called *covariate shift adaptation* for alleviating the influence of lighting condition change. We demonstrated the effectiveness of our proposed method through real-world perceived age prediction experiments.

In the experiments in Section 4.4, test samples were collected from a particular lighting condition, and samples from the same lighting condition were also included in the training set. Although we believe this setup to be practical, it would be interesting to evaluate the performance of the proposed method when no overlap in the lighting conditions exists between training and test data. Following the theoretical study by Cortes et al. (2010) would be a promising direction for further addressing this issue.

In principle, the covariate shift framework allows us to incorporate not only lighting condition change but also various types of environment change such as face pose variation and camera setting change. In our future work, we will investigate whether the proposed approach is still useful in such challenging scenarios.

Recently, novel approaches to importance estimation for high-dimensional problems have been explored (Kanamori et al., 2009; Sugiyama, Kawanabe & Chui, 2009; Sugiyama, Yamada, von Bünau, Suzuki, Kanamori & Kawanabe, 2011; Yamada et al., 2010). In our future work, we would like to incorporating these new ideas into our framework of perceived age estimation, and see how the prediction performance can be further improved. In the context of covariate shift adaptation, the importance weights played a central role for systematically adjusting the difference of distributions in the training and test phases. Beyond covariate shift adaptation, it has been shown recently that the ratio of probability densities can be used for solving various machine learning tasks (Sugiyama, Kanamori, Suzuki, Hido, Sese, Takeuchi & Wang, 2009; Sugiyama, Suzuki & Kanamori, 2012). This novel machine learning framework includes multi-task learning (Bickel et al., 2008; Simm et al., 2011), privacy-preserving data mining (Elkan, 2010), outlier detection (Hido et al., 2011), conditional density estimation (Sugiyama et al., 2010), and probabilistic classification (Sugiyama, 2010). Furthermore, mutual information—which plays a central role in information theory (Cover & Thomas, 2006)—can be estimated via density ratio estimation (Suzuki et al., 2008; Suzuki, Sugiyama & Tanaka, 2009). Since mutual information is a measure of statistical independence between random variables, density ratio estimation can be used also for variable selection (Suzuki, Sugiyama, Kanamori & Sese, 2009), dimensionality reduction (Suzuki & Sugiyama, 2010), independent component analysis (Suzuki & Sugiyama, 2011), and causal inference (Yamada & Sugiyama, 2010). In our future work, we will apply those novel machine learning tools in perceived age prediction.

6. References

Amari, S. (1967). Theory of adaptive pattern classifiers, *IEEE Transactions on Electronic Computers* EC-16(3): 299–307.

Bickel, S., Bogojeska, J., Lengauer, T. & Scheffer, T. (2008). Multi-task learning for HIV therapy screening, *in* A. McCallum & S. Roweis (eds), *Proceedings of 25th Annual International Conference on Machine Learning (ICML2008)*, pp. 56–63.

Boyd, S. & Vandenberghe, L. (2004). *Convex Optimization*, Cambridge University Press, Cambridge, UK.

Cortes, C., Mansour, Y. & Mohri, M. (2010). Learning bounds for importance weighting, *in* J. Lafferty, C. K. I. Williams, R. Zemel, J. Shawe-Taylor & A. Culotta (eds), *Advances in Neural Information Processing Systems 23*, pp. 442–450.

Cover, T. M. & Thomas, J. A. (2006). *Elements of Information Theory*, 2nd edn, John Wiley & Sons, Inc., Hoboken, NJ, USA.

Ding, C. & He, X. (2004). K-means clustering via principal component analysis, *Proceedings of the Twenty-First International Conference on Machine Learning (ICML2004)*, ACM Press, New York, NY, USA, pp. 225–232.

Duda, R. O., Hart, P. E. & Stork, D. G. (2001). *Pattern Classification*, Wiley, New York.

Elkan, C. (2010). Privacy-preserving data mining via importance weighting, *ECML/PKDD Workshop on Privacy and Security Issues in Data Mining and Machine Learning (PSDML2010)*.

FG-Net Aging Database (n.d.).
URL: *http://sting.cycollege.ac.cy/ alanitis/fgnetaging/*

Fishman, G. S. (1996). *Monte Carlo: Concepts, Algorithms, and Applications*, Springer-Verlag, Berlin, Germany.

Fu, Y., Xu, Y. & Huang, T. S. (2007). Estimating human age by manifold analysis of face pictures and regression on aging features, *Proc. of IEEE Multimedia and Expo* pp. 1383–1386.

Geng, X., Zhou, Z., Zhang, Y., Li, G. & Dai, H. (2006). Learning from facial aging patterns for automatic age estimation, *Proc. of ACM International Conf. on Multimedia* pp. 307–316.

Guo, G., Fu, Y., Dyer, C. & Huang, T. S. (2008). Image-based human age estimation by manifold learning and locally adjusted robust regression, *IEEE Trans. on Image Processing* 17(7): 1178–1188.

Guo, G., Mu, G., Fu, Y., Dyer, C. & Huang, T. (2009). A study on automatic age estimation using a large database., *International Conference on Computer Vision in Kyoto (ICCV 2009)* pp. 1986–1991.

Hido, S., Tsuboi, Y., Kashima, H., Sugiyama, M. & Kanamori, T. (2011). Statistical outlier detection using direct density ratio estimation, *Knowledge and Information Systems* 26(2): 309–336.

Hoerl, A. E. & Kennard, R. W. (1970). Ridge regression: Biased estimation for nonorthogonal problems, *Technometrics* 12(3): 55–67.

Jolliffe, I. T. (1986). *Principal Component Analysis*, Springer-Verlag, New York, NY, USA.

Kanamori, T., Hido, S. & Sugiyama, M. (2009). A least-squares approach to direct importance estimation, *Journal of Machine Learning Research* 10: 1391–1445.

Lanitis, A., Draganova, C. & Christodoulou, C. (2004). Comparing different classifiers for automatic age estimation, *IEEE Trans. on Systems, Man, and Cybernetics Part B* 34(1): 621–628.

Lanitis, A., Taylor, C. J. & Cootes, T. F. (2002). Toward automatic simulation of aging effects on face images, *IEEE Trans. on Pattern Analysis and Machine Intelligence* 24(4): 442–455.

MacQueen, J. B. (1967). Some methods for classification and analysis of multivariate observations, *Proceedings of the 5th Berkeley Symposium on Mathematical Statistics and Probability*, Vol. 1, University of California Press, Berkeley, CA., USA, pp. 281–297.

Phillips, P. J., Flynn, P. J., Scruggs, W. T., Bowyer, K. W., Chang, J., Hoffman, K., Marques, J., Min, J. & Worek, W. J. (2005). Overview of the face recognition grand challenge., *Proceedings of the IEEE Computer Society Conference on Computer Vision and Pattern Recognition (CVPR 2005)* pp. 947–954.

Ricanek, K. J. & Tesafaye, T. (2006). Morph: A longitudinal image database of normal adult age-progression., Proceedings of the IEEE 7th International Conference on Automatic Face and Gesture Recognition (FGR 2006) pp. 341–345.

Schölkopf, B. & Smola, A. J. (2002). Learning with Kernels, MIT Press, Cambridge.

Shimodaira, H. (2000). Improving predictive inference under covariate shift by weighting the log-likelihood function, *Journal of Statistical Planning and Inference* 90(2): 227–244.

Simm, J., Sugiyama, M. & Kato, T. (2011). Computationally efficient multi-task learning with least-squares probabilistic classifiers, *IPSJ Transactions on Computer Vision and Applications*, 3: 1–8.

Sindhwani, V., Belkin, M. & Niyogi, P. (2006). The geometric basis of semi-supervised learning, *Semi-Supervised Learning*, MIT Press, Cambridge.

Stone, M. (1974). Cross-validatory choice and assessment of statistical predictions, *Journal of the Royal Statistical Society, Series B* 36: 111–147.

Sugiyama, M. (2010). Superfast-trainable multi-class probabilistic classifier by least-squares posterior fitting, *IEICE Transactions on Information and Systems* E93-D(10): 2690–2701.

Sugiyama, M., Kanamori, T., Suzuki, T., Hido, S., Sese, J., Takeuchi, I. & Wang, L. (2009). A density-ratio framework for statistical data processing, *IPSJ Transactions on Computer Vision and Applications* 1: 183–208.

Sugiyama, M. & Kawanabe, M. (2011). *Covariate Shift Adaptation: Toward Machine Learning in Non-Stationary Environments*, MIT Press, Cambridge, MA, USA. to appear.

Sugiyama, M., Kawanabe, M. & Chui, P. L. (2009). Dimensionality reduction for density ratio estimation in high-dimensional spaces, *Neural Networks* 23(1): 44–59.

Sugiyama, M., Krauledat, M. & Müller, K.-R. (2007). Covariate shift adaptation by importance weighted cross validation, *Journal of Machine Learning Research* 8: 985–1005.

Sugiyama, M., Suzuki, T. & Kanamori, T. (2012). *Density Ratio Estimation in Machine Learning: A Versatile Tool for Statistical Data Processing*, Cambridge University Press, Cambridge, UK. to appear.

Sugiyama, M., Suzuki, T., Nakajima, S., Kashima, H., von Bünau, P. & Kawanabe, M. (2008). Direct importance estimation for covariate shift adaptation, *Annals of the Institute of Statistical Mathematics* 60(4): 699–746.

Sugiyama, M., Takeuchi, I., Suzuki, T., Kanamori, T., Hachiya, H. & Okanohara, D. (2010). Least-squares conditional density estimation, *IEICE Transactions on Information and Systems* E93-D(3): 583–594.

Sugiyama, M., Yamada, M., von Bünau, P., Suzuki, T., Kanamori, T. & Kawanabe, M. (2011). Direct density-ratio estimation with dimensionality reduction via least-squares hetero-distributional subspace search, *Neural Networks*, 24(2): 183–198.

Suzuki, T. & Sugiyama, M. (2010). Sufficient dimension reduction via squared-loss mutual information estimation, *in* Y. W. Teh & M. Tiggerington (eds), *Proceedings of the Thirteenth International Conference on Artificial Intelligence and Statistics (AISTATS2010)*, Vol. 9 of *JMLR Workshop and Conference Proceedings*, Sardinia, Italy, pp. 804–811.

Suzuki, T. & Sugiyama, M. (2011). Least-squares independent component analysis, *Neural Computation* 23(1): 284–301.

Suzuki, T., Sugiyama, M., Kanamori, T. & Sese, J. (2009). Mutual information estimation reveals global associations between stimuli and biological processes, *BMC Bioinformatics* 10(1): S52.

Suzuki, T., Sugiyama, M., Sese, J. & Kanamori, T. (2008). Approximating mutual information by maximum likelihood density ratio estimation, *in* Y. Saeys, H. Liu, I. Inza, L. Wehenkel & Y. V. de Peer (eds), *Proceedings of ECML-PKDD2008 Workshop on New Challenges for Feature Selection in Data Mining and Knowledge Discovery 2008 (FSDM2008)*, Vol. 4 of *JMLR Workshop and Conference Proceedings*, Antwerp, Belgium, pp. 5–20.

Suzuki, T., Sugiyama, M. & Tanaka, T. (2009). Mutual information approximation via maximum likelihood estimation of density ratio, *Proceedings of 2009 IEEE International Symposium on Information Theory (ISIT2009)*, Seoul, Korea, pp. 463–467.

Tivive, F. H. C. & Bouzerdoumi, A. (2006a). A gender recognition system using shunting inhibitory convolutional neural networks, *Proc. of International Joint Conf. on Neural Networks* pp. 5336–5341.

Tivive, F. H. C. & Bouzerdoumi, A. (2006b). A shunting inhibitory convolutional neural network for gender classification, *Proc. of International Conf. on Pattern Recognition* 4: 421–424.

Ueki, K., Miya, M., Ogawa, T. & Kobayashi, T. (2008). Class distance weighted locality preserving projection for automatic age estimation, *Proc. of IEEE International Conf. on Biometrics: Theory, Applications and Systems* pp. 1–5.

Ueki, K., Sugiyama, M. & Ihara, Y. (2010). A semi-supervised approach to perceived age prediction from face images, *IEICE Transactions on Information and Systems* E93-D(10): 2875–2878.

Ueki, K., Sugiyama, M. & Ihara, Y. (2011). Lighting condition adaptation for perceived age estimation, *IEICE Transactions on Information and Systems* E94-D(2): 392–395.

Yamada, M. & Sugiyama, M. (2010). Dependence minimizing regression with model selection for non-linear causal inference under non-Gaussian noise, *Proceedings of the Twenty-Fourth AAAI Conference on Artificial Intelligence (AAAI2010)*, The AAAI Press, Atlanta, Georgia, USA, pp. 643–648.

Yamada, M., Sugiyama, M., Wichern, G. & Simm, J. (2010). Direct importance estimation with a mixture of probabilistic principal component analyzers, *IEICE Transactions on Information and Systems* E93-D(10): 2846–2849.

17

Cell Biometrics Based on Bio-Impedance Measurements

Alberto Yúfera[1,4], Alberto Olmo[2], Paula Daza[3] and Daniel Cañete[4]
[1]*Microelectronics Institute of Seville (IMSE)*
Microelectronics National Center (CNM-CSIC),
[2]*Department of Applied Physics III,*
[3]*Department of Cell Biology,*
[4]*Department of Electronic Technology,*
[2,3,4]*University of Seville,*
Spain

1. Introduction

Many biological parameters and processes can be sensed and monitored using their impedance as marker (Grimmes, 2008), (Beach. 2005), (Yúfera, 2005), (Radke, 2004), with the advantage that it is a non-invasive, relatively cheap technique. Cell growth, cell activity, changes in cell composition, shapes or cell location are only some examples of processes which can be detected by microelectrode-cell impedance sensors (Huang, 2004) (Borkholder, 1998). The electrical impedance of a biological sample reflects actual physical properties of the tissue. In frequency dependent analyses, the β-dispersion ranging from kilohertzs to hundreds of megahertzs (Schwan, 1957) is mainly affected by the shape of the cells, the structure of the cell membranes, and the amount of intra and extra cellular solution. Electrical bio-impedance can be used to assess the properties of biological materials (Ackmann, 1993) involved in processes such as cancer development (Giaever, 1991), (Blady, 1996), (Aberg, 2004); because the cells of healthy tissues and cancer are different in shape, size and orientation, abnormal cells can be detected using their impedance as a marker.
Among Impedance Spectroscopy (IS) techniques, Electrical Cell-substrate Impedance Spectroscopy (ECIS) (Giaever, 1986), based on two-electrode setups, allows the measurement of cell-culture impedances and makes it possible to determine the biological condition (material, internal activity, motility and size) of a cell type and its relationship with the environment; for example, the transfer flow through the cell membrane (Wang, 2010). One of the main drawbacks of the ECIS technique is the need to use efficient models to decode the electrical results obtained. To efficiently manage bio-impedance data, reliable electrical models of the full system comprising electrodes, medium and cells are required. Several studies have been carried out in this field (Giaever, 1991), (Huang, 2004), (Borkholder, 1998), (Joye, 2008), (Olmo, 2010), some of them employing Finite Element simulation (FEM) for impedance model extraction. These models are the key for matching electrical simulations to real system performances and hence for correctly decoding the results obtained in experiments.

The use of FEM analysis with programs such as FEMLAB (Huang, 2004) considers that the DC mode can be employed for sinusoidal steady-state calculation by assigning complex conductivity. This works because the Poisson equation has the same form as the Laplace equation in the charge-free domain. In this chapter we will describe an alternative method of performing FEM simulations of electrode - cell interfaces based on COMSOL. The quasistatic mode of COMSOL is used, which also takes into account magnetic fields to calculate electrical impedance. The models obtained are successfully applied as loads for full electrical system modelling and simulation. In this sense, any biological sample could be electrically modelled and simulated, facilitating reliable information about integrated circuit design for impedance measuring (Yúfera, 2010b). This work includes several improvements to the model in (Huang, 2004), both to the cellular membrane and to the cell-electrode gap region. Impedance changes on small electrodes (32 μm square) caused by 3T3 mouse fibroblasts are simulated in order to validate the model and characterize the microelectrode sensor response to cell size and growth.

The knowledge acquired from the electrode-cell model can be used to create a set of applications useful in cell culture biometry and for improving efficiency biology lab tasks. One use is the detection of cell sizes by characterizing the model in terms of cell-area overlap. Impedance sensor sensitivity curves with cell size will be presented. As an extension of the application for the aforementioned sensitivity curves, a technique for measuring the cell index (CI) coefficient in cell-culture growth processes will be presented and we will show how, as a consequence of this, cell toxicity experiments can be monitored in real-time. An Analog Hardware Description Language (AHDL) model (SpectreHDL) for the mixed-mode simulation of full (electronic and biological) systems will also be described.

By applying the developed model to ECIS curves obtained experimentally, it will be possible to determine cell density and toxin-caused cell death rates. Moreover, cell modelling has recently been applied to cell imaging or bioimpedance microscopy. In (Linderholm, 2006) Electrical Impedance Tomoghaphy (EIT) techniques were reported for cell motility detection. The models proposed can also be applied to decode impedance measurements obtained from cell culture measurements, producing a two dimensional cell location map; that is, a microscopy image based on bio-impedance measurements (Yúfera, 2011).

Having explained the objectives of the work, the proposed contents of this chapter are as follows. The second section gives a brief overview of electrode solution models useful for cell-electrode characterization. The third section presents a useful method for generating cell-electrode electrical models based on COMSOL multiphysics software. The fourth section describes the finite element simulations performed. Processes for extracting useful models are included in the fifth section, which also illustrates cell size detection simulations on a simplified system. AHDL models are presented in sixth section. The seventh section covers the real time monitoring of the cell under cultivation and the application of the proposed model in dosimetric experiments. Finally, in the eighth section, a two dimensional approach to bioimpedance microscopy is described, based on the models previously developed. Conclusions are given in the ninth section.

2. The electrode – electrolyte electrical model

The impedance of electrodes in ionic liquids has been researched quite extensively (Robinson, 1968), (Schwan, 1963), (Simpson, 1963), (Schwan, 1992), (Onaral, 1982) and (Onaral, 1983). When a solid (including metals, semiconductors, and insulators) is immersed in an ionic solution, ions in the solution can react with the electrode and the solid ions from the electrode can enter the solution, leading to complex reactions at the interface. An electrified interface or double layer develops at the interface of the two phases. Eventually, electrochemical equilibrium is established at the interface: the current flowing into the electrode is equal but opposite in sign to that flowing out from the electrode. The net result is the development of a charge distribution at the interface with an associated electric potential distribution. The Helmholtz-Gouy-Chapman-Stern model is the commonly accepted model for describing the charge distribution at the electrode interface (Bockris, 1970).

The negatively charged electrode attracts hydrated ions with positive charges to the surface but repels negatively charged ions away from the surface, yielding the profiles of cation and anion concentration C+ and C-, respectively. The water dipoles are also reoriented under the electric fields. Some ionic species that are not obstructed by their primary hydration sheath, such as some anions, can make their way to and come into contact with the electrode. Most cations have a water sheath due to their lower dissolvation energy (Bockris, 1970) and a smaller contact angle with water dipoles. The charge distribution extends to the bulk solution thanks to thermal motion, forming an ion cloud-like diffusion layer and a charge spatial distribution. The profile of the diffuse zone depends on the Debye length, which in turn depends on the gas constant, the temperature, the ion charge number and the ion concentration of the bulk solution. (Borkholder, 1998).

When a sufficiently small sinusoidal current is applied to the electrode at equilibrium, the electrode potential will be modulated by a sinusoidal overpotential (Schwan, 1968). In the range of linear behavior, the phasor ratio of the output overpotential to the input current defines the AC polarization impedance. During the small current perturbation, charge transfer due to chemical reactions and mass diffusion all occur at the electrode surface. The rate determining step will dictate the electrode polarization impedance.

In the following paragraph, we will first discuss an equivalent circuit representing all the phenomena occurring at the electrode-solution interface and then we will explain each component in the circuit. The electrified interface can be considered as the series connection of two parallel-plate capacitors with the thicknesses of a compact layer and a diffuse layer respectively and with a water dielectric. This is the electrode-solution interfacial capacitance of the electrified double layer. Apart from the double layer capacitance C_I, the electrode-solution interface has faradic impedance representing a barrier to current flow between the electrode and the solution, including the finite rate of mass transport and electron transfer at the electrode surface. These phenomena are modelled in the equivalent circuit in Fig. 1, in which the faradic impedance is in parallel with the double layer capacitance.

The current flowing through the electrified interface will encounter a resistance R_{ct} caused by the electron transfer at the electrode surface and Warburg impedance Z_W due to limited mass diffusion from the electrode surface to the solution. As a result, in the equivalent, the electron transfer resistance R_{ct} is in series with the mass diffusion limited impedance Z_W. As the current spreads to the bulk solution, the electrode has a solution conductivity-determined series resistance, represented as spreading resistance R_S in the equivalent circuit.

2.1 Double layer capacitance (C_I)

The region between the electrode surface and the Outer Helmhotz Plane (OHP) consists mostly of water molecules (Borkholder, 1998). The thickness of the OHP layer is x_H (distance from the metal electrode to OHP), and consequently the capacitance of the Helmholtz layer C_H is given by,

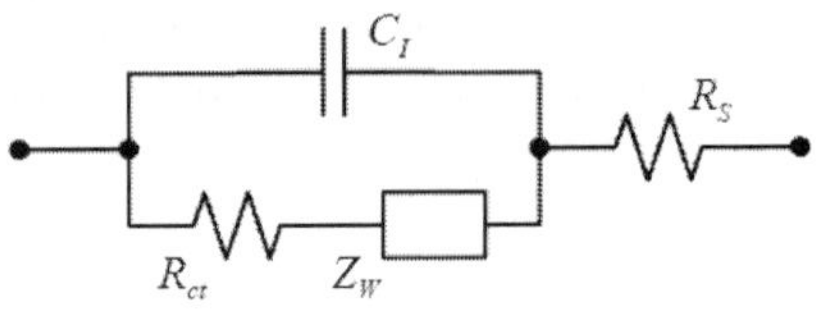

Fig. 1. Equivalent circuit for electrode solution interface. C_I is the double layer capacitance, Faradic impedance includes Z_w, the Warburg impedance and R_{ct}, the charge-transfer resistance. R_s is the spreading resistance.

$$C_H = \varepsilon\varepsilon_0 / x_H \tag{1}$$

The capacitance of the diffuse layer can be derived by differentiating the surface charge with respect to the potential. The total interfacial capacitance C_I of the electrified double layer consists of the series combination of the Helmholtz compact layer and the diffuse layer.

$$1 / C_I = 1 / C_H + 1 / C_G \tag{2}$$

where C_H is the capacitace of the Helmholtz layer described earlier and C_G is the Gouy-Chapman capacitance due to the diffuse ion cloud.This double layer capacitance has been studied in many works (Borkholder, 1998), (Schwan, 1968), (Simpson, 1980), (De Boer, 1978) although most of them agree on the value of C_I. (approximately 15μF/cm^2).

2.2 Warburg impedance (Z_w)

Warburg impedance is related to the mass diffusion process occurring in the electrode-electrolyte interface (Warburg, 1899). In AC measurements, in response to the sinusoidally varying potential, the ion concentration gradient at the interface increases with frequency and the ions diffuse less as frequency increases. The Warburg impedance follows this expression,

$$Z_w = \frac{\omega^{-1/2}.K_w}{A_e.(1+j)} \tag{3}$$

where A_e is electrode area and K_w [$\Omega \cdot sec^{-1/2} \cdot cm^2$] is a constant determined by the electrochemistry and mobility of the ions involved in the charge transfer reaction. It is difficult to find a theoretical value for K_w, which in some works seems to be a parameter that is included in the model to adapt the model to experimental measurements (Huang, 2004). In other works, this impedance is not taken into account, because it is considered negligible for the materials and frequency range used in electrophysiological experiments (Joye, 2008). D. A. Borkholder, in his thesis, gives some reference values for Z_W for circular bare platinum electrodes of different sizes (Borkholder, 1998).

2.3 Charge transfer resistance (R_{ct})

Charge transfer resistance is determined by the electron transfer rate at the interface. Electrodes made of noble metal electrodes such as platinum, gold etc., in physiological saline solution act as catalytic surfaces for the oxygen redox reactions (De Rosa, 1977), (Bagotzky, 1970). The reaction rates for the anodic and cathodic processes are not the same. In the state of equilibrium, the interface current due to oxidation is equal but of the opposite sign to that caused by reduction. Additional potential applied to the electrode will cause a net current flow. In the range comparable to the equilibrium current, the electrode behaves as a linear resistive component the resistance of which is referred to as the charge transfer resistance. As the voltage increases, the excess current increases exponentially and the charge transfer resistance decreases exponentially with the applied voltage. The charge transfer resistance can be described as,

$$R_{ct} = \frac{V_t}{J_o.z} \tag{4}$$

where J_o is the exchange current density (A/cm^2), V_t is the thermal voltage (KT/q) and z is the valence of the ion involved in the charge transfer reaction.

2.4 Spreading resistance (R_s)

The final circuit element which must be included in the basic electrode/electrolyte model is the spreading resistance. As the name implies, this resistance models the effects of the spreading of current from the localized electrode to a distant counter electrode in the solution. It can be calculated by integrating the series resistance of solution shells moving outward from the electrode, where the solution resistance (R in Ω) is determined by

$$R_s = \rho L/A \tag{5}$$

where ρ is the resistivity of the electrolyte (Ω*cm), L is the length between sensing and counting electrodes (cm) and A is the cross-sectional area (cm^2) of the solution through which the current passes. A similar expression is used in most models (Borkholder, 1998), (Joye, 2008).

3. Finite Element Model (FEM)

3.1 Cell electrode model

We first explored the work performed by Huang et al., making use of the computational advantages offered by COMSOL (http://www.comsol.com) over FEMLAB. Our objective was to obtain a model for the impedance changes caused by cell growth on electrodes similar in size to the cell. The cells modeled in the simulation were 3T3 mouse fibroblasts, which closely attach to surfaces and which typically have a cell-surface separation 0.15μm. The cells are about 5 μm in height and, seen from above, are irregularly shaped and approximately 30–50 μm in extent. A circular cell 30 μm in diameter centered on a 32x32 μm square sensing electrode was considered. (see Fig. 2). The sensing electrode was surrounded by a counter electrode with a considerably greater area.

3T3 mouse fibroblasts consist of a thin (about 8 nm), poorly conducting membrane surrounding a highly conductive interior (Giebel, 1999). The cell culture medium simulated by Huang et al. is highly ionic and possesses a conductivity of approximately 1.5 S/m. The cell culture medium fills the cell-electrode gap and forms an electrical double layer (Helmholtz plus diffuse layer) approximately 2 nm thick between the bulk of the medium and the electrode.

Some approximations were made by Huang et al. to address the problem using FEMLAB. Only one quarter of the electrode was simulated. As the problem involves a wide range of distance scales, it was difficult to solve by finite-element techniques, so the following adjustments were made:

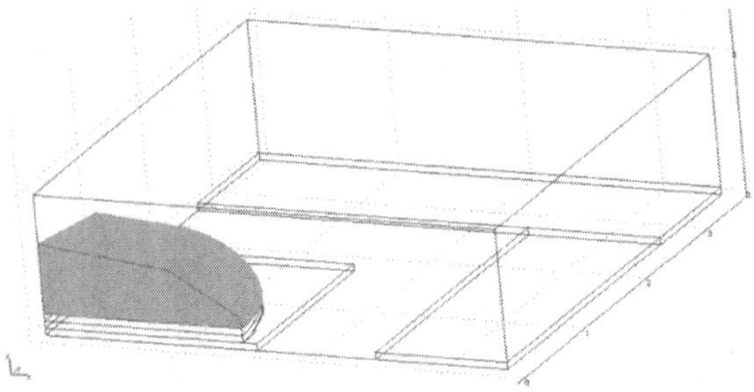

Fig. 2. Geometry of the model simulated in COMSOL.

- The electrical double layer modelling the electrode-solution equivalent circuit was replaced with a 0.5 μm thick region with the same specific contact impedance.

$$\sigma_{dl} + j2\pi f \varepsilon_{dl} = t.\left[\frac{(2\pi f)^{0.5}}{K_w} + j\frac{(2\pi f)^{0.5}}{K_w} + jC_I 2\pi f\right] \quad (6)$$

  where σ_{dl} and ε_{dl} are the conductivity and dielectric permittivity of the double layer, t is the thickness of the region, C_I is the interfacial capacitance per unit area, comprising the series combination of the Helmholtz double layer and the diffuse layer, and K_w is a constant related to the Warburg impedance contribution.
- The cell membrane was replaced by a 0.5 μm thick region with the same capacitance per unit area,

$$c_m = t.C_m \quad (7)$$

  where C_m is the membrane capacitance per unit area and t = 0.5 μm.
- The electrode-cell gap was replaced with a 0.5 μm thick region with the same sheet conductivity, that is

$$\sigma_{gap} = \frac{t_{cell-electrode}}{t}.\sigma_{medium} \quad (8)$$

  where $t_{cell-electrode}$ is the gap thickness and t is again 0.5 μm.

In our study we adopted the geometry of their simulation (see Fig. 2), and the values for the conductivity and permittivity of the electrical double layer were calculated following the same expression as shown eq. (6), with the same values for K_w and C_I as those mentioned in (Huang, 2004). Conductivity of the cell and the medium was also set to 1.5 S/m in our work. However, the model used by X. Huang et al. for the electrode-cell gap and the cellular membrane (eqs. 7 and 8) was refined as shown in the following section.

3.2 Model enhancement

Several modifications were made in the model in order to simulate cell impedance measurements more accurately and obtain a more complex model that reflected real experiments in a more realistic way. These modifications were made in the following areas.

3.2.1 Cell membrane

The equivalent circuit of the attached membrane was modeled as a resistance R_m in parallel with a capacitance C_m, in a similar way to that reported in (Borkholder, 1998). These parameters are defined as,

$$R_m = \frac{1}{g_m . A_e}$$
$$C_m = c_m . A_e \tag{9}$$

where A_e is the area of the attached membrane (in our case A=706.86x10^{-12} m^2), g_{mem} = 0.3 mS/cm^2 is the local membrane conductivity and c_{mem} (1 μF/cm^2) is the membrane capacity per unit area. We can calculate the conductivity and permittivity of the cellular membrane from the impedance using the following expression

$$Z = \frac{1}{K.(\sigma + j\omega\varepsilon)} \tag{10}$$

where K is the geometrical factor (K = area / length). In our case a value of 5 μm was taken as the length. (This value corresponds to the thickness of the membrane layer represented in COMSOL). The value obtained for K was 1.413.10^{-3}, and the values obtained for conductivity and permittivity were σ =1.5 μS/m and ε = 5.001 nF/m (ε_r=565).

3.2.2 Cell membrane-electrolyte interface capacitance

This capacitance was not considered in Huang´s model, but may also be important, as it models the charge region (also called the electrical double layer) which is created in the electrolyte at the interface with the cell. The capacitance C_{hd} is defined as the series of three capacitances,

$$C_{h1} = \frac{\varepsilon_0 \varepsilon_{IHP}}{d_{IHP}} . A_{ce}$$
$$C_{h2} = \frac{\varepsilon_0 \varepsilon_{OHP}}{d_{OHP} - d_{IHP}} . A_{ce} \tag{11}$$
$$C_d = \frac{q\sqrt{2\varepsilon_0 \varepsilon_d K T z^2 n^0 N}}{KT} . A_{ce}$$

where A_{ce} is the area of the attached membrane, ε_0 is the dielectric permittivity of free space; ε_{IHP} and ε_{OHP} are, respectively, the Inner and Outer Helmholtz Plane relative dielectric constant; d_{IHP} is the distance from the Inner Helmholtz Plane to the membrane; d_{OHP} is the distance from the Outer Helmholtz Plane to the membrane; ε_d is the diffuse layer relative dielectric constant; K_B is Boltzmann's constant; T is the absolute temperature; q is the electron charge; z is the valence of ions in the solution; n_0 is the bulk concentration of ions in the solution; and N is the Avogadro constant.

For C_{hd}, the values given in (Joye, 2008) are considered. In particular, it is assumed that ε_{IHP} = 6, ε_{OHP} =32, d_{IHP} = 0.3 nm, d_{OHP} = 0.7 nm, z = 1, T = 300 K, and n_0 =150 mM. The area of the attached membrane is in our case A_{ce}=706.86 μm^2 and ε_d is set to 1. The following values were obtained: C_{h1}= 0.125pF; C_{h2}=0.5pF; C_d=2.22pF, and the total series capacitance was

C_{hd}=0.1pF. Comparing the impedance equivalent to this capacitance with the same expression as before (eq. 10), and remodelling this layer as a 5 μm thick layer with K =1413.10^{-6}, we obtained ε=0.0011 uF/m, which corresponds to ε_r = 124.29, the value that was loaded into COMSOL.

4. COMSOL Simulations

As can be seen in Fig. 2, only one quarter of the electrodes and the cell was simulated. Electrodes were modeled with no thickness. The first layer modeled on top of the electrode surface was the 0.5 μm thick electrical double layer, which can be seen in the figure. On top of the electrical double layer, the cell-electrode gap was modeled with another 0.5 μm layer. In our simulation this layer included the cell membrane-electrolyte interface capacitance. Finally, on top, we have the cell membrane, also modeled as another 0.5 μm layer, and the rest of the cell. For each layer, it is necessary to load the conductivity and permittivity values calculated earlier into COMSOL. All surfaces had an insulating boundary condition (n*J=0) with the exception of the surfaces separating the different layers and sub-domains within the model, which were set to continuity ($n.(J_1-J_2) = 0$), and the bottom surfaces of the two electrodes, which were set to an electric potential of 1V and 0V.
The Quasi-statics module of COMSOL was used to perform the finite element simulations. In this mode, it is possible to obtain the solution for the electric potential for different frequencies. The simulations were performed on a 2.26 GHz Intel(R) Core(TM)2 DUO CPU. Solution times varied with the frequency but ranged from 3 to 6 minutes. In Fig. 3 we can see the solution for the electric potential at the determined frequency of 100 Hz.

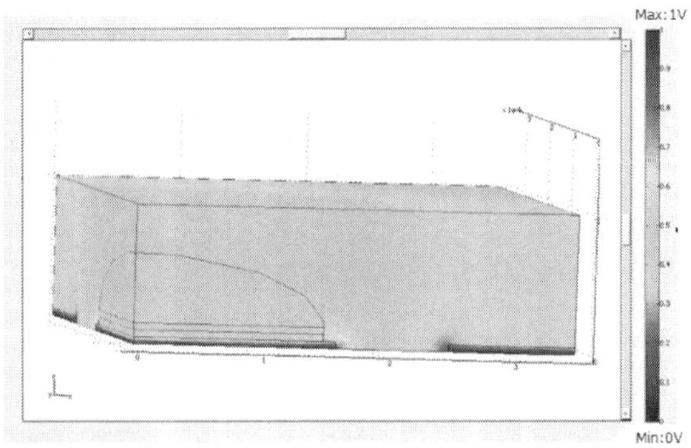

Fig. 3. Electric potential solution at 100Hz.

Two series of simulations, at frequencies ranging from 10^2 Hz to 10^6 Hz, were made with and without the presence of the cell. Once the solution for the electric potential had been found by COMSOL, Boundary Integration was used to find the electric current through the counter electrode. With that value the electric impedance was calculated, taking into account that the voltage difference between electrodes was 1V and that impedance had to by divided by 4 (as only one quarter of the electrodes was being simulated.) The values obtained are shown in Fig. 4.

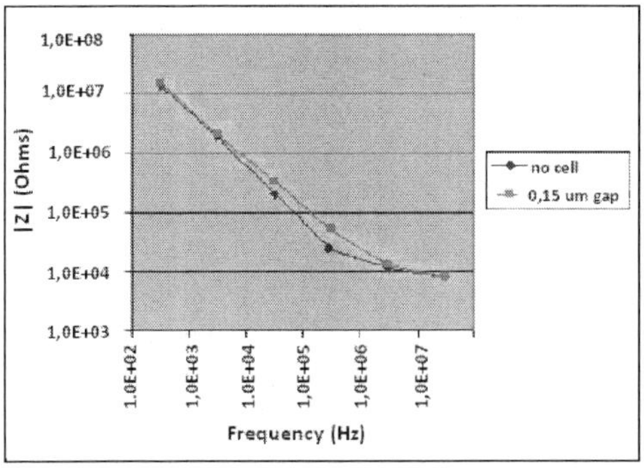

Fig. 4. Impedance magnitude of the microelectrode system with (red line) and without (blue line) the cell.

The measured impedance changes by several orders of magnitude over the simuated frequency range, in accordance with previous works (Schwan 1992), (Onaral, 1982). It can be appreciated how the presence of the cell changes the measured impedance, with the biggest change recorded at a frequency near to 10^5 Hz. This also corroborates the study carried out by (Huang, 2004). Another way of representing the impedance magnitude is to observe the impedance changes in the system with the cell on top with respect to the microelectrode system without cell. This can be done by plotting the normalized impedance change with respect to the cell-less system, defined as

$$r = \frac{Z_c - Z_{nc}}{Z_{nc}} \tag{12}$$

where Z_c and Z_{nc} are the impedance magnitudes with and without cell, respectively. The normalized impedance changes in the system with the 30 μm-diameter cell modelled earlier is plotted in Fig. 4 (red line). Figure 5 shows the normalized impedance parameter, r, as a function of frequency. It can be seen how the 100 kHz frequency seems to be the optimal value for sensing the cell, since sensor sensitivity is maximum.

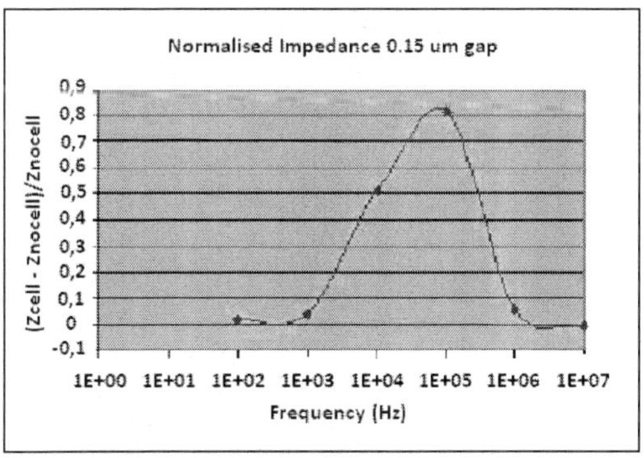

Fig. 5. Simulated normalised impedances of the system, for a 30 μm cell diameter.

5. Model extraction process

Figure 6 shows an example of a two-electrode impedance sensor useful for the ECIS technique: e_1 is called the sensing electrode and e_2 the reference electrode. Electrodes can be manufactured in CMOS process with metal layers (Huang, 2004) or using post-processing steps (Manickam, 2010). The cell location on top of e_1 must be detected. Circuit models developed to describe electrode-cell interfaces (Huang 2004, Joye 2008) contain technology process information and take the cell-electrode overlap area as their main parameter. The correct interpretation of these models gives information for: a) *electrical simulation*: parametrized models can be used to update the electrode circuit in terms of its overlap with cells, b) *imaging reconstruction*: electrical signals measured with sensors can be associated with a given overlap area, making it possible to obtain the actual covering on the electrode from experimental results.

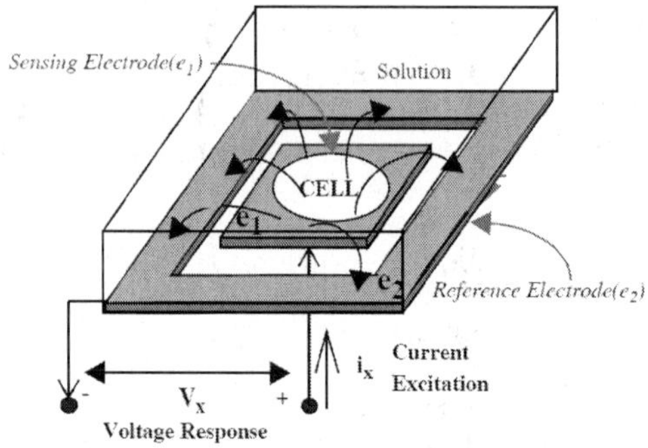

Fig. 6. Basic concept for measuring with the ECIS technique using two electrodes: e_1 (sensing electrode) and e_2 (reference electrode). AC current i_x is applied between e_1 and e_2, and voltage response V_x is measured from e_1 to e_2, including the effect of e_1, e_2 and sample impedances.

This work employs the electrode-cell model reported in (Huang 2004, Olmo 2010), obtained using finite element method simulations. The model in Fig. 7 considers that the sensing surface of e_1 could be total or partially filled by cells. For the two-electrode sensor in Fig. 6, with e_1 sensing area A, $Z(\omega)$ is the impedance by unit area of the empty electrode (without cells on top). When e_1 is partially covered by cells in a surface A_c, $Z(\omega)/(A-A_c)$ is the electrode impedance associated with the area not covered by cells, and $Z(\omega)/A_c$ is the impedance of the area covered. R_{gap} models the current flowing laterally in the electrode-cell interface, which depends on the electrode-cell distance at the interface (in the range of 15-150 nm). R_s is the spreading resistance through the conductive solution. For an empty electrode, the impedance model $Z(\omega)$ is represented by the circuit in Fig. 1. For e_2, not covered by cells, the model in Fig 7a was considered. The e_2 electrode is typically large and grounded, and its resistance is small enough to be rejected. Figure 8 represents the impedance magnitude, Z_c, for the sensor system in Fig. 6, considering that e_1 could be either empty, partially or totally covered by cells. The parameter *ff*, called the *fill factor*, can be zero for $A_c = 0$ (e_1 electrode empty), and 1 for $A_c = A$ (e_1 electrode full). Z_c (*ff*=0)= Z_{nc} is defined as the impedance magnitude of the sensor without cells.

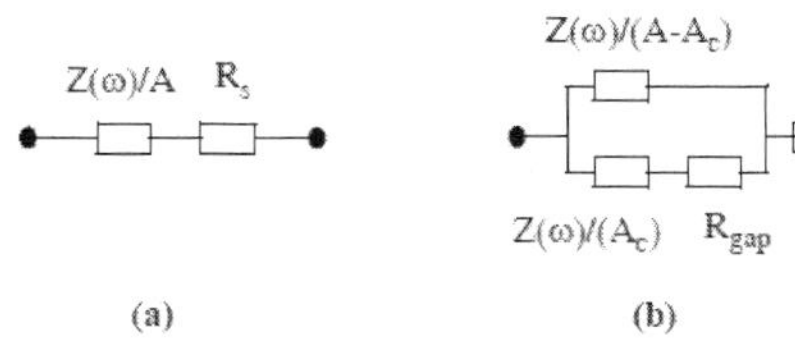

Fig. 7. Proposed model for an electrode-solution-cell model with area A, uncovered by cells (a) and covered over area A_c (b).

The relative changes in impedance magnitude, defined in eq. (12), inform more accurately from these variations: r is the change in impedance magnitude for the two-electrode system with cells (Z_c) as compared to the system without cells (Z_{nc}). Figure 9 shows the *r* versus frequency graph plotted for cell-to-electrode coverage ff from 0.1 to 0.9 in steps of 0.1, using a R_{gap} = 90 kΩ. The size of the electrode is 32 x 32 μm². Again we can identify the frequency range where the sensitivity to cells is high [10 kHz, 1 MHz], represented by r increments. For a given frequency, each value of the normalized impedance r can be linked with its ff, allowing cell detection and estimation of the covered area A_c. Electrodes can be manufactured (in terms of technology and size) according to r-sensitivity curves to improve sensor response.

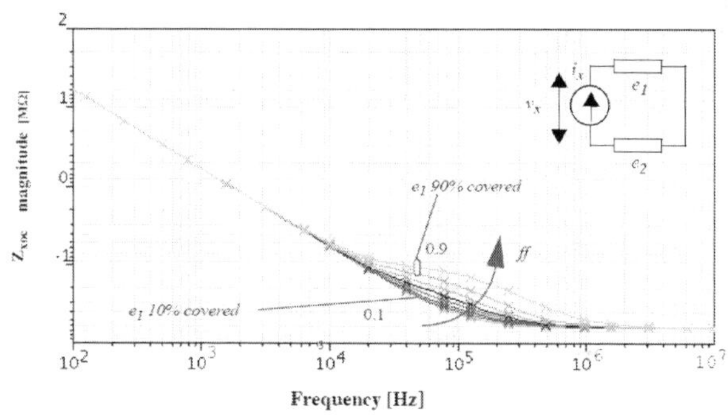

Fig. 8. Impedance evolution when fill factor increases 32 x 32 μm² for a 32 x 32 μm² electrode.

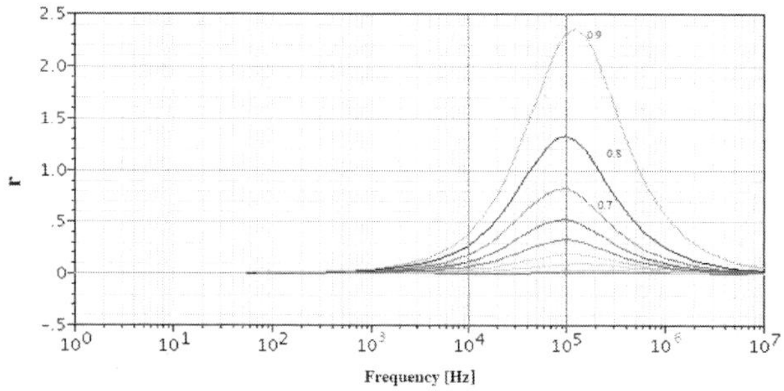

Fig. 9. Normalized impedance r parameter evolution versus frequency. Each line corresponds to values of the fill factor (or A_c covered by cells) in the range of 0.1 (nearly empty) to 0.9 (nearly full).

From Fig. 9, it can be deduced that electrode-cell electrical performance models can be used to derive the overlapping area in a cell-electrode system, an useful resource for biological studies. It can be observed how the curve matches with the frequency range, placing the maximum r value at around 100 kHz, as predicted in the COMSOL simulations. A value of R_{gap} = 90 kΩ was selected for this curve, representing a maximum value of the r curve with ff = 0.69, which in turn represents the ratio (A_c/A), for the cell size of 30 μm diameter shown in the figure obtained using FEM simulations. Other cell sizes, with different diameters, can be selected from the r curves in Fig. 9.

6. AHDL model for simulation

Electrical simulations are commonly employed by electrical engineering designers to obtain useful information and predictions about circuit performance before circuits are sent to the factory. Simulations of front-end sensor data acquisition circuits must be trained versus real sensor models, so a reliable representation of a sensor is a prerequisite for a correct acquisition circuit design. The modelling process for an impedance sensor, as described in this work, can be performed using Analog Hardware Description Language (AHDL), which can easily be incorporated into mixed-mode simulators as SpectreHDL. Another advantage of using AHDL modelling is the possibility of incorporating non-linear circuit element performance, in our case the frequency square root function in the Warburg impedance. This makes AHDL models widely applicable, allowing accurate electrode-solution model simulation, and enabling their incorporation into mixed-mode simulators.

6.1 AHDL model

The proposed model has three main parameters: the electrode area (A), the fill-factor (ff) or percentage of the electrode area covered by cells, and the resistance of the gap region (R_{gap}). Technological data is also included, defining the physical properties of the sensor material and the solution. The HDL model directly implements the equations for circuit elements described in section 2: capacitor double layer, transfer resistance, Warburg resistance and spreading resistance, considering technology information (Borkholder, 1998) and parameters A, ff and R_{gap}. An example with the R_{ct} resistance is described in the following lines:

```
// Resistor transfer: Rct
// Spectre AHDL for AHDL_ELEC, resistor_transfer, ahdl
module resistor_transfer(pin,nin) (A,FF,T)
node [V,I] pin, nin;
parameter real A = 2500p;
parameter real FF = 0.0;
parameter real T = 309;
{
real curr_density = 2.0e-5;                    /* Amp/m2 Au-sol reaction*/
real z_ions = 1.0;                             /* Valence ions */
real q_e =1.6e-19;                             /* Charge e- */
real K   =1.3806e-23;                          /* Constant Boltz. J/°K */
real v_thermal = 0.0;                          /* Thermal V. */
real rt = 0.0 ;                                /* ohm.m2 */
analog {
       v_thermal = K*T/q_e;
       rt = (v_thermal/(curr_density*z_ions));
       rt = rt / A*(1.0-FF);
```

```
        I(pin,nin) <- (V(pin,nin)/rt);
        }
}
```

Analog Hardware Description Language (AHDL) offers an easy way to describe equivalent circuits for the electrode-solution-cell model in Figure 1. Each circuit element is connected using its node description. Inside, the corresponding equations are incorporated to describe its electrical performance, technology data, geometry parameters (A) and model values (R_{gap}). This cell-electrode description can be used to carry out electrical simulations.

6.2 SpectreHDL simulations

In this section we present some simulation results obtained when modelling a commercial electrode: ECIS 8W10E, from Applied Biophysics Inc. (http://www.biophysics.com). A photograph of the complete system is shown in Fig. 10. Eigth wells can be seen, each one containing ten circular gold microelectrodes, of 250 μm diameter. This impedance sensor will be employed for the cell culture experiments described in section 7. For electrical simulations, ten identical electrodes will be considered in parallel. The HDL model for each one is described using the circuit equations from section 2 and the model in Fig. 7b.

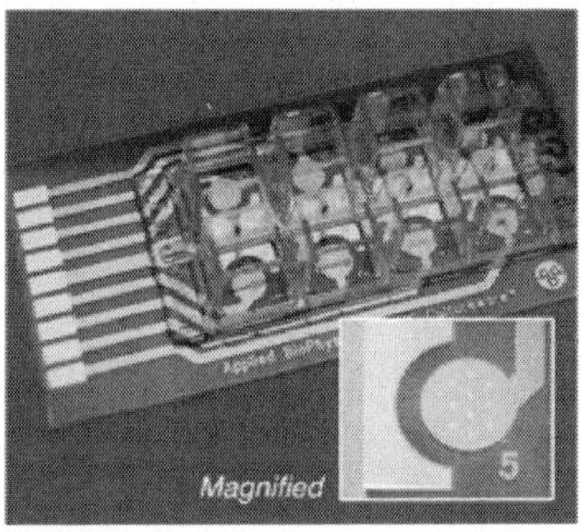

Fig. 10. Electrode employed for impedance measurements.

The expected performance of ten circular electrodes is illustrated in the following simulations. Ten sensing electrodes were used (e_1 in Fig. 6), with just one common reference electrode (e_2), much larger than the sensing electrodes. Also shown are the electrode area dependence (A), the fill factor dependence (ff), and the resistance gap dependence (R_{gap}), to illustrate the model's flexibility in different environments. Fig. 11 represents the normalized impedance r-values expected for these electrodes, for R_{gap} = 22 kΩ, when the fill factor changes from a situation of electrodes uncovered by a cell (ff = 0.1) to nearly fully covered (ff = 0.9). Different cell densities or sizes can be detected using these sensitivity curves. Several values of R_{gap} can also be used to match the models to observed performance. In Fig. 12, several R_{gap} values were set for ff = 0.9, resulting in large changes in r-values. These parameters are used to apply the proposed model to a specific cell line or cell culture experiment. Finally, the electrode area for R_{gap} = 22 kΩ and ff = 0.9 was also changed. In this way, electrode frequency response can be set for the experiment's given electrode geometry and technology parameters (Fig. 13). From these simulated results, it can be observed that optimal working frequency is close to the frequency proposed by the electrode factory (around 4 kHz), and that the electrode area covered by cells can be simulated using the fill factor parameters, making it possible to study cell density and size from the proposed model and simulations.

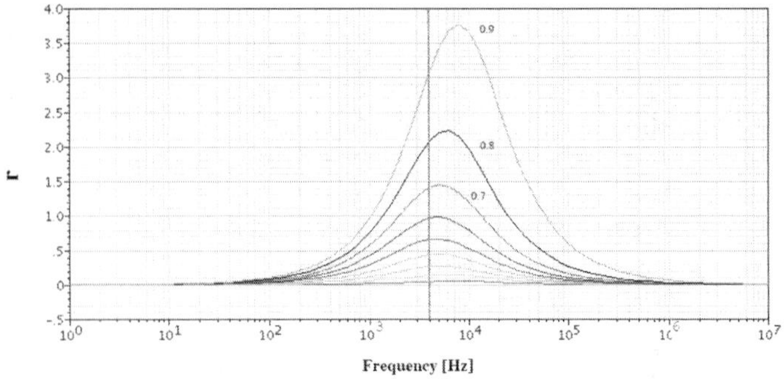

Fig. 11. The r versus frequency curves obtained for ff ∈ [0.1,0.9] and R_{gap} = 22 kΩ, using the data from 8W10E electrodes.

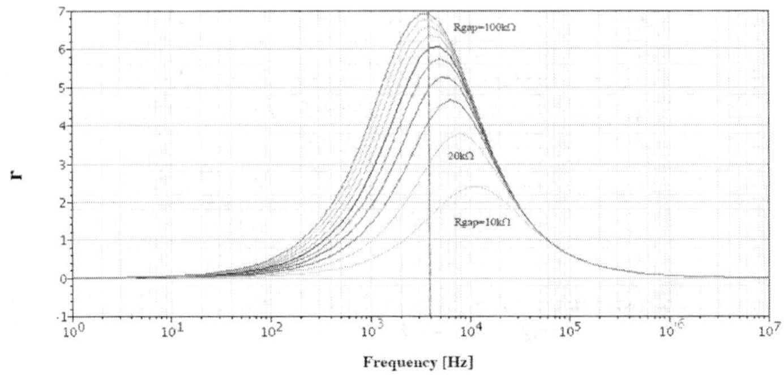

Fig. 12. The r versus frequency curves obtained for R_{gap} ∈ [10 kΩ,100 kΩ] in steps of 10kΩ, for ff = 0.9, using the data from electrodes 8W10E.

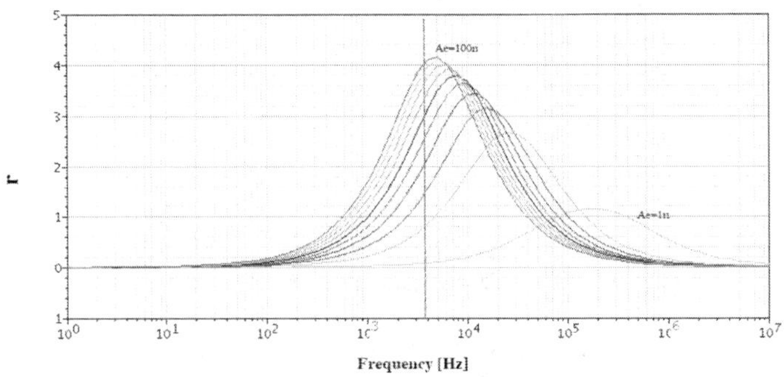

Fig. 13. The r versus frequency curves obtained for R_{gap} = 22 kΩ and ff = 0.9, for different values of the electrode area: 49n (49.10^{-9} m^2) corresponds to a circular electrode with a 250 µm diameter.

6.3 Model extraction

The performance curves obtained above can be used to match experimental results to model parameters and obtain relevant biometric characteristics like cell size, cell culture density and gap resistance. Specific cell parameters such as size and cell density can be obtained from the fill factor values. Gap resistance value provides information about the cell type, because it is related to cell adhesion and cell membrane electrical performance. Models can thus be predicted from FEM simulations performed using the modelling procedure mentioned in section 2. This means that the experimental performance observed in labs can be matched to fit proposed model for the real time monitoring and interpretation of biological experiments, especially cell culture protocols.

7. Cell culture applications

In this section, we present some applications for the cell-electrode electrical model developed earlier. The objective is to employ the information obtained from model electrical simulations to interpret cell culture experiments and to provide information about cell size, cell culture growth processes and cell dosimetric characterization.

7.1 Size definition

Cell size can be measured directly from the fill-factor parameter using the sensor curve sensitivities in Figures 11 to 13. There, maximum sensitivities to cell size (maximum *r* values) are obtained at several frequency ranges (10 kHz to 1 MHz in Fig. 9). Cell size testing should be performed after detection of the optimum frequency range. Another consideration is that the electrode size should be of the same order as the cell size to be detected using the maximum sensitivity delivered by the impedance sensor. This figure demonstrates that cells of different sizes can be detected at a given working frequency by measuring their corresponding normalized impedance value.

7.2 Cell growth

In normal conditions, a cell generates one replica (divides in two equal cells) in a cell cycle, specific to each cell line, which may take several hours. Impedance sensors can be used to measure the number of cells in a cell culture experiment, because the cells are attached to the bottom electrode when they are not dividing. As the number of cells increases, the total bottom area covered by cells increases too, increasing the number of cells placed on top of an electrode and so in time, raising the impedance measured from the electrodes. In normal protocols, growth rate is measured by a tedious process that requires seeding a number of dishes equal to the number of time points to be represented, and counting the number of cells in each dish everyday. Fig. 14a shows a growth curve we measured over seven days using 8W10E sensors in a similar setup to that described in (Giaever, 1986). An initial number of approximately 5000 AA8 cells from chinese hamsters was seeded. From this curve, impedance dynamic range is around 1220 Ω, from 380Ω to 1600 Ω. Considering the initial cell number of 5000 cells very low, we can take the initial impedance as due to the no-cell impedance value (Z_{nc}). At t = 6000 min, the medium was changed, and the confluent phase was achieved at t = 8500 min, approximately. The maximum experimental value given in eq. (12) is around r = 3.1, as illustrated in Fig. 14b. If we consider that in our model the electrodes are almost fully covered by cells (ff = 0.9), the best matching value of R_{gap} is 22 kΩ. System response corresponds to the r-values illustrated in figures 11 and 12. From these curves, the fill factors at different times can be obtained. Table I summarizes the relative normalized impedance values r at several times. Using Fig. 15 to represent the sensor response, it is possible to obtain the corresponding fill factor at every instant. The number of cells in real time can be estimated by considering that the number on top of each electrode can be measured in the order of 500 to 1000 cells. For a well area of 0.8 cm^2, the maximum number of cells goes from 0.8 10^6 to 1.6 10^6. The number of cells, n_{cell} in Table I, is obtained from 0.8×10^6, the expected final cell value. Initially, 5000 cells from the AA8 cell line were seeded. It was considered that a maximum 90% (ff=0.9) of the well surface was covered by the cells. A value of Z_{nc} = 380 Ω for r calculus at eq. (12) was also considered.

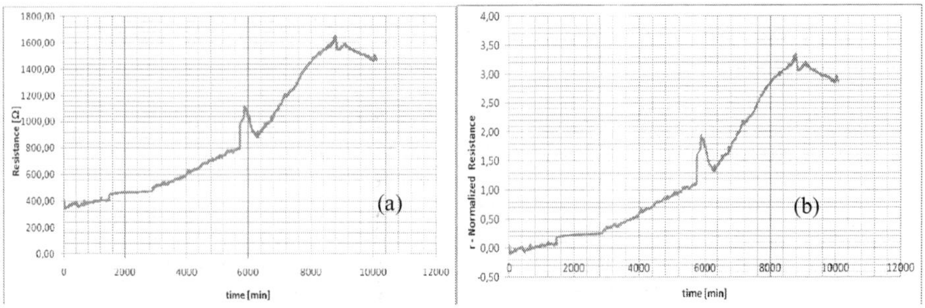

Fig. 14. (a) Impedance evolution in the cell growth experiment. (b) Estimated normalised impedance r evolution.

Table I: Number of cells (ncell) obtained from the impedance Zc measure in Fig. 14a, and using sensor curves proposed for 8W10E electrodes.

time	r	ff	n_{cell}
0	0	-	5000
500	0.024	0.020	18000
1000	0.050	0.050	44000
1500	0.072	0.070	63000
2000	n. a.	n.a	n.a.
2500	n.a.	n.a.	n.a.
3000	0.374	0.362	322000
3500	0.437	0.395	351000
4000	0.615	0.475	422000
4500	0.777	0.530	471000
5000	0.903	.581	516000
5500	1.033	0.602	535000
6000	1.074	0.620	551000
6500	1.507	0.710	631000
7000	1.970	0.775	689000
7500	2.353 --->	0.810	720000
8000	2.837	0.860	764000
8500	3.113	0.890	791000
9000	3.134	0.900	800000
9500	3.010	0.875	778000
10000	2.857	0.864	768000

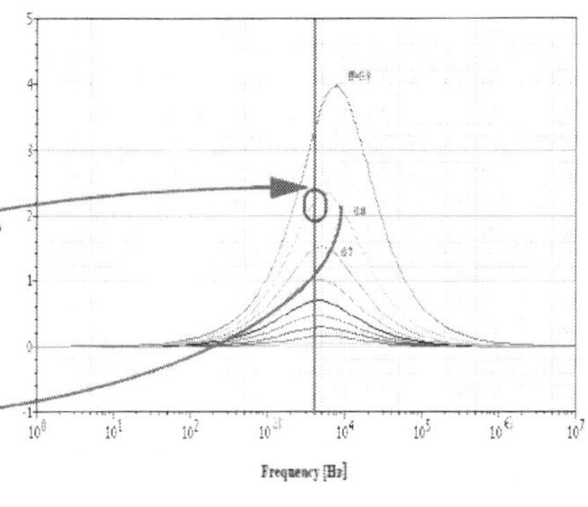

Fig. 15. Table I shows the normalized impedance r-values obtained experimentally from sensor curves in Fig. 14. ff values in Table I are obtained from experimental r-values, using curves in this figure. For example, at t = 7500 min, the measured r value is 2.353, which corresponds to ff = 0.810 on the impedance sensor curves, at the frequency of 4 kHz, and using R_{gap} = 22 kΩ.

The approximate cell number can be deduced from the last column, allowing a real time cell count based on the r-curves associated with the sensors. If different types of cells are considered, for each one the R_{gap} parameter can be calculated, producing r-curves specific to each cell line. This process allows cell lines to be identified and differentiated, for example to match the impedance measure of normal cells and cancer cells (Aberg, 2004).

7.3 Dosimetry analysis

Experiments to characterize the influence of certain drugs on cell growth were also developed. These are usually known as dosimetry protocols, and consist on determining the response of the cell growth to several drug doses. The objective is to demonstrate that the proposed model for a cell-electrode system allows cells to be selected and counted under different conditions. In our case, we again considered the AA8 cell line and, as drugs, we used six different doses of MG132 for growth inhibition, from 0.2 μM to 50 μM. After 72 hours of normal cell growth, the medium was changed and different doses of the drug were: 0.2, 0.5, 1, 5, 10 and 50 μM for wells 3 to 8 respectively. Well 2 was the control.
The experimental impedances obtained for the 8 wells are shown in Fig. 17, for a working frequency of 4 kHz. At the end of the experiment it could be observed that impedance decreases as drug dosage increases. The control well (W2) was full of cells with the maximum impedance, while the well with the maximum dosage (W8) had the lowest resistance, at the bottom. The black line (W1) represents electrode-solution impedance. In this case, after the medium change (t = 4000 min), impedance was seen inexplicably to drop below the initial baseline level (400 Ω). As in cell cultivation experiments, a resistance value was taken based on a starting value of Z_{nc}, as a value representative of electrode-solution impedance. Final impedance values at 8000 min, Z_c,, were considered as the final response, in Table II. Considering Z_{nc} and Z_c, r values are calculated in the third column. The r versus frequency curves in Fig. 15, can be used to obtain the estimated ff values from the proposed model. The number of cells at the end of the experiment was also counted, and is shown in the last column for each well. Considering ff_{max} = 0.9 for an experimentally measured cell number of 8.06×10^5, it is possible to calculate the rest of the expected ff values.
The same data is summarized in Table III for 2, 4 and 10 kHz respectively. The best match is obtained at 4 kHz in fill factor (ff parameter). The impedance baseline, Z_{nc}, for r calculus can be seen to decrease with frequency, due probably to electrode impedance dependence. For medium resistance (W1) and high drug concentration wells (W7,W8), the resistance measured is below Z_{nc}, so eq. (12) cannot be applied for r calculation.

8. Cell imaging impedance-based

Another application of the proposed model for testing cell culture impedance sensing is the implementation of an impedance based imaging system (Yúfera, 2011)(Yúfera, 2010a). This can be done when, as was commented before, cell sizes are of the same order as microelectrode sensors. In this example, we have only considered a simplified model for electrode-solution-cell electrical performance.

8.1 Cell location applications

The cell-electrode model: Fig. 19 shows the model employed for a two-electrode sensor as shown in Fig. 6. For the empty electrode, the impedance model $Z(\omega)$ has been chosen. The

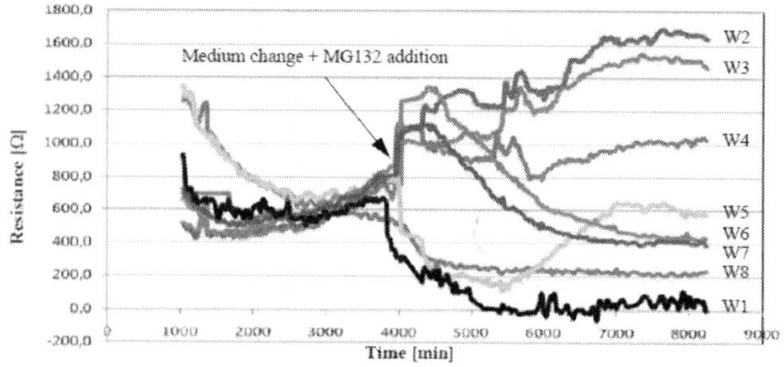

Fig. 16. Impedance obtained in 8 wells (W1 to W8) at 2 kHz frequency. W1: Medium. W2: Control. W3: 0.2μM. W4: 0.5 μM. W5: 1 μM. W6: 5 μM. W7: 10 μM. W8: 50 μM.

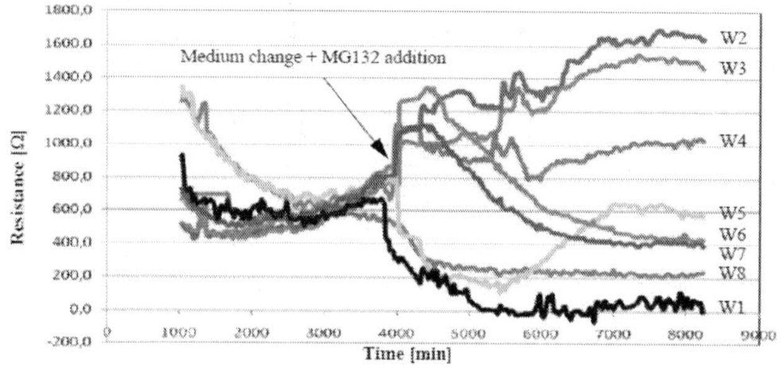

Fig. 17. Impedance obtained in 8 wells at 4 kHz frequency. (a) W1: Medium (b) W2: Control (c) W3: 0.2μM (d) W4: 0.5 μM (e) W5: 1 μM (f) W6: 5 μM (g) W7: 10 μM and (h) W8: 50 μM.

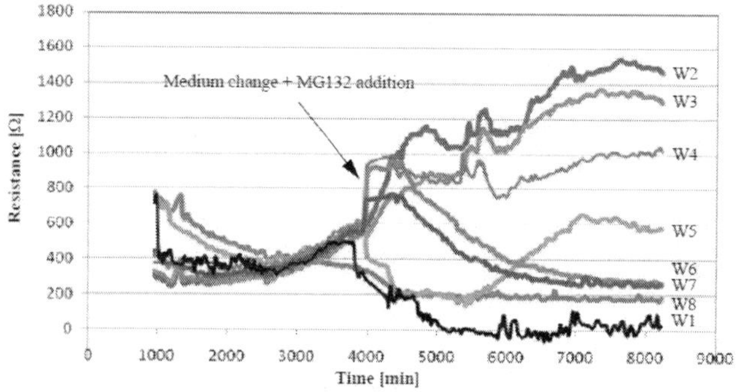

Fig. 18. Impedance obtained in 8 wells at 10 kHz frequency. (a) W1: Medium (b) W2: Control (c) W3: 0.2μM (d) W4: 0.5 μM (e) W5: 1 μM (f) W6: 5 μM (g) W7: 10 μM and (h) W8: 50 μM.

Well	Z_c at t= 8000 min	r from Z_c and Z_{nc}	ff estimated from model	ff expected	N° cells measured
1	259.2	-	-	-	Medium
2	1631.7	3.1	0.90	0.900	8.06×10^5
3	1454.7	2.6	0.85	0.690	6.13×10^5
4	1030.6	1.5	0.72	0.610	5.41×10^5
5	625.8	0.5	0.44	0.410	3.60×10^5
6	417.4	0.05	0.037	0.036	3.20×10^4
7	406.8	0.015	0.016	0.024	2.10×10^4
8	99.6	< 0	-	0.005	4.00×10^3

Table II. Experimental values for relative impedance (r) and fill-factor (ff). $Z_{nc} = 400\ \Omega$. Frequency = 4 kHz.

	Z_c [Ω] at t= 8000 min			r from Z_c and Z_{nc}			ff Estimated from model			
Well	2kHz	4kHz	10kHz	2kHz	4kHz	10kHz	2kHz	4kHz	10kHz	ff expected
1	212	259.2	178	-	-	-	-	-	-	Medium
2	1650	1631.7	1500	2.43	3.1	3.76	0.98	0.90	0.90	0.900
3	1524	1454.7	1336	2.18	2.6	3.24	0.94	0.85	0.88	0.690
4	1042	1030.6	1012	1.17	1.5	2.21	0.82	0.72	0.82	0.610
5	580	625.8	580	0.21	0.5	0.84	0.32	0.44	0.44	0.410
6	432	417.4	282	-	0.05	-	-	0.037	-	0.036
7	419	406.8	267	-	0.015	-	-	0.016	-	0.024
8	84	99.6	43	-	-	-	-	-	-	0.005

Table III. Experimental values for relative impedance (r) and fill-factor (ff) at different frequencies. $Z_{nc} = 480\Omega$, 400Ω, and 315Ω for 2, 4 and 10 kHz working frequency respectively.

circuit is shown in Fig. 19c, where C_p, R_p and R_s are dependent on both electrode and solution materials. For e_2 the model in Fig 19a, uncovered by cells, was considered. Usually, the reference electrode is common for all sensors, because its area is much higher than e_1. Figure 19d shows the relative impedance magnitude, r, for the sensor system, using a cell-to-electrode coverage ff from 0.1 to 0.9 in steps of 0.1. Again, the frequency range where the sensitivity to cells is high, represented by *r* increments, can be readily identified. For imaging reconstruction, the study in (Yúfera, 2011) proposes a new CMOS system to measure the r parameter for a given frequency, and to detect the corresponding covering area on each electrode according to the sensitivity shown in Fig 19d.

8.2 2D image applications

The case simulated was that of an 8x8 two-electrode array. The sample input to be analysed was a low density MCF-7 epithelial breast cancer cell culture, as shown in Fig. 20a. In this image some areas are covered by cells and others are empty. Our objective was to use the area parametrized electrode-cell model and the proposed circuits to detect their location. The selected pixel size was 50 µm x 50 µm, similar to cell dimensions. Figure 20a shows the grid selected and its overlap with the image. We associated a squared impedance sensor,

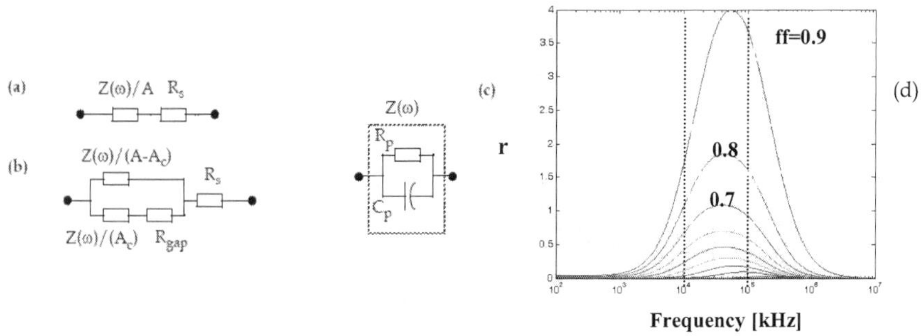

Fig. 19. Electrical models for (a) e_1 electrode without cells and, (b) e_1 cell-electrode. (c) Model for Z(ω). (d) Normalized magnitude impedance *r* for *ff*= 0.1 to 0.9 in steps of 0.1. C_P=1nF, R_P=1MΩ, R_s=1kΩ and R_{gap}=100 kΩ.

similar to the one described in Fig. 6, with each pixel in Fig. 20a to obtain a 2D system description valid for electrical simulations. An optimum pixel size can be obtained using design curves for normalized impedance *r* and its frequency dependence. Each electrical circuit associated to each e_1 electrode in the array was initialized with its corresponding fill factor (*ff*) producing the matrix in Fig 20b. Each electrode or pixel was associated with a number in the range [0,1] (ff) depending on its overlap with cells on top. These numbers were calculated to an accuracy of 0.05 from the image in Fig.20(a). The *ff* matrix represents the input of the system to be simulated. Electrical simulations of the full system were performed at 10 kHz to obtain the impedance corresponding to each electrode using the AHDL model proposed in section 6. Pixels were simulated by rows, from the bottom left hand corner (pixel 1) to the top right hand corner (pixel 64) (Yúfera, 2011).
To have a 2D graph image of the fill factor (the area covered by cells) in all pixels, values or r were obtained from the measurements taken and the sensor curves are used in Fig. 19. The results are shown in Fig 21, which represents the 8x8 ff-maps. In the maps, each pixel has a

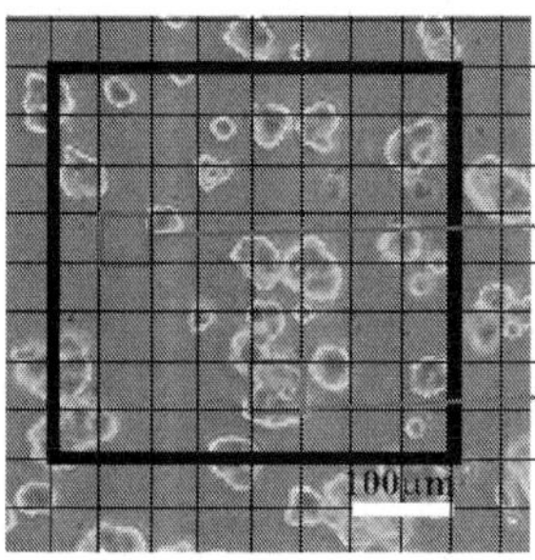

pixel 64

0.30	0.40	0.15	0	0.10	0.15	0	0
0.20	0	0	0.20	0.50	0.55	0.35	0.40
0.35	0.10	0.15	0.30	0	0.05	0.10	0.10
0	0	0.20	0 20	0.40	0.30	0.30	0.15
0	0	0.05	0.10	0.60	0.50	0	0.25
0.40	0	0.05	0.30	0.50	0.30	0	0.05
0.80	0.05	0	[illegible]	0.90	0.30	0	0.45
0.90	0.20	0	0.50	0.10	0	0.05	0.15

pixel 1

(a) (b)

Fig. 20. (a) 8x8 pixel area selection in epithelial breast cancer cell culture. (b) Fill factor map (*ff*) associated with each electrode (pixel).

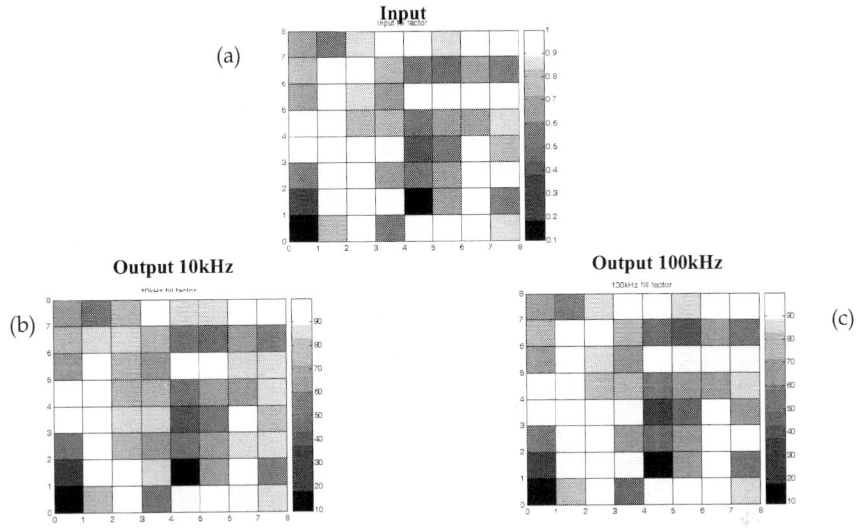

Fig. 21. 2D diagram of the fill factor maps for 8x8 pixels: (a) ideal input. Image reconstructed from simulations at (b) 10 kHz and (c) 100 kHz.

grey level depending on its fill factor value (white is empty and black full). Specifically, Fig. 21(a) shows the ff-map for the input image in Fig. 20b. Considering the parametrized curves in Fig. 19 at 10 kHz frequency, the fill factor parameter was calculated for each electrode and the results are shown in Fig. 21b. The same simulations were performed at 100 kHz, producing the *ff*-map in Fig. 21c. As Fig. 19 predicts, the best match with the input is found at 100 kHz, since normalized impedance is more sensitive and the sensor has a higher dynamic range at 100 kHz than at 10 kHz. For reported simulations, 160 ms and 16 ms per frame are required working at 10 kHz and 100 kHz, respectively. This frame acquisition time is enough for real time monitoring of cell culture systems.

9. Conclusions and discussion

The main objective of the work described in this chapter is to develop alternative methods for measuring and identifying cells involved in a variety of experiments, including cell cultures. To this end, we have focussed on obtaining models of the sensor system employed for data acquisition, and on using them to extract relevant information such as cell size, density, growth rate, dosimetry, etc.

First of all, the impedance parameter was selected as an excellent indicator of many biological processes, and Electical Cell-substrate Impedance Spectroscopy (ECIS), a technique currently considered very promising, was employed. Microelectrode impedance based sensors were analyzed, and finite element simulations carried out to model the electrical performance of both electrode-solution and electrode-solution-cell systems. This modelling process, starting with basic descriptions such as cell size, impedance, etc., enabled the use of different cells.

A practical circuit for electrode-solution and electrode-solution-cell simulation was employed, with the capacity to incorporate modelling information derived from FEM simulations. An Analog Hardware Description Model was proposed to incorporate this model to mixed-mode simulations at SpectreHDL, allowing non-linear performance characterization for the model proposed and the use of mixed-mode simulators.
Commercial 8W10E electrodes supplied by Applied Biophysics (AB) were modelled using the proposed five element circuit, obtaining good matches. Optimal measurement frequency was identified near 4 kHz, which is the frequency usually recommended by AB. Relative impedance changes were also related to the cell density thanks to fill factor parameters.
A set of experiments were conducted to match the proposed models with the performance observed at cell cultures. The first proposal was to study cell growth evolution based on 8W10E electrodes. Curves obtained experimentally, using a basic set-up, allowed real time growth to be monitored. An estimation of the number of cells was obtained using sensor curves calculated from the proposed electrical model.
Dosimetry experiments reproduce conditions similar to those of cell growth, but in this case, a growth inhibitor was added at different doses. In the experimental data, it was observed that the higher the dose, the more the measure impedance decreased, in according with the expected performance. We still do not fully understand the experimental results obtained from the measurements. Impedance inexplicably fell below the expected baseline (Z_{nc}). However, impedance curves for the control and for small drug doses were perfectly aligned up to this impedance level. A proposed model was set up with R_{gap} = 22 kΩ to explain the experimental data. Deviations from the data expected were in the order of 10% and 20% in fill factor calulations (ff), and were more precise at 4 kHz.
A final application was illustrated in the field of microscopy for cell cultures through the correct decoding of impedance response. By calculating the fill factor, it is possible to define the area occupied over one microelectrode.
As mentioned earlier, deviation in the fill factor determination was seen to be large, and the results are not accurate; in the future it will be necessary to analize the influence of certain error sources to increase the system performance. Firstly, data was collected using a simple set-up, in an attempt to reproduce the procedures followed in the experiments done by Giaever and Keese (1986). In these measurements, serial load resistance was large (1 MΩ), to limit the current through the cell to 1 μA. Signal-to-Noise Ratio (SNR) was low due to the large noise contribution from high resistance values. Other, improved setups should be used in the future. Secondly, the proposed model has the advantage that it needs only one parameter (R_{gap}), in comparison with another reported model using three parameters (Giaever, 1991). This makes easy to handle experimental data, but also introduces inaccuracies. The possibility of adding more parameters to the model should be considered in the future. Finally, finite element simulations must be developed for different cell lines to really test the cell identity in biometric applications.

10. Acknowledgements

This work is in part supported by the funded Project: Auto-calibración y auto-test en circuitos analógicos, mixtos y de radio frecuencia: Andalusian Government project P0-TIC-5386, co-financed with the FEDER program. We would also like to thank Citoquimica Ultraestructural Group (BIO132) of the Cell Biology Department, Seville University, for its experimental help in conducting cell culture experiments, and Blanca Buzón for her support on finite element simulation.

11. References

Åberg, P., Nicander, I., Hansson, J., Geladi, P., Holmgren, U. and Ollmar, S. (2004). Skin Cancer Identification Using Multi-frequency Electrical Impedance: A Potential Screening Tool, *IEEE Transaction on Biomedical Engineering*. vol. 51, n° 12, pp: 2097-2102.

Ackmann, J. (1993). Complex Bioelectric Impedance Measurement System for the Frequency Range from 5Hz to 1MHz. *Annals of Biomedical Engineering*, Vol. 21, pp. 135-146.

Applied Biophysics. http://www.biophysics.com/.

Beach, R.D. et al, (2005). Towards a Miniature In Vivo Telemetry Monitoring System Dynamically Configurable as a Potentiostat or Galvanostat for Two- and Three-Electrode Biosensors, *IEEE Transaction on Instrumentation and Measurement*, vol. 54, n° 1, pp: 61-72.

Bagotzky, V.S., Vasilyev, Y.B., Weber, J. and Pirtskhalava, N.. (1970). Adsorption of anions on smooth platinum electrodes. *J. Electroanal. Chem. & Interfacial Electrochem.*, Elsevier sequoia A.A., Netherlands, 27: 31-46 (1970).

Blady, B. & Baldetorp, B.. (1996). Impedance spectra of tumour tissue in comparison with normal tissue; a possible clinical application for electrical impedance tomography. *Physiol. Meas.*, vol. 17, suppl. 4A, pp. A105–A115.

Borkholder, D. A., (1998). Cell Based Biosensors Using Microelectrodes, *PhD Thesis*. Stanford.

Bockris, J.O'M. & Reddy, A.K.N.. (1970). Modern electrochemistry. *Plenum Press*, New York.

COMSOL, The use of the Conductive Media DC Application Mode for time-harmonic problems, solution 904, support knowledge base, from http://www.comsol.com.

De Boer, R.W. & van Oosterrom, A. (1978). Electrical properties of platinum electrodes: Impedance measurements and time domains analysis, *Med. Biol. Eng. Comput.*, 16: 1-9.

DeRosa, J.F. & Beard, R.B.. (1977). Linear AC electrode polarization impedance at smooth noble metal interfaces, *IEEE Transactions on Biomedical Engineering*, BME-24(3): 260-268.

Giaever, I. et al. (1986). Use of Electric Fields to Monitor the Dynamical Aspect of Cell Behaviour in Tissue Cultures. *IEEE Transaction on Biomedical Engineering*, vol. BME-33, No. 2, pp. 242-247.

Giaever, I. & Keese, C. R. (1991). Micromotion of mammalian cells measured electrically, *Proc. Nail. Acad. Sci. USA. Cell Biology*, vol. 88, pp: 7896-7900, Sep. 1991.

Giebel, K.F. et al. (1999). Imaging of cell/substrate contacts of living cells with surface plasmon resonance microscopy, *Biophysics Journal*, vol. 76, pp: 509–516.

Grimnes, S. & Martinsen, O. (2008). Bio-impedance and Bioelectricity Basics, *Second edition. Academic Press*, Elsevier.

Huang X. et al. (2004). Simulation of Microelectrode Impedance Changes Due to Cell Growth, *IEEE Sensors Journal*, vol.4, n°5, pp: 576-583.

Joye N. et al. (2008). An Electrical Model of the Cell-Electrode Interface for High-density Microelectrode Arrays, *30th Annual International IEEE EMBS Conference*, pp: 559-562.

Linderholm, P., Braschler. T., Vannod, J., Barrandon, Y., Brouardb, M., and Renaud, P. (2006). Two-dimensional impedance imaging of cell migration and epithelial stratification. *Lab on a Chip*. 6, pp: 1155–1162.

Manickam, A., Chevalier, A., McDermott, M., Ellington, A. D. and Hassibi, A. (2010). A CMOS Electrochemical Impedance Spectroscopy (EIS) Biosensor Array, *IEEE Transactions on Biomedical Circuits and Systems*, vol 4, n° 6. pp: 379-390.

Olmo, A. & Yúfera, A. (2010). Computer Simulation of Microelectrode Based Bio-Impedance Measurements with COMSOL, *Third International Conference on Biomedical Electronics and Devices, BIODEVICES 2010*. pp: 178-182. Valencia, Spain.

Onaral, B. & Schwan, H.P. (1982). Linear and nonlinear properties of platinum electrode polarization. Part I: Frequency dependence at very low frequencies, *Med. Biol. Eng. Comput.*, 20: 299-306.

Onaral, B. & Schwan, H.P. (1983). Linear and nonlinear properties of platinum electrode polarization. Part II: time domain analysis, *Med. Biol. Eng. Comput.*, 21: 210-216.

Radke, S.M & Alocilja, E.C. (2004). Design and Fabrication of a Microimpedance Biosensor for Bacterial Detection, *IEEE Sensor Journal*, vol. 4, n° 4, pp: 434-440.

Robinson, D.A. (1968). The electrical properties of metal microelectrodes, *Proceedings of the IEEE*, 56(6): 1065-1071.

Schwan, H. P. (1957). Electrical properties of tissue and cell suspensions. *Advances in Biological and Medical Physics. New York, Academic press*, vol. 5, pp. 147–224.

Schwan, H.P. (1963), "Determination of biological impedance", in Physical in Biological Research, W.L. Nastuk, *Ed. New York: Academic press*, 6(ch.6) (1963).

Schwan, H.P. (1968). Electrode polarization Impedance and Measurement in Biological Materials, *Ann. N.Y. Acad. Sci.*, 148(1): 191-209.

Schwan, H.P. (1992). Linear and Nonlinear electrode polarization and biological materials, *Annal. Biomed. Eng.*, 20: 269-288.

Schwan, H.P. (1992), Linear and Nonlinear electrode polarization and biological materials, *Annal. Biomed. Eng.*, 20: 269-288.

Simpson, R.W., Berberian, J.G. and Schwan, H.P. (1963). Nonlinear AC and DC polarization of platinum electrodes, *IEEE Transactions on Biomedical Engineering*. BME-27(3): 166-171.

SpectreHDL Reference Manual, Cadence Design Systems Inc.

Yúfera, A. et al. (2005). A Tissue Impedance Measurement Chip for Myocardial Ischemia Detection, *IEEE Transaction on Circuits and Systems: Part I*. vol. 52, n°.12 pp: 2620-2628.

Yúfera A. & Rueda, A. (2010a). A Close-Loop Method for Bio-Impedance Measurement with Application to Four and Two-Electrode Sensor Systems, *Chapter 15 in New Developments in Biomedical Engineering, IN-TECH*, Edited by: Domenico Campolo, pp: 263-286.

Yúfera, A. & Rueda, A. (2010b). Design of a CMOS closed loop system with applications to bio-impedance measurements, *Microelectronics Journal*. Elsevier. vol. 41, pp: 231-239.

Yufera, A. & Rueda, A. (2011). A Real-Time Cell Culture Monitoring CMOS System Based on Bio-impedance Measurements. *Kluwer Academic Pub.: Analog Integrated Circuits and Signal Processing, Special Issue on ICECS 2009*. (Accepted for publication)

Wang, P & Liu, Q. editors (2010). Cell-Based Biosensors: Principles and Applications, Artech House Series.

Warburg, E. (1899). Ueber das verhalten sogenannter unpolarisirbarer elektroden gegen wechselstrom, *Physik & Chemie*, 3: 493-499.

18

Hand Biometrics in Mobile Devices

Alberto de Santos-Sierra[1], Carmen Sanchez-Avila[1], Javier Guerra-Casanova[1] and Aitor Mendaza-Ormaza[2]
[1]Group of Biometrics, Biosignals and Security
Centro de Domótica Integral
Universidad Politécnica de Madrid
[2]Universidad Carlos III de Madrid, Leganés
Spain

1. Introduction

New trends in biometrics are inclined to adapt both identification and verification process to mobile devices in order to provide real scenarios and applications with a more secure frame. In fact, upcoming applications related to electronic commerce are demanding more trustworthy and reliable techniques to ensure their operations and transactions Van Thanh (2000), for instance. In other words, biometrics are requested to provide an appropriate alternative to current pin codes and passwords.
Furthermore, commercial biometric systems normally have no constraints in terms of computational cost or involved hardware but they do aim the highest accuracy in personal identification. In contrast, applying biometrics to mobile devices requires a reconsideration of previous lack of constraints since a mobile device is at present far from being comparable to current biometric systems in terms of hardware.
Based on these concerns, this document presents a biometric system based on hand geometry oriented to mobile devices, since hand images were acquired with mobile devices. This approach offers the possibility of identifying individuals easily with a non-intrusive acquisition procedure, using a picture taken with the mobile phone and avoiding the use of a flat surface to place the hand, providing this system with a non-contact characteristic. Moreover, the hand can be acquired without constraints in orientation, distance to camera or illumination, since the proposed technique within this paper is invariant to previous changes. This property provides an increase in the acceptance of the biometric technique by the final user, together with the fact that no removal of rings, watches and the like is required for image acquisition.
In contrast, such lack of constraints in acquisition demands a more challenging solution in relation to segmentation and feature extraction. The former operation must be able to isolate completely hand from background, regardless what is behind the hand. In case of feature extraction, the template must be independent from which hand is considered for identification (left or right hand) and invariant to changes in orientation, position, distance to camera and the like. In addition, the proposed template considers finger widths and lengths and, besides, information from four fingers (index, middle, ring and little/pinky) is considered, instead of global features from the whole hand.

The proposed system has been tested with three databases collected in different environments, with two mobile phones and therefore different cameras.
First database was created to evaluate the proposed algorithm in terms of detection accuracy, containing samples of 120 individuals from a population with different ages, gender and races, taken with an HTC, throughout a period of four months. Segmentation algorithm is exclusively tested by images in second database, a collection of synthetic hand images, based on first database, but with different environments (soil, grass, tiles, walls and the like), so that real scenarios can be simulated. Finally, third database was collected to evaluate to what extent segmentation and feature extraction algorithm were invariant to different degrees of hand opening, distance to camera and rotation. This latter database was completed using a Sony Ericsson w380i mobile.
The achieved results provide an Equal Error Rate of $4.1\pm$ 0.2 % by using 60 features (15 features in each finger) and seven training samples for template extraction, being able to obtain an EER of 3.8 ± 0.1 % when increasing the number of training samples to ten, by using a Support Vector Machine linear classifier.
The layout of the paper is arranged as follows: Section 2 provides a literature review in hand biometrics, drawing attention to hand geometry approaches. Sections 3 and 4 describe both the segmentation procedure and how features are extracted. Before presenting the results in Section 6, a description of the databases involved to evaluate the biometric system is provided in Section 5. Finally, this document ends with conclusions and future work in Section 7.

2. Related work

The distinctive characteristics within the human hand have inspired different identification techniques based mainly on geometric aspects Sanchez-Reillo et al. (2000); Zheng et al. (2007), texture patterns Kong et al. (2009) and hand vein templates Shahin et al. (2008). Considering geometric aspects, there exist several previous works based on a wide variety of topics Singh et al. (2009); Zheng et al. (2007): fingers and hand measurements Sanchez-Reillo et al. (2000); Singh et al. (2009), hand contour de Santos Sierra et al. (2009); Yoruk et al. (2006), 3D geometric representation Kanhangad et al. (2009), graph description Gross et al. (2007); Rahman et al. (2008) and so forth. Furthermore, research lines in hand biometrics based on geometric aspects consider a fusion among different characteristics leading to an enhancement in verification and identification Varchol et al. (2008); Wang et al. (2009); Yang & Ma (2007).
An aspect of relevance regards how the hand is acquired concerning not only the acquisition devices but also to what extent hand background is under control. Generally, CCD Cameras are the most common device to acquire hand images Covavisaruch & Prateepamornkul (2006); Sanchez-Reillo et al. (2000); Yu et al. (2009) providing with a wide variety of images resolutions depending on the camera. In addition, scanners are also considered as an adequate alternative to CCD devices Hashemi & Fatemizadeh (2005); Varchol et al. (2008). For the sake of a precise acquisition, hand is usually located on a flat surface provided with guiding pegs ensuring that hand is exactly placed on the same position. However, some problems arise from this approach which concern shape deformation, an increase in the device acquisition complexity and, more recently, contact-based acquisition devices can be considered controversial regarding hygiene and public-health issues Zheng et al. (2007).
On the basis of this fact, peg-free biometric systems tackles with this problem although many approaches still preserve the flat surface to locate the hand. Some works propose an acquisition procedure avoiding completely any contact with surfaces de Santos Sierra et al. (2009); Zheng et al. (2007). However, these contact-free approaches cope with the problem

of isolating the hand from a more complicated background, since previous works based on contact or semi-contact devices had a controlled background. In other words, contact-free biometric systems provide less invasiveness in acquisition at expense of an increase in the computational cost of the feature extraction and segmentation algorithm.

Regarding invasiveness, most of previous works require a removal of rings, bracelets and similar complements Kukula et al. (2007), although many trends tend to extract hand features and descriptors without requiring any removal de Santos Sierra et al. (2009); Yoruk et al. (2006).

Different illumination scheme have been proposed. Normally, a gray-scale image provides enough information concerning not only geometric aspects but also palmprint or fingerprint texture information Arif et al. (2006); Yang & Ma (2007). In contrast, color images provide more information on skin color and therefore more useful information for contact-less approaches. In addition, several color spaces have been also proposed to facilitate the procedure of segmentation, although most common used space is RGB Tan et al. (2009).

Several authors have proposed an infra-red illumination environment Ferrer et al. (2009); Shahin et al. (2008) based on the fact that infra-red illumination allows to extract hand contour easily since infra-red light highlights that region closer to the focus, and therefore, background is rarely illuminated. However, these acquisition systems require both a special illumination and an infra-red camera, difficult to be embedded on daily devices like mobiles and smartphones, for instance.

Hand biometric acceptation was assessed in Kukula & Elliott (2005; 2006); Kukula et al. (2007) evaluating the performance of the biometric system in relation to the number of attempts in accessing the system. In fact, the repeated used of the device provides an increase in the identification accuracy of participants. Therefore, the individuals get easily habituated to hand biometric devices, although many users required more restricted instructions when facing the system. Similar conclusions were obtained in de Santos Sierra et al. (2009) where hand images were acquired in a free space.

New trends in biometrics tend to adapt current systems to mobile devices. However, not every biometric technique is suitable for this adaptation. Furthermore, mobile devices imply certain limitations in relation to computational cost and performance efficiency and accuracy. Obviously, mobile security is not so demanding as, for instance, an application in an international airport. In the literature, there exist previous approaches concerning biometrics and mobiles involving different biometric characteristics: Face Recognition on Symbian OS Abeni et al. (2006); Ijiri et al. (2006), Voice Recognition Shabeer & Suganthi (2007), Keystroke Dynamics Saevanee & Bhatarakosol (2008), Hand de Santos Sierra et al. (2009), Palmprint Han et al. (2007) or Finger Pressure McLoughlin & Naidu (2009); Saevanee & Bhatarakosol (2008); Shabeer & Suganthi (2007). All previous work coincide on the same conclusions: mobile devices imply limitations for biometric accuracy and efficiency, but provide a high degree of security in daily applications.

3. Segmentation

As presented in the literature review (Section 2), segmentation in hand biometrics was almost a trivial operation, since the background is completely uniform and different in color and intensity to hand texture Boreki & Zimmer (2005); Sanchez-Reillo et al. (2000); Yu et al. (2009). However, the acquisition procedure proposed within this document requires a more demanding segmentation procedure able to isolate entirely and precisely hand from background. Notice, that this background is unknown and there is no prior information about it, since images could be acquired at any place regardless the environment.

Images were acquired in a RGB color space, which is a standard format for comercial mobiles. However, the proposed segmentation is carried out in a different color space, since RGB provides not enough information to distinguish properly hand from background. In order to obtain an adequate accuracy in segmentation, CIELAB (CIE 1976 L*,a*,b*) was selected due to its ability to describe all visible colors by the human eye Gonzalez & Woods (1992); Mojsilovic et al. (2002); Tan et al. (2009).

This color space transformation facilitates enormly the segmentation operation by offering a representation in which pixels corresponding to skin texture are separated in terms of L*a*b* intensities from rest pixels. Selecting which layer contains more distinctive information for segmentation is in fact a crucial matter Albin et al. (2002); Gonzalez & Woods (1992); Recky & Leberl (2010); Wang et al. (2010). The proposed method makes use of the Entropy of an image, H, to select which layer contains more unique or distinguishing information Luengo-Oroz et al. (2010).

Normally, experiments show that layer a provides more distinctive information.

After selecting the proper layer, pixels must be divided into two groups: a group containing pixels corresponding to hand, and a second group gathering those pixels describing background. This classification is carried out by a k-means algorithm, which provides a suitable clustering solution for segmentation problem Recky & Leberl (2010), gathering in a unique cluster those pixels corresponding to hand texture.

Although a deep explanation of k-means procedure is far beyond the scope of this article Gonzalez & Woods (1992), the segmentation problem can be stated as follows: given an image I, the aim of this k-means algorithm is to divide the image I into k clusters, minimizing the intracluster square sum (Eq. 1):

$$\arg\min_{\mathbf{S}} \sum_{i=1}^{k} \sum_{\mathbf{x}_j \in S_i} \left\| \mathbf{x}_j - \boldsymbol{\mu}_i \right\|^2 \tag{1}$$

where $\mathbf{S}$ corresponds to the segments in which the image I is divided, and μ_i represents the ith clustering distribution mean.

Classification is based not only in colour information, but also in the position within image. This is essential for avoiding the effects of rings and small ornaments on hands, since they are considered as part of the hand, despite of slightly deforming the hand. However, the aim of this procedure is twofold: to ensure fingers not to be splitted from hand, remaining the hand as a unique solid blob, and to keep simple the segmentation algorithm (the most time consuming step in hand recognition, Section 6), considering both the fact that the procedure could be implemented in mobiles and that ignoring measures extracted from regions associated to rings is easier than correcting the error provided by the ring.

A deeper understanding of the effects produced on the template and system accuracy remains as future work, together with an adequate processing to avoid this effect.

Obviously, this fact affects posterior measures, and therefore, the effects of rings in feature extraction will be explained under Section 4.

In order to obtain a binary image (those pixels belonging to hand represented by a high value, and thus background represented by zero), k is set to $k = 2$. In addition, ensuring which group corresponds to hand ('1' values) or background ('0' values) is easily carried out by analyzing which group is more isolated from the outside boundary (image border). Reader may notice that this assumption implies that individuals have colaborated with the system in locating the hand within the camera visual field, and therefore within the boundaries of the image. Otherwise, in case of hand being too close to camera (and therefore not being confined within image boundaries) or hand not appearing completely in the image, a correct segmentation will not be carried out (hand does not appear completely within image), implying that image must be rejected and requiring a new acquisition.

Due to illumination and background, the result provided by the k-means algorithm could be slightly inaccurate in the boundary, and therefore a morphological operation must be performed to soften that contour. The selected operation is a morphological opening Gonzalez & Woods (1992); Luengo-Oroz et al. (2010), with a disk structural element of small dimension (5 pixels of radius), since such a structural element suites adequately hand geometry, based on the rounded shape of a hand, without any sharp contour.

4. Template extraction

This section defines the features to be extracted from hand in order to reduce the biometric information contained within the hand to more comparable and measurable parameters. These features must describe and define the hand uniquely and univocally, and must remain invariant to changes of size, distance to camera, rotation and similar variations in acquisition. Some previous works provide similar templates based on width fingers and distances extracted from hand Boreki & Zimmer (2005); Sanchez-Reillo et al. (2000), and others consider free-space acquisition Ferrer et al. (2009); Zheng et al. (2007), but without considering a high degree of freedom in hand changes and mobile devices acquisition.

Before extracting features, tips and valleys are detected according to previous work de Santos Sierra et al. (2009); Munoz et al. (2010), based on the difference of pixels in the hand contour and hand centroid.

The proposed method extracts features by dividing the finger from the basis to the tip in m parts. Each of these former parts measures the width of fingers, based on the euclidean distance between two pixels. Afterwards, for each finger, the m components are reduced to n elements, with $n < m$, so that each n component contains the average of $\lfloor \frac{m}{n} \rfloor$ values, gathering mean value, μ and standard deviation σ. In other words, template is extracted based on an average of a finger measures set, being more reliable and precise than one single measure (Section 6). This approach provides a novelty if compared to previous works in literature (Section 2), where single measures were considered.

Furthermore, each n component is normalized by the corresponding finger length, in an attempt to provide independence on distance to camera.

Therefore, the template can be mathematically described as follows. Let $F = \{f_i, f_m, f_r, f_l\}$ be the set of possible fingers, namely index, middle, ring and little, respectively. Let $\Lambda = \{\lambda_i, \lambda_m, \lambda_r, \lambda_l\}$ be the set of distances for the corresponding finger.

Each finger f_k is divided into m parts from basis to top, resulting in the set of widths $\Omega_{f_k} = \{\omega_1, \ldots, \omega_m\}$. From set Ω, the template is represented by $\Delta_{f_k} = \left\{ \frac{\delta_1^{f_k}}{\lambda_{f_k}}, \ldots, \frac{\delta_n^{f_k}}{\lambda_{f_k}} \right\}$, where each $\delta_t^{f_k}$ is defined as the average value of at least $\lfloor \frac{m}{n} \rfloor$ components in Ω_{f_k}. Notice that this division could imply that last element δ_n could be the average of more than $\lfloor \frac{m}{n} \rfloor$ components in order to ensure that every element in Ω_{f_k} is considered to create Δ_{f_k}.

Features are not extracted in thumb finger due to its variability in terms of movement, position and direction, and thus, none sufficient distinctive information can be extracted, despite of

normalization. Therefore, the biometric template representing a hand is composed of a total of $4 \times n$ elements. This relation will be studied in detailed within results section (Section 6).
In order to compare templates among individuals, this paper proposes (Support Vector Machines, SVM Kumar & Zhang (2006; 2007)) with linear kernel functions as an adequate and accurate classifier, which has provided the best results when compared to other classifiers and kernel functions.
The number of samples to create the template in order to train the SVM properly is studied in Section 6.3.

5. Database acquisition

This biometric method is oriented to mobile applications and therefore, the algorithm must be tested with images acquired from mobile devices. The databases differ in the mobile device involved to acquire images, number of individuals, images sizes and the like. First database is used to evaluate (train, validate and test) the whole system considering identification efficiency. Second was created synthetically based on first database to evaluate only the performance of segmentation, with the aim of assessing the implemented algorithms in different environments and scenarios. Finally, third database was collected to evaluate feature changes in rotation, hand opening and distance to camera. These databases are available at *http://www.gb2s.es*.

5.1 First database

This database is the most complete containing hand captures of 120 different individuals of an age range from 16 to 60 years old, gathering males and females in similar propotion.
Furthermore, considering a contact-less approach for biometric hand recognition, every hand image was acquired without placing the hand in any flat surface neither requiring any removal of rings, bracelets or watches. Instead, the individual was required to open his/her hand naturally, so the mobile device (an HTC) could take a photo of the hand at 10-15 cm of distance with the palm facing the camera.
This acquisition implies no severe constraints on neither illumination nor distance to mobile camera and every acquisition was carried out under natural light. These approach combines several current challenges in hand biometric recognition with the limitation of mobile devices. Therefore, it is a database with a huge variability in terms of size, skin color, orientation, hand openness and illumination conditions.
In order to ensure a proper feature extraction, independently on segmentation, acquisitions were taken on a defined blue-coloured background, so that segmentation can be easily performed, focusing on hands. This background can be easily replaced by another texture like soil, tiles and the like, as it will be seen in Sections 5.2 and 6.1.
Some samples of this first database are provided in Figure 1.
Both hands were taken, in a total of two sessions: During the first session, 10 acquisitions from both hands are collected; second session is carried out after 10-15 minutes, collecting again 10 images per hand.
The image size provided by the device is 648x338 pixels.

5.2 Second database

Second database is entirely aimed to evaluate segmentation, assessing to what extent the segmentation algorithm can satisfactory perform a hand isolation from background on real scenarios.

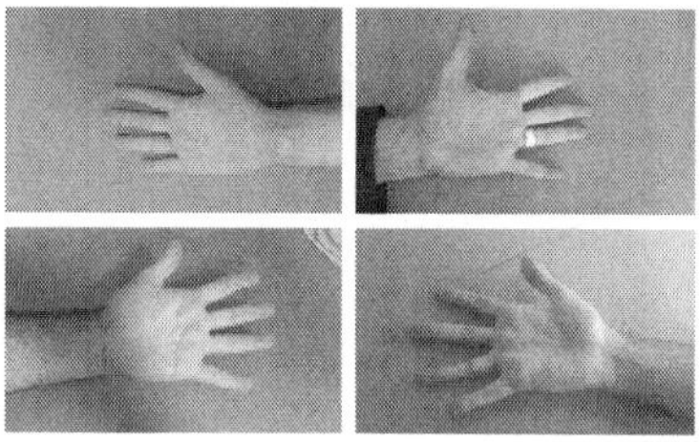

Fig. 1. Samples of first database, with blue-coloured background.

In order to simulate that hand is located over different backgrounds, that region considered as background in the segmentation procedure carried out for images in first database is replaced by different textures. Afterwards, an opening morphological operator (with a disk structural element of radius 5) for colour images Gonzalez & Woods (1992) is considered to avoid possible edges separating hand and the latter texture, providing a more realistic environment. Different backgrounds are considered in an attempt to cover all possible real scenarios, containing textures from carpets, fabric, glass, grass, mud, different objects, paper, parquet, pavement, plastic, skin and fur, sky, soil, stones, tiles, tree, walls and wood. Five different images from every texture were considered to ensure more realistic environments. All previous texture backgrounds were taken from *http://mayang.com/textures/*.
Some examples of second database can be seen in Figure 2.

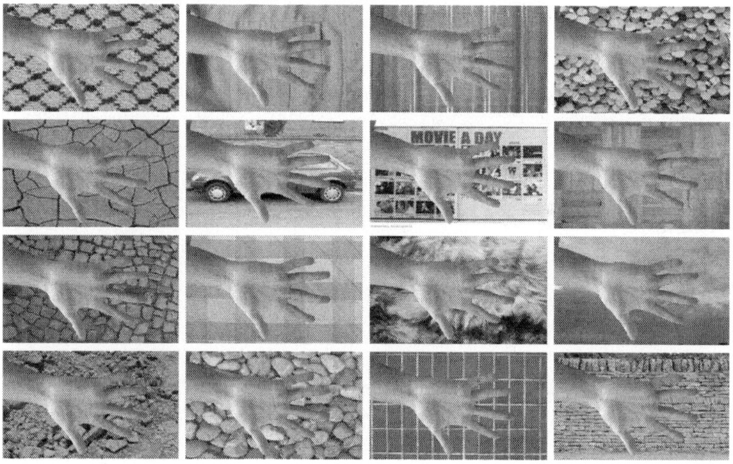

Fig. 2. Samples of second database in different backgrounds for a given acquisition taken from first database.

For each image on first database, a total of 5×17 (five images and 17 textures) images are created. Therefore, second database collects a total of $120 \times 2 \times 20 \times 5 \times 17 = 408000$ images (120 individuals, 2 hands, 20 acquisitions per hand, five images and 17 textures) to properly evaluate segmentation on real scenarios.

5.3 Third database

Finally, third database is collected by a Sony Ericcson w380i mobile devices, with a camera of 1.3 Megapixels and image dimensions of 1280x1024. This database contains images from 10 individuals with the aim of measuring three different aspects: rotation, 50 images per angle were taken (5 angles: 0°, 45°, 90°, 135° and 180°. All these angles are referred to the vertical line formed by individual's head and feet, placing the hand with the palm facing the individual with the fingers pointing to the top), for each individual (a total of 2500 images); hand openness, 50 images per position (three possibilities: normal, not very open and very open) and individual which makes a total of 1500 images; and distance to camera, 50 images per distance (two distances: 15 cm and 30 cm) and individual (10 users) making a total of 1000 images. Regarding openness degree, normal degree is considered when the surface of the palm is totally flat, with a openness radius of infinity. Not very open means to have a slightly concave curvature of the palm (a big positive radius of curvature), and very open means to have a slightly convex curvature of the palm (a big radius of curvature, but opposite to the previous one).

Therefore, the database contains 5000 images of 10 individuals. It must be pointed out that angles in this database are not precisely measured but approximated, similarly to the distance to the camera and the hand openness degree. Several samples of this database are provided in Figure 3.

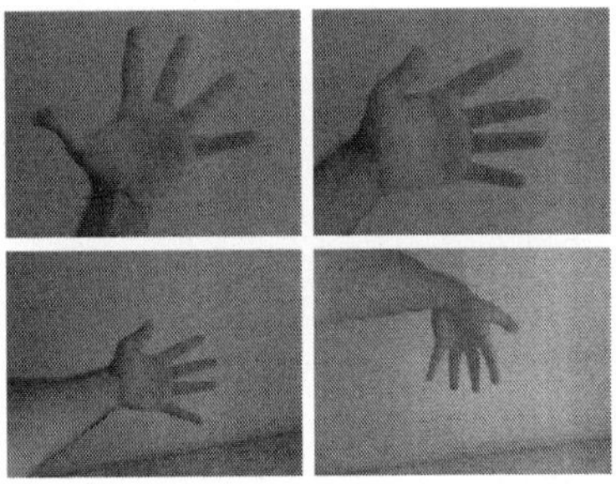

Fig. 3. Samples of third database, with different rotation angles, distance to camera and openness degree.

6. Results

Evaluation in hand biometrics involves assessing how the segmentation procedure isolates hand from background, to what extent features are invariant to changes (position, scale or orientation) and the accuracy in identifying and verifying individuals given a database. Therefore, this section will be divided into three different parts corresponding to each aspect to be assessed.

6.1 Segmentation evaluation

Concerning segmentation evaluation, a supervised evaluation method Munoz et al. (2010); Zhang et al. (2008) was considered, comparing the segmentation result to a ground-truth solution obtained based on the segmentation carried out for first database. This first database contains hand acquisitions with a known background, becoming relatively easy to extract

Texture	$F(\%)$	Texture	$F(\%)$	Texture	$F(\%)$
Carpets	80.7±0.3	Paper	86.2±0.1	Stones	82.4±0.1
Fabric	83.2±0.1	Parquet	76.1±0.2	Tiles	89.2±0.2
Glass	92.5±0.2	Pavement	88.3±0.3	Tree	98.7±0.1
Grass	91.6±0.1	Skin and Fur	86.7±0.1	Wall	87.3±0.3
Mud	82.6±0.2	Sky	89.3±0.1	Wood	83.8±0.2
Objects	88.2±0.3	Soil	71.1±0.2		

Table 1. Segmentation evaluation by means of factor F in a synthetic database with 17 different background textures.

precisely hand from background. This segmentation will be considered as ground-truth for this evaluation.

In order to measure to what extent the result is similar to ground-truth, factor F Alpert et al. (2007) provides a suitable manner to assess segmentation. Factor F is defined by Eq. 2:

$$F = \frac{2RP}{R+P} \tag{2}$$

where R (Recall) represents the number of true positives (true segmentation, i.e. classify a pixel corresponding to hand as hand) in relation to the number of true positives and false positives (false background segmentation, i.e. consider a pixel corresponding to background as hand) and P (Precision) represents the number of true positives in relation to the number of true positives and false negatives (false hand segmentation, i.e. consider a pixel corresponding to hand as background).

The results of factor F obtained for second database are presented in Table 1. Notice that those textures similar in color and textures to hand (like mud, wood, skin and parquet) decrease the performance of the segmentation algorithm.

In addition, we present the segmentation result within Figure 4, where first row provides some examples of both left and right hands from first database, together with their segmentation results in second row, representing the ground-truth segmentation. Besides, some examples from the synthetic database were taken to compare segmentation results between an under-control background and their corresponding synthetic images, with a random background (third and four row). A complete understanding of the effects of these backgrounds on identification rates will be a future work aim.

6.2 Feature invariance evaluation

Providing information on the evaluation regarding feature extraction represents a difficult task, since the assessment would consist of comparing each feature in different situations for each database. However, a small sample of features (representative of all possible features) is considered and are compared using the fifth database. These features correspond to those more close to tip, i.e. $\frac{\delta_n^{f_k}}{\lambda_{f_k}}$, assuming that the variation of these features is similar to the rest of features along fingers.

Therefore, the evaluation will consider only three aspects: different degrees of hand opening, distance to camera and rotation.

Firstly, this study only considers three degrees of opening: Original position (standard position), a not very open hand where the individual is indicated to close slightly the hand

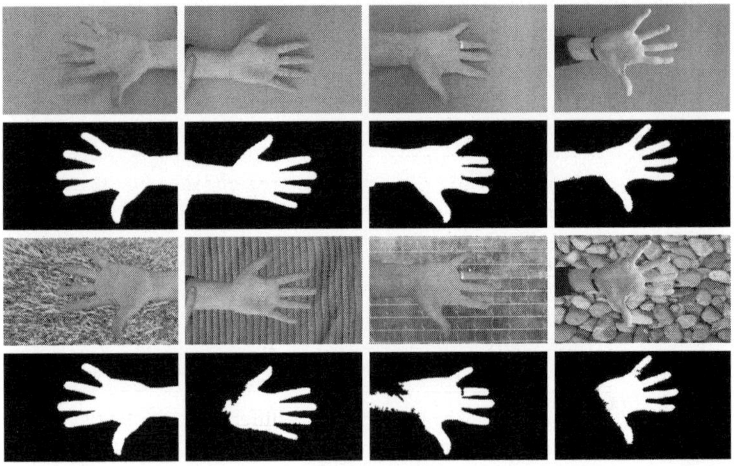

Fig. 4. Segmentation Evaluation: First row, original first database; Second row, the corresponding segmentation result (ground-truth); Third row, the associated synthetic images with different backgrounds; Fourth row, the segmentation result for synthetic images.

Feature	Original	Not very open	Very open
Index	1.2±0.1	1.1±0.2	1.5±0.1
Middle	1.4±0.2	1.3±0.1	1.7±0.2
Ring	1.3±0.1	1.4±0.2	1.6±0.1
Little	1.7±0.1	1.7±0.1	1.9±0.1

Table 2. Variation of a certain feature in each finger for different openness degrees. The results correspond to the mean difference value and their dispersion.

and a very open hand degree where the subject is supposed to open entirely the hand, Figure 3.

Table 2 contains the difference in terms of pixels between extracted features to a feature reference set, containing on the first column the differences among hand acquisitions with no changes in hand openness. This experiment is repeated 1000 times selecting randomly the feature reference set, presenting thus the main statistics (difference mean and deviation) in Table 2. This procedure is the same in posterior Tables 3 and 4.

These results highlight that although there is no significative variation in terms of difference average and deviation, there is a slight variation when the hand is entirely open. This is due to the fact that opening the hand extensively can deform to some extent the geometry of the fingers, but this variation is not significative if compared to original values (first column).

Secondly, the variation of the features is studied in relation to the distance between hand and mobile. According to fifth database, only two distances were considered: standard distance (15 cm approx.) and far distance (30 cm approx.). Notice that a very short distance to the mobile camera makes the hand not fit the mobile screen. Table 3 shows how the distance affects moderately the error between features, although the variation is comparable to original deviation, and it is possible to affirm that extracted features are invariant to distance to camera.

Feature	Original (15 cm approx.)	30 cm approx.
Index	1.2±0.1	1.5±0.3
Middle	1.4±0.2	1.8±0.2
Ring	1.3±0.1	1.9±0.3
Little	1.7±0.2	2.1±0.2

Table 3. Variation of a certain feature in each finger for two distances to camera. The results correspond to the mean difference value and their dispersion.

Feature	Original	45°	90°	135°	180°
Index	1.2±0.1	1.4±0.2	1.3±0.1	1.1±0.1	1.4±0.1
Middle	1.4±0.2	1.5±0.1	1.2±0.2	1.3±0.2	1.3±0.2
Ring	1.3±0.1	1.5±0.2	1.6±0.1	1.4±0.1	1.5±0.2
Little	1.7±0.2	1.8±0.1	1.5±0.2	1.8±0.1	1.7±0.1

Table 4. Variation of a certain feature in each finger for different rotation angles. The results correspond to the mean difference value and their dispersion.

Finally, Table 4 provides information on the property of invariance for extracted features. There is no significative change in feature difference when compared to original, which means that features are invariant to rotation.

Moreover, a practical manner of assessing whether features are invariant to changes is indicated by the global Equal Error Rate (EER) provided in next subsection (Section 6.3). Notice that first database contains a wide range of cases with different values in position, orientation and distance to camera.

Finally, there exist other factors worthy of study, and which remain as future work, like blur effects in image, since it is very common that images acquired by a mobile phone are blurred due to small movements of the camera when obtaining the picture.

6.3 System accuracy, EER

Previous sections have provided an evaluation in terms of segmentation and feature extraction. However, the most important aspect regards the capability of the biometric system to identify or verify individuals. The evaluation of the biometric accuracy involves again a wide number of elements such as the database, the number of samples and features used to train the system. Thereby, a deep understanding of these former factors is required to obtain the best results in identification/verification.

In contrast, this section will only consider two aspects covering those main problems that general biometric systems cope with: 1) The relation between accuracy and number of features; 2) The dependency of the whole biometric system (in terms of Equal Error Rate, EER Sanchez-Reillo et al. (2000)) in relation to the number of samples required to train the system.

The first study is carried out by using the first database (Section 5), fixing the number of training samples ($T = 7$) and testing samples ($U = 13$), being assessed by a K-fold cross validation approach. Samples from first session in the database were used as training samples, using acquisitions from second session as testing samples. For simplicity sake, five values for n were considered: $\{5, 7, 10, 12, 15\}$. Changes due to smaller variations in n are negligible.

Furthermore, only one hand is considered (left or right) in identification, selected by the individual from which hand was taken. A fusion of both hands recognition could improve

	$n = 5$	$n = 7$	$n = 10$	$n = 12$	$n = 15$
EER (%) with Δ_n	16.4 ±0.1	14.1±0.2	8.3±0.3	5.7±0.3	4.1±0.1
EER (%) with $\Omega_{m=n}$	18.3 ±0.2	15.2 ±0.1	9.1±0.2	7.2±0.2	6.8±0.1

Table 5. Variation of Equal Error Rate (EER, %) in relation to the number of features, n.

Training Samples	$T = 3$	$T = 5$	$T = 7$	$T = 10$
EER (%) with Δ_n	14.2±0.4	8.4±0.2	4.1±0.2	3.8±0.1
EER (%) with $\Omega_{m=n}$	15.4±0.2	10.1±0.3	6.8±0.1	6.1±0.2

Table 6. Variation of Equal Error Rate (EER, %) in relation to the number of training samples, T.

the overall accuracy, but lacks of interest for a final application in mobiles. Nonetheless, this system allows users to provide any of both hand for identification, so that individuals should not remember with which hand were enrolled.

The results obtained under this experimental layout are presented in Table 5, where EER (%) is provided in relation to n for both the proposed approach based on average values (Δ_n) and the traditional approach, based on single fingers width ($\Omega_{m=n}$).

Besides, reader may notice that the same number of samples are extracted from each finger, although it could be possible that some fingers contribute differently to the final pattern. A deeper understanding of this idea remains as future work, (Section 7).

Similarly, the relation between EER and the number of training samples is of interest, since a compromise must be achieved between this two previous parameters. Notice that an application based on a high number of training samples will cause a rejection from final users due to its obvious inconvenience. To this end, Table 6 is provided, employing samples from first session in the database (ten samples $T = 10$) to train and samples from second session to test ($U = 10$ samples, concretely). The experimental result is obtained setting the feature extraction parameters to $n = 15$, employing a K-fold cross validation approach. Obviously, the higher the number of elements in training, the higher the system accuracy. However, a modest variation in terms of EER is observed with $T \geq 7$, being $T = 7$ the selected value which gathers a compromise between accuracy and comfortability (number of training samples). In addition, a comparison to a traditional approach ($\Omega_{m=n}$) is also provided in Table 6.

6.4 Mobile implementation

The presented system has been implemented on two different arquitectures: a MATLAB implementation to be run in a PC computer @2.4 GHz Intel Core 2 Duo with 4GB 1067 MHz DDR3 of memory and a JAVA-Android implementation oriented to an HTC Desire with @1GHz and 576 MB RAM. Reader can notice obvious differences in hardware, and therefore the implementation of this approach must be tackled with different perspectives in each situation.

Table 7 provides a comparative study of the speed performance of each implementation. Although HTC implementation is more time-consuming, it takes less than 3 seconds to identify individuals, which is very suitable for daily applications. Temporal values in former Table 7 were obtained by measuring both implementations average performance by using first database.

Process	PC @2.4GHz (seconds)	HTC @1GHz (seconds)
Acquisition	< 0.1	< 0.1
Segmentation	0.5 ± 0.07	1.3 ± 0.1
Feature Extraction	0.3 ± 0.1	0.7 ± 0.1
Matching	0.1 ± 0.02	0.4 ± 0.06

Table 7. Comparative temporal study of implementations in PC (second column) and HTC mobile (third column) measured in seconds

7. Conclusions and future work

This document presents a biometric system based on hand geometry oriented to mobile devices. This system incorporates some novel and challenging aspects since images are acquired without no severe constraints in terms of illumination, position, distance to camera and orientation; acquisitions were taken with cameras embedded on commercial mobile devices, providing thereby low resolution images lacking in details and precision; and no flat surface is required to locate the hand or pegs to force a certain position to the hand.
Due to all these previous characteristics, a non-invasive biometric system comes up gathering not only comfortability to the final user (take a hand picture with the mobile) but also reliance on the performance of the biometric system, being able to identify individuals with an EER of $4.1 \pm 0.2\%$ with seven training samples and a total of 60 features ($4 \times n$ with $n = 15$) and seven training samples. Moreover, an EER of $3.8 \pm 0.1\%$ can be obtained by increasing the number of training samples to ten images.
In addition, this biometric system has been seriously evaluated covering every main aspect in a biometric system: segmentation, feature extraction and identification rate. The evaluation relies on three databases, which are publicly available on *http://www.gb2s.es* collecting a wide range of samples with the purpose of assessing previous aspects, considering different devices, environment conditions, situations, backgrounds, population and the like.
The obtained results come up with an important conclusion: the proposed extracted features yield to an independence to changes in image acquisitions.
Furthermore, a study concerning invariance to blur operations will be contemplated. Blur and fuzzy effects deserve special attention since they simulate the behavior of a moved acquisition, something very common in mobile acquisitions due to the low quality acquisition system. Despite of building a new database, this effects will be reproduced with different image processing algorithms. In addition, a deeper understanding on the contribution of each individual feature in relation to final accuracy will be also considered, together with a fusion scheme with palmprint.
With the aim of a mobile device application, several details must be improved. First of all, it is desirable to reduce the number of training samples, preserving the accuracy. Secondly, an adaptive SVM is supposed to decreased EER throughout time, decreasing the number of false rejections (situations that exasperate the final user). Thirdly, a PCA algorithm could obtain the principal components in the extracted templates, reducing the number of features within the pattern. Furthermore, a study on the device independence of the biometric system will lead to make possible the fact of enrolling an individual with one device and accessing with other, yielding to multiple applications.
In addition, reader may notice that this system entails, at least, more than one individual in order to carry out a comparison. This situation barely happens on a mobile device, since they are not shared by more than one individual. Therefore, how would it be possible to keep

rates without knowing others bimetric data? This biometric system provides a solution for a centralized access system, where the accesing devices are mobile apparatus. Individuals can have access to their corresponding mobiles, by being verified using this biometric system.
However, if this biometric system is used offline, i.e. without accessing previous centralized system, the biometric algorithm must be able to identify the individual without being compared to others. This can be achieved by storing fake templates in mobile, or providing a one-class SVM. In any case, this situation involves a final scenario and a final implementation and, therefore, it has little relation to biometric topics, despite of being a challenging problem regular to all biometric systems applied to daily applications.
Finally, an adaptation to current biometric standards ISO/IEC JTC1/SC37 will be also considered.

8. References

Abeni, P., Baltatu, M. & D'Alessandro, R. (2006). Nis03-4: Implementing biometrics-based authentication for mobile devices, *Global Telecommunications Conference, 2006. GLOBECOM '06. IEEE*, pp. 1 -5.

Albin, S., Rougeron, G., Peroche, B. & Tremeau, A. (2002). Quality image metrics for synthetic images based on perceptual color differences, *Image Processing, IEEE Transactions on* 11(9): 961 - 971.

Alpert, S., Galun, M., Basri, R. & Brandt, A. (2007). Image segmentation by probabilistic bottom-up aggregation and cue integration, *IEEE Conference on Computer Vision and Pattern Recognition, 2007. CVPR '07.*, pp. 1–8.

Arif, M., Brouard, T. & Vincent, N. (2006). Personal identification and verification by hand recognition, *Engineering of Intelligent Systems, 2006 IEEE International Conference on*, pp. 1–6.

Boreki, G. & Zimmer, A. (2005). Hand geometry: a new approach for feature extraction, *Automatic Identification Advanced Technologies, 2005. Fourth IEEE Workshop on*, pp. 149–154.

Covavisaruch, N. & Prateepamornkul, P. (2006). Personal identification system using hand geometry and iris pattern fusion, *Electro/information Technology, 2006 IEEE International Conference on*, pp. 597–602.

de Santos Sierra, A., Guerra Casanova, J., Sánchez Ávila, C. & Jara Vera, V. (2009). Silhouette-based hand recognition on mobile devices, *43rd Annual 2009 International Carnahan Conference on Security Technology, 2009.*, pp. 160–166.

Ferrer, M., Fabregas, J., Faundez, M., Alonso, J. & Travieso, C. (2009). Hand geometry identification system performance, *Security Technology, 2009. 43rd Annual 2009 International Carnahan Conference on*, pp. 167–171.

Gonzalez, R. C. & Woods, R. E. (1992). *Digital Image Processing*, Addison-Wesley Longman Publishing Co., Inc., Boston, MA, USA.

Gross, R., Li, Y., Sweeney, L., Jiang, X., Xu, W. & Yurovsky, D. (2007). Robust hand geometry measurements for person identification using active appearance models, *Biometrics: Theory, Applications, and Systems, 2007. BTAS 2007. First IEEE International Conference on*, pp. 1–6.

Han, Y., Tan, T., Sun, Z. & Hao, Y. (2007). Embedded palmprint recognition system on mobile devices, *ICB*, pp. 1184–1193.

Hashemi, J. & Fatemizadeh, E. (2005). Biometric identification through hand geometry, *Computer as a Tool, 2005. EUROCON 2005.The International Conference on*, Vol. 2, pp. 1011–1014.

Ijiri, Y., Sakuragi, M. & Lao, S. (2006). Security management for mobile devices by face recognition, *Mobile Data Management, 2006. MDM 2006. 7th International Conference on*, pp. 49 – 49.

Kanhangad, V., Kumar, A. & Zhang, D. (2009). Combining 2d and 3d hand geometry features for biometric verification, *Computer Vision and Pattern Recognition Workshops, 2009. CVPR Workshops 2009. IEEE Computer Society Conference on*, pp. 39–44.

Kong, A., Zhang, D. & Kamel, M. (2009). A survey of palmprint recognition, *Pattern Recogn.* 42(7): 1408–1418.

Kukula, E. & Elliott, S. (2005). Implementation of hand geometry at purdue university's recreational center: an analysis of user perspectives and system performance, *Security Technology, 2005. CCST '05. 39th Annual 2005 International Carnahan Conference on*, pp. 83–88.

Kukula, E. & Elliott, S. (2006). Implementation of hand geometry. an analysis of user perspectives and system performance, *Aerospace and Electronic Systems Magazine, IEEE* 21(3): 3–9.

Kukula, E., Elliott, S., Gresock, B. & Dunning, N. (2007). Defining habituation using hand geometry, *Automatic Identification Advanced Technologies, 2007 IEEE Workshop on*, pp. 242–246.

Kumar, A. & Zhang, D. (2006). Personal recognition using hand shape and texture, *Image Processing, IEEE Transactions on* 15(8): 2454–2461.

Kumar, A. & Zhang, D. (2007). Hand-geometry recognition using entropy-based discretization, *Information Forensics and Security, IEEE Transactions on* 2(2): 181–187.

Luengo-Oroz, M. A., Faure, E. & Angulo, J. (2010). Robust iris segmentation on uncalibrated noisy images using mathematical morphology, *Image Vision Comput.* 28: 278–284. URL: *http://portal.acm.org/citation.cfm?id=1663651.1663753*

McLoughlin, I. & Naidu, N. (2009). Keypress biometrics for user validation in mobile consumer devices, *Consumer Electronics, 2009. ISCE '09. IEEE 13th International Symposium on*, pp. 280 –284.

Mojsilovic, A., Hu, H. & Soljanin, E. (2002). Extraction of perceptually important colors and similarity measurement for image matching, retrieval and analysis, *Image Processing, IEEE Transactions on* 11(11): 1238 – 1248.

Munoz, A.-C., de Santos Sierra, A., Ávila, C., Casanova, J., del Pozo, G. & Vera, V. (2010). Hand biometric segmentation by means of fuzzy multiscale aggregation for mobile devices, *Emerging Techniques and Challenges for Hand-Based Biometrics (ETCHB), 2010 International Workshop on*, pp. 1 –6.

Rahman, A., Anwar, F. & Azad, S. (2008). A simple and effective technique for human verification with hand geometry, *Computer and Communication Engineering, 2008. ICCCE 2008. International Conference on*, pp. 1177–1180.

Recky, M. & Leberl, F. (2010). Windows detection using k-means in cie-lab color space, *Pattern Recognition (ICPR), 2010 20th International Conference on*, pp. 356 –359.

Saevanee, H. & Bhatarakosol, P. (2008). User authentication using combination of behavioral biometrics over the touchpad acting like touch screen of mobile device, *Computer and Electrical Engineering, 2008. ICCEE 2008. International Conference on*, pp. 82 –86.

Sanchez-Reillo, R., Sanchez-Avila, C. & Gonzalez-Marcos, A. (2000). Biometric identification through hand geometry measurements, *Pattern Analysis and Machine Intelligence, IEEE Transactions on* 22(10): 1168–1171.

Shabeer, H. & Suganthi, P. (2007). Mobile phones security using biometrics, *Conference on Computational Intelligence and Multimedia Applications, 2007. International Conference on*, Vol. 4, pp. 270 –274.

Shahin, M., Badawi, A. & Rasmy, M. (2008). A multimodal hand vein, hand geometry, and fingerprint prototype design for high security biometrics, *Biomedical Engineering Conference, 2008. CIBEC 2008. Cairo International*, pp. 1–6.

Singh, A., Agrawal, A. & Pal, C. (2009). Hand geometry verification system: A review, *Ultra Modern Telecommunications Workshops, 2009. ICUMT '09. International Conference on*, pp. 1–7.

Tan, W., Wu, C., Zhao, S. & Chen, S. (2009). Hand extraction using geometric moments based on active skin color model, *Intelligent Computing and Intelligent Systems, 2009. ICIS 2009. IEEE International Conference on*, Vol. 4, pp. 468–471.

Van Thanh, D. (2000). Security issues in mobile ecommerce, *Database and Expert Systems Applications, 2000. Proceedings. 11th International Workshop on*, pp. 412 –425.

Varchol, P., Levicky, D. & Juhar, J. (2008). Multimodal biometric authentication using speech and hand geometry fusion, *Systems, Signals and Image Processing, 2008. IWSSIP 2008. 15th International Conference on*, pp. 57–60.

Wang, W.-C., Chen, W.-S. & Shih, S.-W. (2009). Biometric recognition by fusing palmprint and hand-geometry based on morphology, *Acoustics, Speech and Signal Processing, 2009. ICASSP 2009. IEEE International Conference on*, pp. 893–896.

Wang, X., Huang, X. & Fu, H. (2010). A color-texture segmentation method to extract tree image in complex scene, *Machine Vision and Human-Machine Interface (MVHI), 2010 International Conference on*, pp. 621 –625.

Yang, F. & Ma, B. (2007). Two models multimodal biometric fusion based on fingerprint, palm-print and hand-geometry, *Bioinformatics and Biomedical Engineering, 2007. ICBBE 2007. The 1st International Conference on*, pp. 498–501.

Yoruk, E., Konukoglu, E., Sankur, B. & Darbon, J. (2006). Shape-based hand recognition, *IEEE Transactions on Image Processing* 15(7): 1803–1815.

Yu, P., Xu, D., Li, H. & Zhou, H. (2009). Fingerprint image preprocessing based on whole-hand image captured by digital camera, *Computational Intelligence and Software Engineering, 2009. CiSE 2009. International Conference on*, pp. 1–4.

Zhang, H., Fritts, J. E. & Goldman, S. A. (2008). Image segmentation evaluation: A survey of unsupervised methods, *Computer Vision and Image Understanding* 110(2): 260 – 280.

Zheng, G., Wang, C.-J. & Boult, T. (2007). Application of projective invariants in hand geometry biometrics, *Information Forensics and Security, IEEE Transactions on* 2(4): 758–768.